Principles of Oceanography

Principles of Oceanography

Edited by Theodore Roa

SYRAWOOD
PUBLISHING HOUSE
New York

Published by Syrawood Publishing House,
750 Third Avenue, 9th Floor,
New York, NY 10017, USA
www.syrawoodpublishinghouse.com

Principles of Oceanography
Edited by Theodore Roa

International Standard Book Number: 978-1-68286-444-9 (Hardback)

Cataloging-in-publication Data

Principles of oceanography / edited by Theodore Roa.
    p. cm.
Includes bibliographical references and index.
ISBN 978-1-68286-444-9
1. Oceanography. 2. Marine sciences. I. Roa, Theodore.
GC11.2 .P75 2017
551.46--dc23

Printed in the United States of America.

TABLE OF CONTENTS

PREFACE

The objective of this book is to give a general view of the different areas of oceanography and its applications. It is the study of oceans borrowing from varied fields like biology, physics, climatology, geology, etc. Different approaches, evaluations, methodologies and advanced studies on oceanography have been included in this book. It is a compilation of topics which include the major principles of oceanography along with its branches like biological oceanography, chemical oceanography, geological oceanography and physical oceanography, to name few. It aims to serve as a resource guide for students and experts alike and contribute to the growth of the discipline. A number of latest researches have been included to keep the readers up-to-date with the global concepts in this area of study.

This book is a comprehensive compilation of works of different researchers from varied parts of the world. It includes valuable experiences of the researchers with the sole objective of providing the readers (learners) with a proper knowledge of the concerned field. This book will be beneficial in evoking inspiration and enhancing the knowledge of the interested readers.

In the end, I would like to extend my heartiest thanks to the authors who worked with great determination on their chapters. I also appreciate the publisher's support in the course of the book. I would also like to deeply acknowledge my family who stood by me as a source of inspiration during the project.

Editor

Cholera and Shigellosis: Different Epidemiology but Similar Responses to Climate Variability

Benjamin A. Cash[1]*, **Xavier Rodó**[2,3], **Michael Emch**[4], **Md. Yunus**[5], **Abu S. G. Faruque**[5], **Mercedes Pascual**[6,7]

1 Center for Ocean-Land-Atmosphere Studies, Fairfax, Virginia, United States of America, 2 Institut Català de Ciències del Clima (IC3), Barcelona, Catalunya, Spain, 3 Institució Catalana de Recerca i Estudis Avançats, Barcelona, Catalunya, Spain, 4 University of North Carolina Chapel Hill, Chapel Hill, North Carolina, United States of America, 5 International Centre for Diarrheal Disease Research, Dhaka, Bangladesh, 6 Department of Ecology and Evolutionary Biology University of Michigan, Ann Arbor, Michigan, United States of America, 7 Howard Hughes Medical Institute, Chevy Chase, Maryland, United States of America

Abstract

Background: Comparative studies of the associations between different infectious diseases and climate variability, such as the El Niño-Southern Oscillation, are lacking. Diarrheal illnesses, particularly cholera and shigellosis, provide an important opportunity to apply a comparative approach. Cholera and shigellosis have significant global mortality and morbidity burden, pronounced differences in transmission pathways and pathogen ecologies, and there is an established climate link with cholera. In particular, the specific ecology of *Vibrio cholerae* is often invoked to explain the sensitivity of that disease to climate.

Methods and Findings: The extensive surveillance data of the International Center for Diarrheal Disease Research, Bangladesh are used here to revisit the known associations between cholera and climate, and to address their similarity to previously unexplored patterns for shigellosis. Monthly case data for both the city of Dhaka and a rural area known as Matlab are analyzed with respect to their association with El Niño and flooding. Linear correlations are examined between flooding and cumulative cases, as well as for flooding and El Niño. Rank-correlation maps are also computed between disease cases in the post-monsoon epidemic season and sea surface temperatures in the Pacific. Similar climate associations are found for both diseases and both locations. Increased cases follow increased monsoon flooding and increased sea surface temperatures in the preceding winter corresponding to an El Niño event.

Conclusions: The similarity in association patterns suggests a systemic breakdown in population health with changing environmental conditions, in which climate variability acts primarily through increasing the exposure risk of the human population. We discuss these results in the context of the on-going debate on the relative importance of the environmental reservoir vs. secondary transmission, as well as the implications for the use of El Niño as an early indicator of flooding and enteric disease risk.

Editor: Wei-Chun Chin, University of California, Merced, United States of America

Funding: This work was funded by the United States National Oceanic and Atmospheric Administration grant # F020704. The authors further acknowledge the support of core donors that provide unrestricted support to the International Centre for Diarrheal Disease Research, Bangladesh (icddr,b) for its operations and research. Current donors providing unrestricted support include the Australian International Development Agency, the government of the People's Republic of Bangladesh, the Canadian International Development Agency, the Swedish International Development Cooperative Agency, and the Department for International Development, United Kingdom. The authors gratefully acknowledge these donors for their support and commitment to icddr,b's research efforts. M. Pascual is an investigator of the Howard Hughes Medical Institute. The funders had no role in study design, data collection and analysis, decision to publish, or preparation of the manuscript.

Competing Interests: The authors have declared that no competing interests exist.

* Email: bcash@gmu.edu

Introduction

Comparative studies of associations between climate variability and human health across different infectious diseases are largely lacking. Potential climate influences are typically addressed only for specific infections and locations for which sufficiently long retrospective records exist. Such comparison studies, however, are instrumental in identifying systemic breakdowns in human health at the population level that do not depend on specific pathogen ecologies or epidemiology. Similarities in responses to climate forcing instead signal common mechanisms and, in so doing, identify key pathways of action for intervention. In addition, associations between disease risk and slowly varying features of the climate system allow for the possibility of developing early warning systems for disease risk in the affected areas.

Diarrheal illnesses provide one important disease class for the application of a comparative approach. Cholera, arguably the paradigmatic water-borne diarrheal disease, is caused by the bacteria *Vibrio cholerae* and has been responsible for numerous pandemics throughout history, including the current era [1]. Infection generally results from ingesting water contaminated with the bacteria, and left untreated leads to mortality rates as high as 50% [2]. Global mortality rates for cholera are estimated at $\sim$3% [3]. Mortality for shigellosis, another widely spread diarrheal illness that is caused by bacteria of the *Shigella* genus can reach 5–15% for some strains in areas with poor medical care [4–5]. The infectious dose for shigellosis is also substantially lower than for cholera and the disease is responsible for significant mortality and morbidity in the developing world [6]. Given their global mortality and morbidity burden, the potentially pronounced differences in transmission pathways and pathogen ecologies, and the existing evidence for a role of climate forcing on cholera, cholera and shigellosis represent excellent candidates for a comparative study on their associations with climate.

Endemic cholera has been prominent in studies on the influence of climate variability on diarrheal diseases. In particular, the El Niño-Southern Oscillation (ENSO), widely recognized as one of the most significant modes of interannual climate variability, has been shown to affect the severity of seasonal outbreaks in Bangladesh [7–12]. Favorable conditions for increased cholera risk in the fall have been shown to be associated with positive ENSO (El Niño) events in the preceding winter, thus establishing ENSO as a potential early-warning indicator for Bangladesh cholera outbreaks. At the local level increased rainfall and associated flooding have also been shown to increase fall cholera cases, and represent the means by which the remote influence of ENSO in transmitted to Bangladesh [11], [13].

The particular attention paid to cholera is due in no small part to the extended record of cases available for Bangladesh, advances on the microbial ecology of *Vibrio cholerae*, and the recognition that the pathogen survives outside the human host in aquatic environments such as estuaries and brackish water [14–16]. Climate conditions favoring the population growth or survival of the pathogen would exacerbate transmission through environmental aquatic reservoirs. This paradigm has led to an emphasis on mechanisms that are pathogen specific in mediating the effect of climate variability. An alternative, although not necessarily exclusive view, is one in which the vulnerability of the human host plays a key role and at seasonal and interannual time scales is modified by anomalous climate conditions such as extreme floods [17–19].

The relationship between water-borne bacterial diseases in Bangladesh and the monsoon rains is of particular interest, given that Bangladesh lies at the confluence of three major rivers (Ganges/Padma, Brahmaputra, Megna) in an extensive estuarine region that is considered the hearth of cholera. Bangladesh is also a low-lying country that experiences some of the highest rainfall totals in the Indian monsoon region and is inundated to a greater or lesser degree on an annual basis. While a certain degree of flooding is necessary to maintain the fertility of the delta system and associated agriculture, major flooding events have resulted in serious loss of life and destruction of infrastructure throughout Bangladesh's history [20]. Given the evidence linking flooding and post-monsoon cholera outbreaks [17–18], [22–24], it is of particular interest to consider what role the severity of annual flooding might play for shigellosis in that same region and season, particularly in the context of potential remote forcing by ENSO.

The extensive surveillance program of the International Center for Diarrheal Diseases Research, Bangladesh (icddr,b) makes it possible for us to revisit here the known associations between cholera and climate variability, and to specifically inquire about the possibility of similar but unknown patterns for shigellosis. In the results presented below, we show that there are indeed similar associations between cholera and shigellosis cases in Bangladesh and flooding, as well as for cholera and shigellosis cases and sea surface temperatures in the tropical Pacific. These similar associations have significant implications for the mechanisms behind the influence of climate on enteric disease outbreaks in Bangladesh, as well as for the use of ENSO as an early warning indicator of outbreak risk.

We indeed find a strong similarity of association patterns for post-monsoon outbreaks, whose implications for the mechanisms behind the influence of climate forcing we discuss.

Data and Methodology

Ethics Statement

The Diarrheal Disease Surveillance System (DDSS) of icddr,b is a routine ongoing activity of the Dhaka and Matlab Hospital, which has been approved by the Research Review Committee (RRC) and Ethical Review Committee (ERC) of icddr,b. At the time of enrollment, verbal consent was taken from the adult patients and caregivers or guardians in case of the children patients, with the information to be stored in the hospital database and used for conducting research. This verbal consent was documented by keeping a check mark in the questionnaire, which was again shown to the adult patients or the parents/guardians of children patients. Patients or parents/guardians were assured about the non-disclosure of information collected from them, and were also informed about the use of data for improving patient care activities as well as scientific research and publication without disclosing the name or identity of the patients. ERC was satisfied with the voluntary participation, maintenance of the rights of the participants and confidential handling of personal information by the hospital physicians and has approved this consent procedure.

Data

Flood Affected Area (FAA), expressed as the percentage of the total area of Bangladesh inundated, was obtained from the annual reports of the Bangladesh Flood Forecasting and Warning Center (FFWC; http://www.ffwc.gov.bd). Values range from 0.2% in 1994 to a remarkable 68% during the severe flooding of 1998. Although FAA is only available as annual values, flooding in Bangladesh is dominated by the rainy season (June-September; JJAS). The reported FAA for each year was taken to be representative of this period.

The NINO34 index is a standard and widely used measure of the strength and state of ENSO, the dominant mode of climate variability on interannual timescales. Values were provided by the National Oceanic and Atmospheric Administration (NOAA) Climate Prediction Center. NINO34 is defined as the area-averaged sea surface temperature (SST) over the region (5°S-5°N, 120°W-170°W) and is presented here as 3-month running mean anomalies calculated relative to the 1971-2000 base period. Gridded SST values used to calculate the rank correlation maps (see below) were taken from the Hadley Centre Sea Ice and Sea Surface Temperature (HadISST) v1.1 data set [25].

The disease data analyzed in this work was taken from two surveillance programs overseen by the icddr,b. In one program, based in icddr,b's Dhaka Hospital, stool specimens from every 25th patient were tested for cholera and shigellosis from 1979–1995, and from every 50th patient from 1996 onwards. The second surveillance program is located in Matlab, Bangladesh, in a rural area approximately 57 km southeast of the capital of Dhaka. In contrast to the Dhaka hospital, patients attended to at the Matlab hospital are mostly those presenting symptoms of severe diarrheal illness. Stool specimens of all diarrhea patients from the Health and Demographic Surveillance System (HDSS) area were tested for shigellosis and cholera. In both locations cases were separated by causative species and strain, and in the analysis presented here we focus exclusively on cases due to the dominant form of the each disease for our period of study, which is 1983–2010. For cholera this was this El Tor biotype, which replaced the former Classical biotype in the past decades and is now the dominant strain both in Bangladesh and around the world. For shigellosis *S. flexneri* represent the dominant form of the disease in Bangladesh.

Time series of the monthly cases (Fig. 1a) and the magnitude of the fall peak (Fig. 1b) for the two diseases in both Matlab and Dhaka show that, in addition to significant interannual variability for both diseases and locations, cholera cases exceed those of shigellosis and that there is a clear downward trend in Dhaka shigellosis cases for this period.

Methodology

Our focus in this work is on interannual variations in disease outbreaks, as opposed to longer-term trends, and as such all data was detrended prior to analysis. As a check on the sensitivity of our results to the removal of the long-term trends we first analyzed data detrended using a simple linear regression. We then repeated the analysis using data detrended via Singular Spectrum Analaysis (SSA), which allows for a more flexible and nonlinear definition of the trend [26]. Our results were not sensitive to the choice of detrending technique (See File S1) and as such results reported here are based on the simpler linear method.

To quantify the magnitude and geographical extent of the link between the occurrence of water borne disease cases and potential environmental drivers, such as SST, we calculate Spearman rank-correlations between reported cases, flooding, and SST (see also [11], [27]). We employed rank-correlation, rather than the more common Pearson's correlation, because it is a non-parametric method more suitable for use with the non-normally distributed flooding and disease data. The rank-correlation, ρ, is calculated as

$$\rho = \frac{n^3 - \frac{D_1 + D_2}{2} - 6\sum_{i=1}^{n} d_i^2}{\sqrt{(n^3 - D_1)(n^3 - D_2)}} \qquad (1)$$

in the case where the data includes ties [28], where n is the number of months, d_i is the difference in rank for each month, and D_1 and D_2 are the sum of the cubes of the sizes of the ties in the first and second samples, respectively, where the size of a tie is defined as the number of months with the same value. For $n > 20$ we can test for the significance of the above quantity using the transformed, normally distributed variable

$$\omega = (\rho + \frac{6}{n(n^2 - 1)})\sqrt{n-1} \qquad (2)$$

Results and Discussion

Time series of monthly cases (Fig. 1a) and the magnitude of the fall peak (Fig. 1b) of cholera and shigellosis in Matlab and Dhaka show that, in addition to the significant interannual variability for

both diseases and locations, there is a downward trend in Dhaka shigellosis cases for this period that is significant at the 95% level (as determined by SSA; see File S1 for details). The significance of the trend is further emphasized by the fact that only for Dhaka shigellosis do results made using the detrended and raw case data differ.

As noted in the Introduction, previous studies have identified a clear statistical association, as well as a physical link, between post-monsoon cholera cases in Matlab and ENSO events in the preceding winter. Building on this previous work (e.g., [11–12]), we calculate the rank correlation between shigellosis cases during the 'fall peak' (August-December; ASOND) and SST at every grid point for the preceding boreal winter (December-February; DJF). Calculating the correlation for each grid point, rather than for a single index defined for a limited and predetermined region, allows for a fuller description of the association between disease outbreaks and global SST. Calculating correlation maps for multiple diseases and locations allows for a richer comparison of their associations with SST as well.

The correlation map for ASOND cholera in Matlab and DJF SST (Fig. 2a) produces a pattern that closely resembles the one for Matlab shigellosis (Fig. 2b), particularly through the tropical regions. This visual similarity is confirmed by the very high pattern correlation of 0.88 between the two maps. While there are minor differences between the patterns for cholera and shigellosis in the Indian Ocean, the diagnosed relationship with the tropical Pacific is nearly identical. Moreover, the maps also closely resemble the pattern of SST anomalies that mark the mature phase of a warm ENSO event, and both series are significantly correlated with the NINO34 index (0.42 for cholera, 0.55 for shigellosis; both p<0.05). This apparent association with El Niño was explored previously for Matlab cholera [11], in which experiments with a general circulation model confirmed the physical significance of the correlation pattern and its correspondence to ENSO. This analysis was essential in establishing that the patterns identified by the rank-correlation analysis do indeed represent ENSO, and do not resemble the ENSO pattern merely by chance. The close resemblance between the correlation patterns for cholera and shigellosis strongly implies that shigellosis is responding in a similar manner to the remote forcing from ENSO.

Applying this same correlation map analysis to Dhaka cholera (Fig. 3a) and shigellosis (Fig. 3b) cases, we again find strong similarities in the association patterns for the two diseases. While the Dhaka cholera map closely resembles the two Matlab maps, the highest correlations values are shifted towards the central Pacific for shigellosis. This shift results in a lower correlation between the two Dhaka patterns (0.72) than we find for Matlab, and a lower value of the correlation with the NINO34 index (0.54 for cholera, 0.21 for shigellosis; cholera correlation p<<0.01, shigellosis correlation not significant).

The fact that Dhaka shigellosis shows a statistically significant correlation with SST across a broad section of the tropical Pacific, while at the same time showing a relatively low correlation with the NINO34 index, highlights the advantages of analyzing correlation patterns rather than focusing on single indices. This shift in the region of significant correlations between Dhaka shigellosis and SST also suggests a closer association with the central, rather than eastern, Pacific component of ENSO. Thus, despite the dramatic differences in ecology between the two pathogens and the environments of the two study sites, we find a clear association between post-monsoon cases and ENSO in the preceding winter for both diseases and locations.

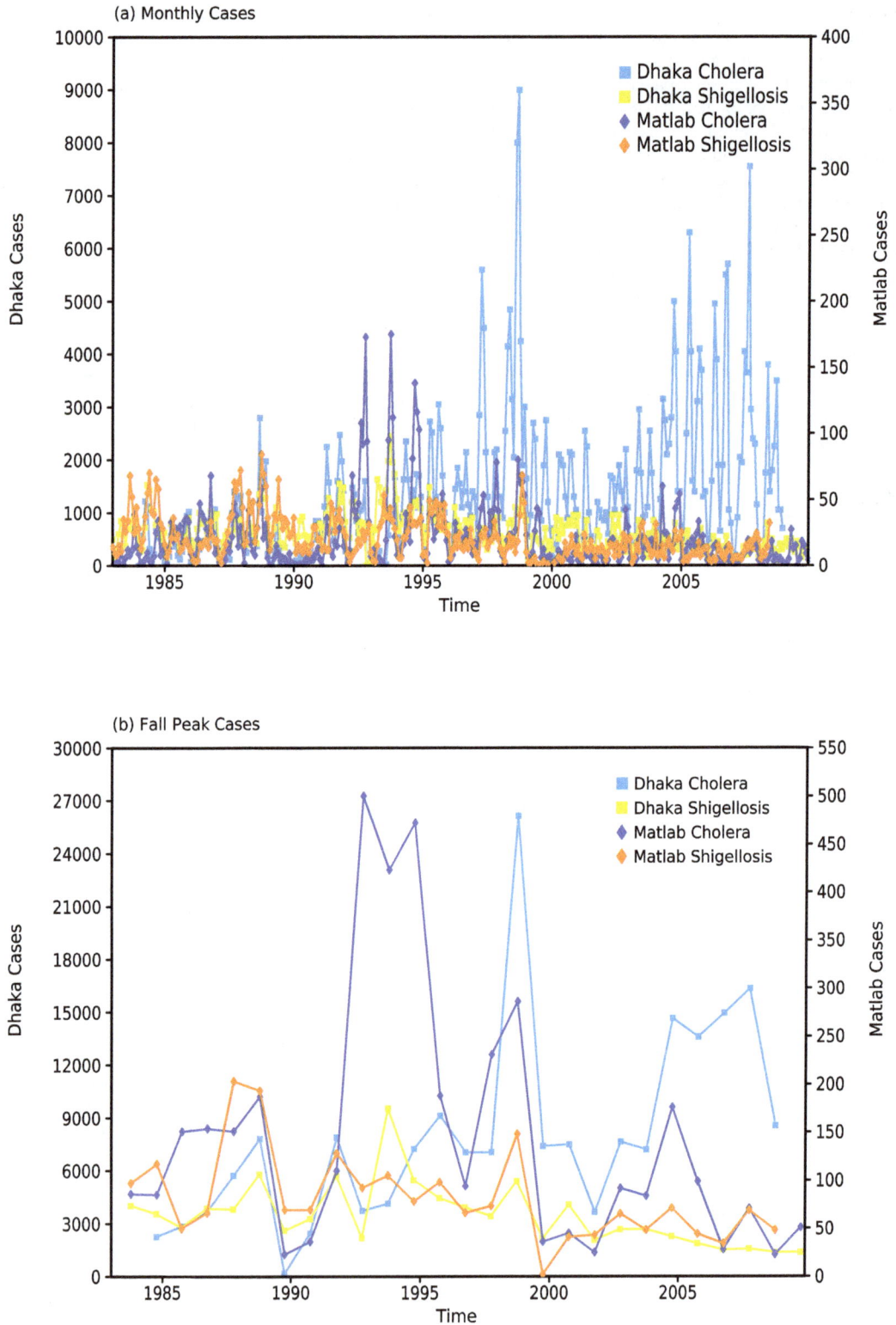

Figure 1. Time series of (a) monthly *S. flexneri* shigellosis and *V. cholerae* El Tor cholera cases and (b) cases during the fall peak (August-December; ASOND) from 1983–2009 for Dhaka and Matlab.

Earlier work (e.g., [11–12]) advanced the theory that enhanced local rainfall and subsequent flooding is the mediating factor between tropical SST anomalies and cholera cases in Bangladesh. Enhanced local rainfall linked to remote SST anomalies has also been associated with increased malaria cases in northwest India [27]. Consistent with this mechanism, we find strong positive

(a) Matlab: Fall peak cholera and winter SST

(b) Matlab: Fall peak shigellosis and winter SST

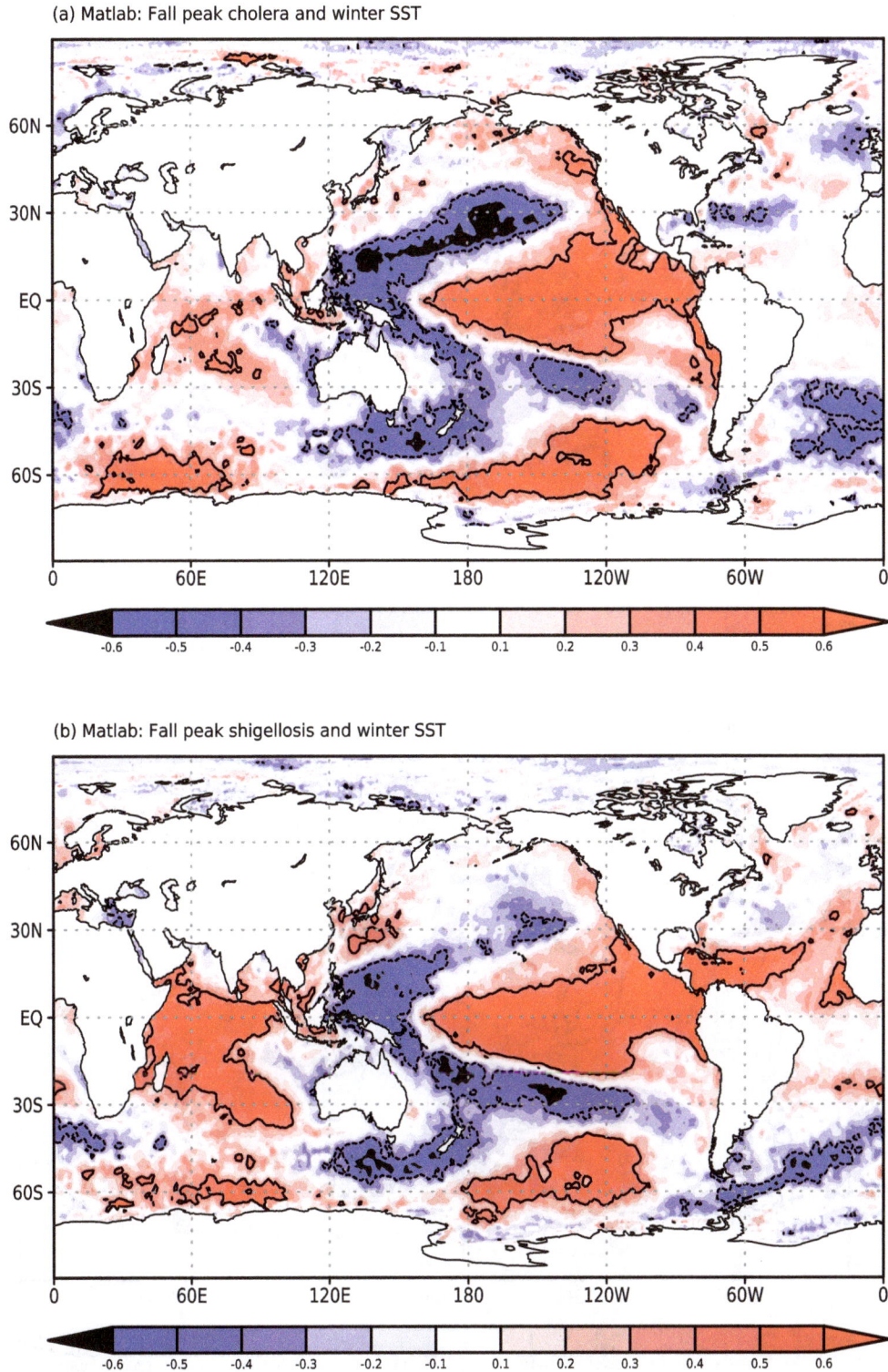

Figure 2. Rank-correlation maps for Matlab (a) fall peak (August-December; ASOND) El Tor cholera and (b) ASOND *S. flexneri* **shigellosis cases and preceding December-February sea surface temperature (SST).** Maps are correlated at a value of 0.88 for the subregion (35S, 35N).

associations between FAA and both cholera (Fig. 4) and shigellosis (Fig. 5) cases in both locations. The relationship is particularly clear for the Dhaka fall peak in cholera cases (Fig. 4a), with no notable outliers and a rank correlation coefficient of 0.74 (p<<

0.01). In contrast, for the Matlab data the period from 1992–1994 stands out sharply as an outlier, with 2–3 times the number of cases compared to other years with similar FAA. If this period is excluded the rank correlation coefficient with FAA is 0.54 (p<

(a) Dhaka: Fall peak cholera and winter SST

(b) Dhaka: Fall peak shigellosis and winter SST

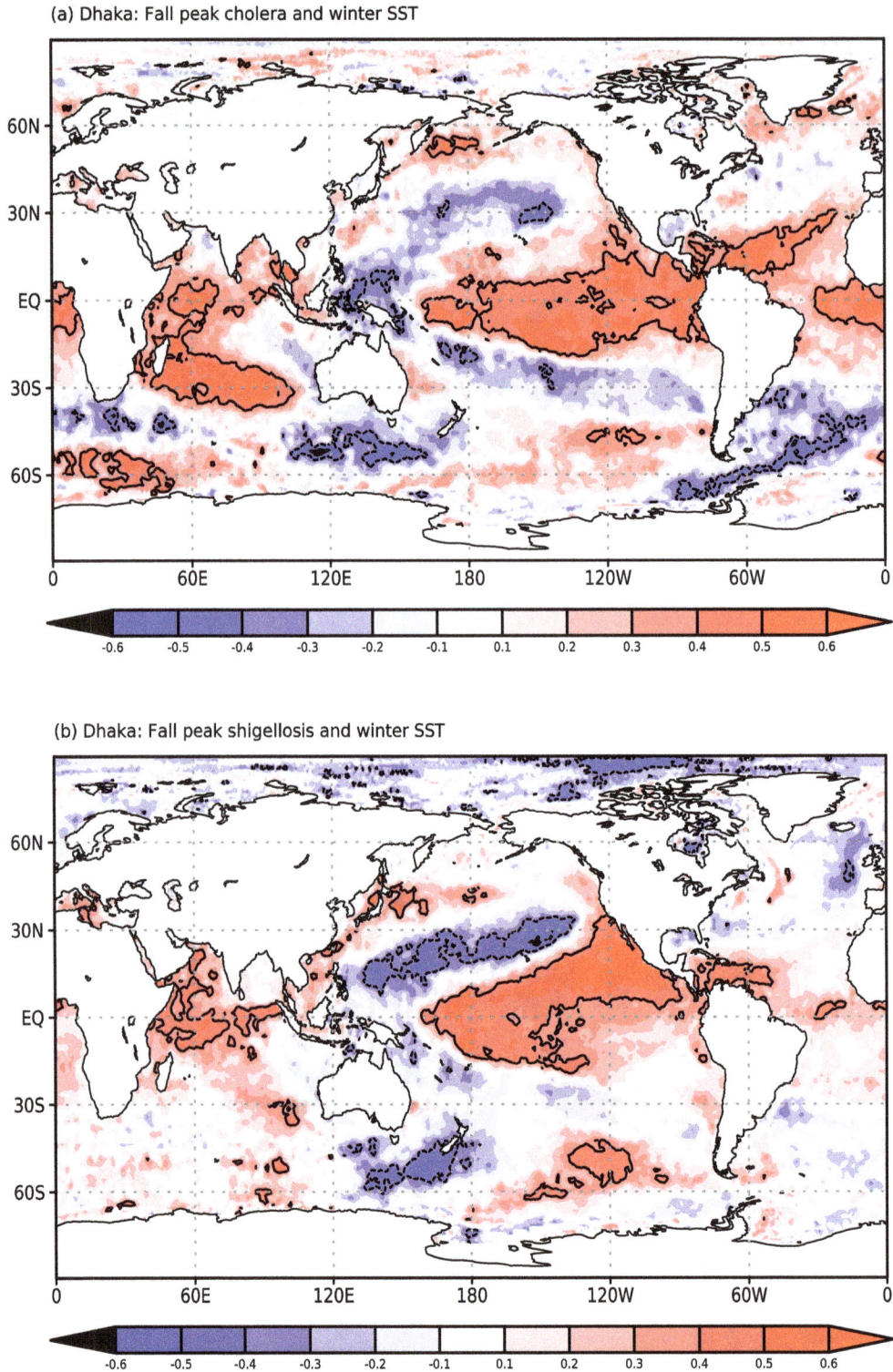

Figure 3. Rank-correlation maps for Dhaka (a) fall peak (August-December; ASOND) El Tor cholera and (b) ASOND *S. flexneri* **shigellosis cases and preceding December-February sea surface temperature (SST).** Maps are correlated at a value of 0.72 for the subregion (35S, 35N).

0.01; correlation is negligible if 1992–1994 are included). We do not find similar elevated values for this period in Dhaka. In addition, while 2007 was a year of heavy flooding and increased

cholera prevalence in Dhaka, cases are relatively low in the Matlab record.

It is not known whether the inconsistencies in the cholera records for the two locations reflect sampling differences between

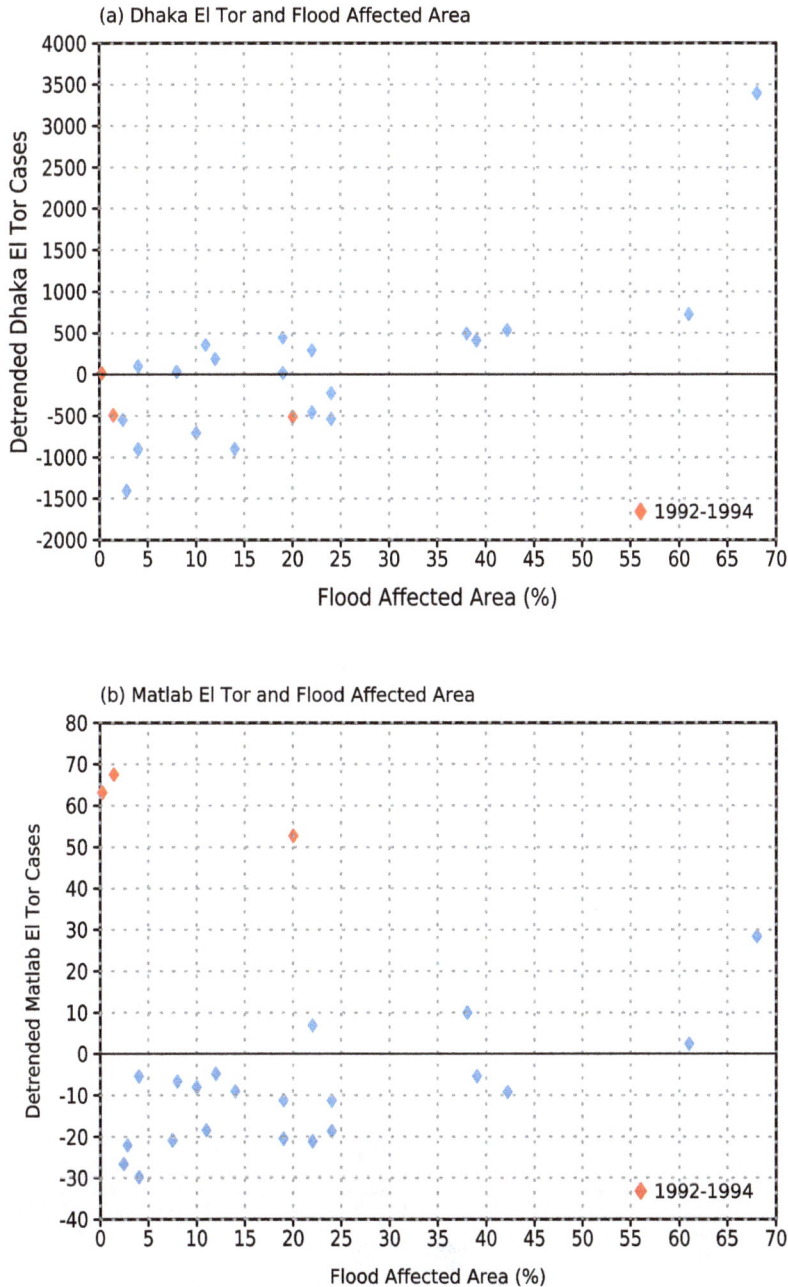

Figure 4. (a) Flood Affected Area (FAA) and linearly detrended fall peak (August-December; ASOND) Matlab El Tor cholera cases. Rank correlation of 0.56 when 1992–1994 removed. (b) FAA and linearly detrended ASOND Dhaka El Tor cholera incidence. Rank correlation of 0.74.

the large, urban environment of Dhaka and the smaller, more rural environment of Matlab, or if they reflect a degree of localization in outbreaks. One difference in the sampling methodology is that in Matlab only patients who come from the Health and Demographic Surveillance System (HDSS) area are included in the study, while in Dhaka any patient who arrives at the hospital is included. It should also be noted that 1992–1994 is the period in which 0139 Bengal emerged in Bangladesh, and this may be related to the unusual values from Matlab.

For shigellosis there is somewhat greater scatter at lower levels of FAA (Fig. 5) than for cholera, perhaps reflecting the fact that shigellosis requires a much smaller infectious dose than cholera

[6]. Overall the correlation with FAA is comparable to that of cholera (0.74 and 0.50 for Matlab and Dhaka, respectively, p<< 0.01). Interestingly, the 1992–1994 period is unremarkable for shigellosis in Matlab but is a partial outlier for Dhaka (note that correlation between Dhaka shigellosis and NINO34 increases from 0.21 to 0.41 if these years are removed), further suggesting that enteric disease in Bangladesh during these three years merits additional investigation.

The above analysis establishes that the annual flood affected area provides a clear predictor for the number of diarrheal disease cases in Bangladesh during the fall peak. However, it does so at very short leads. The fall peak (ASOND) overlaps with the end of

(a) Dhaka Shigellosis and Flood Affected Area

(b) Matlab Shigellosis and Flood Affected Area

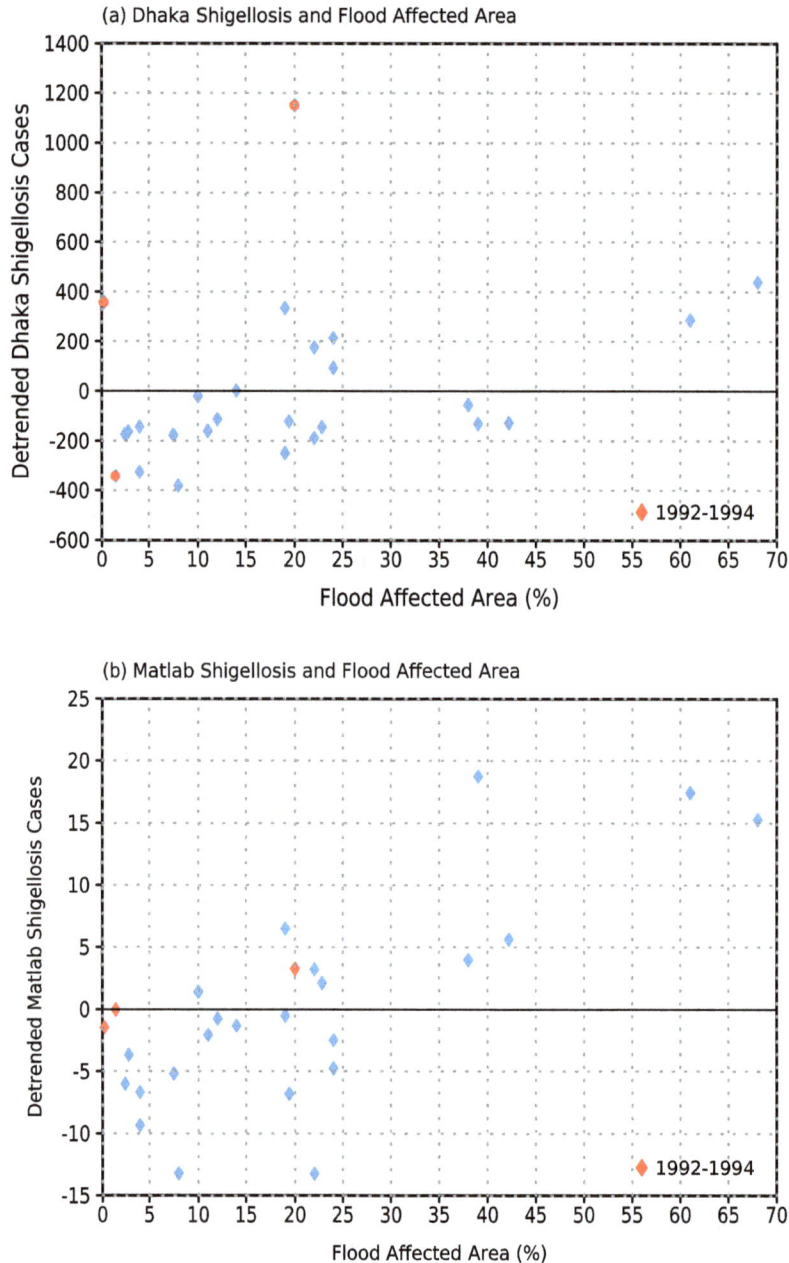

Figure 5. (a) Flood Affected Area (FAA) and linearly detrended fall peak (August-December; ASOND) Matlab *S. flexneri* shigellosis cases. Rank correlation of 0.74. (b) FAA and linearly detrended ASOND Dhaka *S. flexneri* shigellosis incidence. Rank correlation of 0.50.

the monsoon period (JJAS), providing only limited advanced warning for public health officials. In order to increase the lead-time of risk forecasts, we further explore the link between disease cases and ENSO (Figs. 2 and 3) in light of the association between disease cases and flooding (Figs. 4 and 5). Years with FAA below 30% are not strongly associated with the value of the NINO34 index (Fig. 6) for any of the three preceding seasons considered (December-January-February, DJF; March-April-May, MAM; June-July-August, JJA). FAA values above 30%, however, are not seen following negative DJF NINO34 values (Fig. 6a). For MAM (Fig. 6b) and JJA (Fig. 6c), these more severe floods can be associated with either positive or negative values of the NINO34 index. Hence, ENSO indices for the spring and summer seasons

cannot be used to discriminate between strong and weak flooding years and thus high and low risk disease seasons.

Summary and Conclusions

Cholera and shigellosis are two prominent diarrheal illnesses that together are responsible for significant mortality and morbidity throughout the developing world, including Bangladesh. The causative organisms, *V. cholerae* and *S. flexneri*, differ significantly in their ecology, transmission pathways, and infectious doses, among other features. Despite these differences, our analysis of case data taken from two separate surveillance sites in Bangladesh demonstrates that interannual variations in the severity of the fall outbreaks of both diseases are closely and

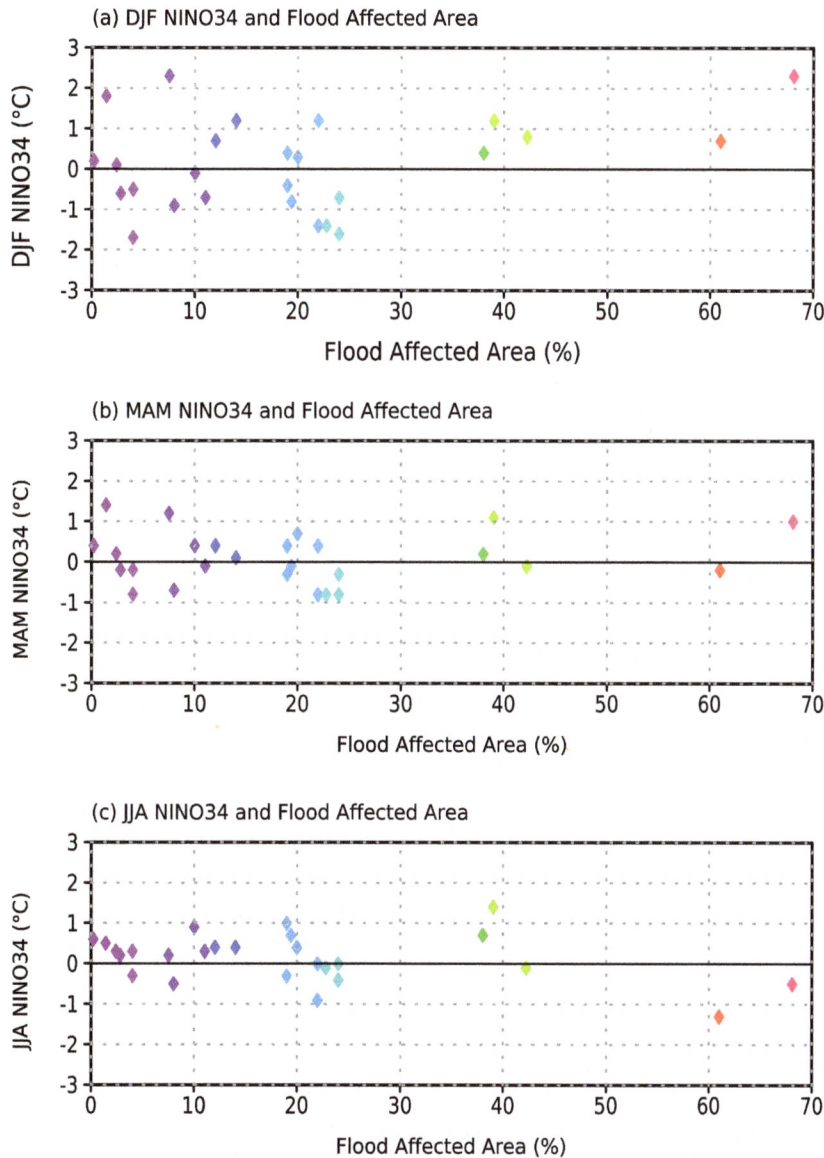

Figure 6. Flood Affected Area (FAA) versus (a) December-February (DJF) NINO34 index, (b) March-May (MAM) NINO34 index, and (c) June-August (JJA) NINO34 index.

similarly associated with the areal extent of the annual monsoon floods. Of potentially greater significance, particularly from the perspective of risk forecasting, we also find that both the monsoon floods and post-monsoon disease outbreaks are significantly correlated with ENSO activity in the preceding winter.

This association between winter ENSO and Bangladesh flooding is consistent with previous work [29], which identified a linear relationship between flooding and DJF NINO34 values. However, that study did not address the asymmetry between positive and negative ENSO events, or the variations in flooding between positive events. Recent work [30] suggests that the lag of two seasons between ENSO during DJF and the summer monsoon flooding is related to the excitation of a coupled air-sea mode of variability in the Indian Ocean region, which allows the impact of DJF ENSO anomalies to persist into summer. It is interesting to note that the magnitude of the flooding is not strongly correlated

with the magnitude of the index, indicating that more than just ENSO is playing a role.

The striking similarity in the association between cholera and shigella and climate variability, despite the significant differences in the ecology of the two pathogens, strongly suggests that the two diseases share a similar transmission pathway in the post-monsoon period. Flooding inevitably increases the potential for exposure to contaminated water, an increase in risk likely to be exacerbated by the attendant overcrowding and breakdown in sanitary infrastructure. Flooding also, inevitably, subjects individuals to heightened stress, which is known to affect immune response and increases susceptibility to infectious disease [31]. While the relationship between flooding and El Tor cases in Bangladesh has been noted before (e.g., [13], [22–23], as has as an increase in shigellosis for the large flooding events or 1988, 1998 and 2004 [17], this general association between flooding, cholera, and shigellosis has not been explored previously.

The association between flooding and shigellosis, in conjunction with the association between flooding and cholera is of particular interest because it demonstrates that the association between post-monsoon cholera cases and flooding is not unique. Rather it suggests that it is the general increase in exposure risk of the human population that follows flooding, rather than the specific ecology of any one pathogen, which underlies the impact of climate variability on diarrheal disease in Bangladesh. In particular our analysis demonstrates that a long-lived environmental reservoir connected to primary transmission, which is known to exist for cholera but not for shigellosis, is not a necessary condition for flooding to drive post-monsoon disease cases in. It instead suggests that climate variability acts by modulating secondary transmission, where the latter can involve different pathways, through water or food or person to person contact, and is characterized by a dependency on previous levels of infection in the population [16].

Our findings have focused on the post-monsoon season in part because of the established role of ENSO during the preceding winter on cholera during this part of the year, through the pathway of increased precipitation and flooding. It is worth mentioning however that the disease exhibits an additional, pre-monsoon, peak and the overall seasonal pattern is known to vary regionally, in particular with respect to the relative importance of the two seasons [22], [24]. Comparisons with other diarrheal diseases at this regional level would be informative. Similarly, comparison of spatio-temporal patterns within a large urban environment such as Dhaka would further elucidate similarities and differences in epidemiology, especially given the reported heterogeneity within the city itself in the response to ENSO [18].

Linking cholera and shigellosis to flooding provides a physical mechanism for environmental influence on the interannual variability of the two diseases; linking flooding to ENSO provides a potential mechanism for forecasting risk. Our analysis shows that the most severe flooding events in Bangladesh follow warm winter conditions in the central and eastern tropical Pacific; none follow cold conditions. This relationship allows for an initial prediction of flooding and disease risk to be made as early as the end of the boreal winter; the risk of severe flooding and associated disease outbreaks should be substantially reduced when ENSO is in a negative state. Understanding why severe flooding follows warm winter Pacific SST in only a subset of years, as well as why the relationship is not linear with the magnitude of the SST anomalies is critical to improving the accuracy of disease risk forecasts, and is the subject of ongoing research. Likewise, expanding comparative analysis to include other water-borne and vector-borne infections should further refine our understanding of the fundamental processes underlying the connections between climate variability and disease outbreaks.

Author Contributions

Conceived and designed the experiments: BAC MP. Performed the experiments: BAC. Analyzed the data: BAC XR ME MY AF MP. Contributed reagents/materials/analysis tools: BAC XR ME MY AF MP. Contributed to the writing of the manuscript: BAC XR ME MY AF MP.

References

1. Kohn GC (1995) Encyclopedia of plague and pestilence. New York: Wordsworth Reference, 408 pp.
2. Cook GC (1996) Management of cholera: the vital role of rehydration. In: Cholera and the Ecology of *Vibrio Cholerae*, (B. S. Drasar and B. D. Forrest, Eds) Suffolk: Chapman and Hall, 54–94.
3. Ali M, Lopez AL, You YA, Kim YE, Sah B, et al. (2012) The global burden of cholera. Bull World Health Organ 90: 209–218, doi:10.2471/BLT.11.093427
4. Rahaman MM, Khan MM, Aziz KMS, Islam MS, Kibriya AK (1975) An outbreak of dysentery caused by *Shigella dysenteriae* type 1 on a Coral Island in the Bay of Bengal. The Journal of Infectious Diseases 132: 15–19.
5. Kotloff KL, Winickoff JP, Ivanoff B, Clemens JD, Swerdlow DL, et al. (1999) Global burden of *Shigella* infections: implications for vaccine development and implementation of control strategies. Bull World Health Organ, 77: 651–666.
6. Emch M, Ali Md, Yunus Md (2008) Risk areas and neighborhood-level risk factors for *Shigella dysenteriae 1* and *Shigella flexneri*. Health & Place 14: 96–105.
7. Pascual M, Rodó X, Ellner SP, Colwell RR, Bouma MJ (2000) Cholera dynamics and El Niño-Southern Oscillation. Science 8: 1766–1769.
8. Pascual M, Bouma MJ, Dobson AP (2002) Cholera and climate: revisiting the quantitative evidence. Microbes Infect 4: 237–245.
9. Rodó X, Pascual M, Fuchs G, Faruque ASG (2002) ENSO and cholera: A nonstationary link related to climate change? PNAS 99: 12901–12906.
10. Koelle K, Pascual M (2004) Disentangling extrinsic from intrinsic factors in disease dynamics: A nonlinear time series approach with an application to cholera. American Naturalist 163: 901–913.
11. Cash BA, Rodó X, Kinter III JL (2008) Links between tropical Pacific SST and cholera incidence in Bangladesh: Role of the eastern and central tropical Pacific. J Climate 21: 4647–4663.
12. Cash BA, Rodó X, Kinter III JL, Yunus Md (2010) Disentangling the impact of ENSO and Indian Ocean variability on the regional climate of Bangladesh: Implications for cholera risk. J Climate 23: 2817–2831.
13. Koelle K, Rodó X, Pascual M, Yunus Md, Mostafa G (2005) Refractory periods and climate forcing in cholera dynamics. Nature 436: 696–700.
14. Colwell RR (1996) Global climate and infectious disease: The cholera paradigm. Science 274: 2025–2031.
15. Lipp EK, Huq A, Colwell RR (2002) Effects of global climate on infectious disease: the cholera model. Clinical Microbiology Reviews 15: 757–770.
16. Codeço CT, Coelho FC (2006) Trends in Cholera Epidemiology. PLoS Med 3: e42; doi:10.1371/journal.pmed.0030042 [Online 31 January 2006].

17. Schwartz BS, Harris JB, Khan AI, Larocque RC, Sack DA, et al. (2006) Diarrheal epidemics in dhaka, bangladesh, during three consecutive floods: 1988, 1998, and 2004. Am J Trop Med Hyg 74: 1067–1073.
18. Reiner RC Jr, King AA, Emch M, Yunus Md, Faruque ASG, Pascual M (2012) Highly localized sensitivity to climate forcing drives endemic cholera in a megacity. PNAS 109: 2033–2036.
19. Rinaldo A, Bertuzzo E, Mari L, Righetto L, Blokesch M, et al. (2012) Reassessment of the 2010–2011 Haiti cholera outbreak and rainfall-driven multiseason projections. PNAS 109: 6602–6607, doi:10.1073/pnas.1203333109.
20. Hossain Md S (2009) Annual flood report 2009. 76 pp, Bangladesh Flood Warning and Forecast Center, Dhaka, Bangladesh.
21. Koelle K, Pascual M, Yunus Md (2005) Pathogen adaptation to seasonal forcing and climate change. Pro. Royal Soc, B 272: 971–977.
22. Akanda AS, Jutla AS, Islam S (2009) Dual peak cholera transmission in Bengal Delta: A hydroclimatological explanation. Geophys Res Lett 36: L19401.
23. Akanda AS, Jutla AS, Alam M, de Magny GC, Siddique AK, et al. (2011) Hydroclimatic influences on seasonal and spatial cholera transmission cycles: Implications for public health intervention in the Bengal Delta. Water Resour. Res 47: doi:10.1029/2010WR009914 [Online 10 May 2011].
24. Bertuzzo E, Mari L, Righetto L, Gatto M, Casagrandi R (2012) Hydroclimatology of dual-peak annual cholera incidence: Insights from a spatially explicit model. Geophys Res Lett 39: doi:10.1029/2011GL050723 [Online 10 March 2012].
25. Rayner NA, et al. (2003) Global analyses of sea surface temperature, sea ice, and night marine air temperature since the late nineteenth century. J Geophys Res 108: doi:10.1029/2002JD002670 [Online 17 July 2003].
26. Ghil M., et al. (2002) Advanced spectral methods for climatic time series. Reviews of Geophysics 40: 1.1–1.41.
27. Cash BA, Rodó X, Ballester J, Bouma MJ, Baeza A, et al. (2013) Malaria epidemics and the influence of the tropical South Atlantic on the Indian monsoon. Nature Climate Change 3: 502–507.
28. Kraft CH, van Eden C (1968) *A nonparametric introduction to statistics*. The Macmillan Company.
29. Chowdhury MR, Ward MN (2007) Seasonal flooding in Bangladesh – variability and predictability. Hydro Proc 21: 335–347.
30. Kosaka Y, Xie S-P, Lau N-C, Vecchi G (2013) Origin of seasonal predictability for summer climate over the Northwestern Pacific. PNAS 110: 7574–7579.
31. Marsland AL, Bachen AH, Cohen S, Rabin B, Manuck SB (2002) Stress, immune reactivity and susceptibility to infectious disease. Physiology and Behavior 711–716.

2

Elevated Temperature Alters the Lunar Timing of Planulation in the Brooding Coral *Pocillopora damicornis*

Camerron M. Crowder[1]*, **Wei-Lo Liang**[2], **Virginia M. Weis**[1], **Tung-Yung Fan**[2,3]*

1 Department of Integrative Biology, Oregon State University, Corvallis, Oregon, United States of America, **2** Institute of Marine Biology, National Dong Hwa University, Pingtung, Taiwan, R.O.C., **3** National Museum of Marine Biology and Aquarium, Pingtung, Taiwan, R.O.C.

Abstract

Reproductive timing in corals is associated with environmental variables including temperature, lunar periodicity, and seasonality. Although it is clear that these variables are interrelated, it remains unknown if one variable in particular acts as the proximate signaler for gamete and or larval release. Furthermore, in an era of global warming, the degree to which increases in ocean temperatures will disrupt normal reproductive patterns in corals remains unknown. *Pocillopora damicornis*, a brooding coral widely distributed in the Indo-Pacific, has been the subject of multiple reproductive ecology studies that show correlations between temperature, lunar periodicity, and reproductive timing. However, to date, no study has empirically measured changes in reproductive timing associated with increased seawater temperature. In this study, the effect of increased seawater temperature on the timing of planula release was examined during the lunar cycles of March and June 2012. Twelve brooding corals were removed from Hobihu reef in Nanwan Bay, southern Taiwan and placed in 23 and 28°C controlled temperature treatment tanks. For both seasons, the timing of planulation was found to be plastic, with the high temperature treatment resulting in significantly earlier peaks of planula release compared to the low temperature treatment. This suggests that temperature alone can influence the timing of larval release in *Pocillopora damicornis* in Nanwan Bay. Therefore, it is expected that continued increases in ocean temperature will result in earlier timing of reproductive events in corals, which may lead to either variations in reproductive success or phenotypic acclimatization.

Editor: Mónica Medina, Pennsylvania State University, United States of America

Funding: This research was supported by a Ministry of Science and Technology grant NSC-101-2611-M0291-002 awarded to T-Y.F., a National Science Foundation East Asian and Pacific Summer Institute grant awarded to C.M.C., and a National Science Foundation OISE award (#1042509) awarded to V.M.W. The funders had no role in study design, data collection and analysis, decision to publish, or preparation of the manuscript.

Competing Interests: The authors have declared that no competing interests exist.

* Email: Crowderc@onid.oregonstate.edu (CMC); tyfan@nmmba.gov.tw (T-YF)

Introduction

Reproductive timing is a critical factor in coral reproductive success and has been correlated to multiple environmental variables including those associated with seasonality such as temperature, solar irradiance, tidal cycles, nocturnal illumination associated with lunar periodicity, and light-dark cycles corresponding with diel fluctuations [1–8]. Previous studies have effectively demonstrated correlations between environmental variables and coral reproduction; however, examining the direct causality of individual variables on the timing of reproduction is critical to understanding the mechanisms controlling reproductive timing. Determining which variables are vital for the coordination of reproductive events will reveal information underlying coral reproductive function. Corals are both the bio-engineers and foundational primary producers of coral reef ecosystems and therefore, understanding how environmental variables affect timing of reproduction is essential for predicting future impacts of climate change on coral reef ecosystem stability.

Temperature has been shown to be a critical variable affecting coral reproductive success. Corals of all life stages are negatively affected by increasing sea-surface temperature attributed to global climate change [9–11]. While significant effort has been dedicated to describing the effects of temperature on the physiology and ecology of corals, less attention has focused on the effects of elevated temperature on coral reproductive timing and success [9,12]. Recent studies have revealed that temperatures exceeding tolerance thresholds reduce polyp fecundity, gametic quality [13,14] and the number of reproductive events in corals [15]. While it is evident that elevated temperatures can impair reproductive processes, predictable seasonal fluctuations in temperature might be a key component controlling the timing of reproduction. Multiple field studies have observed correlations between temperature and reproductive timing [16,17]. This correlation could demonstrate that some corals are displaying reproductive plasticity or alterations in the timing of release to adapt to change, which could be an important mechanism for larval survival and fitness with rising ocean temperatures associated with climate change. Understanding how temperature affects the timing of reproduction will provide information as to how continued increases in sea-surface temperature may further alter reproductive processes and patterns in corals.

Approximately 15% of coral species brood internally fertilized larvae (planulae) that are released during a process known as planulation. Planulation typically occurs multiple times annually and in some cases on a monthly basis [18,19]. Planulae are often buoyant and have the capacity for wide dispersal, but are also able to quickly settle upon release leading to fast rates of colonization

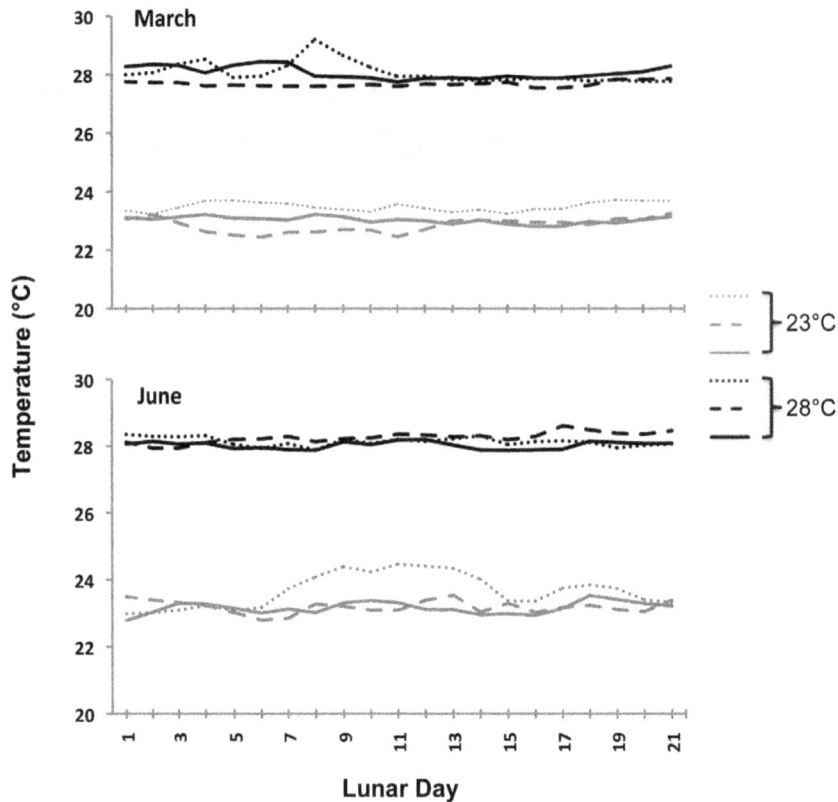

Figure 1. Average daily tank temperatures for both 23 and 28°C temperature treatments in lunar March and June 2012. Three lines represent temperature values in the three separate treatment tanks. Measurements were taken every 10 minutes and averaged for each lunar day.

[2,20]. The timing of planulation has a direct influence on larval survival, dispersal, and recruitment [2,19]. Reproductive events optimally timed to environmental variables such as temperature and light, can affect larval survival particularly in larvae containing symbiotic dinoflagellates [21–24].

The widely studied brooding coral *Pocillopora damicornis*, has been the subject of multiple studies investigating reproduction, particularly reproductive synchrony associated with lunar periodicity [19,20,25–27]. Collectively, these observations show variability in planulation patterns. *P. damicornis* has been shown to release planulae at every lunar phase during monthly lunar reproductive cycles with some consistencies observed within similar geographical locations. This diversity in the timing of planulation between different geographical regions may be the result of phenotypic plasticity that increases fitness [28]. However, to date, little is known about phenotypic plasticity of coral reproductive timing, and specifically how this plasticity may be driven by environmental variables, such as rising ocean temperatures.

In the context of the ongoing debate as to the effect of individual environmental variables on the timing of planulation, it is timely to ask if changes in temperature can directly alter planulation patterns in *P. damicornis*. The purpose of this study was to determine the effect of increased seawater temperature on the timing of planulation within a single lunar cycle and between lunar cycles of different seasons. To investigate this, *P. damicornis* colonies were monitored in temperature-controlled tanks exposed to natural lunar cycles with seawater temperature set to either 23°C (low) or 28°C (high). For both seasons, the timing of planulation was found to be plastic, with elevated temperature

treatments resulting in significantly earlier peaks of planula release. This suggests that temperature modifies the lunar timing of larval release in *Pocillopora damicornis* in Nanwan Bay.

Materials and Methods

Coral Collection

Twelve adult colonies of *P. damicornis* were randomly collected with permission from the Kenting National Park, Taiwan from a 10-meter area at Hobihu reef (21°56.799'N, 120°44.968'E) in Nanwan Bay, Taiwan on two separate occasions in lunar March and June 2012. Colonies ranged in size from 7–15 cm in diameter and were removed from reefs 4–5 meters in depth. Within an hour after collection, colonies were transported back to The National Museum of Marine Biology and Aquarium (NMMBA) in Checheng, Taiwan.

Experimental Design

Six 150 liter tank mesocoms situated within a larger (30 ton) aquarium system as described by Mayfield et al. (2013) were used in this study. Three of the tanks were maintained at 23°C (mean winter ocean temperature) and three were maintained at 28°C (mean summer ocean temperature). Two coral colonies were randomly selected and assigned to each tank. Tanks were covered with shade cloth and exposed to natural outdoor sunlight and received a constant supply of sand-filtered seawater. Corals were allowed to acclimate to the experimental temperature conditions for one week prior to monitoring of planulation.

Monitoring of planulation began on lunar day 1 (new moon), March 22, 2012 (lunar March) and July 19, 2012 (lunar June).

Every evening during the lunar cycle, corals from each tank were individually placed into flow-through containers surrounded by 100 μm mesh plankton netting. Flotation devices were placed inside the netting to prevent nets from touching coral colonies. Each morning, nets were removed and the total number of planulae inside the netting was counted. Tank temperatures were recorded every 10 minutes using a HOBO temperature logger throughout lunar March and June 2012 (Figure 1).

Data Analysis

The percent of the total number of planulae released each day was calculated as the total number of planulae released per individual colony per day (mean ± standard deviation, n = 6 colonies per temperature treatment) divided by the total number of planulae released by each colony during the monthly reproductive cycle. Raleigh's test [29] was used to test the null hypothesis that planula release occurred uniformly throughout the lunar cycle. If the null hypothesis was rejected, then the mean lunar day (MLD) and angular deviation of planulation were calculated using circular statistics [19,27,29] to determine lunar timing of planulation. Circular statistic calculations were completed directly in an excel spreadsheet based on the equations and methods described in Zar et al. (1999).

A 1-way ANOVA was performed to compare MLD and angular deviation (AD) for each tank (n = 3), with average MLD and AD calculated for the two corals within each tank, within temperature treatments. A general linear mixed model with repeated measures was completed using a Poisson distribution with colony size treated as a covariate. Statistical analyses were completed using the statistical package R (R Foundation for Statistical Computing Vienna, Austria).

Results

Raleigh's test for uniform distribution indicated that planula release did not occur uniformly throughout the lunar cycle for either temperature treatment over both lunar March and June (p<0.001 for all tests) (Table 1). Differences were observed in the timing of planulation, measured as the percentage of planulae released per day with corals at 28°C releasing planulae earlier in the lunar cycle than those at 23°C for both lunar March and June (Figure 2). In lunar March, peak percentage of release occurred on lunar day 8 for the 28°C treatment, compared to lunar day 19 for the 23°C treatment. In lunar June, peak percentage of release was observed as two smaller peaks on lunar day 6 and 10 for the 28°C

treatment, compared to a more significant peak at lunar day 12 for the 23°C treatment.

MLD of planulation closely resembled patterns in percentage of planulae released whereas angular deviation, a measure of dispersion, was not significant in lunar March or lunar June, where there was a larger spread in days of release (Table 1). In lunar March, the average MLD and angular deviation of release were 17.5±1.1 and 23.7±5.4 for the 23°C treatment and 8.4±0.7 and 16.1±1.9 for the 28°C treatment, respectively. In lunar June, the average MLD and angular deviation of release were 12.5±1.0 and 23.8±8.6 for the 23°C treatment and 7.7±1.7 and 26.5±10.4 for the 28°C treatment, respectively. One-way nested ANOVAs showed significant differences in MLD (p<0.001) between temperature treatments in lunar March and significant differences in MLD (p = 0.014), between temperature treatments in lunar June (Table 2).

Additionally, our results indicate that there were substantially more planulae released in lunar March (1,454±890 at 23°C and 3,593±1,359 at 28°C) compared to lunar June (186.3±55 at 23°C and 127±93 at 28°C) (Table 1). Significant differences in the total number of planulae released per lunar day were found to be associated with lunar day in both lunar March (p<0.001) and June (p<0.001) and colony size was not found to have a significant effect within individual lunar months (Table 3). Additionally, significant differences in the total number of planulae released per day were associated with temperature for lunar March (p = 0.001) but not lunar June (Table 3).

Discussion

Influence of Seasonality

In this study we observed a substantial difference in the total number of planulae released between lunar March and June 2012. This observed difference in the total number of planule released is likely due to seasonality. This hypothesis is consistent with previous findings showing that there were differences in planulae abundance between seasons in Pocilloporid corals in the Northwestern Philippines with dry seasons (March–May) having higher numbers of planulae released then wet seasons in (June–October) [20]. Differences in total number of planulae released in our study could also be attributed to colony size, since corals used in lunar March were, on average, 1 cm larger than those used in lunar June (Table 1). However, differences in planulation abundances are not expected to affect our results because timing of planulation was examined individually for each month.

Table 1. Results of Raleigh's test for uniform distribution of planula release by P. damicornis colonies during March and June lunar cycles.

	Lunar March		Lunar June	
	23°C	28°C	23°C	28°C
MLD	17.5±1.1	8.4±0.7	12.5±1.0	7.74±1.7
Angular Deviation	23.7±5.4	16.1±1.9	23.8±8.6	26.5±10.4
z	386–7261	2133–4638	93–208	11–291
p	<0.001	<0.001	<0.001	<0.001
Planulae	1,454±890	3593±1359	186.3±55	127±93
Colony Size (cm)	10.4±1.3	11.1±1.8	9.7±1.1	9.8±1.4

Mean lunar day (MLD) ± standard deviation, angular deviation of release ± standard deviation, range of Raleigh's test statistic (z), p-value for Raleigh's test (p), mean number of planulae released per tank ± standard deviation, and average colony diameter (cm) ± standard deviation.

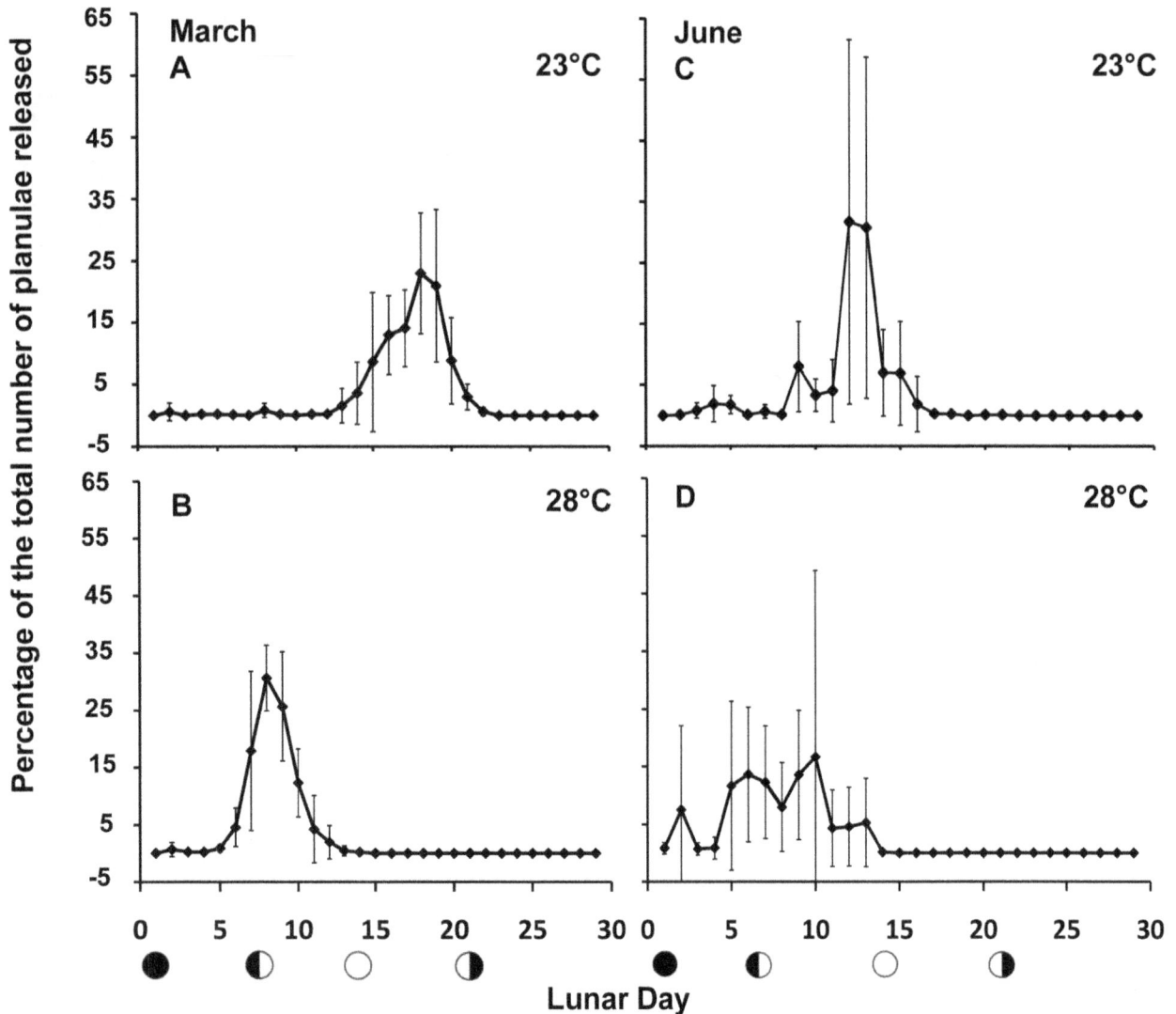

Figure 2. Percent of *P. damicornis* planulae released each day during lunar March (A and B) and Lunar June (C and D) 2012 reproductive cycles for colonies incubated at 23 (A and C) and 28°C (B and D). Points represent means of six colonies ± SE. Moon symbols represent lunar phases (new, 1st quarter, full, and last quarter).

Timing of Planulation

Lunar periodicity has been correlated with the timing of reproduction for multiple coral species [2,18] especially *P. damicornis*, where the timing of planulation is consistently linked to lunar phases [19,20,26,27]. While lunar periodicity likely does play a role in the timing of reproduction, in some cases with good predictability, other environmental factors may be able to disrupt these cycles. In this study we directly tested the influence of low (23°C) and high (28°C) temperatures on reproductive timing, and show that changes in temperature have the capacity to significantly alter the timing of planulation. Clear differences were observed in the timing of planulation, shown as the percentage of planulae released each day (Figure 2), and temperature was found to have a significant impact on the MLD of planulation for both lunar March and June (Table 1 & 2). The observed variations in lunar day of release between temperature treatments suggests that elevated temperature modifies the affect of other cues, such as lunar periodicity, to drive timing of release.

Early release in the high temperature treatment suggests that temperature affects the reproductive physiology of the adult coral and/or the developing planulae, resulting in an acceleration of release [30]. This hypothesis is supported by a previous study on a broadcast spawning coral, *Echniopora lamellosa*, in Taiwan showing that reproductive processes, such as gametogenesis and spawning are plastic and can be accelerated by increasing seawater temperature [31]. Another study on Caribbean corals, within the genus *Madracis*, found that maturation of gametes was positively correlated with increases in seawater temperature, indicating that observed changes in timing could be attributed to internal cues associated with gametogenesis [32].

Alternatively, shifts in the timing of planulation could be a result of negative changes in planula physiology. This hypothesis is supported by multiple studies that have observed decreases in larval survival with elevated temperature. Larvae of the Hawaiian coral *Fungia scutaria* exposed to 27, 29, and 31°C showed gradual decreases in survivorship with animals incubated at 31°C having the highest rates of mortality [24]. Another study conducted on *P*.

Table 2. Results from a 1-way ANOVA reporting the mean lunar day and angular deviation of planula release for lunar March and June.

Mean Lunar Day	Lunar March					Lunar June				
	DF	SS	MS	F	P	DF	SS	MS	F	P
Temperature	1	125.68	125.68	175.80	1.90 E-04	1	29.22	29.22	17.66	1.40E-02
Residuals	4	2.86	0.71			4	6.62	1.65		
Angular Deviation										
Temperature	1	85.43	85.43	5.93	7.2 E-02	1	11.00	10.96	0.107	0.76
Residuals	4	57.62	14.41			4	409.20	102.29		

Table 3. Results from a general linear mixed model with repeated measures using a Poisson distribution with colony size treated as a covariate for the total number of planulae released per lunar day.

	Lunar March				Lunar June			
	Estimate	SE	Z	P	Estimate	SE	Z	P
Intercept	2.866	2.735	1.050	0.295	5.762	3.576	1.611	0.107
Lunar Day	-0.062	0.001	-61.030	2.00E-16	-0.082	0.004	-19.050	2.00E-16
Temperature	0.256	0.078	3.250	0.001	-0.116	0.093	-1.251	0.211
Colony Size	-0.400	0.292	-1.370	0.171	-0.026	0.262	-0.101	0.919

damicornis larvae in Taiwan found 28°C to be the thermal threshold for maximum respiration in planulae, with higher temperatures leading to reduced respiration and likely metabolic depression [33]. Additionally, a 5-fold decrease in survival was observed with a 1.5°C increase in temperature in *Acropora palmata* embryos and larvae from the Caribbean [11]. Although success of planulae after release was not assessed in this study, it has been shown that early-released planulae have lower settlement rates than those released later within a single reproductive cycle [19].

Reproductive Plasticity

Our results reveal that temperature can act as a driver for plasticity in reproductive timing. Reproductive plasticity can enhance individual success in harsh or fluctuating environments [28]. While our study did not examine reproductive plasticity specifically in an adaptive context, we postulate that the phenotypic plasticity observed in this study may suggest capacity for an adaptive response to elevated temperatures in corals. This ability to shift reproductive timing in high temperature environments may indicate that climate change induced increases in ocean temperature may not be detrimental to reproduction, but rather simply alters its timing. Our findings also indicate that such shifts in timing can occur relatively quickly, as reproductive plasticity was observed over a single reproductive cycle. This is similar to findings observed in the coral *Echinopora lamellosa*, that show early spawning, when transplanted from colder northern to warmer southern Taiwan [19]. Understanding reproductive plasticity in corals is important because plasticity may provide corals the flexibility they need to be successful in a changing climate. However, many questions remain about how reproductive plasticity will influence the fate of corals, and their ecosystems, in the long term.

Conclusions

Our results provide empirical evidence that a 5°C increase in temperature, accelerates the timing of planula release in *P. damicornis* in Nanwan Bay, Taiwan.

It important to note that the corals sampled for this study are of unknown genetic origin, and therefore could be clones of the same genotype. If so, this could decrease the variability in temperature response. Nonetheless, our findings reveal that there is plasticity in the timing of reproduction and these changes can occur rapidly, within a single lunar reproductive cycle. These results highlight the reality that increases in ocean temperature have the capacity to disrupt patterns of planulation in corals. Depending on how shifts in the timing of planulation correlate with other environmental variables and conditions, these alterations could lead to increased planulae mortality and decreased recruitment success, or may be indicative of the potential for adaptation to warming ocean temperatures. Understanding the affect of temperature on reproductive timing in *P. damicornis* provides information that can be used to predict patterns in reproductive success and colonization in a future of rapidly changing ocean climate conditions.

Acknowledgments

We thank those who provided field support and feedback on analysis and writing: Dr. Li-Hsueh Wang, Steve Doo, Crystal McRae, Chih-Jui Tan, Sylvia Zamudio, Lorenzo Bramanti, Joleen Tseng, Robbin Chen, Alex Smith, Zach Haber, Peter Greene, Julia Stevens, Sarah Morton, Luna Sun, Nate Kirk, Sheila Kitchen, Angela Poole, Emily Weis, Hannah Tavalire, Jessica Reimer, Stevan Arnold, Eli Meyer, Dee Denver, and Patrick Chappell.

Author Contributions

Conceived and designed the experiments: T-YF W-LL CMC. Performed the experiments: CMC W-LL. Analyzed the data: CMC W-LL T-YF. Contributed reagents/materials/analysis tools: T-YF VMW. Contributed to the writing of the manuscript: CMC VMW T-YF.

References

1. Babcock R, Bull G, Harrison P, Heyward A, Oliver J, et al. (1986) Synchronous spawnings of 105 scleractinian coral species on the Great Barrier Reef. Marine Biology 90: 379–394.

2. Harrison P, Wallace C (1990) Reproduction, dispersal and recruitment of scleractinian corals. Ecosystems of the world 25: 133–207.

3. McGuire M (1998) Timing of larval release by *Porites astreoides* in the northern Florida Keys. Coral Reefs 17: 369–375.

4. Jokiel PL, Guinther EB (1978) Effects of temperature on reproduction in the hermatypic coral *Pocillopora damicornis*. Bulletin of Marine Science 28: 786–789.

5. Brady AK, Hilton JD, Vize PD (2009) Coral spawn timing is a direct response to solar light cycles and is not an entrained circadian response. Coral Reefs 28: 677–680.

6. Villanueva RD, Baria MVB, dela Cruz DW, Dizon RM (2011) Diel timing of planulation and larval settlement in the coral *Isopora cuneata* (Scleractinia: Acroporidae). Hydrobiologia 673: 273–279.

7. Goodbody-Gringley G, de Putron S (2009) Planulation patterns of the brooding coral *Favia fragum* (Esper) in Bermuda. Coral Reefs 28: 959–963.

8. Goodbody-Gringley G (2010) Diel planulation by the brooding coral *Favia fragum* (Esper, 1797). Journal of Experimental Marine Biology and Ecology 389: 70–74.

9. Hoegh-Guldberg O, Mumby P, Hooten A, Steneck R, Greenfield P, et al. (2007) Coral reefs under rapid climate change and ocean acidification. Science 318: 1737–1742.

10. Donner SD (2009) Coping with commitment: projected thermal stress on coral reefs under different future scenarios. PLoS One 4: e5712.

11. Randall CJ, Szmant AM (2009) Elevated temperature affects development, survivorship, and settlement of the elkhorn coral, *Acropora palmata* (Lamarck 1816). The Biological Bulletin 217: 269–282.

12. Mayfield A, Fan T-Y, Chen C-S (2013) Physiological acclimation to elevated temperature in a reef-building coral from an upwelling environment. Coral Reefs 32: 909–921.

13. Michalek-Wagner K, Willis BL (2001) Impacts of bleaching on the soft coral *Lobophytum compactum*. I. Fecundity, fertilization and offspring viability. Coral Reefs 19: 231–239.

14. McClanahan TR, Weil E, Cortes J, Baird A, Ateweberhan M (2009) Consequences of coral bleaching for sessile reef organisms. Berlin: Springer-Verlag.

15. Howells EJ, Berkelmans R, van Oppen MJ, Willis BL, Bay LK (2013) Historical thermal regimes define limits to coral acclimatization. Ecology 94: 1078–1088.

16. De Putron SJ, Ryland JS (2009) Effect of seawater temperature on reproductive seasonality and fecundity of *Pseudoplexaura porosa* (Cnidaria: Octocorallia): latitudinal variation in Caribbean gorgonian reproduction. Invertebrate Biology 128: 213–222.

17. Nozawa Y, Harrison PL (2007) Effects of elevated temperature on larval settlement and post-settlement survival in scleractinian corals, *Acropora solitaryensis* and *Favites chinensis*. Marine Biology 152: 1181–1185.

18. Harrison PL (2011) Sexual reproduction of scleractinian corals. Coral reefs: an ecosystem in transition. Springer. 59–85.

19. Fan T-Y, Li J-J, Ie S-X, Fang L-S (2002) Lunar periodicity of larval release by pocilloporid corals in southern Taiwan. Zoological Studies-Tapei 41: 288–294.

20. Villanueva RD, Yap HT, Montano MNE (2008) Timing of planulation by pocilloporid corals in the northwestern Phillipines. Marine Ecology Progress Series 370: 111–119.

21. Fan T-Y, Lin K-H, Kuo F-W, Soong K, Liu L-L, et al. (2006) Diel patterns of larval release by five brooding scleractinian corals. Marine Ecology Progress Series 321: 42.

22. Edmunds PJ, Gates RD, Leggat W, Hoegh-Guldberg O, Allen-Requa L (2005) The effect of temperature on the size and population density of dinoflagellates in larvae of the reef coral *Porites astreoides*. Invertebrate Biology 124: 185–193.

23. Yakovleva IM, Baird AH, Yamamoto HH, Bhagooli R, Nonaka M, et al. (2009) Algal symbionts increase oxidative damage and death in coral larvae at high temperatures. Marine Ecology Progress Series 378: 105–112.

24. Schnitzler CE, Hollingsworth LL, Krupp DA, Weis VM (2012) Elevated temperature impairs onset of symbiosis and reduces survivorship in larvae of the Hawaiian coral, *Fungia scutaria*. Marine Biology 159: 633–642.

25. Jokiel PL, Ito RY, Liu PM (1985) Night irradiance and synchronization of lunar release of planula larvae in the reef coral *Pocillopora damicornis*. Marine Biology 88: 167–174.

26. Richmond RH, Jokiel PL (1984) Lunar periodicity in larva release in the reef coral *Pocillopora damicornis* at Enewetak and Hawaii. Bulletin of Marine Science 34: 280–287.

27. Tanner JE (1996) Seasonality and lunar periodicity in the reproduction of pocilloporid corals. Coral Reefs 15: 59–66.

28. Via S, Gomulkiewicz R, De Long G, Scheiner S, Schlichting C, et al. (1995) Adaptive phenotypic plasticity: consensus and controversy. Trends in Ecology and Evolution 10: 212–217.

29. Zar JH (1999) Biostatistical Analysis. Prentice-Hall, New Jersey. 422–469.

30. Schmidt-Nielsen K (1997) Animal physiology: adaptation and environment. Cambridge University Press.

31. Fan T-Y, Dai C-F (1999) Reproductive plasticity in the reef coral *Echinopora lamellosa*. Marine Ecology Progress Series 190: 297–301.

32. Vermeij MJA (2004) The reproductive biology of closely related coral species: gametogenesis in *Madracis* from the southern Caribbean. Coral Reefs 24: 206–214.

33. Edmunds PJ, Cumbo V, Fan T-Y (2011) Effects of temperature on the respiration of brooded larvae from tropical reef corals. The Journal of Experimental Biology 214: 2783–2790.

3

Seagrass Canopy Photosynthetic Response Is a Function of Canopy Density and Light Environment: A Model for *Amphibolis griffithii*

John D. Hedley[1]*, Kathryn McMahon[2], Peter Fearns[3]

1 Environmental Computer Science Ltd., Tiverton, Devon, United Kingdom, **2** School of Natural Sciences and Centre for Marine Ecosystems Research, Edith Cowan University, Joondalup, Western Australia, **3** Department of Imaging and Applied Physics, Curtin University of Technology, Perth, Western Australia

Abstract

A three-dimensional computer model of canopies of the seagrass *Amphibolis griffithii* was used to investigate the consequences of variations in canopy structure and benthic light environment on leaf-level photosynthetic saturation state. The model was constructed using empirical data of plant morphometrics from a previously conducted shading experiment and validated well to *in-situ* data on light attenuation in canopies of different densities. Using published values of the leaf-level saturating irradiance for photosynthesis, results show that the interaction of canopy density and canopy-scale photosynthetic response is complex and non-linear, due to the combination of self-shading and the non-linearity of photosynthesis versus irradiance (P-I) curves near saturating irradiance. Therefore studies of light limitation in seagrasses should consider variation in canopy structure and density. Based on empirical work, we propose a number of possible measures for canopy scale photosynthetic response that can be plotted to yield isoclines in the space of canopy density and light environment. These plots can be used to interpret the significance of canopy changes induced as a response to decreases in the benthic light environment: in some cases canopy thinning can lead to an equivalent leaf level light environment, in others physiological changes may also be required but these alone may be inadequate for canopy survival. By providing insight to these processes the methods developed here could be a valuable management tool for seagrass conservation during dredging or other coastal developments.

Editor: Kay C. Vopel, Auckland University of Technology, New Zealand

Funding: This research was funded by an Edith Cowan University (ECU) Industry Collaboration Scheme with BMT Oceanica Pty. Ltd., the industry partner, awarded to KM, JH, and PF. ECU Faculty Visiting Fellow Scheme supported JH for travel to ECU. KM was supported by the ECU Collaborative Research Network. Environmental Computer Science Ltd. provided support in the form of salary for author JH. The funders had no role in the study design, data collection and analysis, decision to publish, or preparation of the manuscript. The specific roles of these authors are articulated in the "author contributions" section.

Competing Interests: This work was partly funded by BMT Oceanica. The author J Hedley is an employee of Environmental Computer Science Ltd.

* Email: j.d.hedley@envirocs.com

Introduction

Seagrass meadows are a dominant habitat of most coastal environments and provide important ecosystem services such as primary production, nutrient cycling, sediment stabilization, food and habitat for other organisms and trophic transfers to adjacent habitats [1]. Globally, these ecosystem services have been valued at an approximated US$ 19000 ha^{-1} yr^{-1} [2] but emerging understanding of the carbon storage capability of seagrass meadows implies this may be an underestimate [3]. Despite these recognized values, the area of seagrass is reducing world-wide at an increasing rate. Waycott et al. [4] estimated 29% of the known areal extent has disappeared since seagrass areas were initially recorded in 1879, and the rate of decline has accelerated in the last two decades.

The key anthropogenic pressures impacting seagrass meadows at local scales are urban, industrial and agricultural runoff, infrastructure development and dredging [5]. These pressures impact seagrasses directly via physical removal or indirectly through the introduction of pollutants such as nutrients, or suspended sediments that result in a reduction of light reaching seagrass meadows. Seagrasses are sensitive to light reduction as they are typically adapted to high light environments [1].

Increasing research is being undertaken to improve the management and conservation of seagrass meadows through improved understanding of the risks they face (e.g. [6]), developing bioindicators of the pressures they are exposed to [7] and thresholds of stressors such as light reduction which may differentiate sub-lethal effects from permanent loss of seagrass [8,9]. In general, leaf-level photosynthetic activity in response to irradiance follows a 'photosynthesis versus irradiance curve', which is linear for subsaturating irradiances but becomes non-linear, as progressively increasing irradiance causes saturation of the photosynthetic electron transport chain, and finally attains a plateau phase, which is defined as maximal photosynthesis rate (P_{max}) [10,11]. A key physiological parameter that represents a species response to a given light level is E_k, defined as the intersection between the initial linear slope and P_{max} on a P-I curve. E_k is frequently referred to as the 'saturating irradiance' [12,13] although technically it is slightly below the irradiance at which full photosynthetic saturation occurs, and above the irradiance at which saturation starts to cause deviation from linearity. E_k can be empirically determined and for each species may vary over a restricted range due to physiological acclimatization or factors such as temperature [14].

Various light threshold analyses have been proposed as having predictive capability for seagrass mortality. Dependent on available data, light levels can be assessed with respect to different factors or components of the environment, including the water column light attenuation coefficients [15] or Secchi disk depths [16]; light at the top of the seagrass canopy expressed as percentage of surface irradiance [17,18]; instantaneous or mean daily irradiance [8,19] or the number of hours of irradiance above E_k per day, H_{sat}, [8,20]. These thresholds can also be integrated over time, which is relevant to management when pressures persist over particular durations, e.g., dredging or flood plumes. The percentage of days below a particular mean daily irradiance [8] or the sum of the hours of irradiance below E_k compared to reference conditions [9] are two examples for which thresholds have been proposed to predict the onset of seagrass mortality.

One important component that all of these thresholds do not consider is the interaction of the seagrass canopy itself with the benthic light field, since it is the amount of light reaching individual leaves of a seagrass that governs the plants photosynthetic response [21]. The photosynthetic activity in turn influences how the seagrass meadow responds to the changes in light [12] and overall plant productivity [22]. Canopy structure of seagrass meadows can also vary markedly due to natural variations in light [23] or in response to light perturbations [9]. Due to canopy self-shading, light levels at the top of the canopy may be very different to light levels within the canopy, and will vary throughout the canopy in a manner dependent on the incident benthic light field, canopy structure and bending angle of the leaves, which vary under water motion [24,25,26]. Therefore, a mechanistic explanation of how light levels affect canopy sustainability must include the interaction of the canopy structure with the incident light field.

In this study we developed a 3D model of a complex seagrass canopy (*Amphibolis griffithii*) of varying structure, from low to high leaf area index (LAI), by adapting the model described in [25] and [27]. We modeled the exposure of these virtual canopies to a number of environmentally relevant levels of light reduction to assess the amount of light reaching each leaf surface and how this varies under different canopy densities and positions due to movement associated with water motion. Finally, we assessed the canopy saturation state by relating the light each leaf receives to values of leaf-level E_k for *A. griffithii* found in the literature. The modeling scenarios were based on empirically quantified canopy structures from specific plant morphologies, and were designed to be comparable to a shading experiment that was conducted on *A. griffithii* in 2005 [9].

In summary, the objectives were:

- To develop a 3D canopy model for a seagrass species with a complex canopy, hence demonstrating an advance in technical capability with respect to the simple *Thalassia* morphology model of Hedley and Enríquez [25].

- To understand the consequences on within-canopy light capture and canopy saturation state of 1) canopy position: upright vs. moving under high wave action, 2) canopy structure: low to high LAI (1.27 to 7.65), and 3) light reduction: 0–95% shading

- To identify potential descriptors of canopy light levels which could have use for the management of seagrass beds under light reduction events such as dredging or coastal pollution.

Methods

Canopy structures

The modelling experiment was designed to mirror aspects of a previously published empirical shading manipulation experiment [9,28]. The empirical study utilised an extensive (>6 ha) meadow of *Amphibolis griffithii* in 4.5 m water depth at Jurien Bay, Western Australia (30° 18′ 34″ S, 115° 00′ 26″ E; WGS84 datum). A control plus two-treatment shading experiment was conducted, the first phase of which ran from 10th March to 14th June 2005. Before and during the experiment individual *A. griffithii* plants were sampled and characterised in terms of stem and branch lengths, internodal distances, and number and dimensions of terminal leaves (Fig. 1a).

In the computer model, ten sets of individual plant data from the initial control sampling were replicated as vector mesh structures (Fig. 1f). The model plants were assembled into five canopies of leaf area index (LAI) from 1.27 to 7.65, by varying the choice and number of plants in a 20 cm×20 cm segment of substrate (Table 1, Fig. 2). The leaves and stems of the vector mesh structures were modelled as a point-mass and force system according to methods typically used for modelling cloth in the computer graphics industry [29]. A dynamic numerical integrator modelled the plant structures flexing naturalistically under a simple wave-action force model. Two wave actions, 'high' and 'low', were employed. In the dynamic model the low wave energy treatment plants were allowed to assume a typical upright position with no wave induced movement (Fig. 2). Under high wave energy plants underwent a vigorous cycle of forward and backward motion (Figs. 2f–i). From these dynamic models canopy structure treatments were extracted as instantaneous snapshots for each of the five LAI treatments: 1) a single snapshot for low wave action, 2) 14 snapshots through a cycle of movement for high wave action (Fig. 2). The 14 snapshots for high wave action were individually passed to the optical model (see below) and the results were averaged, thereby assuming the canopies undergo this movement continuously and photosynthetic response is the mean of the responses at any instant in time.

Water column optical model and shading

The canopy structures were input to the optical model for estimating diurnal leaf-incident irradiance. The model framework, previously described in [27] and [25], propagates sky radiance distributions through the canopy to give leaf incident irradiance in 17 wavebands of 20 nm width from 400–740 nm. Spectral irradiance can then be reduced to photosynthetically available radiation (PAR) at leaf level, and related to leaf tissue photosynthetic saturating irradiance, approximated by E_k (Fig. 1). To parameterise the model, hourly clear sky radiance distributions were produced using libRadtran and a directional radiance model [30,31] corresponding to the Jurien Bay site on 27th April; the middle of the post summer 3 month trial phase in Lavery et al. [9] (Fig. 1c).

The sky radiance distributions were input to PlanarRad (http://www.planarrad.com), a plane parallel water column model functionally similar to HydroLight [32,33] to estimate the hourly top of canopy radiance distribution (Fig. 1e). The model provides directional radiance tabulated over a hemisphere of zenith and azimuth angles, but to remove any dependency on sun azimuth and canopy orientation downwelling irradiances were azimuthally averaged to have only a zenith angle dependency (Fig. 1c, e). The water column utilised a library set of spectral inherent optical properties (IOPs, for details see [34]) which when input to the model produced a diffuse attenuation of planar PAR irradiance,

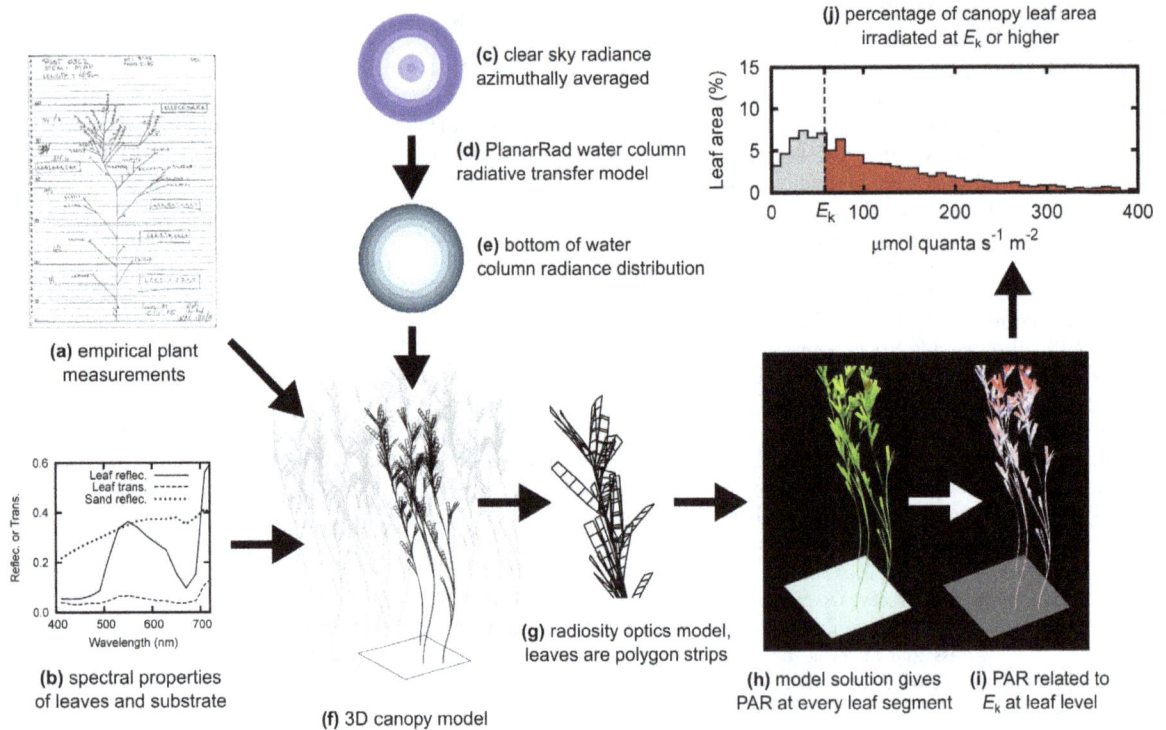

Figure 1. Overview of modelling system. (a, b) empirical data informs construction of 3D canopy model (f), (c, d, e) A plane-parallel model estimates directional radiance incident on the top of the canopy, (f, g, h) a geometric optical model handles radiative transfer to and between leaf segments, (i, j) PAR distribution over leaf area is reduced to the percentage of the canopy irradiated above leaf-level photosynthetic saturation, E_k.

k_d, of approximately 0.2 m^{-1}. In comparison, k_d values measured at the time of the empirical shading experiment ranged from 0.07 to 0.19 over a four month period but were 0.19 in April (Department of Parks and Wildlife, unpublished data). A set of nine modelled shading treatments were implemented by taking the top of canopy radiance distribution and reducing the values by 10%, 20%, etc. up to 95% (Table 1). Hence shading was spectrally neutral as was the shade cloth used in Lavery et al. [9], where the shading treatments were equivalent to 81–87% and 89–95% in our notation. The empirical study therefore represented quite a strong shading effect with respect to the modelled range. The water column optical model was additionally evaluated by comparing modelled top of canopy daily PAR irradiance against *in-situ* measurements from the associated study [28].

Canopy structure optical model

The top of canopy irradiance was propagated through a geometrical optics model [25,27] that accounts for inter-reflection and transmission between leaf segments. The spectral reflectance and transmittance of *A. griffithii* leaves was taken from the paper of Durako [35]. In this study we did not attempt to capture inter or intra-plant variability in leaf absorptance. This can be done [25], but the data collection requirements are onerous. All surfaces were considered Lambertian reflectors and transmitters. The underlying substrate reflectance was set from a library sand spectral reflectance that had a mean value of 0.33. The 20×20 cm modelled canopy segment was repeated periodically horizontally so the modelled canopy was of uniform LAI and has no edge (Fig. 1f).

Empirical measurements of PAR irradiance close to midday at both the canopy top and on the substrate underneath canopies of *A. griffithii* of differing LAIs were available for validation of the

canopy optical model from the study of McMahon and Lavery [28]. Canopy transmission was measured in control and treatment plots of varying but known LAI through measuring the instantaneous photosynthetic photon flux density (PPFD, μmol $\text{m}^{-2} \text{ s}^{-1}$) at the top and base of the canopy. The light sensor (Odyssey PAR sensor) was calibrated against a standard calibration light source (Quartz Tungsten Halogen Reference Lamp operated at 3150°K from a LI-1800-02 Optical Radiation Calibrator). The low wave energy structure model treatments at the hour closest to midday were used to perform this validation, and a number of additional runs with different LAIs to those in Table 1 were added to further populate the validation data. An additional quality assurance protocol for the canopy optical model is to set the within-canopy water absorptance to zero and then verify energy conservation between the top of canopy incident and exitant irradiances and energy absorbed by all surfaces in the model [25]. This was performed for a subset of the runs in Table 1.

Relation to photosynthetic properties

The model solution provided incident PAR at every point on every leaf at a resolution of approximately 0.5 cm^2. This was then related to the leaf level saturation irradiance for *A. griffithii*, approximated by E_k. Masini and Manning [12] evaluated E_k in *A. griffithii* as ranging from 25 to 55 μmol quanta $\text{m}^{-2} \text{ s}^{-1}$ for temperatures of 13°C to 23°C, of which the upper value is closer to the conditions of the empirical data from the associated study here. While Masini and Manning [12] did not assess physiological variation in E_k in *A. griffithii*, in the same study *Posidonia sinuosa* was shown to have E_k that varied from 50.4 to 39.1 μmol quanta $\text{m}^{-2} \text{ s}^{-1}$ in depths of 4 m and 12–15 m respectively. Therefore to accommodate a realistic variation in E_k in the absence of data, we

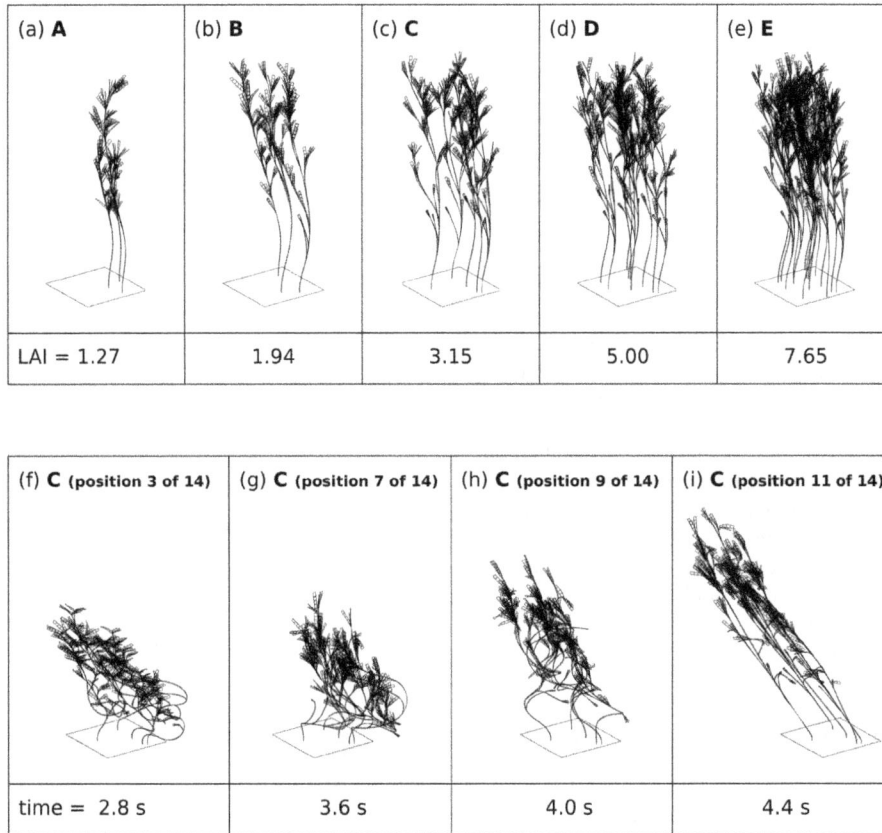

(a) **A**	(b) **B**	(c) **C**	(d) **D**	(e) **E**
LAI = 1.27	1.94	3.15	5.00	7.65

(f) **C** (position 3 of 14)	(g) **C** (position 7 of 14)	(h) **C** (position 9 of 14)	(i) **C** (position 11 of 14)
time = 2.8 s	3.6 s	4.0 s	4.4 s

Figure 2. Example canopy structures and positions used in the model treatments. (a)–(e) low wave action canopy position for the five LAI treatments A to E. (f)–(i) subset of time sequence positions under high wave action for canopy C, all 14 positions were used in the optical model. In all cases the canopy structure is notionally repeated in all horizontal directions such that the square substrate section tessellates.

have used the comparable range of 45–55 μmol quanta m^{-2} s^{-1} to put all of our results into the context of potential physiological variation in E_k. To produce plots including the range and mid-point specific values of 45, 50 and 55 were used. Based on the value of E_k of 45 or 55 μmol quanta m^{-2} s^{-1}, the model can report the instantaneous proportion of the leaf area of the canopy that is irradiated at or above the saturation irradiance. To interpret these results, for each canopy structure and shading treatment, a value termed H^A_{sat} was calculated as the time integral of the percentage leaf area above saturation in a 24 hour period, with units % leaf area×hour. This measure is discussed later, but was intended to be analogous to H_{sat}, the daily top of canopy irradiated hours above saturation [9,20] but also factoring in the canopy self-shading.

Table 1. Modelling experiment design.

LAI	Structure	Shading (%)	Hour
A −1.27	low wave energy (1 position)	0	×12
B −1.94	high wave energy (14 positions)	10	
C −3.15		25	
D −5.00		40	
E −7.65		50	
		60	
		70	
		85	
		95	

A fully-factored set of model runs were performed for each of five LAI treatments, 15 canopy structures and nine shading treatments over 12 hourly diurnal intervals, a total of 8100 runs.

Results

Optical model validation

The sky radiance and water column model produced a daily top of canopy PAR dose of 11.0 mol quanta m^{-2}, whereas the comparable *in-situ* measured average daily PAR irradiance over 3 months was 19.0 mol quanta m^{-2} [9]. Since our study was primarily concerned with the relative effect of the shading treatments and LAI this discrepancy is not of great importance, but could be due to: 1) the accuracy of the libRadtran sky radiance model (no validation data available); 2) the accuracy of the Odyssey PAR sensors, which can have issues in long-term stability (Slivkoff, pers. comm.), or; 3) the model water column k_d(PAR), which was at the upper range compared to measurements taken during the empirical study (0.2 vs. 0.070.19). This deviation in k_d(PAR) does provide an almost exact explanation for the discrepancy, but since k_d(PAR) is a wavelength-integrated output of the model parameterised on spectral IOPs for absorption and backscatter it is not trivial to set an arbitrary value of k_d(PAR). In the scope of this study, using the closest IOP set from actual measured data [34] was considered adequate. In reality the daily measured PAR was sometimes above and sometimes below the model value, so all things considered the modelled canopy PAR dose was reasonable and the discrepancy is inconsequential to the subsequent interpretation of the results.

The percentage of the incident top of canopy PAR irradiance transmitted to the substrate, as a function of leaf area index, validated well against empirical data (Fig. 3). The empirical data showed wide variation, but the modelled transmitted irradiances corresponded very closely to the upper bound of the empirical data. This is to be expected, since some of the real canopies contained free standing and epiphytic macroalgae which would have reduced the transmission beyond that described by the *A. griffithii* LAI alone. The upper bound points most likely represent the most monospecific *A. griffithii* canopies and correspond best to the model. An exponential function fit to the model data ($n = 28$) gave r^2 of 0.96, the fit of all 27 empirical data points to that same function gives an r^2 of 0.73. However, if only four outliers are removed (Fig. 3) the empirical data r^2 rises to 0.90. The model and empirical data therefore compare well, especially given the practical difficulty in making accurate within-canopy light measurements.

The performance of the model in terms of energy conservation was demonstrated in the subset of runs for which water absorption was set to zero. For the majority of runs energy losses were less than 2% and for all runs they were less than 3%. In practice, when water absorption is non-zero, energy conservation performance would be better than these figures suggest. The current model implementation requires water absorption to be set to zero for energy accounting, but this in itself removes a damping effect on the multiple scattering and increases energy losses through numerical errors. Therefore the model solutions for leaf incident irradiance can be considered, at worst, slight underestimates by around 2%.

Effect of canopy structure and position on leaf level PAR

As expected, the distribution of leaf level PAR irradiance became increasingly skewed to lower values as LAI increased (Figs. 4a, e, i, m, q). In low LAI canopies the distribution of PAR over leaf area was almost flat: leaf tissue received a wide range of PAR with almost equal probability, and much of it was above saturating irradiance at mid-day under the model conditions of clear sky and moderately clear water. In denser canopies the leaf level light distribution had a long high-end tail: many leaves

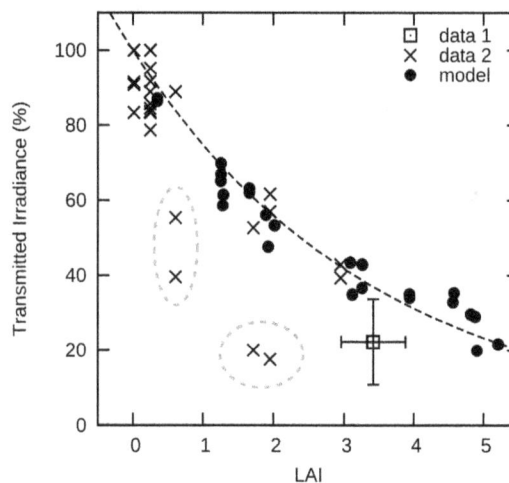

Figure 3. Percentage of downwelling top of canopy PAR irradiance reaching the substrate, as a function of LAI. Results for empirical in-situ measurements (data 1 and 2) and modelled estimates (model) are shown. Curve fit to model data points is y = 100 * exp(−0.29×LAI).×(data 2) - are 26 individual coincident measurements of LAI and irradiance above and below the canopy. ⊡ (data 1) is a single point based on a site mean of 13 LAI determinations from 2.84 to 4.09 and a set of associated but not spatially co-incident light measurements, error bars are one S.D. The four encircled points are the outliers referred to in the text. Transmittance data collection is described in McMahon and Lavery 2014 and corresponding LAI is unpublished data from Lavery et al. 2009. ● are 25 model runs including both those described in detail and some additional runs. In the models runs solar zenith angle was approximately 28°.

received light below E_k, but a few leaves received very high light (Fig. 4i). Overall the pattern was clearly linked to the relative openness or self-shading within the canopies. The range of E_k of 45 to 55 μmol quanta m^{-2} s^{-1} was generally small compared to range of irradiances the leaves experienced, but this was more true for the lower LAI and unshaded treatments (Fig. 4).

The treatments of canopy position of upright or moving under wave action appeared to have little effect on leaf level PAR irradiance (e.g. Fig. 4a vs. 4b). Numerically the canopy movement slightly reduced the daily integrated percentage of saturated leaf area for all but the lowest LAI (Fig. 5). However overall there was not a statistically significant difference at either E_k of 45 or 55 μmol quanta m^{-2} s^{-1} (paired value *t*-test, $p>0.05$). Therefore there is no evidence from our data that canopy movement affects time-integrated light capture. However the instantaneous light capture has a high variation under movement. While the standard deviation was 10–20% of the mean (Fig. 5) at some individual time points the saturated leaf area was up to 50% more or less than the mean. As expected, shading scaled the *x*-axis position of leaf-level irradiance distribution plots by the corresponding factor. That is, halving the top of canopy irradiance halved the leaf level incident irradiance at every leaf (Figs. 4b, d, f, h, etc.).

Diurnally accumulated saturated leaf area

The accumulated percentage of leaf area above saturation over a twelve-hour day showed a complex relationship between both shading and LAI (Fig. 6). While, as expected, increasing either shading or LAI monotonically decreased the accumulated percentage of leaf area above saturation, the shape of the function was non-linear and there was an interaction between shading and LAI (Fig. 6a). The contour lines in Figure 6a make clear the trade-

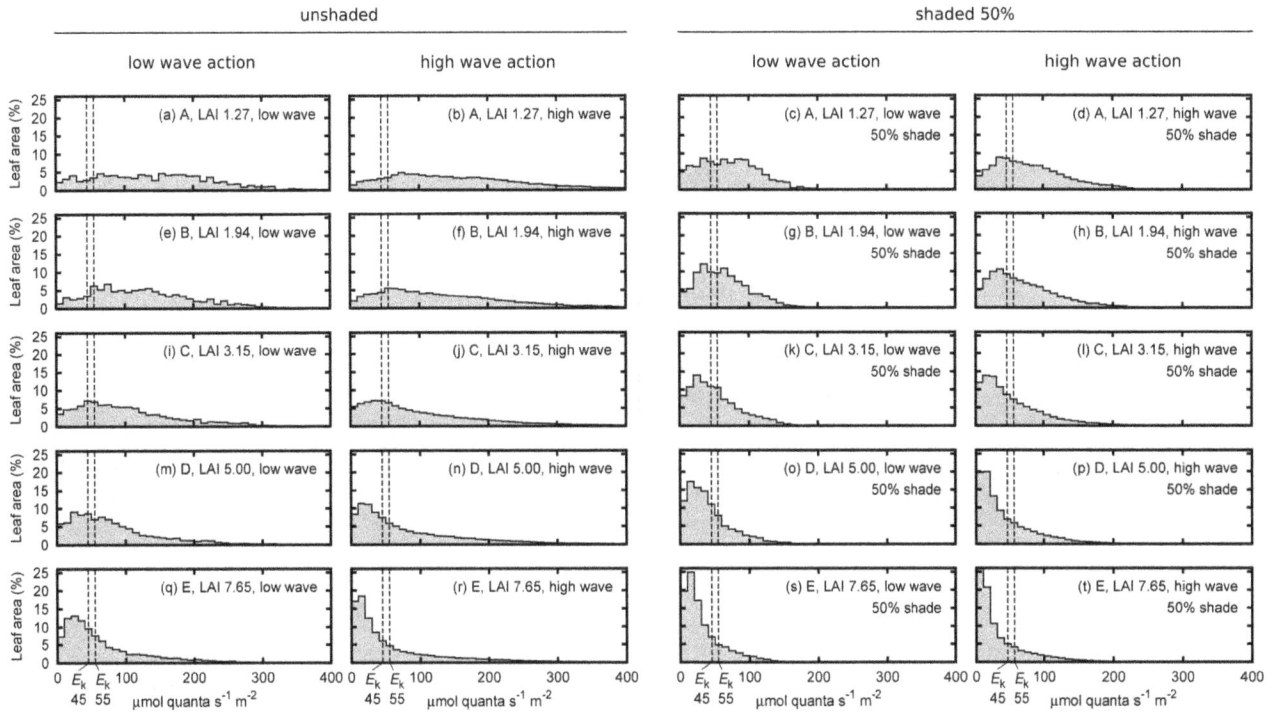

Figure 4. Distribution of leaf saturation state in the canopy in terms of percentage of leaf area at midday. Treatments are upright low wave action canopy positions and the average over the high wave action movement positions (left columns), and the same for 50% shading (right columns), for the five LAI treatments. The estimated photosynthetic saturation irradiances, E_k, of 45 and 55 μmol quanta m^{-2} s^{-1} inferred from Masini and Manning (1997) are shown as vertical dotted lines.

off between leaf area index and light with respect to the saturation state of the canopy. These lines show equal points in the LAI–shading function space, so for example a canopy of LAI 5.5 with no shading was equivalent to LAI 2.0 with 50% shading, with respect to the diurnally accumulated saturation of relative leaf area. The potential acclimation range of E_k from 45 to 55 μmol quanta m^{-2} s^{-1} (assuming water temperature at approx. 23°C) added a degree of freedom to the LAI-shading relationship approximately equivalent to 1 unit of LAI at low shading (e.g. along the x-axis of Fig. 6a), but this increased as shading increased to 60% or more (Fig. 6a). Therefore at low shading modifying E_k

over the suggested range is equivalent to changing LAI by plus or minus one half.

For the high wave action treatment there is a small qualitative difference in the position of the contour lines in low LAI and low shading region as compared to the upright low wave action treatment (Fig. 6b vs. 6a). However a sensitivity analysis of the data tables underlying Figures 6a and 6b showed that these differences are equivalent to an error in the shading percentage of only 6 points or less. In other words, if in a practical application shading were quantified at discrete levels of 0, 5, 10, 15% etc. then the difference between upright and moving canopies would be negligible.

Discussion

Geometric optical modelling of seagrass canopies and validation

In terms of the geometrical optical modelling of seagrass canopies, the results presented here corroborate those of Hedley and Enríquez [25], showing that it is possible to construct a physical three dimensional model of a seagrass canopy and obtain acceptable validation against *in-situ* light measurements. Through-canopy transmission was estimated accurately for pure *Amphibolis* canopies, but the importance of considering epiphytes or other canopy constituents was underlined by the high variability of the empirical data, which in some cases had lower light penetration than the model predicted based on *Amphibolis* LAI alone.

With respect to morphological complexity, *A. griffithii* is on the more complex end of the spectrum in comparison to strap leaf morphologies of *Thalassia* and many other seagrass species [36], to which this modelling framework was previously confined.

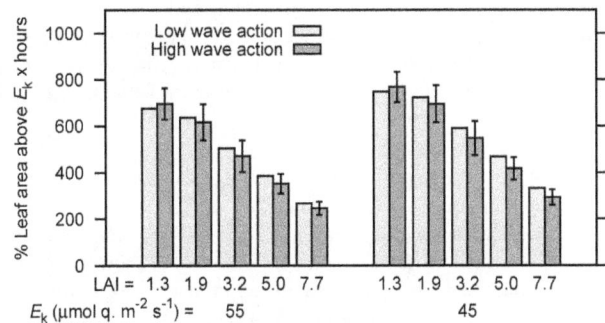

Figure 5. Time accumulated percentage of leaf area irradiated above photosynthetic saturation irradiance for low wave action and high wave action treatments, for E_k of 45 and 55 μmol quanta m^{-2} s^{-1}. Error bars on high wave action treatment are the standard deviation over all 14 movement positions and hence indicate the range in the instantaneous canopy saturation state.

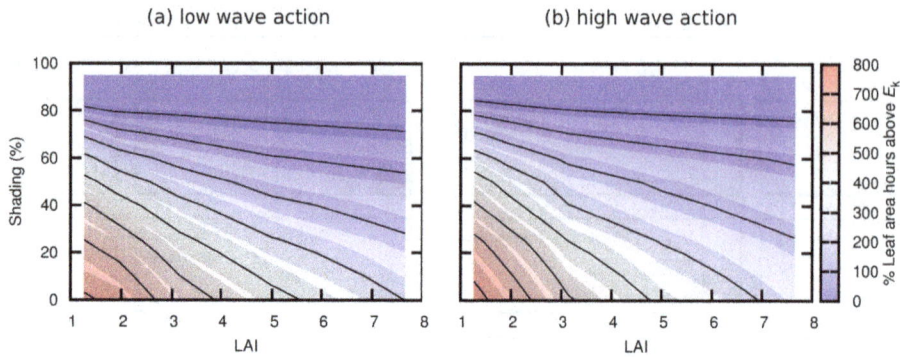

Figure 6. Time accumulated percentage of leaf area irradiated above photosynthetic saturation irradiance. The colour scale shows H^A_{sat}, the "percentage leaf area hours" above E_k, as a function of LAI and shading. (a) is for upright low wave action canopy structures, (b) is for the average over the high wave energy canopy positions. Contour lines are isoclines based on the mid value of E_k equal to 50 μmol quanta m^{-2} s^{-1} while the surrounding greyed region shows the limits for E_k of 45 to 55 μmol quanta m^{-2} s^{-1}. The isoclines are located at H^A_{sat} of 50, 100, 200, and then steps of 100 up to 800% leaf area hours.

Therefore the potential for future models of other seagrass species is good as these two examples capture the range in canopy complexity. Hedley and Enríquez [25] used profiles of light through the canopy to derive a diffuse attenuation profile, k_d, for validation. In this study only light measurements at the top and bottom of the canopy were available. However, this simpler validation may be preferable and adequate. In practice, empirical measurements of light profiles within canopies are difficult to make, and rarely fit well to exponential attenuation with depth. The measurements at the top and bottom of the canopy are the strongest "signal" for within-canopy attenuation and can be used to derive k_d if canopy height is known. So in future empirical work to which such modelling may be subsequently applied, we recommend measuring downwelling irradiance at the top and bottom of canopies, together with canopy height and LAI.

Influence of shading, LAI and position on diurnal leaf photosynthetic saturation

The previous empirical shading study on *A. griffithii* [9] quantified the change in H_{sat} induced by shading, i.e. the total number of hours of top-of-canopy irradiance that was above photosynthetic saturation, as compared to the unshaded treatments. This quantity, summed over time, was demonstrated as a good indicator of changes in canopy biomass and capacity for subsequent recovery. However H_{sat} relates the top of canopy irradiance to leaf-level photosynthetic saturation irradiance and so ignores canopy self-shading and other structural factors, which therefore introduce an additional degree of freedom. Here, we factor in the canopy structure by considering the percentage of the canopy leaf area above saturating irradiance accumulated over time, H^A_{sat}, with units of % leaf area×hour. This descriptor extends H_{sat} by reducing to a single number the interaction of the duration of saturating irradiance and the canopy self-shading. It can be roughly interpreted as the daily ratio of saturated photosynthesis to leaf area at canopy scale, and ranges from 0 to 1200 for plants completely saturated for 12 hours of daylight.

Considering the variation in H^A_{sat} with leaf area index and shading, as expected LAI has a strong effect on diurnal leaf saturation state (Fig. 6.). A change in LAI from 1 to 7 has as much effect as 60% shading (Fig. 6a), so the ambient light field cannot be treated independently from the canopy structure when photosynthetic processes at leaf level are of interest. Furthermore, the relationship between LAI and both shading and leaf saturation

state is a non-linear interaction; Figure 6 represents a curved surface in both axes of shading and LAI. This occurs because while leaf level irradiance is a linear function of canopy level irradiance, the leaf level photosynthetic response is not a linear function of irradiance when the irradiance approaches or exceeds E_k. As leaf level irradiance approaches and exceeds E_k the photosynthetic response levels off at P_{max}. In general, since photosynthesis versus irradiance curves are non-linear in the region of the saturating irradiance any derived measure of leaf level photosynthetic activity will have a complex relationship with LAI unless all leaves are well below saturating irradiance.

Within this study there is no statistically significant evidence to support the statement that canopy movement effects light capture and photosynthetic response. Qualitatively it is interesting that under movement the lowest LAI canopy experienced an increase in daily saturation whereas the higher LAI canopies were systematically lower (Fig. 5). To test the statistical significance of this observation would require substantial further modelling effort and was outside the scope of this study, however at low LAIs sideways movement may serve to enhance light collection by spreading leaves out horizontally and making them insensitive to the directionality of incident light. Under wave action such flattening is intermittent and as we have shown here is not a factor of great photosynthetic significance. Additionally, under these conditions the optical consequences of surface waves and sediment resuspension should also be considered and may be more significant [37]. However, in other systems and species, canopy flattening can be a result of shallow water depth or tidal or estuarine flow [38]. In this case the semi-constant flattening of the canopy may be of optical significance.

Potential for LAI modification as an acclimation response

Figure 6 indicates that modification of LAI is a possible response to maintain the saturation state of the canopy under reduced or enhanced light conditions. This role of morphological plasticity has been demonstrated in a number of experimental studies (e.g. [39]) and hypothesized as a regulatory mechanism in *Thalassia* [13]. From our model data (Fig. 6), if a canopy of LAI 7 is observed to reduce to LAI 4 after a period of 40% reduction in light, this loss of biomass might be interpreted as a trajectory of canopy decline but alternative interpretation is that of an acclimation response to maintain the leaf level photosynthetic state. This interpretation is independent of the mechanism by which it occurs. Leaf mortality might be considered just a by-product of inadequate light to maintain respiration, but if the net effect is a return to an unstressed leaf-level light regime then the distinction between a compensatory morphological adjustment and a decline is at best ambiguous. This argument is of course dependent on the definition of 'decline'; it does not apply if net productivity per unit area rather than biomass is the criteria. Here, by 'decline' we mean an implied trajectory toward canopy eradication.

Under the interpretation of potential acclimation a key question is whether the reduction in LAI remains on the isocline for canopy saturation state (i.e. the contour lines in Fig. 6). A canopy that moves on a trajectory through LAI–shading 'space' such that it stays on a contour line is experiencing the same time integrated percentage of its leaf area at saturating irradiance (see A to E in Fig. 7). That is, it experiences the same daily photosynthetic saturation in relation to its leaf biomass. We might therefore hypothesise that if a canopy can sustainably exist at one point on an H^A_{sat} isocline, canopies can also sustainably exist at other points on that line, all other things being equal. Movement along an isocline can occur purely by modification of the LAI, alternative acclimation responses such as modification of E_k at the leaf-level will enable movement perpendicular to the isoclines, illustrated by the range around the isoclines in Figures 6 and 7 delimited by E_k of 45 to 55 µmol quanta m^{-2} s^{-1}. Ignoring the latter possibility, and focussing on LAI modification alone, the prospects for long term survival of a canopy under a change in light environment can be estimated by following the isocline from its current location in LAI-shading space to the new light environment (y-axis) location. If at this location the LAI is greater than zero (judged by extrapolation), then the canopy could survive by thinning out to this LAI, at which point it will have the same relative photosynthetic saturation of its leaf area. In the following section we use this concept to interpret the results from the previous empirical shading experiment [9].

Canopy trajectories in LAI and shading space

In the post-summer treatment of the shading experiment of Lavery et al. [9] canopies with an LAI of ~4 had reduced to an LAI of ~2 after three months of 84% shading. This change in LAI and shading can be represented as a trajectory on the H^A_{sat} map: Figure 7, point A to B. At 6 months of shading the LAI had reduced to one and by 9 months the canopy was almost eradicated

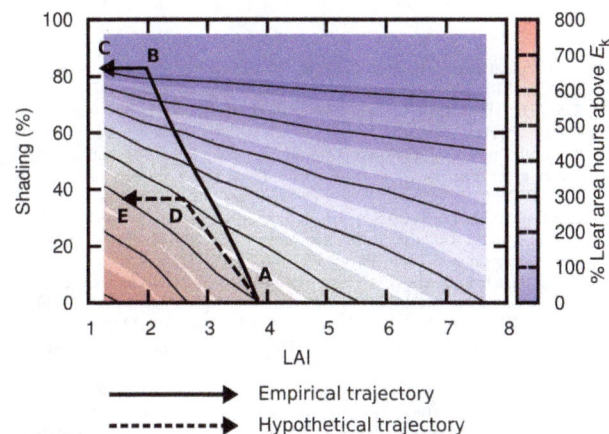

Figure 7. Trajectories of canopies from empirical shading experiment and hypothetical example. Underlying plot is as described in the caption of Figure 6a.

and did not subsequently recover (Fig. 7, point C). Assuming for the moment that H^A_{sat}, the accumulated percentage area of leaf saturation above E_k, is a measure relevant to canopy sustainability then Figure 7 indicates that such a measure could have predictive power for canopy survival. In the previous section we postulated that canopies can move along the isoclines by modification of LAI alone. The initial reduction of LAI in the empirical data (Fig. 7, point A to B) occurred in the first three months in response to 84% shading. The trajectory cuts across the isoclines because there is a time lag as the canopy cannot become thinner instantaneously. At three months (Fig. 7, point B) the LAI has reduced to 2 but the shading is extreme so at the leaf level the light environment is still very much reduced. There can be two response pathways, either physiological changes may allow the canopy to exist on the new isocline, such as the mobilisation of stored reserves [40] or reductions in the saturating irradiance [23], or if such processes cannot bring about sufficient change then further LAI reduction is required in an attempt to return to an isocline closer to the original. In this case following the isocline to the right and extrapolating to the intercept with the x-axis it is clear that the light environment is equivalent to a canopy with huge LAI of at least 20+ in the original un-shaded situation. For the canopy to survive would require physiological changes that would permit canopies of these high LAIs to exist normally in this environment. Such canopies did not exist, hence physiological changes are insufficient (it is clear from Fig. 7 that variation in E_k is inadequate), hence the LAI continued to decrease and eventually the canopy was eradicated (Fig. 6, Point C).

The previous example is a straightforward case of severe light limitation, but with moderate shading (for example 40%, Fig. 7) the situation is more complicated. The empirical data of Lavery et al. [9] only contained shading at a minimum of 81% so the example of 40% in Figure 7 is hypothetical. If a situation of 40% shading is introduced, assuming the validity of H^A_{sat}, it is clear that the canopy could survive by reducing LAI from 4 to 1 (moving along the isocline from point A to point E, Fig. 7). However, because of the time lag in reducing LAI, an initial trajectory in which LAI partially reduces (A to D, Fig. 7) is realistic and is likely to include physiological responses in tandem. For example E_k could decrease, but at LAI around 3 the range of E_k from 45 to 55 only allows accommodation of up to 20% shading at the most (Fig. 7). The existence of a time lag is supported by studies on different seagrass species that have incorporated less extreme shading treatments over short time-scales; physiological changes such as increases in chlorophyll and reduction in LAI occur after longer durations of reduced light [39]. At point D in Figure 7 the canopy lies on an isocline that represents a canopy of LAI 5.5 in the unshaded environment. If such canopies can sustainably exist, at the same depth, water clarity etc. then the canopy may survive at LAI of 2 to 3 (point D), rather than reducing LAI to 1 (point E). Either way, Figure 7 has predictive power for the canopy response in that if it is anticipated a 40% shading event may occur, e.g., from dredging activities, then it is clear that a canopy that is sustainable at LAI of 4 could reduce to LAI of 1, or, could induce physiological changes to maintain an LAI higher than 1, but in the latter case only if canopies greater than LAI of 4 currently exist in that environment.

The possible trajectories in LAI-shading space of Figure 7 are dependent on the capability and time constants of other physiological acclimation mechanisms. These mechanisms could include adjustments to photosystem kinetics to increase the efficiency (i.e. lower E_k), increases in chlorophyll content and a:b ratio to enhance light capture, or mobilisation of stored carbohydrates for maintenance and growth of the existing leaf biomass [7]. To our knowledge there is no published data on photo-acclimation in *A. griffithii* under changing light conditions. However, unpublished data by co-author McMahon shows that under high levels of shading there are reductions in the saturating irradiance and other photo-acclimation responses, which maintain electron transport rates at unshaded values, but there is a time-limit over which this photo-acclimation is maintained of around 21 days. Therefore, the model as we have developed here is very relevant for predicting impacts associated with longer term reductions in light of over three weeks or more.

Other measures of canopy scale photosynthetic response to light

In the previous discussion we have assumed that the concept of isoclines of equal light environment' with respect to H^A_{sat} is valid. Alternative measures may be more appropriate but this does not affect the primary concept that canopy self-shading can be equivalent to environmental shading, and that there are two mechanisms of photosynthetic acclimation: physiological and via canopy structure. Any alternative measure of the photosynthetically relevant light environment would likely have a similar form to that of Figures 6 and 7. The interaction of self-shading and the non-linearity of leaf-level photosynthesis must inevitably result in a complex canopy scale response to LAI and the light environment for all canopies that are subjected to irradiances above photosynthetic saturation. Another candidate measure would be the integration of photosynthesis over time, i.e. to propagate the leaf-level light through a photosynthesis versus irradiance (P-I) curve to give an integrated photosynthesis measure equivalent to μmol O_2 evolution. In addition, plots of actual top of canopy PAR light levels may have greater descriptive power than percentage shading (Figures 6 and 7). Lavery et al. [9] observed different canopy changes at similar shading levels and interpreted these as being due to differences in the absolute light levels. In this study we suggested H^A_{sat} as a simple extension of the top-of canopy H_{sat}, since that measure has been demonstrated to have predictive power for canopy sustainability [8,9] and has been used in management contexts, and percentage shading was employed as a mirror of the empirical treatments. Clearly, there are many opportunities for further experiments and modelling to determine the most relevant measure of canopy photosynthetic response, the key point being that that measure needs to include within-canopy light propagation.

Conclusions

Three dimensional canopy modelling of *Amphibolis griffithii* has revealed that the interaction of light levels and canopy density on canopy-scale photosynthetic activity is complex and non-linear, in particular due to the non-linearity of leaf-level photosynthesis at saturating irradiance. The accumulated percentage area of leaf saturation above saturating irradiance, H^A_{sat}, was proposed as a measure relevant to canopy sustainability, based on extension of the equivalent top-of-canopy measure H_{sat} that has previously proved useful. The available empirical data were not sufficient to evaluate the efficacy of H^A_{sat} due to lack of lower shading treatments. Evaluating this measure and other candidates such as integrated leaf-level photosynthesis requires further experimental work. Nevertheless the principle has been demonstrated that plots of equal light environment' (Fig. 6) produced for different seagrass species, water depths, and water column optical properties could have practical management applications for predicting and interpreting canopy changes under light reduction events. Reduction in seagrass density in response to shading must be interpreted in terms of the leaf-level light environment. While physiological responses are also important, existing canopies in the same environment can provide information of the limits of physiological acclimation, and indicate if change in light levels will induce a trajectory to steady state sustainability, or to eradication. An important future step is to understand the time constants in change and recovery trajectories, to determine how long shading events can be tolerated and the required recovery periods. This information will be invaluable to coastal management.

Acknowledgments

J Hedley wishes to acknowledge Susana Enríquez, whose ideas and insight have informed the work described here. K McMahon acknowledges the collaboration with P S Lavery, which provided background for this research, L Twomey and M Westera from BMT Oceanica Pty Ltd for insights contributing to this work and N Dunham for canopy measurements.

Author Contributions

Conceived and designed the experiments: JH KM PF. Performed the experiments: JH KM. Analyzed the data: JH KM PF. Contributed reagents/materials/analysis tools: JH KM. Contributed to the writing of the manuscript: JH KM PF.

References

1. Orth RJ, Carruthers TJB, Dennison WC, Duarte CM, Forqurean JW, et al. (2006) A global crisis for seagrass ecosystems. Bioscience 56: 987–996.
2. Costanza R, d'Arge R, de Groot R, Farber S, Grasso M, et al. (1997) The value of the world's ecosystem services and natural capital. Nature 387: 253–260.
3. Fourqurean JW, Duarte CM, Kennedy N, Marbà N, Holmer M, et al. (2012) Seagrass ecosystems as a globally significant carbon stock. Nature Geoscience 5: 505–509.
4. Waycott M, Duarte CM, Carruthers TJB, Orth RJ, Dennison WC, et al. (2009) Accelerating loss of seagrasses across the globe threatens coastal ecosystems. Proceedings of the National Academy of Sciences 106: 12377–12381.
5. Grech A, Chartrand-Miller K, Erftemeijer P, Fonseca M, McKenzie L, et al. (2012) A comparison of threats, vulnerabilities and management approaches in global seagrass bioregions. Environmental Research Letters 7: 024006.
6. Grech A, Coles R, Marsh H (2011) A broad-scale assessment of the risk to coastal seagrasses from cumulative threats. Marine Policy 35: 560–567.
7. McMahon K, Collier C, Lavery PS (2013) Identifying robust bioindicators of light stress in seagrasses: A meta-analysis. Ecological Indicators 30: 7–15.
8. Collier CJ, Waycott M, McKenzie LJ (2012) Light thresholds derived from seagrass loss in the coastal zone of the northern Great Barrier Reef, Australia. Ecological Indicators 23: 211–219.
9. Lavery PS, McMahon K, Mulligan M, Tennyson A (2009) Interactive effects of timing, intensity and duration of experimental shading on Amphibolis griffithii. Marine Ecology Progress Series 394: 21–33.
10. Lambers H, Chapin FS, Pons TL (1998) Plant physiological ecology. New York: Springer. 540 pp.
11. Kirk JTO (1994). Light and photosynthesis in aquatic ecosystems. Cambridge: Cambridge Press. 528 pp.
12. Ralph PJ, Durako MJ, Enríquez S, Collier CJ, Doblin MA (2007) Impact of light limitation on seagrasses. Journal of Experimental Marine Biology and Ecology 350: 176–193.
13. Cayabyab NM, Enríquez S (2007) Leaf photoacclimatory responses of the tropical seagrass Thalassia testudinum under mesocosm conditions: a mechanistic scaling-up study. New Phytol 176: 108–123.
14. Masini RJ, Manning CR (1997) The photosynthetic responses to irradiance and temperature of four meadow-forming seagrasses. Aquatic Botany 58: 21–36.
15. Duarte CM, Marba N, Krause-Jensen D, Sanchez-Camacho M (2007) Testing the predictive power of seagrass depth limit models. Estuaries and Coasts 30: 652–656.
16. O'Brien KR, Grinham A, Roelfsema CM, Saunders MI, Dennison WC (2011) Viability criteria for the presence of the seagrass Zostera muelleri in Moreton Bay, based on benthic light dose. In: MODSIM 2011: International Congress on Modelling and Simulation, Proceedings. Modelling and Simulation Society of Australia and New Zealand (MODSIM 2011), Perth, Australia, 12–16 December 2011. 4127–4133.
17. Kemp WM, Batiuk R, Bartleson R, Bergstrom P, Carter V, et al. (2004) Habitat requirements for submerged aquatic vegetation in Chesapeake Bay: Water quality, light regime, and physical-chemical factors. Estuaries 27: 363–377.

18. Dennison WC, Orth RJ, Moore KA, Court Stevenson J, Carter V, et al. (1993) Assessing water quality with submersed aquatic vegetation: habitat requirements as barometers of Chesapeake Bay health. BioScience 43: 86–94.

19. Gacia E, Marba N, Cebrian J, Vaquer-Sunyer R, Garcias-Bonet N, et al. (2012) Thresholds of irradiance for seagrass *Posidonia oceanica* meadow metabolism. Mar Ecol Prog Ser 466: 69–79.

20. Dennison WC, Alberte RS (1985) Role of daily light period in the depth distribution of *Zostera marina* (eelgrass). Mar Ecol Prog Ser 25: 51–61.

21. Enríquez S, Merino M, Iglesias-Prieto R (2002) Variations in the photosynthetic performance along the leaves of the tropical seagrass *Thalassia testudinum*. Marine Biology 140: 891–900.

22. Fourqurean JW, Zieman JC (1991) Photosynthesis, respiration and whole plant carbon budget of the seagrass *Thalassia testudinum*. Mar Ecol Prog Ser 69: 161–170.

23. Collier CJ, Lavery PS, Masini RJ, Ralph PJ (2007) Morphological, growth and meadow characteristics of the seagrass *Posidonia sinuosa* along a depth-related gradient of light availability. Mar Ecol Prog Ser 337: 103–115.

24. Carruthers TJB, Walker DI (1997) Light climate and energy flow in the seagrass canopy of *Amphibolis griffithii* (J.M.Black) den Hartog. Oecologia 109: 335–341.

25. Hedley JD, Enríquez S (2010) Optical properties of canopies of the tropical seagrass *Thalassia testudinum* estimated by a three-dimensional radiative transfer model. Limnology and Oceanography 55: 1537–1550.

26. Zimmerman R (2006) Light and photosynthesis in seagrass meadows. In: Larkum AWD, Orth RJ, Duarte C, editors. Seagrasses: Biology, ecology and conservation. Springer. 303–321.

27. Hedley JD (2008) A three-dimensional radiative transfer model for shallow water environments. Optics Express 16: 21887–21902.

28. McMahon K, Lavery PS (2014) Canopy-scale modifications of the seagrass *Amphibolis griffithii* in response to and recovery from light reduction. Journal of Experimental Marine Biology and Ecology 455: 38–44.

29. House DH, Breen DE (2000) Cloth modeling and animation. Massachusetts: A. K. Peters. 344 p.

30. Mayer B, Kylling A (2005) The libRadtran software package for radiative transfer calculations – description and examples of use. Atmospheric Chemistry and Physics 5: 1855–1877.

31. Grant RH, Heisler GM, Gao W (1996) Photosynthetically-active radiation: Sky radiance distributions under clear and overcast conditions. Agricultural and Forest Meteorology 82: 267–292.

32. Mobley CD (1994) Light and Water. San Diego: Academic Press. 608 p.

33. Mobley CD, Sundman L (2000) HydroLight 4.1 user's guide. Sequoia Scientific. Available: http://www.sequoiasci.com/products/Hydrolight.aspx.

34. Hedley JD, Roelfsema CM, Phinn SR, Mumby PJ (2012). Environmental and sensor limitations in optical remote sensing of coral reefs: Implications for monitoring and sensor design. Remote Sensing 4: 271–302.

35. Durako MJ (2007) Leaf optical properties and photosynthetic leaf absorptances in several Australian seagrasses. Aquatic Botany 87: 83–89.

36. Green EP, Short F (2004) World atlas of seagrasses. Berkeley: University of California Press. 320 p.

37. Pedersen TM, Gallegos CL, Nielsen SL (2012) Influence of near-bottom re-suspended sediment on benthic light availability. Estuarine, Coastal and Shelf Science 106: 93–101.

38. Koch E, Gust G (1999) Water flow in tide and wave dominated beds of the seagrass *Thalassia testudinum*. Mar Ecol Prog Ser 184: 63–72.

39. Collier CJ, Waycott M, Ospina AG (2012b) Responses of four Indo-West Pacific seagrass species to shading. Marine Pollution Bulletin 65: 342–354.

40. Brun FG, Vergara JJ, Hernádez I, Pérez-Lloréns JL (2003) Growth, carbon allocation and proteolytic activity in the seagrass *Zostera noltii* shaded by *Ulva* canopies. Funct Plant Biol 30: 551–560.

4

Abiotic versus Biotic Drivers of Ocean pH Variation under Fast Sea Ice in McMurdo Sound, Antarctica

Paul G. Matson[1]*, **Libe Washburn**[2], **Todd R. Martz**[3], **Gretchen E. Hofmann**[1]

1 Department of Ecology, Evolution, and Marine Biology, University of California Santa Barbara, Santa Barbara, California, United States of America, 2 Department of Geography, University of California Santa Barbara, Santa Barbara, California, United States of America, 3 Geosciences Research Division, Scripps Institution of Oceanography, University of California San Diego, La Jolla, California, United States of America

Abstract

Ocean acidification is expected to have a major effect on the marine carbonate system over the next century, particularly in high latitude seas. Less appreciated is natural environmental variation within these systems, particularly in terms of pH, and how this natural variation may inform laboratory experiments. In this study, we deployed sensor-equipped moorings at 20 m depths at three locations in McMurdo Sound, comprising deep (bottom depth>200 m: Hut Point Peninsula) and shallow environments (bottom depth ~25 m: Cape Evans and New Harbor). Our sensors recorded high-frequency variation in pH (Hut Point and Cape Evans only), tide (Cape Evans and New Harbor), and water mass properties (temperature and salinity) during spring and early summer 2011. These collective observations showed that (1) pH differed spatially both in terms of mean pH (Cape Evans: 8.009±0.015; Hut Point: 8.020±0.007) and range of pH (Cape Evans: 0.090; Hut Point: 0.036), and (2) pH was not related to the mixing of two water masses, suggesting that the observed pH variation is likely not driven by this abiotic process. Given the large daily fluctuation in pH at Cape Evans, we developed a simple mechanistic model to explore the potential for biotic processes – in this case algal photosynthesis – to increase pH by fixing carbon from the water column. For this model, we incorporated published photosynthetic parameters for the three dominant algal functional groups found at Cape Evans (benthic fleshy red macroalgae, crustose coralline algae, and sea ice algal communities) to estimate oxygen produced/carbon fixed from the water column underneath fast sea ice and the resulting pH change. These results suggest that biotic processes may be a primary driver of pH variation observed under fast sea ice at Cape Evans and potentially at other shallow sites in McMurdo Sound.

Editor: Erik V. Thuesen, The Evergreen State College, United States of America

Funding: This research was supported by (1) U.S. National Science Foundation (NSF) award ANT-0944201 to GEH from the Antarctic and Organisms and Ecosystems Program, (2) NSF grant OCE (OTIC) 0844394 to TRM (in support of the SeaFET development), and (3) by funds from the University of California in support of a multi-campus research program, Ocean Acidification: A Training and Research Consortium, (http://oceanacidification.msi.ucsb.edu/) to GEH and TRM. The funders had no role in study design, data collection and analysis, decision to publish, or preparation of the manuscript.

* Email: pmatson@gmail.com

Introduction

Information regarding natural environmental variation is crucial to understanding how marine populations may respond to future changes in ocean climate. Recent technological advances, such as the development of deployable pH sensors [1], have enabled marine scientists to begin to explore natural variation in ocean pH at much higher temporal frequencies and provided a glimpse at how natural pH variation differs across ecosystems [2]. The magnitude of this pH variation between ecosystems is readily apparent, with coastal Antarctic and offshore oligotrophic ecosystems exhibiting much lower magnitudes of variation than seen in coastal ecosystems in temperate and tropical regions [2]. However, we have less understanding of how pH variation may differ within geographic regions and the relative strength of abiotic (*e.g.*, surface air/sea mixing, subsurface water mass mixing, heat flux) and biotic (*e.g.*, photosynthesis and respiration) processes in driving these patterns. Such spatial heterogeneity in pH variation

could lead to some areas functioning as either hot-spots for adaptation to dynamic environmental conditions or refugia from future conditions.

Polar ecosystems are expected to be the first to experience the impacts of anthropogenic induced climate change, as well as to experience the greatest relative change in environmental conditions [3]. Antarctic marine ecosystems are generally recognized for their environmental stability; for instance, ocean temperature in the southern Ross Sea remains at or near $-1.8°C$ for the majority of the year [4,5]. This homogeneous environment is reflected in the physiology of Antarctic species (reviewed by [6]), with many being stenothermal and demonstrating poor abilities to acclimate to elevated temperatures [7] and some having even lost the ability to generate a heat shock response [8]. While the majority of ecophysiological research with Antarctic marine species has focused on temperature stress, much less is known regarding how these species may respond to fluctuations in the pH environment (see [9–13]). However, recent data have suggested that some critical species may respond in a deleterious fashion [14,15]. Since marine organisms are adapted to local conditions [16,17], it is crucial to improve our understanding of natural ocean pH variation within Antarctic ecosystems in order to more effectively predict how these species may respond to a changing ocean climate.

Given the remoteness and logistical difficulties inherent to polar research, substantially less is known of the environmental variability of coastal Antarctic seas compared to those in temperate regions. There is a critical need to increase oceanographic observations within the Southern Ocean to improve our understanding of how it will respond to global change [18]. In the southern Ross Sea, McMurdo Sound has received consistent study of its physical and biological oceanography since first discovered by James Ross in 1841. More recent oceanographic investigations have identified two primary water masses present within McMurdo Sound: High Salinity Surface Water (HSSW) and Ice Shelf Water (ISW). These water masses can be identified based on their physical characteristics, with HSSW having relatively warmer temperatures and higher salinity than the colder, fresher ISW [19]. Circulation patterns in the eastern Sound are complex with variation between southward flowing HSSW along Ross Island and northward flowing ISW from under the Ross Ice Shelf, though flow is generally southward [5,20–24] (Figure 1). In the western Sound, current flow is northward and primarily composed of ISW [20,21,24,25]. The physical oceanography within the Sound has created distinct benthic communities on the western and eastern sides of the Sound [21], with greater overall benthic primary production in the east [21,26]. Coastal bathymetry in the eastern Sound tends to be steeper than in the west, with more abundant hard substrate in the form of basaltic gravel and larger bedrock as opposed to soft glacial dust in the western Sound [27]. Physical scouring by surface sea ice, as well as formation of anchor ice, results in strong zonation patterns with longer-lived species occurring at greater depths [28]. Despite the high latitude of McMurdo Sound (77°S), benthic communities may contain up to three species of macroalgae [27]. These species appear to be adapted to conditions of low temperature and long periods with little or no available light and remain photosynthetically active and capable of rapid responses in production once light levels increase during austral spring [29,30]. In addition, a diverse microalgal community is present within the bottom of the sea ice [31]. This community may contribute 20–65% of the primary productivity in areas covered by sea ice [32,33] but is patchily distributed across space and time [34]. While many of these studies have quantified *in situ* photosynthetic efficiency and production by algal groups in McMurdo Sound, their contribution to ocean pH variation is not currently known.

A combination of mooring-based ocean observations and modeling was used to explore the natural dynamics of ocean pH during spring and early summer within McMurdo Sound and the potential abiotic and biotic drivers of that variation. To identify patterns of pH variation and potential abiotic drivers, oceanographic sensors were deployed on moorings at multiple locations in McMurdo Sound to observe pH and water mass properties during spring. To identify the potential for a biological driver of variation, specifically under-ice algal photosynthesis, a mechanistic model was developed based on parameters of light availability, photosynthetic efficiency, and estimates of biomass for three algal groups found at Cape Evans on the west coast of Ross Island. This approach helped shed light on the natural dynamics of ocean pH under Antarctic sea ice and what mechanisms may be driving the observed patterns of variation across space.

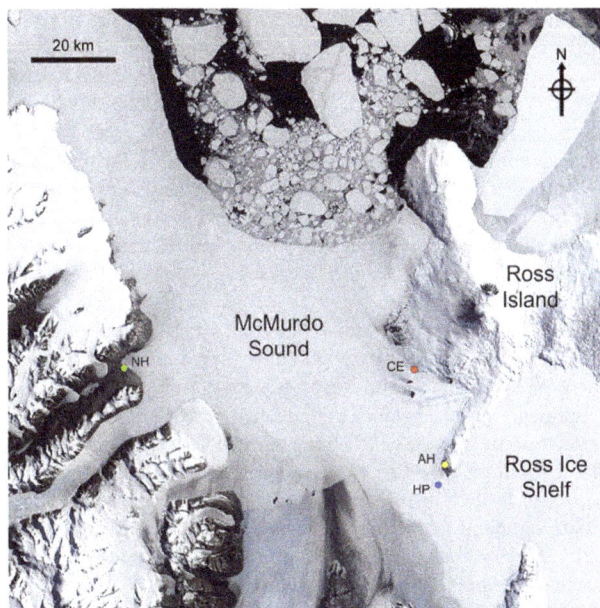

Figure 1. Map showing study location in McMurdo Sound, Antarctica. Colored circles indicate mooring (Hut Point "HP", blue; Cape Evans "CE", red; New Harbor "NH", green) and irradiance sensor locations (Arrival Heights "AH", yellow). This image is based on Landsat satellite images collected during a previous year and does not reflect sea ice extent from the 2011/2012 season.

Methods

Study sites

Three oceanographic moorings were deployed within McMurdo Sound during austral spring (October–December) in 2011. Two locations, Cape Evans (S 77° 38.060′, E 166° 24.918′) and Hut Point (S 77° 52.425′, E 166° 35.164′), were located on the eastern side of the sound, while a third location was located on the western side of the sound at New Harbor (S 77° 34.576′, E 163° 31.702′) (Figure 1). This research was conducted under the auspices of the U.S. Antarctic Program in accordance with environmental regulations laid out in the Antarctic Treaty. No deployments were made in protected areas and no permissions or permits were required. These locations were selected in order to observe the two primary water masses within McMurdo sound: the southerly flowing water mass along the eastern side (Cape Evans), the northward flowing water mass coming from under the Ross Ice Sheet (New Harbor), as well as a location where the two water masses may mix at the southern end of Ross Island (Hut Point). The three locations differed in depth, with Cape Evans and New Harbor being shallower coastal sites (<30 m depth) while the Hut Point site was a deeper site (>200 m depth). Benthic moorings were suspended above the seafloor by a subsurface buoy at Cape Evans and New Harbor while the mooring at Hut Point was suspended from the surface through a hole in the ice by a steel cable within a wooden hut. All sensors were deployed at 20 m depth, with the exception of pressure sensors at Cape Evans and New Harbor, which were attached to the mooring anchor.

Sensor arrays

Moorings were instrumented with a suite of sensors to record time series of temperature, salinity, pH, and tidal height. Temperature and salinity were measured using a non-pumped conductivity-temperature (CT) MicroCAT sensor (SBE-37 SM;

Sea-Bird Electronics) that sampled at 5-min intervals. pH was measured using an autonomous data logger based on a Honeywell Durafet pH sensor [1] and sampled at 1-hr intervals. Tidal heights were measured using water level loggers (Hobo U20-001-03-Ti; Onset) that recorded water pressure at 10-min intervals. All oceanographic data was processed using a 1-hr low pass filter and then sampled at 1-hr intervals. All three locations were equipped with CT and pressure sensors while pH sensors were only deployed at Cape Evans and Hut Point; the pH sensor intended for New Harbor was damaged during shipping and rendered inoperable.

Calibration of pH sensors required a discrete water sample collected *in situ*. This single point calibration approach is justified when the sensor obeys the Nernst equation and the temperature component of the standard potential has been previously characterized; both of which have been repeatedly demonstrated for these sensors [1]. The water sample was collected adjacent to the sensor by SCUBA divers (Cape Evans) or by lowering a 5 L Niskin sampling bottle from the surface (Hut Point) prior to retrieval. From this sample, a 500 mL water sample was returned to the laboratory for CO_2 analysis modified from Standard Operating Procedures (SOP) for spectrophotometric pH (SOP 6b) and Total Alkalinity (TA, SOP 3b) [35] as reported in [36]. *In situ* pH was then calculated using CO_2calc [37] using the constants of [38] as refit by [39]. Due to the calibration approach used, sensor accuracy depends mostly upon collection of a representative discrete sample. Based on experience, there is an expectation that the data presented here accurately represent pH variability with a finite yet unquantified error in accuracy dominated by sampling errors. Past experience suggests that sampling errors lead to vicarious calibration errors of ~0.01 pH or less. Second order errors due to extending the fit of temperature dependent equilibrium constants in CO_2calc and temperature dependent sensor calibration coefficients for the SeaFET sensor, both fit to data above zero, introduces additional unquantified error; yet this error is most likely smaller than the aforementioned discrete sampling error [1].

Photosynthesis-Irradiance Model

A simple mass balance model was constructed to estimate the potential contribution of photosynthesis by sea ice and benthic algae to the pH variation observed at Cape Evans. The control volume for this model comprised four depth bins of equal length and width but varying depths (10 m, 15 m, 20 m, and 25 m) to account for the sloping bathymetry at Cape Evans [27]. These depths determined both the light transmitted to the benthos (z_w) as well as the volume of the water column. While no measurements of flow from Cape Evans were collected during this study, the water column is generally well-mixed during spring in McMurdo Sound [5,20] and at Cape Evans in particular [27], likely driven by diurnal tides in the region [40,41]. Both O_2 and $CO_{2(aq)}$ were assumed to be uniformly distributed at all times within each bin and no diffusion between bins.

Levels of photosynthetically active radiation (PAR, in units of μmol photons m^{-2} s^{-1}) at the surface (E_s) were measured nearby at the NSF UV Monitoring Station (http://uv.biospherical.com/) at McMurdo Station on Arrival Heights (77°50′ S, 166°40′ E). From these data, we identified the day of maximum PAR observed (November 11, 2011; 1495 μmol photons m^{-2} s^{-1}) and used this to create a synthetic hourly surface PAR (E_s) dataset for a 24 h period using the equation:

$$E_s(t) = 900 + \left(600\left[-cos\left(\left[\frac{\pi}{12}\right]t + 0.1\right)\right]\right) \quad (1)$$

where t is hour of the day. PAR at different depths $E(z,t)$ was calculated at water depths z below the base of the sea ice at times t over a 24 h period using the following equation from [42]:

$$E(z,t) = E_s(t)exp\left\{\frac{-[(k_i \cdot z_i) + (k_m \cdot z_m) + (k_w \cdot z)]}{cosj(t)}\right\} \quad (2)$$

where k_i, k_m, and k_w are the attenuation coefficients of the sea ice, sea ice-associated microbial layer, and water column, respectively; z_i and z_m are the thicknesses of the sea ice and sea ice associated microbial community, respectively. Attenuation of light by the sea ice-associated microbial layer is assumed to be driven primarily from light absorption by sea ice algae; therefore k_m and z_m quantify the absorption and concentration of chlorophyll *a* (chl*a*), respectively. Cape Evans has been noted as having little to no snow accumulation at the surface [27,29,43] or platelet ice present beneath the sea ice [27,44], therefore neither were included in the equation. The fraction of incident light transmitted through the sea ice was approximated by:

$$cosj(t) = [1 - 0.563 \, cos^2 h_s(t)] \quad (3)$$

where h_s is the solar elevation angle above the horizon [45,46].

Using these estimates of light availability at depth, oxygen production by an algal community composed of sea ice algae, benthic macroalgae, and crustose coralline algae was calculated. For each algal group, we estimated net primary production of oxygen based on the hyperbolic tangent equation of [47]:

$$PP_{net}(z,t) = \left[P_{max} \cdot tanh\left(\frac{\alpha E(z,t)}{P_{max}}\right) - R\right]b \quad (4)$$

where P_{max} is the maximum production at saturating irradiance expressed as oxygen production per unit biomass [μmol O_2 (g biomass)$^{-1}$ h^{-1}], α is the rate of photosynthetic efficiency at non-saturating irradiance per unit biomass [μmol O_2 (g biomass)$^{-1}$ h^{-1} (μmol photons m^{-2} s^{-1})$^{-1}$], R is algal respiration (μmol O_2 (g biomass)$^{-1}$ h^{-1}) per unit biomass, and b is the biomass density (g biomass m^{-2}). P_{max}, α, R, and b for benthic algae used in this model were derived from *in situ* measurements of oxygen production at Cape Evans [29,30]. P_{max}, α, R, and b for sea ice algae, originally reported in terms of C [42], were converted to units of O_2 by multiplying PP_{net} (in units of C) by the photosynthetic quotient, estimated as 1.03 (mol O_2/mol C) [48]. For all taxa, the amount of oxygen produced was divided by the photosynthetic quotient to estimate the total carbon removed from the system. Additionally, all algae in this system were assumed to rely on diffusive uptake of $CO_{2(aq)}$, which has no effect on alkalinity. While the carbon concentration mechanisms for these species are not explicitly known, diffusive uptake of $CO_{2(aq)}$ has been shown to be common for many species of red marine macroalgae from the class *Florideophyceae* (of which *P. antarctica* is a member) [49], as well as in some Antarctic sea ice algae when [$CO_{2(aq)}$]>5 μM [50].

Carbonate System Estimation

Carbonate system parameters were estimated using CO2SYS for Matlab [51] for both mooring data sets and estimating the effect of photosynthesis on pH. Total alkalinity (TA) was estimated

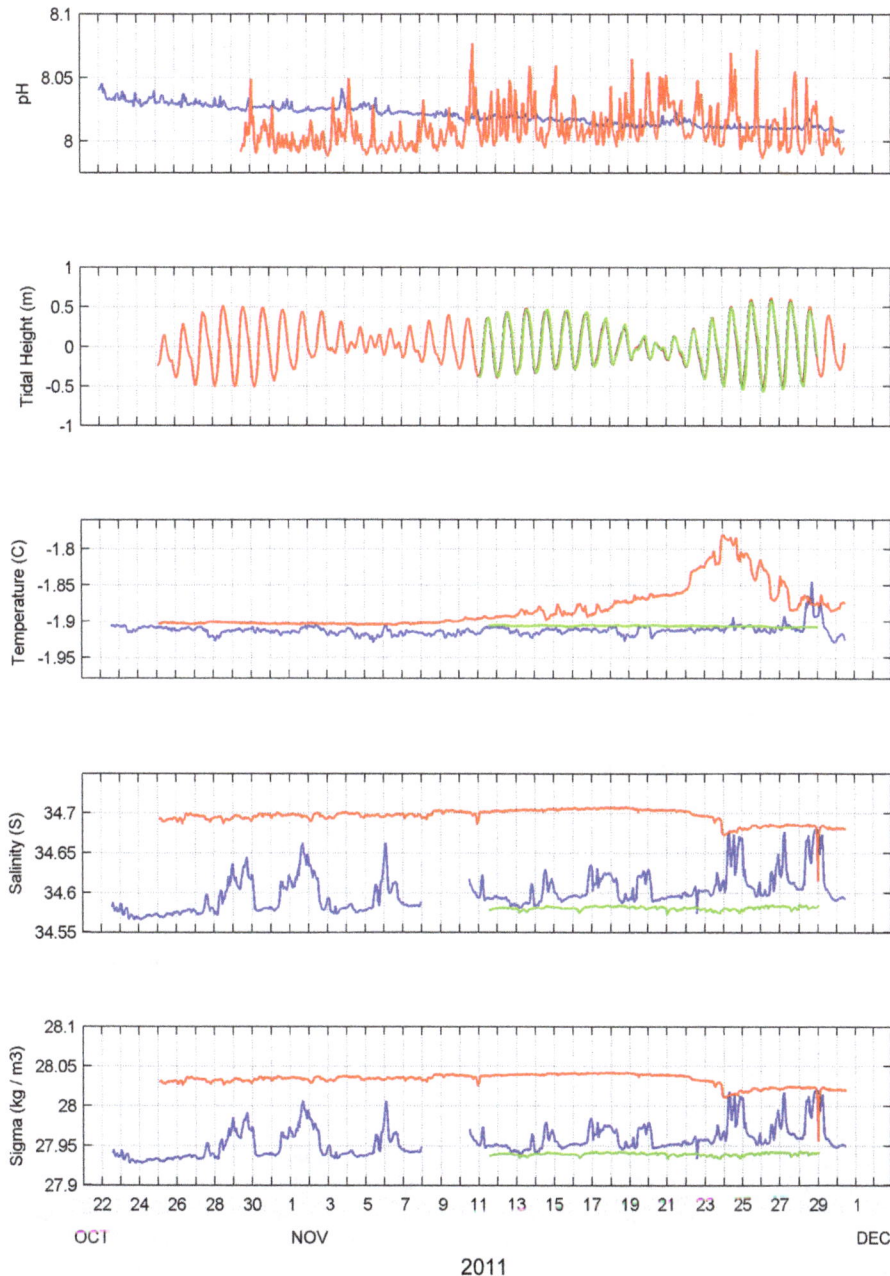

Figure 2. Water mass properties observed during spring in McMurdo Sound. Locations are indicated by color: Hut Point (blue), Cape Evans (red), and New Harbor (green).

from *in situ* salinity and temperature measurements from Hut Point (2336.5±1.1 μmol kgSW^{-1}) and Cape Evans (2341.2±0.5 μmol kgSW^{-1}) using the equations of [52] for the Southern Ocean. These estimates were in agreement with measurements of alkalinity collected during this study at Hut Point (2335.7±11.8 μmol kgSW^{-1}; $n = 35$) and Cape Evans (2342.5±11.3 μmol kgSW^{-1}; $n = 6$). To estimate the photosynthesis-irradiance model, a value for total inorganic carbon (C_T) of 2247.5 μmol kgSW^{-1} was used as a baseline level for the system without photosynthesis based on the lowest pH values recorded at Cape Evans (7.987) during this study and the estimated alkalinity. This value falls in line with winter end-member values of C_T from other studies of the carbonate system in Antarctic waters [53–56].

Throughout these results, pH is reported on the total hydrogen ion scale. All calculations within CO2SYS use the dissociation constants of [38], refit by [39]. Estimates of total silicate (56 μmol kgSW^{-1}) and total phosphate (2.35 μmol kgSW^{-1}) were based on values reported values for spring within this region of McMurdo Sound [20,57]. Salinity and temperature values were 34.7 and −1.9°C, respectively.

Results

Mooring Observations

Overall patterns of pH variation during spring appeared to vary between locations (Figure 2; Table 1). At Hut Point, pH exhibited

Table 1. Descriptive statistics of oceanographic observations from each location.

Variable	Location	Median	Mean	s.d.	Min	Max
pH	Hut Point	8.012	8.020	0.007	8.006	8.042
	Cape Evans	8.005	8.009	0.015	7.987	8.077
Temperature (°C)	Hut Point	−1.913	−1.912	0.007	−1.929	−1.845
	Cape Evans	−1.890	−1.879	0.030	−1.905	−1.781
	New Harbor	−1.906	−1.906	0.001	−1.909	−1.904
Salinity	Hut Point	34.596	34.600	0.023	34.566	34.679
	Cape Evans	34.699	34.697	0.009	34.615	34.708
	New Harbor	34.581	34.581	0.002	34.573	34.585
Density (kg m^{-3})	Hut Point	27.952	27.955	0.019	27.928	28.019
	Cape Evans	28.035	28.033	0.008	27.957	28.042
	New Harbor	27.940	27.940	0.002	27.943	27.933

The number of observations (n) varies between sites (Hut Point: for Temperature and pH, $n = 935$, for Salinity and Density, $n = 875$; Cape Evans: $n = 768$; New Harbor: $n = 420$).

very low hourly variation with an overall trend of decreasing pH over the deployment duration. Mean pH was 8.020 ± 0.007 ($\pm$ s.d.) and ranged from 8.006 to 8.042 and the distribution of these values had a low level of skewness (Pearson's median skewness: $\gamma = 0.07$). At Cape Evans, pH was initially lower but observed hourly variation was much greater than at Hut Point. Mean pH was 8.009 ± 0.015 and ranged from 7.987 to 8.077 with the distribution showing a higher level of skewness towards higher pH values (Pearson's median skewness: $\gamma = 0.25$). The mean absolute rate of change in pH also differed between the two locations, with a greater mean rate of change observed at Cape Evans (0.007) versus Hut Point (0.001).

During the course of the deployment, three spring tides and two neap tides were observed, with a maximum tidal range of ~1 m (Figure 2). Water temperature was relatively constant at all locations from late October to late November, with the exception of a warming event at Cape Evans that occurred over a 7 day period in which temperature increased ~0.12°C (Figure 2; Table 1). Salinity differed on the east and west sides of McMurdo Sound, with greater salinity at Cape Evans versus New Harbor, while salinity at Hut Point fluctuated intermittently throughout the deployment period (Figure 2; Table 1). These fluctuations appear to be mixing between the water masses observed at New Harbor and Cape Evans, as indicated by the relationship between temperature and salinity at the three locations (Figure 3). Seawater density in this region is primarily driven by differences in salinity. Assuming that median values of salinity at Cape Evans and New Harbor represent end-member values for the upper 20 m, calculations indicate that the water mass at Hut Point was composed of ~14% of the southward flowing water mass mixed with the water mass emerging from under Ross Ice Shelf off Hut Point during this time period. A density discontinuity was observed at Cape Evans on 23–24 November that may indicate advection of a new water mass into the study area or a separate region-specific process. This event appeared to have had little effect on pH at either Hut Point or Cape Evans, with no apparent changes in pH coinciding with density fluctuations at either location. These results suggest that physical processes inherent to water mass advection and mixing do not drive the observed pH variation in McMurdo Sound.

To explore how the calcification environment in McMurdo Sound may vary through time, the carbonate system at both Hut Point and Cape Evans was estimated. Patterns of estimated pCO_2 fluctuation were inversely related to observed pH, with low levels of variation at Hut Point and higher levels of variation at Cape Evans (Figure 4). At Hut Point, mean pCO_2 was estimated at 413 ± 8 µatm ($\pm$ s.d) with a range of 391 to 427 µatm and mean $\Omega_{calcite}$ and $\Omega_{aragonite}$ of 1.98 ± 0.03 and 1.24 ± 0.02, respectively. At Cape Evans, mean pCO_2 was estimated at 426 ± 16 µatm with a range of 358 to 450 µatm and mean $\Omega_{calcite}$ and $\Omega_{aragonite}$ of 1.95 ± 0.06 and 1.22 ± 0.04, respectively. At no time was $\Omega_{aragonite} < 1$ at either location.

Comparisons of pH spectra between Hut Point and Cape Evans indicate different dominant frequencies at each location. A large diel peak was present in the pH spectrum at Cape Evans, but not at Hut Point (Figure 5A); both exhibited variance increases at frequencies of 0.2 cpd and lower. Of the total variance within the pH time series at Cape Evans (0.00021), ~25% occurs on a diel frequency (0.000052; variance integrated between 0.84 and 1.27 cycles per day). A similar diel peak is found in spectra of the tide and solar irradiance at Cape Evans (Figure 5B and 5C). While a relationship between pH and tide height was not apparent (Figure 6A), variation in pH appeared to increase marginally as tide exchange rate became more negative (Figure 6B). pH (mean $\pm$ s.d.) was greater during periods of ebb than flood flow (pH$_{ebb}$: 8.0119 ± 0.0161, $n = 397$; pH$_{flood}$: 8.0059 ± 0.0135, $n = 371$; Wilcoxon Rank Sum test: $W = 125314$, $p < 0.001$; Figure 6C).

Photosynthetic model outputs

Based on observed values of surface PAR, Equation 1 was used to simulate hourly surface PAR for a 24 h period during late austral spring (mid November) with a mean of 900 ± 433 µmol photons m^{-2} s^{-1} (mean $\pm$ s.d.), a maximum of ~1500 µmol photons m^{-2} s^{-1}, and a minimum of ~300 µmol photons m^{-2} s^{-1} (Figure 7A). Equation 2 calculated 3.53% ± 0.59 (mean $\pm$ s.d.) of surface PAR passed through 1.75 m of sea ice to be available for photosynthesis by sea ice algae. Based on an assumed sea ice algal biomass corresponding to 125 mg chl$a$ m^{-2} (Table 2), light was further attenuated such that mean percentage of surface PAR available to benthic algae at depth over a 24 hr period was 0.060 ± 0.023, 0.032 ± 0.013, 0.017 ± 0.0075, and 0.0092 ± 0.0043, at 10 m, 15 m, 20 m, and 25 m, respectively.

Model estimates of oxygen production rates differed greatly between the three algal groups (Figure 7). Based on parameters of

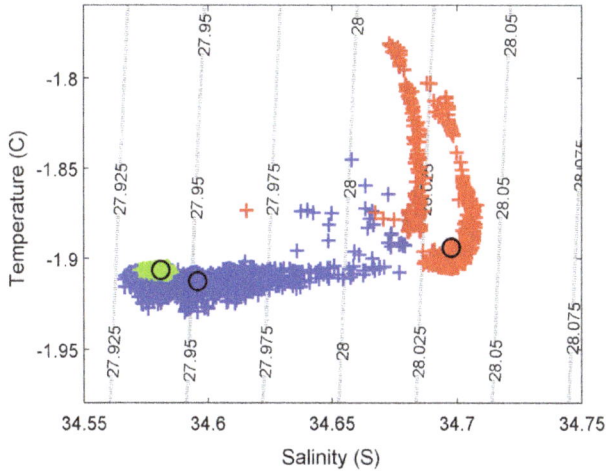

Figure 3. Temperature-Salinity plot showing relationships between temperature, salinity, and density for Hut Point (blue), Cape Evans (red), and New Harbor (green). Individual sampling points are represented by crosses while location-specific median values are represented by filled circles.

algal biomass (Table 2), the maximum oxygen production rate by sea ice algae (5353 μmol O_2 m^{-2} h^{-1}; Figure 7D) was approximately 4 times greater than the highest estimated rate for foliose red algae (1345 μmol O_2 m^{-2} h^{-1}; Figure 7B) and 44 times greater than for crustose coralline red algae (121 μmol O_2 m^{-2} h^{-1}; Figure 7C). When averaged across depths (10 to 25 m), these oxygen production rates translate to a carbon drawdown of 5.34 μmol C $kgSW^{-1}$ d^{-1} from the water column via algal photosynthesis. This removal of inorganic carbon, assumed to be entirely in the form of CO_2 (aq), would result in an increase of 0.016 pH units. When sea ice algal biomass was varied between 0 (no sea ice algae) and 400 mg chl$a$ m^{-2} (theoretical maximum possible biomass [58]), pH increased between 0.013 to 0.046 units, respectively (Figure 8). While this pH effect range accounts for 14–51% of the maximum pH variation observed at Cape Evans, 95% of the diel variation in pH ranged 0.028 units, well within the model estimates. Analyses indicated that pH effects estimated by this model are most sensitive to changes in both light availability and carbonate system parameters (Table 3). Variation in the thickness and attenuation of sea ice doubled the effect on estimated pH due to photosynthesis. The greatest sensitivities were in the carbonate system via changes in alkalinity and total carbon; increasing TA increased the buffering potential within the system while decreases in C_T

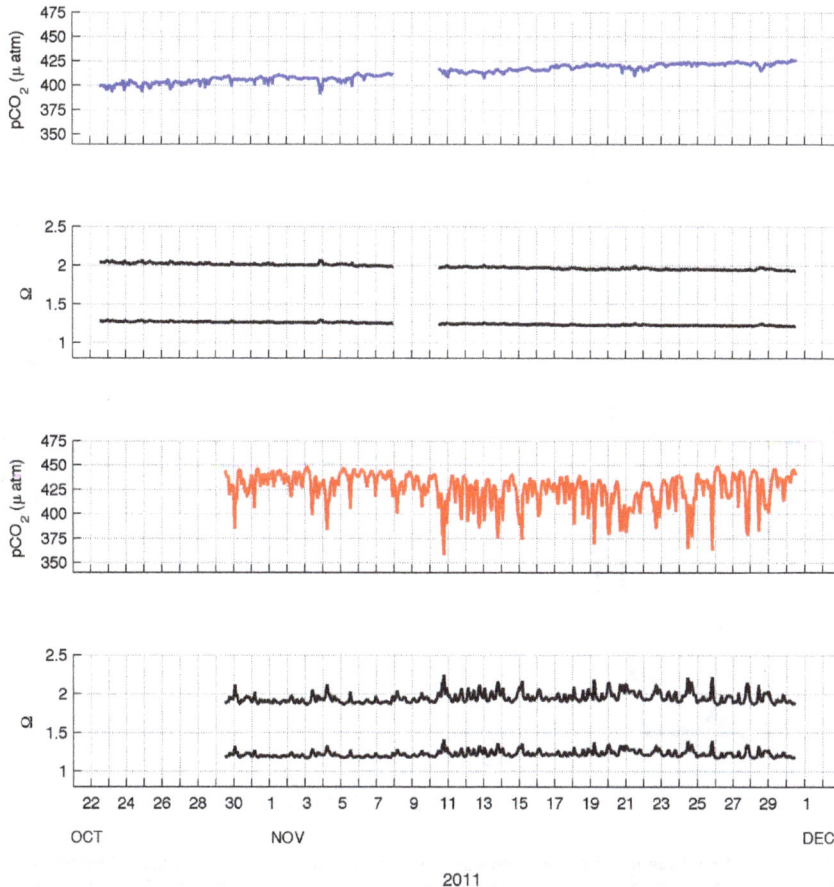

Figure 4. Estimated pCO$_2$ and carbonate ion saturation states (Ω) based on observed pH and estimated alkalinity. Data in the upper two panels refer to Hut Point while the two lower panels refer to Cape Evans. Lines in pCO$_2$ panels are color coded by location (Hut Point, blue; Cape Evans, red) while lines in carbonate saturation panels indicate $\Omega_{calcite}$ (upper black line in panel) and $\Omega_{aragonite}$ (lower black line in panel). Alkalinity estimates were location-specific and based on temperature and salinity measurements using the method of [52].

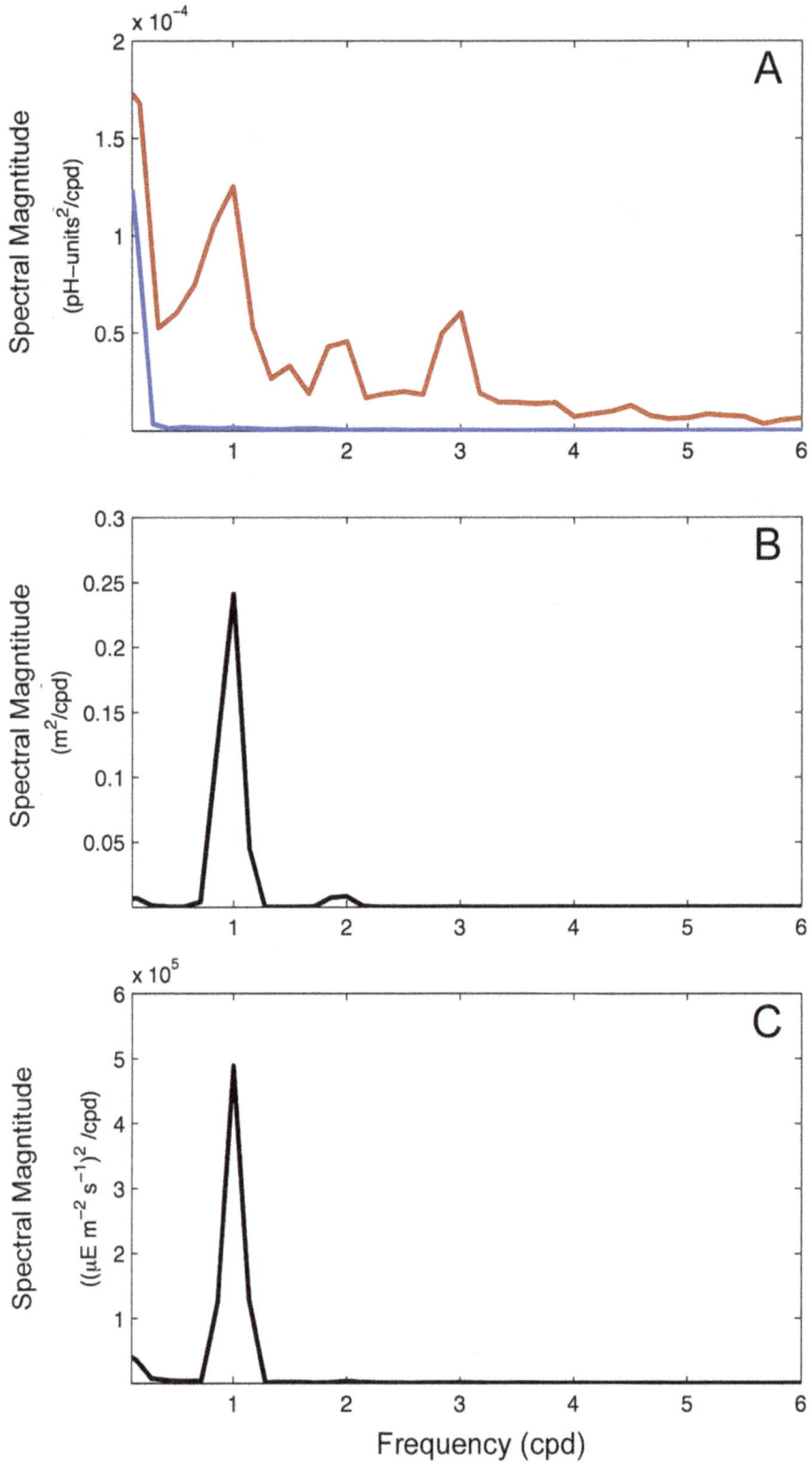

Figure 5. Spectra for pH, tide, and surface irradiance from eastern McMurdo Sound. pH spectra (A) for Hut Point (blue) and Cape Evans (red). Tide spectra (B) are from Cape Evans. Surface irradiance spectra (C) were measured at Arrival Heights. Frequency is in the units of cycles per day.

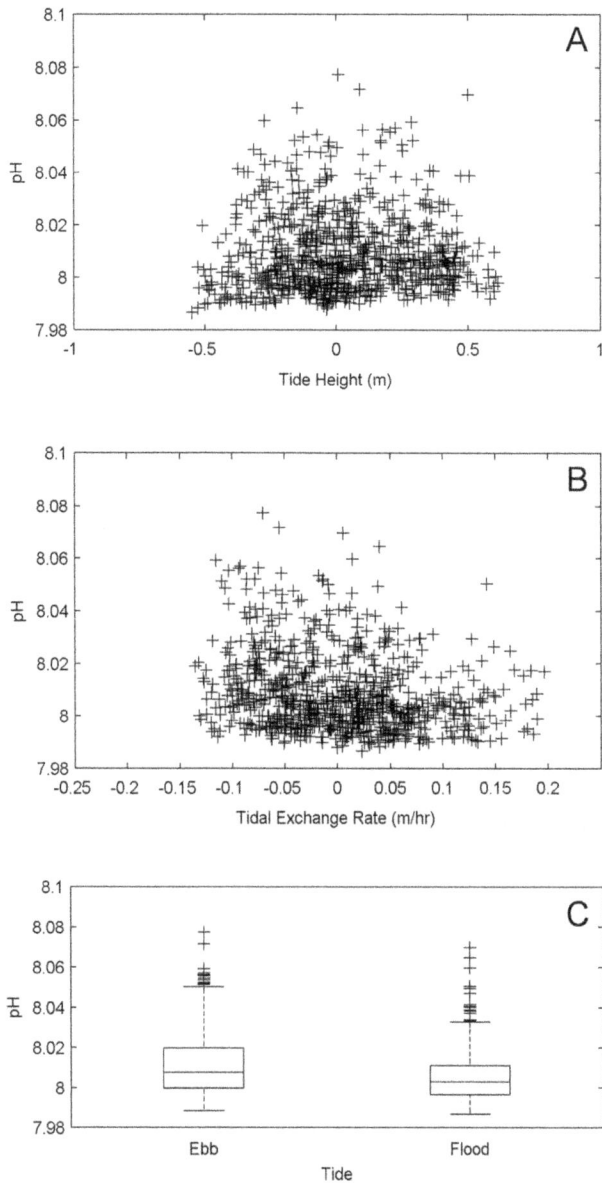

Figure 6. Relationship between pH and tides observed at Cape Evans. Panels show pH and tidal height (A), tidal exchange rate (B), and ebb versus flood stage (C).

reduced the abundance of carbonate ion species (HCO_3^- and CO_3^{-2}).

Discussion

This study deployed oceanographic sensors to observe high frequency variation in pH and water mass properties under fast sea ice at multiple locations within McMurdo Sound in the southern Ross Sea during austral spring, 2010. In addition, the potential for algal photosynthesis to serve as a biological driver of diel pH variation was estimated using a mechanistic model. The salient findings of this study were: (1) near-surface pH variation differed spatially between the two study locations with greater variation observed at the shallow coastal site than at the deeper location: (2) advection and water mass mixing were observed at Hut Point but did not appear to affect pH; and (3) modeling results

suggest that under-ice algal photosynthesis is capable of driving a large portion of the diel pH variance observed at our shallow coastal site.

Based on the data presented here, near-surface waters (20 m depth) in McMurdo Sound experienced low levels of diel pH variation during this time period. The greatest pH range was observed at our shallow coastal location (Cape Evans) with a range of 0.0906 units, which is approximately 4–5 times lower than observed along the California coast [2]. These low levels of variation are similar to those first reported by [59] in McMurdo Sound during austral spring in 2010. Together, these observations support the expectation that Antarctic species living under fast sea ice are less likely to experience large fluctuations in pH over short time scales, unlike species found within upwelling systems [2,60]. However, it is important to note that these observations are limited to spring. The transition from winter to summer is marked by a dramatic change in water column structure [5] as well as the delivery of the phytoplankton bloom from the Ross Sea polynya, north of Ross Island. The bloom, primarily consisting of *Phaeocystis pouchetti*, is transported under the sea ice in eastern McMurdo Sound by the southward flowing current and generally appears off McMurdo Station by early to mid December [61]. This bloom may be responsible for a seasonal shift of 0.3–0.5 pH units between winter and summer [4,62]. Despite the relatively low pH variation, calcification conditions (i.e. saturation state, Ω) were still low, especially compared to both temperate [60] and tropical regions [63]. Longer duration deployments, ideally overwintering, are needed in order to capture the scale of high-frequency variation that occurs during this seasonal shift in productivity.

In addition to developing high-frequency time series of ocean pH under Antarctic sea ice, this study investigated whether abiotic or biotic processes may be driving the observed pH variation. Despite detecting at least three different water masses across our study locations, near-surface mixing events did not appear to contribute to the pH variation observed in McMurdo Sound. Temperature-salinity properties indicate that two water masses with properties similar to HSSW were present at Cape Evans. A water mass with properties similar to ISW was present continuously at New Harbor, while repeated mixing events between water masses were observed at Hut Point. It is possible that these events were driven by changes in near-surface flow direction, which has been shown to vary from southward to northward depending on the tidal phase during this time of year in eastern McMurdo Sound [24]. Further, yearlong mooring-based observations by [5] at Cape Armitage found the highest levels of variation in salinity and flow to occur at shallow depths (50–55 m versus 150–325 m). Salinity values were consistently lower ($\Delta S = 0.2$–0.3) in this study compared to values reported by [20] but comparable to measurements by [5]. This may be attributed to the shallower sensor depth in this study (stationary at 20 m vs. vertical profiles from 0–600 m for [20]); the relative differences in salinity between the western and eastern Sound were consistent with those previously reported [20,24]. The slight warming that occurs between Hut Point and New Harbor is likely due to the proximity of Hut Point to the Ross Ice Shelf [20]. The lack of a pH difference between the two primary water masses is interesting given their history of atmospheric contact. While HSSW flowing south from the Ross Sea polynya past Cape Evans at current speeds of ~2–3 cm/s near the surface [21,24] may have more recent atmospheric contact, ISW has very little, if any, contact with the atmosphere while underneath the Ross Ice Shelf. Modeling efforts estimate that 2.4–3.5 years is required for the formation of ISW from HSSW under the Ross Ice Shelf [64,65],

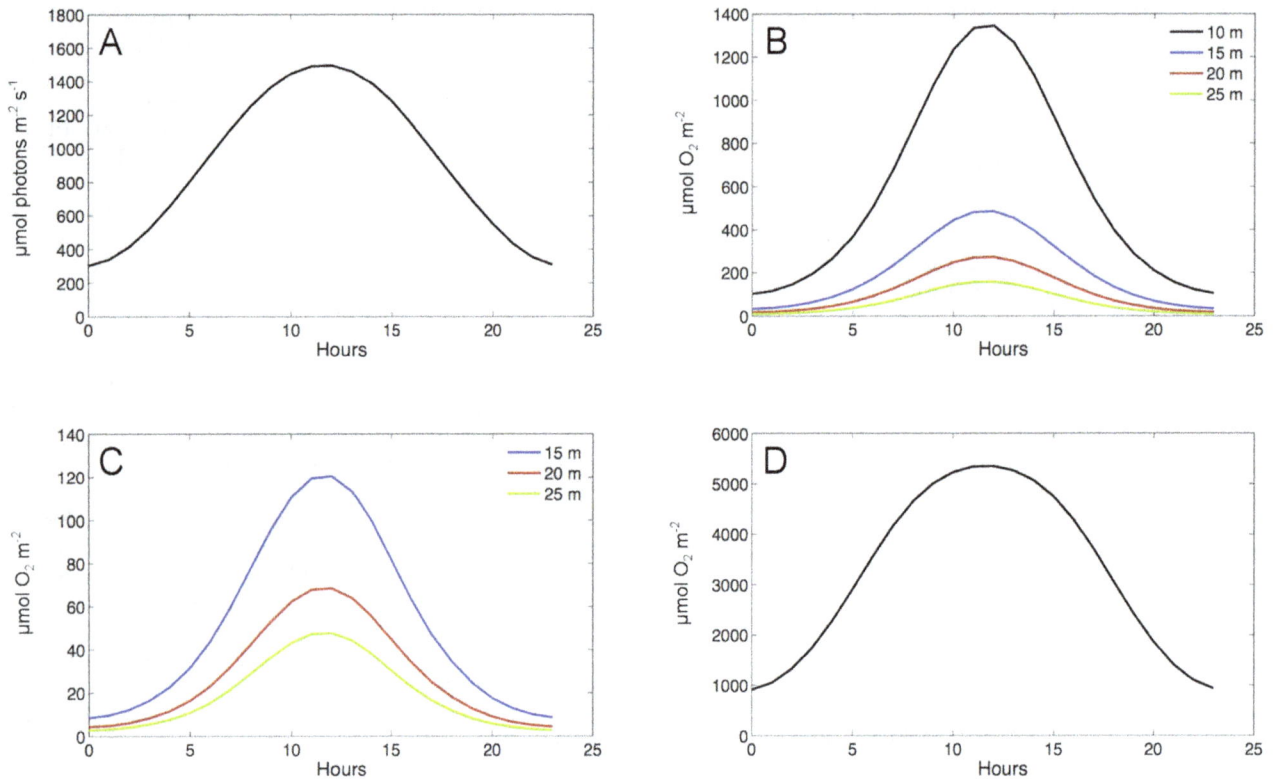

Figure 7. Model estimates of surface PAR (A) and hourly O_2 production rate by algal functional groups over a 24 h period. Algal groups include foliose red (B), crustose coralline (C), and sea ice algae (D). Colored lines indicate estimated O_2 production at different depths.

Table 2. Photosynthesis-irradiance model parameters.

Parameter	Values [Source]
k_l	1.35 m^{-1} [57]
k_m	0.016 m^2 mg^{-1} [57]
k_w	0.09 m^{-1} [29]
z_i	1.75 m
z_m	125 mg chl$a$ m^{-2} (mean value[1])
z_w	10–25 m [27,29]
$P_{max, ia}$	50.6 µmol O_2 mg^{-1} chl$a$ h^{-1} [42]
$P_{max, fr (net)}$	6.3–10 µmol O_2 gFW^{-1} h^{-1} [29]
$P_{max, cc (net)}$	11.4–16.25 µmol O_2 m^{-2} thallus h^{-1} [30]
R_{ia}	3% of P_{max} [42]
α_{ia}	1.03 µmol O_2 mg^{-1} chl$a$ h^{-1} (µmol photons m^{-2} s^{-1})$^{-1}$ [42]
α_{fr}	1.09–1.67 µmol O_2 gFW^{-1} h^{-1} (µmol photons m^{-2} s^{-1})$^{-1}$ [29]
α_{cc}	4.6 µmol O_2 m^{-2} thallus h^{-1} (µmol photons m^{-2} s^{-1})$^{-1}$ [30]
b_{ia}	same as z_m
b_{fr}	518–559 gFW m^{-2} [27,29]
b_{cc}	80–100% cover m^{-2} [27,30]
TA	2341.2 µmol kgSW^{-1} (This study)
C_T	2247.5 µmol kgSW^{-1} (This study)

Algal-specific values are presented for sea ice algae (ia), *Phyllophora antarctica* (fr), and crustose coralline (cc).
[1]Mean value calculated from sea ice algae chlorophyll a measurements by [34,43,44,69] at Cape Evans.

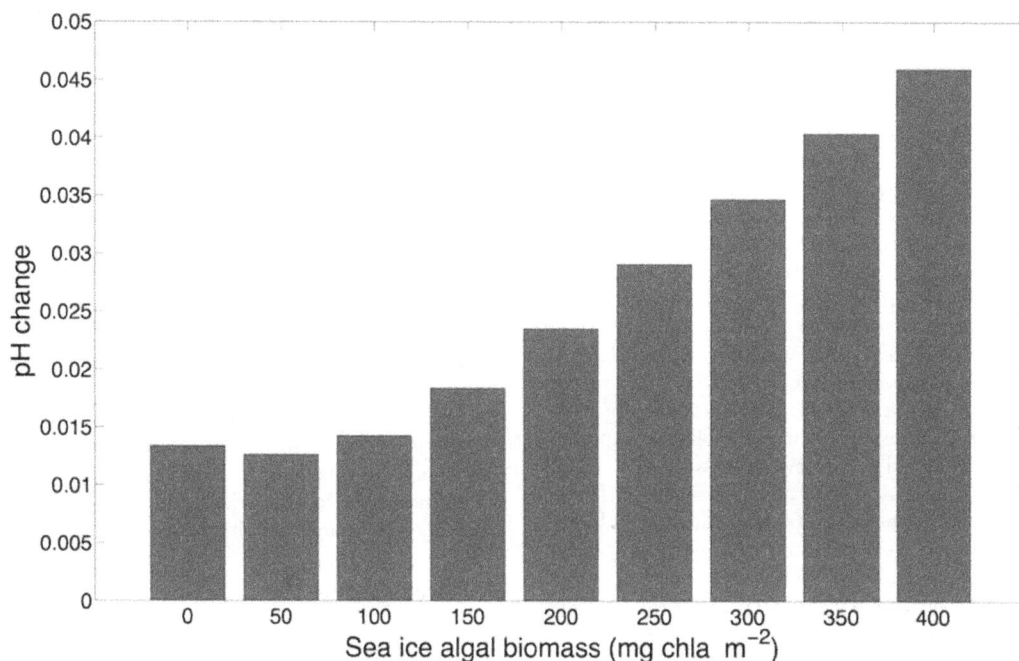

Figure 8. Model estimates of pH effect due to photosynthesis by algal community over a 24 h period in relation to sea ice algal biomass.

which in some places may be up to 700 m thick. Despite this difference in time under the ice between HSSW and ISW, median pH differed by only 0.0146 units between Hut Point and Cape Evans, and was slightly greater at Hut Point. Future deployments of pH sensors and current meters at multiple depths across McMurdo Sound and under the Ross Ice Shelf would add insight regarding processes affecting the carbonate system within these water masses.

Spatial differences in pH within McMurdo Sound appear to be more likely driven by a combination of biological production, bathymetry, and tidal exchange. The combination of the positive diel peak in pH with the increased pH during ebb tide periods at Cape Evans suggests the potential for pH variation in McMurdo Sound to be driven through a combination of photosynthetic production and tidal exchange between shallow and deep water

columns. Sea ice algal biomass appeared to have a greater influence on increasing pH relative to benthic algae under mean levels of biomass, likely due to the higher quantity of available light at that depth. This community of microscopic algae appears to be patchy both across space and time [31,33,34,66] making quantification of standing stock difficult over large areas. However, regional patterns have been reported with biomass tending to increase under thicker sea ice [67] and generally greater biomass in the western versus eastern McMurdo Sound [31]. Composition of the sea ice algal community appears to fluctuate seasonally, with shade-tolerant species being replaced as light levels increase through the summer [33]. The effect of these shifts in species assemblages on carbon drawdown and pH is not known, though increased levels of production would likely increase diel pH changes. Elevated pCO_2 concentrations have been shown to

Table 3. Sensitivity analyses of model parameters for the light environment, algal biomass, and ocean carbonate system.

Parameter	Null+5%	Null−5%
sPAR	3.12	−3.91
k_i	−11.72	11.72
z_i	−11.72	11.72
k_m	−1.56	1.56
k_w	−0.78	0.78
b_{ia}	3.12	−3.13
b_{fr}	0	0
TA	−30.47	14.84
C_T	11.72	−28.13

Values represent the percent change in estimated pH in response to an increase/decrease of each parameter value by 5%. Null value of each parameter is listed in Table 2.

increase growth of sea ice algae present in McMurdo Sound, though this effect is lost when pH drops below 7.6 [13]. Bathymetry may also be an important contributor to spatial difference in pH variation within McMurdo Sound. While the presence of sea ice algae is not depth dependent, their influence on dissolved inorganic carbon in the surrounding water is diluted as water column depth increases. This may help explain the relative lack of high-frequency pH variation observed at Hut Point, though the cause of the low-frequency decrease in pH is less clear. In addition to a deeper water column, Hut Point also had a thick layer of snow on the ice surface (>0.3 m), which would severely reduce the light available to sea ice algae. This spatial relationship between shallow versus deep locations was also seen in 2010, with greater diel variation observed at shallower locations (Cape Evans and Cinder Cones) than at a deeper site north of Hut Point (Erebus Basin) [59]. Tidal exchange may also play an important role in the observed pH variation. Previous work by [5,21,23,24] showed spatial heterogeneity on periodic flow patterns (along-shore versus cross-shore) in areas of McMurdo Sound farther south of Cape Evans. If cross-shore transport were sufficient at Cape Evans, the increased pH variation observed during periods of ebb flow may be a signal of the shallow water moving past the mooring to deeper water. However, given the lack of such measurements in this study, it is not possible to test this hypothesis here. With lower levels of pH variation occurring over areas with greater depths, pelagic species may be less likely to experience natural fluctuations in pH on a regular basis, unlike species found in shallow coastal habitats. Further organismal research will be required to determine if these spatial differences in pH variation have led to differential tolerances and physiological responses in species residing in these two habitats.

This model showed daily photosynthetic production could account for a large proportion of the observed pH variance over a 24-hr period, but was unable to resolve temporal patterns in pH variation across subsequent days. This disconnection resembles that found by [68] when modeling pCO_2 and O_2 dynamics within ice-covered lakes; they suggest this is due to convective mixing under surface ice. The model presented here assumes a well-mixed water column, a common condition at Cape Evans during this time of year [27,34,43]. The magnitude of the pH effect may be reduced due to the formation of a boundary layer along the bottom of the sea ice, which may decrease the photosynthetic exchange of CO_2 and O_2 in the water column. This layer thickens under slower current speeds, reducing gas diffusion away from the sea ice algae [43]. As such, this physical process should be incorporated into future models, as well as additional parameters to account for tidal exchange and non-algal community respiration. Further, given the sensitivity of the model to variation in carbonate system parameters (alkalinity and C_T), the effects of productivity on alkalinity and carbon species uptake should be included as it becomes available. These parameters could be empirically constrained by implementing repeated depth-stratified water sampling at regular intervals during a 24-h period, similar to the approach used by [53] for carbonate system effects of a *Phaeocystis* bloom in the Ross Sea polynya. Despite these limitations, this model offers a starting point to begin to explore a biological driver of pH variation in a unique environment that is threatened by changing global climate.

Acknowledgments

The authors thank members of the U.S. Antarctic Program, particularly R. Robbins and S. Rupp, for diving operations at McMurdo Station, Antarctica. We thank team members of Bravo-134 for their support in the field (Prof. M. Sewell, Dr. G. Dilly, Dr. E. Rivest, Dr. P. Yu, E. Hunter, H. Kaiser III, and L. Kapsenberg). We thank B. Peterson and Y. Takeshita from the Martz Lab at SIO for technical support with pH sensors. We thank Prof. R. Nisbet at UCSB for advice and guidance regarding model development, Dr. D. Reed and Dr. R. Miller at UCSB for insight regarding algal photosynthesis, and Dr. P. Vroom for thoughtful comments regarding this manuscript. We thank C. Gotschalk for assistance and advice regarding programming and data processing.

Author Contributions

Conceived and designed the experiments: PGM GEH. Performed the experiments: PGM GEH. Analyzed the data: PGM GEH LW TRM. Contributed reagents/materials/analysis tools: GEH LW. Contributed to the writing of the manuscript: PGM GEH LW TRM.

References

1. Martz TR, Connery JG, Johnson KS (2010) Testing the Honeywell Durafet for seawater pH applications. Limnol Oceanogr Methods 8: 172–184.

2. Hofmann GE, Smith JE, Johnson KS, Send U, Levin LA, et al. (2011) High-frequency dynamics of ocean pH: a multi-ecosystem comparison. PLoS One 6: e28983.

3. Doney SC, Ruckelshaus M, Duffy JE, Barry JP, Chan F, et al. (2012) Climate change impacts on marine ecosystems. Ann Rev Mar Sci 4: 11–37.

4. Littlepage JL (1965) Oceanographic investigations in McMurdo Sound, Antarctica. Antarct Res Ser 2: 1–38.

5. Mahoney AR, Gough AJ, Langhorne PJ, Robinson NJ, Stevens CL, et al. (2011) The seasonal appearance of ice shelf water in coastal Antarctica and its effect on sea ice growth. J Geophys Res 116.

6. Peck LS (2005) Prospects for survival in the Southern Ocean: vulnerability of benthic species to temperature change. Antarct Sci 17: 497.

7. Peck LS, Morley SA, Clark MS (2010) Poor acclimation capacities in Antarctic marine ectotherms. Mar Biol 157: 2051–2059.

8. Hofmann GE, Lund SG, Place SP, Whitmer AC (2005) Some like it hot, some like it cold: the heat shock response is found in New Zealand but not Antarctic notothenioid fishes. J Exp Mar Biol Ecol 316: 79–89.

9. Cummings V, Hewitt J, Van Rooyen A, Currie K, Beard S, et al. (2011) Ocean acidification at high latitudes: potential effects on functioning of the Antarctic bivalve *Laternula elliptica*. PLoS One 6: e16069.

10. Byrne M, Ho MA, Koleits L, Price C, King CK, et al. (2013) Vulnerability of the calcifying larval stage of the Antarctic sea urchin *Sterechinus neumayeri* to near-future ocean acidification and warming. Glob Chang Biol 19: 2264–2275.

11. Enzor LA, Zippay ML, Place SP (2013) High latitude fish in a high CO_2 world: Synergistic effects of elevated temperature and carbon dioxide on the metabolic rates of Antarctic notothenioids. Comp Biochem Physiol A Mol Integr Physiol 164: 154–161.

12. Yu PC, Sewell MA, Matson PG, Rivest EB, Kapsenberg L, et al. (2013) Growth attenuation with developmental schedule progression in embryos and early larvae of *Sterechinus neumayeri* raised under elevated CO_2. PLoS One 8: e52448.

13. McMinn A, Muller MN, Martin A, Ryan KG (2014) The response of Antarctic sea ice algae to changes in pH and CO_2. PLoS One 9: e86984.

14. Bednaršek N, Tarling GA, Bakker DCE, Fielding S, Jones EM, et al. (2012) Extensive dissolution of live pteropods in the Southern Ocean. Nat Geosci 5: 881–885.

15. Kawaguchi S, Ishida A, King R, Raymond B, Waller N, et al. (2013) Risk maps for Antarctic krill under projected Southern Ocean acidification. Nat Clim Chang 3: 843–847.

16. Sanford E, Kelly MW (2011) Local adaptation in marine invertebrates. Ann Rev Mar Sci 3: 509–535.

17. Kelly MW, Padilla-Gamino JL, Hofmann GE (2013) Natural variation and the capacity to adapt to ocean acidification in the keystone sea urchin *Strongylocentrotus purpuratus*. Glob Chang Biol 19: 2536–2546.

18. Meredith MP, Schofield O, Newman L, Urban E, Sparrow M (2013) The vision for a Southern Ocean Observing System. Curr Opin Environ Sustain 5: 306–313.

19. Jacobs SS, Gordon AL, Ardai JT (1979) Circulation and Melting Beneath the Ross Ice Shelf. Science 203: 439–443.

20. Barry JP (1988) Hydrographic Patterns in McMurdo Sound, Antarctica and Their Relationship to Local Benthic Communities. Polar Biol 8: 377–391.

21. Barry JP, Dayton PK (1988) Current Patterns in McMurdo Sound, Antarctica and Their Relationship to Local Biotic Communities. Polar Biol 8: 367–376.

22. Leonard GH, Purdie CR, Langhorne PJ, Haskell TG, Williams MJM, et al. (2006) Observations of platelet ice growth and oceanographic conditions during the winter of 2003 in McMurdo Sound, Antarctica. J Geophys Res 111.

23. Robinson NJ, Williams MJM, Barrett PJ, Pyne AR (2010) Observations of flow and ice-ocean interaction beneath the McMurdo Ice Shelf, Antarctica. J Geophys Res 115.
24. Robinson NJ, Williams MJM, Stevens CL, Langhorne PJ, Haskell TG (2014) Evolution of a supercooled Ice Shelf Water plume with an actively growing subice platelet matrix. Journal of Geophysical Research: Oceans 119, 3425–3446.
25. Assmann K, Hellmer HH, Beckmann A (2003) Seasonal variation in circulation and water mass distribution on the Ross Sea continental shelf. Antarct Sci 15: 3–11.
26. Dayton PK, Watson D, Palmisano A, Barry JP, Oliver JS, et al. (1986) Distribution patterns of benthic microalgal standing stock at McMurdo Sound, Antarctica. Polar Biol 6: 207–213.
27. Miller KA, Pearse JS (1991) Ecological Studies of Seaweeds in McMurdo Sound, Antarctica. Am Zool 31: 35–48.
28. Dayton PK, Robilliard GA, DeVries AL (1969) Anchor ice formation in McMurdo Sound, Antarctica, and its biological effects. Science 163: 273–274.
29. Schwarz AM, Hawes I, Andrew N, Norkko A, Cummings V, et al. (2003) Macroalgal photosynthesis near the southern global limit for growth; Cape Evans, Ross Sea, Antarctica. Polar Biol 26: 789–799.
30. Schwarz AM, Hawes I, Andrew N, Mercer S, Cummings V, et al. (2005) Primary production potential of non-geniculate coralline algae at Cape Evans, Ross Sea, Antarctica. Mar Ecol Prog Ser 294: 131–140.
31. Palmisano A, Sullivan CW (1983) Sea ice mircobal communities (SIMCO): 1. Distribution, abundance, and primary production of ice microalgae in McMurdo Sound, Antarctica in 1980. Polar Biol 2: 171–177.
32. Legendre L, Ackley SF, Dieckmann GS, Gulliksen B, Horner R, et al. (1992) Ecology of sea ice biota. 2. Global significance. Polar Biol 12: 429–444.
33. McMinn A, Martin A, Ryan K (2010) Phytoplankton and sea ice algal biomass and physiology during the transition between winter and spring (McMurdo Sound, Antarctica). Polar Biol 33: 1547–1556.
34. Ryan KG, Hegseth EN, Martin A, Davy SK, O'Toole R, et al. (2006) Comparison of the microalgal community within fast ice at two sites along the Ross Sea coast, Antarctica. Antarct Sci 18: 583.
35. Dickson AG, Sabine CL, Christian JR, editors (2007) Guide to best practices for ocean CO_2 measurements: PICES Special Publications. 191 p.
36. Fangue NA, O'Donnell MJ, Sewell MA, Matson PG, MacPherson AC, et al. (2010) A laboratory-based, experimental system for the study of ocean acidification effects on marine invertebrate larvae. Limnol Oceanogr: Methods 8: 441–452.
37. Robbins LL, Hansen ME, Kleypas JA, Meylan SC (2010) CO_2calc - A user-friendly seawater carbon calculator for Windows, Mac OS X, and iOS (iPhone). US Geol Survey Open File Report 128: 17.
38. Mehrbach C, Culberson CH, Hawley JE, Pytkowicz RM (1973) Measurement of the apparent dissociation constants of carbonic acid in seawater at an atmospheric pressure. Limnol Oceanogr 18: 897–907.
39. Dickson AG, Millero FJ (1987) A comparison of the equilibrium constants for the dissociation of carbonic acid in seawater media. Deep Sea Res 34: 1733–1743.
40. Goring DG, Pyne A (2003) Observations of sea-level variability in Ross Sea, Antarctica. NZ J Mar Freshwater Res 37: 241–249.
41. Stevens CL, Robinson NJ, Williams MJM, Haskell TG (2009) Observations of turbulence beneath sea ice in southern McMurdo Sound, Antarctica. Ocean Science 5: 435–445.
42. Rivkin RB, Putt M, Alexander SP, Meritt D, Gaudet L (1989) Biomass and production in polar planktonic and sea ice microbial communities: a comparative study. Mar Biol 101: 273–283.
43. McMinn A, Ashworth C, Ryan KG (2000) In situ net primary productivity of an Antarctic fast ice bottom algal community. Aquat Microb Ecol 21: 177–185.
44. Trenerry LJ, McMinn A, Ryan KG (2002) In situ oxygen microelectrode measurements of bottom-ice algal production in McMurdo Sound, Antarctica. Polar Biol 25: 72–80.
45. Kozlyaninov MV, Pelevin VN (1966) On the application of a one-dimensional approximation in the investigation of the propagation of optical radiation in the sea. Dept Commerce. 54 p.
46. Jerlov NG (1976) Marine Optics. New York: Elseiver Press.
47. Jassby AD, Platt T (1976) Mathematical Formulation of the Relationship Between Photosynthesis and Light for Phytoplankton. Limnol Oceanogr 21: 540–547.
48. Satoh H, Watanabe K (1988) Primary Productivity in the Fast Ice Area near Syowa Station, Antarctica, during Spring and Summer 1983/84. J Oceanogr Soc Japan 44: 287–292.
49. Raven JA, Ball LA, Beardall J, Giordano M, Maberly SC (2005) Algae lacking carbon-concentrating mechanisms. Can J Bot 83: 879–890.
50. Gleitz M, Kukert H, Riebesell U, Dieckmann GS (1996) Carbon acquisition and growth of Antarctic sea ice diatoms in closed bottle incubations. Mar Ecol Prog Ser 135: 169–177.
51. Van Heuven S, Pierrot D, Lewis E, Wallace DWR (2009) MATLAB program developed for CO_2 system calculations. ORNL/CDIAC-105b. Carbon Dioxide Information Analysis Center, Oak Ridge National Laboratory, U.S. Department of Energy, Oak Ridge, Tennessee.
52. Lee K, Tong LT, Millero FJ, Sabine CL, Dickson AG, et al. (2006) Global relationships of total alkalinity with salinity and temperature in surface waters of the world's oceans. Geophys Res Lett 33.
53. Bates NR, Hansell DA, Carlson CA, Gordon LI (1998) Distribution of CO_2 species, estimates of net community production, and air-sea CO_2 exchange in the Ross Sea polynya. J Geophys Res 103: 2883–2896.
54. Fransson A, Chierici M, Yager PL, Smith WO (2011) Antarctic sea ice carbon dioxide system and controls. J Geophys Res 116.
55. Roden NP, Shadwick EH, Tilbrook B, Trull TW (2013) Annual cycle of carbonate chemistry and decadal change in coastal Prydz Bay, East Antarctica. Mar Chem 155: 135–147.
56. Shadwick EH, Tilbrook B, Williams GD (2014) Carbonate chemistry in the Mertz Polynya (East Antarctica): Biological and physical modification of dense water outflows and the export of anthropogenic CO_2. J Geophys Res Oceans 119: 1–14.
57. Rivkin RB (1991) Seasonal Patterns of Planktonic Production in McMurdo Sound, Antarctica. Am Zool 31: 5–16.
58. Nielsen ES (1962) One the maximum quantity of plankton chlorophyll per surface unit of a lake or the sea. Int Rev Gesamten Hydrobiol Hydrogr 47: 333–338.
59. Matson PG, Martz TR, Hofmann GE (2011) High-frequency observations of pH under Antarctic sea ice in the southern Ross Sea. Antarct Sci 23: 607–613.
60. Yu PC, Matson PG, Martz TR, Hofmann GE (2011) The ocean acidification seascape and its relationship to the performance of calcifying marine invertebrates: Laboratory experiments on the development of urchin larvae framed by environmentally-relevant pCO_2/pH. J Exp Mar Biol Ecol 400: 288–295.
61. Palmisano AC, SooHoo JB, SooHoo SL, Kottmeier ST, Craft LL, et al. (1986) Photoadaptation in Phaeocystis pouchetii advected beneath annual sea ice in McMurdo Sound, Antarctica. J Plankton Res 8: 891–906.
62. McNeil BI, Tagliabue A, Sweeney C (2010) A multi-decadal delay in the onset of corrosive 'acidified' waters in the Ross Sea of Antarctica due to strong air-sea CO_2 disequilibrium. Geophysical Research Letters 37.
63. Rivest EB, Hofmann GE (2014) Responses of the Metabolism of the Larvae of Pocillopora damicornis to Ocean Acidification and Warming. PLoS ONE 9: e96172.
64. Smethie WM, Jacobs SS (2005) Circulation and melting under the Ross Ice Shelf: estimates from evolving CFC, salinity and temperature fields in the Ross Sea. Deep Sea Res Part I: Oceanogr Res Pap 52: 959–978.
65. Reddy TE, Holland DM, Arrigo KR (2010) Ross ice shelf cavity circulation, residence time, and melting: results from a model of oceanic chlorofluorocarbons. Cont Shelf Res 30: 733–742.
66. McMinn A, Pankowskii A, Ashworth C, Bhagooli R, Ralph P, et al. (2010) In situ net primary productivity and photosynthesis of Antarctic sea ice algal, phytoplankton and benthic algal communities. Mar Biol 157: 1345–1356.
67. McMinn A, Ryan KG, Ralph PJ, Pankowski A (2007) Spring sea ice photosynthesis, primary productivity and biomass distribution in eastern Antarctica, 2002–2004. Mar Biol 151: 985–995.
68. Baehr MM, DeGrandpre MD (2004) In situ pCO_2 and O_2 measurements in a lake during turnover and stratification: observations and modeling. Limnol Oceanogr 49: 330–340.
69. McMinn A, Ryan K, Gademann R (2003) Diurnal changes in photosynthesis of Antarctic fast ice algal communities determined by pulse amplitude modulation fluorometry. Mar Biol 143: 359–367.

5

Recent Warming of Lake Kivu

Sergei Katsev[1,2]*, Arthur A. Aaberg[2], Sean A. Crowe[3], Robert E. Hecky[1,4]

1 Large Lakes Observatory, University of Minnesota Duluth, Duluth, Minnesota, United States of America, **2** Department of Physics, University of Minnesota Duluth, Duluth, Minnesota, United States of America, **3** Department of Earth, Ocean, and Atmospheric Sciences, University of British Columbia, Vancouver, British Columbia, Canada, **4** Department of Biology, University of Minnesota Duluth, Duluth, Minnesota, United States of America

Abstract

Lake Kivu in East Africa has gained notoriety for its prodigious amounts of dissolved methane and dangers of limnic eruption. Being meromictic, it is also expected to accumulate heat due to rising regional air temperatures. To investigate the warming trend and distinguish between atmospheric and geothermal heating sources, we compiled historical temperature data, performed measurements with logging instruments, and simulated heat propagation. We also performed isotopic analyses of water from the lake's main basin and isolated Kabuno Bay. The results reveal that the lake surface is warming at the rate of 0.12°C per decade, which matches the warming rates in other East African lakes. Temperatures increase throughout the entire water column. Though warming is strongest near the surface, warming rates in the deep waters cannot be accounted for solely by propagation of atmospheric heat at presently assumed rates of vertical mixing. Unless the transport rates are significantly higher than presently believed, this indicates significant contributions from subterranean heat sources. Temperature time series in the deep monimolimnion suggest evidence of convection. The progressive deepening of the depth of temperature minimum in the water column is expected to accelerate the warming in deeper waters. The warming trend, however, is unlikely to strongly affect the physical stability of the lake, which depends primarily on salinity gradient.

Editor: Inés Álvarez, University of Vigo, Spain

Funding: The work was supported by the University of Minnesota Office of International Programs Seed Grant to SK. The funder had no role in study design, data collection and analysis, decision to publish, or preparation of the manuscript.

Competing Interests: The authors have declared that no competing interests exist.

* Email: skatsev@d.umn.edu.

Introduction

Deep meromictic lakes are good climate monitors, as changes in heat fluxes across the lake surface become reflected in heat content of the deeper waters [1]. Consistent with recent atmospheric warming, surface temperatures of stably stratified East African Great Lakes increased by about one degree over the last century [1,2]. The temperature rise could be linked to a number of adverse effects, such as increased physical stability of the water column that decreases nutrient upwelling and diminishes ecosystem productivity [1,2].

The 450 m deep Lake Kivu, at the border of Rwanda and the Democratic Republic of the Congo, is unique among the African Great Lakes in that in addition to heat exchanges with the atmosphere its temperature is affected by subterranean heat inputs. The sublacustrine heat comes from a complex geological system that includes two active volcanoes, Nyiragongo and Nyamuragira [3–5], which constitute the lake's northern watersheds. These deep heat sources cause the lake temperature to increase with depth below the surface mixed layer, a feature not observed in other lakes. In addition to these unique features, the lake is best known for the prodigious amounts of carbon dioxide and methane in its deep waters [6,7], which present both a hazard of a catastrophic limnic eruption [8,9] and an economic opportunity of extracting the methane for energy generation. Persistent stratification in Lake Kivu is maintained by a salinity gradient with multiple pycnoclines, the strongest ones located at the depths of 60, 160, 250, and 310 meters. The pycnoclines are believed to be maintained by several sublacustrine inflows, which

preserve the salinity and temperature gradients against dissipation by diffusion [8]. The uppermost pycnocline (60 m) marks the lower boundary of the epilimnion and the seasonal maximum depth of wind-induced mixing. The temperature decreases from surface into the epilimnion and down to the depth of approximately 80 m, but increases below that depth due to sublacustrine heat inputs (Fig. 1). The resultant reverse temperature gradient in deep waters implies an upward transport of heat towards the depth of the temperature minimum. This heat may be removed only through epilimnetic mixing, e.g. if weaker temperature gradients during a dry season [7] allow mixing to that depth, or if a cold inflow enters the lake at that depth.

Past studies [10] have reported surface warming of Lake Kivu of up to 0.5°C over the past 30 years. Monimolimnion warming, on the other hand, has received less attention. Until recently, warming was thought to be dominated by the atmospheric signal and limited to upper 250 m (the depth of the strongest pycnocline [4,10]). Thermal effects of the Nyiragongo lava inflows after a 2002 eruption were deemed insignificant [4,10]. Recently, however, warming below 250 m was also acknowledged [11], though whether the source of the heat was atmospheric or subterranean was not identified. Constraining the rate of the heat transfer in the deep monimolimnion is difficult [12], as the reverse temperature gradient results in double-diffusion structures [13] that are intermittent and often laterally localized within several hundred meters [12]. Non-diffusive transport mechanisms, such as convection generated by localized heat sources or density flows [14] have not been quantified. Despite several attempts to

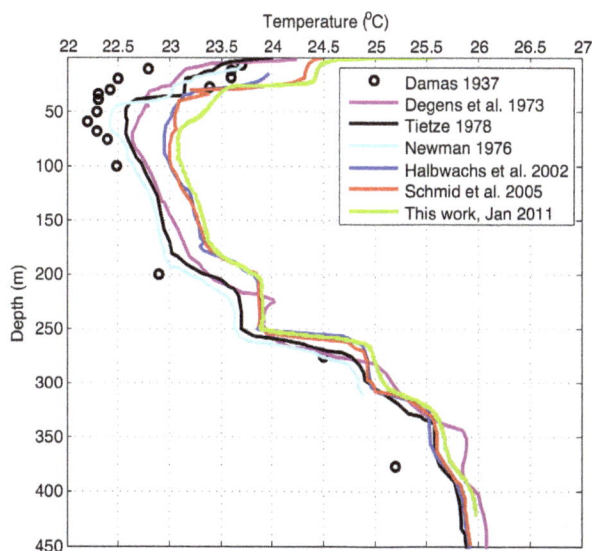

Figure 1. Evolution of temperature profiles in Lake Kivu.

calculate mixing coefficients and fluxes [12,15,16], the rates of vertical mixing remain uncertain.

In this paper, we compile multi-decadal records of lake temperature profiles that show that the temperatures have increased throughout the entire water column of Lake Kivu. We quantify these increases, as well as the extent of seasonal mixing and epilimnetic heat removal. Focusing on multi-year and multi-decadal trends, we discuss the likely causes of warming, sources of heat in the surface mixed layer and deep waters, evidence for convective mixing in deep waters, and implications for lake stability.

Methods

Depth profiles of temperature and conductivity were taken in Lake Kivu using a Sea&Sun CTD 90 M probe. The manufacturer's stated accuracy was 0.002°C and precision was 0.0002°C. A string of Onset U22 temperature recorders was deployed on a rope suspended from the KP1 methane extraction platform (1°43' 56.445" S, 29°14' 34.246" E) from January 2011 to March 2013. The temperature recorders had a precision of 0.02°C, which was sufficient for the purpose of resolving temporal temperature variations at their respective depths. Systematic errors within the recorders' accuracy range (0.2°C) were corrected by comparing their readings with those taken by the CTD. The deployment was verified in November 2011 and October 2012, at which point several faulty recorders were replaced and additional ones installed. A titanium-cased Seabird (SBE39) thermistor (accuracy 0.002, precision 0.0001°C, stability 0.0002°C/month) was deployed at the depth of 355 m on a rope suspended from the Rwanda Energy Corporation (REC) platform (1°43' 55.9" S, 29°14' 34.1" E) between September 2012 and March 2013. A second SBE39 temperature recorder equipped with a pressure sensor was deployed at 68 m depth to monitor for possible changes in deployment depth due to displacement of the platform or elasticity of the suspended rope. Historical temperature profiles in Lake Kivu water column over the past 80 years were obtained by digitization from published literature sources. The pressure-to-depth conversion for the CTD profiles was performed by multiplying the gauge pressure in dbar by 1.0197, consistent with

the conversion used in previous studies in Lake Kivu [17]. This ignores the difficult-to-calculate depth variation of water density in Lake Kivu [4] but results in an error of no more than 0.5%, which is on the order of the CTD's pressure sensor accuracy. Water samples were taken from a selection of depths in both the main basin (in 2011 and 2012) and Kabuno Bay (in 2012) using a Niskin bottle with the pressure valve loosened to allow the exsolving gasses to escape on ascent. The samples were placed in crimped vials with no headspace by overfilling them from the bottom using a tube connected to the Niskin bottle spigot, and subsequently analyzed for the ^{18}O and ^{2}H isotopic composition at the GEOTOP facility at McGill University. Additional samples were collected by hand from a surface stream (1° 42' 14.99" S, 29° 16' 14.71" E), a hot spring near Gisenyi, and a single rain event near Kabuno Bay (1° 38' 28.63" S, 29° 8' 18.56" E, October 1, 2012). Duplicate water column samples taken according to USEPA protocols were analyzed for CFC concentrations by the Tritium Laboratory at the University of Miami, who also calculated the recharge ages (effective time of last contact with the atmosphere, in years before sampling) according to standard procedures. Analytical data have been made publicly available through the IEDA EarthChem data repository. Data are also available upon request to the corresponding author. Research in Rwanda was conducted under permit MINEDUC/S&T/0011/2010.

Results

Stratification and mixing

The temperature (Fig. 1) and conductivity (not shown) distributions in Lake Kivu conform to the salinity-controlled stratification described in multiple previous studies (e.g. [8]). The temperature minimum was observed in 2011–2012 at about 78 m depth (Fig. 1). Excursions in temperature and salinity profiles (not shown here) and comparison of profiles taken at different locations indicated cold sublacustrine inflows at depths 166, 173, and 242 m, and a warm inflow between 260 and 300 m depth, consistent with previous findings [10,18]. Double diffusion "staircase" structures in temperature profiles were detected at depths intervals 170–200 m and 275–340 m. Temperature records in the epilimnion (Fig. 2) indicated that surface mixing during the dry season in 2011 and 2012 reached the depth of approximately 55 m, about 20 m shallower than the depth of the temperature minimum. The temperatures below the mixing depth, recorded by thermistors, increased at an average rate of ~0.02–0.04°C per year ($p > 0.05$) (Fig. 2). The temperature increase recorded by a more precise Seabird recorder at 68 m depth between October 2012 and March 2013 was 0.016°C per year ($p > 0.05$) (Fig. 3).

The temperature time series recorded by the deep-water thermistor at 355 m depth shows temperature variations on the order of 0.01°C. This variance is characteristic of the temporal variability in CTD profiles at this depth, though greater than the variability reported in the deep waters of Lake Matano, a 700 m deep meromictic lake in Indonesia with no known heat sources at depth [19]. Importantly, the time series in Fig. 3 reveals several events with substantially larger temperature excursions, on the order of 0.05°C.

Water isotopic signals

The ^{18}O and ^{2}H isotopic composition (Fig. 4) of Lake Kivu water column falls on a line between two endmembers: surface water where isotopic composition is dominated by the evaporative signal, and deep water where the isotopic signature trends towards the composition of groundwater, as exemplified by the composi-

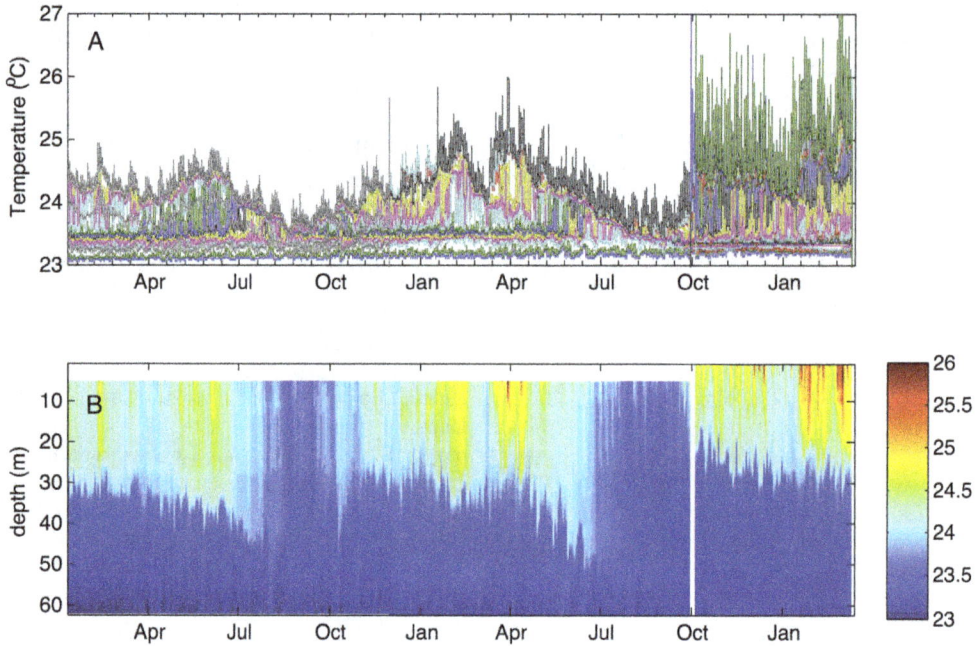

Figure 2. Epilimnion temperatures between January 2011 and March 2013. Temperatures were recoded by a string of moored Onset temperature loggers. A. Records from individual thermistors. B. Same data as a contour plot against depth. Thermistors were recovered and redeployed in November 2011 and October 2012, at which times several loggers were replaced and additional ones added.

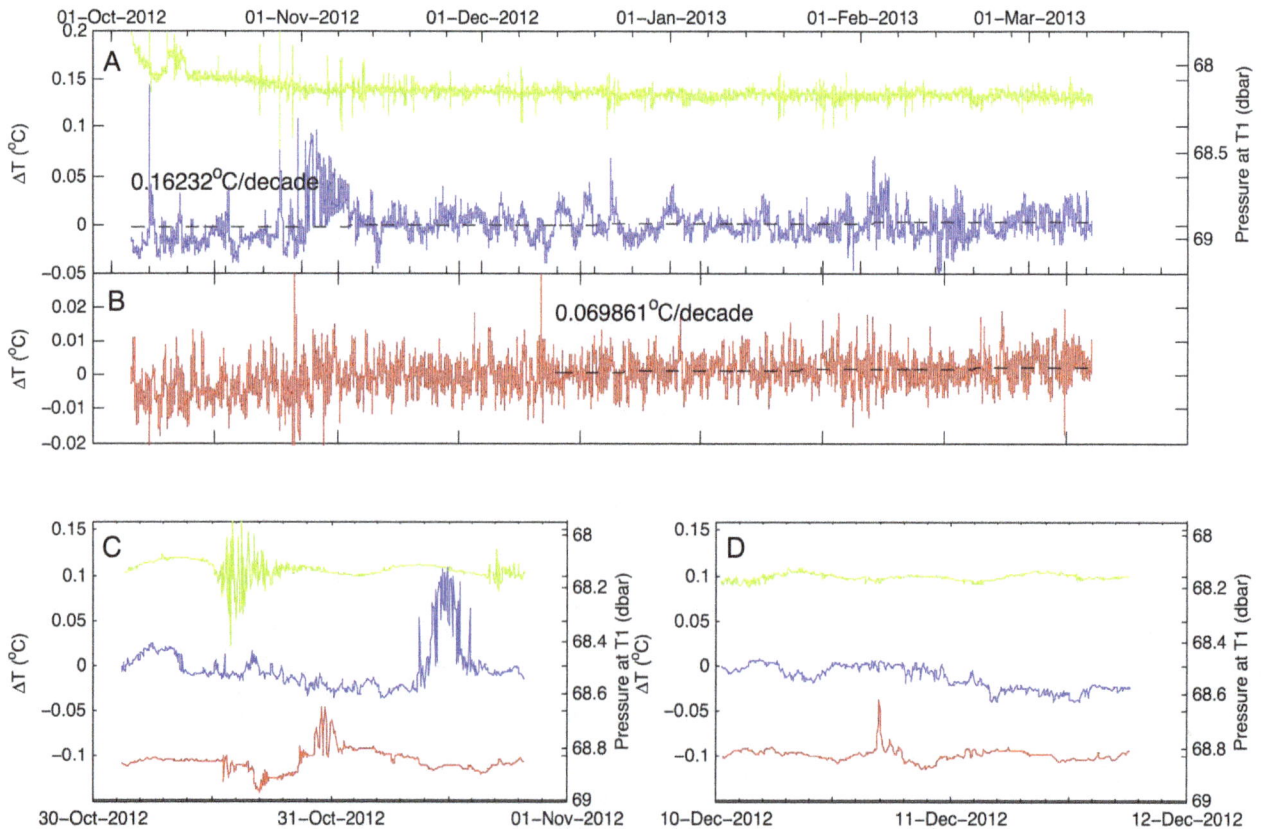

Figure 3. Deep-water temperatures between October 2012 and March 2013. Temperatures were recorded using moored Seabird temperature loggers. A,B: Temperature time series (deviations from the mean) at the depths of 68 m (T1) and 355 m (T2). The green line is pressure recorded at T1. C,D: The same but in more detail for two anomalous events in October and December 2012.

tion at a hot spring in Gisenyi. The waters of Lake Kivu become isotopically lighter with depth, generally following the depth dependence of water density/conductivity (Fig. 4), which suggests that the depth distributions of isotopes are controlled by mixing rates (e.g. [19]). The waters in Kabuno Bay, which is separated from the main basin by a shallow sill, differ radically in their isotopic composition from the main basin. Water below the permanent pycnocline at 10 m in Kabuno Bay is isotopically much lighter than anywhere in the main basin (Fig. 4) but falls along the same line between the evaporative and groundwater endmembers (Fig. 4). This suggests that Kabuno Bay, being smaller, is more affected by sublacustrine inflows, many of which are believed to be located along the north shore of Lake Kivu. Small-scale variations in the depth distribution of ^{18}O isotopes vs. ^{2}H isotopes may reflect the inflow compositions. For instance, the δ^2H profile in Kabuno Bay exhibits a minimum near the persistent pycnocline, at the depth where a cold inflow along the north shore has been detected in our CTD profiles (not shown). Similarly, Fig. 4 suggests inflows in the main basin at around 230 m, 260 m, 170 m, 70 m, and 35 m depth. The trends observed in our data are significantly clearer than those presented previously [5], due to higher spatial and concentration resolutions.

CFC concentrations in the monimolimnion were low. At the depth of 80 m, which approximately corresponds to the temperature minimum, the recharge age was>40 years, indicating that surface mixing did not reach that depth in recent past.

Discussion

The warming trend

A compilation of previously published and recently obtained temperature profiles (Fig. 1) reveals that temperatures in Lake Kivu have been increasing throughout the entire water column. Temperatures at the depth of temperature inversion (~78 m) have increased by approximately 0.5°C since mid-1970s, a change of ~0.12°C per decade. This is matched or exceeded by the warming rates (~0.2°C per decade) recorded by our instruments below the depth of seasonal mixing in 2011–2012 (Figs. 2–3). Temperatures in the upper epilimnion, averaged over seasonal variations, seem

to have increased over the past 40 years by half a degree to a degree (Fig. 1). Waters below 350 m have warmed since the 1970s by about 0.15°C (Fig. 1), though this number is less certain as early profiles exhibit significant variability (Fig. 1). It is worth noting that the profiles by Degens (1973) [20] and Newman (1976) [16] were taken at about the same time, on the same cruise. Degens's modified heat flow recorder likely did not have the same response time and accuracy as Newman's calibrated and rapid thermistors, so Newman's profile is likely to be more accurate. That this profile is similar and only slightly offset from the data of Damas (1937) [21] suggests that most of the warming (at all depths) occurred in the last 40 years. Figure 5b shows the evolution of temperatures at selected depths between major thermoclines where temperature gradients are minimal. The monimolimnion warming is strongest near the mixolimnion and progressively decreases with depth (Fig. 5). The warming rates inferred for the deep monimolimnion (Fig. 5) are consistent with the trend seen in our thermistor data (Fig. 3). Concurrent with warming, the thermal structure of the lake has changed in that the major thermoclines (pycnoclines) moved upward by about 15 m since the 1970s (Fig. 1; see also [14] and [18]), likely as a result of subsurface water inflows. Interestingly, the thermocline upwelling velocities (about 0.1 m y^{-1} for the 370 m thermocline, 0.2 m y^{-1} for the 310 and 250 m thermoclines, and 0.35 m y^{-1} for the 160 m thermocline) roughly match the values for the vertical advection velocities in the "steady state" model of [8] (from 0.15 m y^{-1} below 250 m to around 0.79±0.13 m y^{-1} between 110 and 200 m depth; cited in [22]).

The warming trend in the mixed layer of Lake Kivu is consistent with changes in other large African Lakes. In Lake Malawi, temperatures at 100 m depth are increasing at a rate of 0.06±0.02°C per decade [23]. In Lake Tanganyika, the upper water-column (150 m) has been warming at 0.1±0.01°C per decade since 1913, and deep waters (600 m) by ~0.05 C per decade since 1938 [2,24]. Lake Victoria warmed by 0.3°C between 1960s and 1991 (0.1 C per decade [25]), and Lake Albert warmed by 0.5°C between 1963 and 1990s (~0.15°C per decade [13]). Warming typically results either from an increased amount of absorbed heat (e.g. due to higher air temperatures or

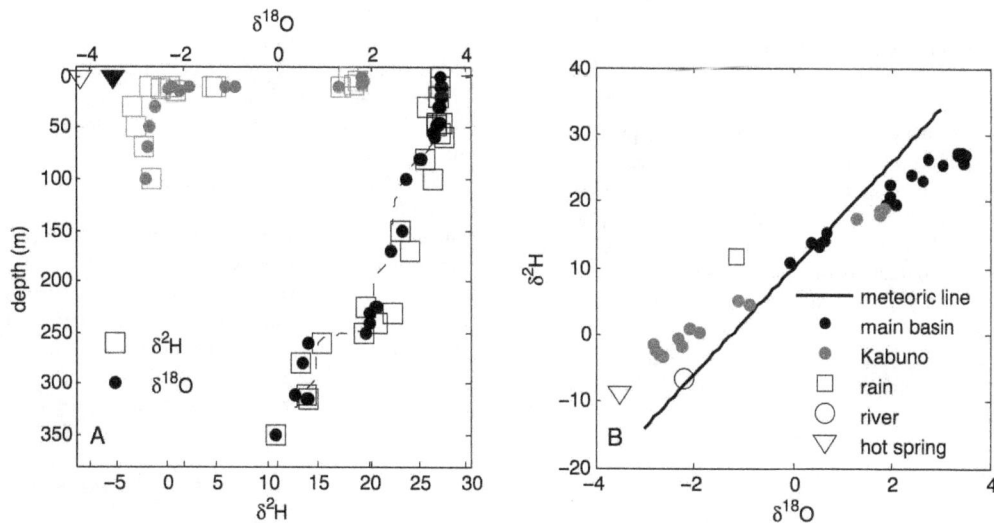

Figure 4. ^{18}O and ^{2}H isotopic signals in Lake Kivu's main basin (black) and Kabuno Bay (grey). Also shown are data for a surface hot spring in Gisenyi (triangles), surface stream in Gisenyi (open circle), and a single rain event (square in panel B). The dashed line in panel A is an inverted and scaled profile of water conductivity, shown here for illustration as an indicator of water column stratification.

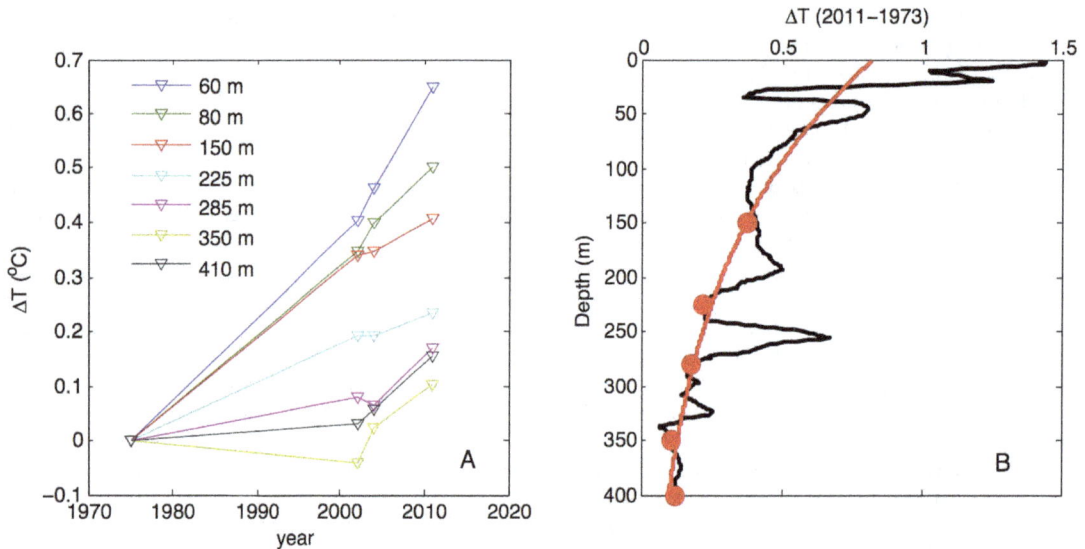

Figure 5. Changes in water column temperature. A) At selected depths against time, relative to 1975 measurements [32]. B) As a function of depth, based on comparison of our January 2011 profile with [32]. The exponential fit line is drawn through points (marked by symbols) between major thermoclines where temperature gradients are minimal and also least affected by the vertical movement of thermoclines.

reduced cloud cover) during the wet season or a decrease in evaporative cooling (e.g. due to an increased air humidity) during the dry season. Long-term trends in air temperatures around Lake Kivu are poorly known. Across the region, however, air temperatures kept pace with the global increase of 0.6±0.2°C over the past century (~0.07 per decade [26]): temperatures increased by ~0.5°C since 1980 in Kenya (0.15°C/decade [27]), 0.4°C since 1930s in winter around Lake Malawi (0.05°C/decade [23]), and 0.5–0.7°C between 1950 and 1995 in Lake Tanganyika (~0.13°C per decade [2]). The Berkeley Earth compilation of meteorological data suggests the warming rate of 0.22°C/decade for Rwanda since the 1970s.

Heat transport, mixing, and causes of warming

The reverse temperature gradient below 80 m depth implies an upward transport of heat from deep waters towards the depth of temperature inversion. The epilimnetic heat fluxes are also directed (downward) towards this depth. Though epilimnion temperature gradients become weaker during the dry season [7], mixing does not normally reach the T inversion depth (Fig. 2), although occasional deeper mixing events have been suggested [18]. The total increase in the main basin's heat content observed below 80 m between 1973 and 2011 (~200 MJ m^{-2}) requires an average excess heat flux through the uppermost chemocline of ~0.18 W m^{-2}. In contrast to the main basin of the lake, the temperature profiles in Kabuno Bay exhibit a strong temperature minimum throughout the year [7], and the temperature minimum is maintained by a cold inflow at that depth (10 m) (Fig. 4 and unpublished temperature data). In the main basin, cold inflows at the depth of the temperature minimum have not been detected and a calculation based on the balance of heat fluxes also suggested the absence of such an inflow [17].

Temperature variations in the deep water column (below 250 m) indicate active energy transfer there (Figs. 1 and 3). In a quiescent deep monimolimnion of a 600 m deep meromictic lake Matano, for example, temperature profiles in the deep waters remain constant within one hundredth of a degree (instrumental accuracy) over decades [19]. In contrast, in Lake Kivu the deep water temperature profiles vary significantly over time (Fig. 1), and possibly laterally among deep sampling locations. Some lateral variability may be expected in the vicinity of subsurface inflows (e.g., Fig. 4); however, the time series in Fig. 3 also indicates significant temporal variability. This suggests that the deep monimolimnion experiences water movements and heat fluxes, perhaps in response to localized heating or seismic events. For example, our temperature recorder at 350 m registered two anomalous events in a six-month period (Fig. 3) that were not obviously correlated to the records higher up in the water column. The mixing intensity in the deep waters therefore cannot be assumed on long time scales to be close to molecular diffusion, as assumed by previous models (e.g., [8,22]) or inferred from measured temperature microstructures [15]. Vertical mixing in Lake Kivu remains poorly quantified, as standard methods are difficult to apply under the nearly steady-state conditions [19] and the effects of episodic events are difficult to take into account. Past modeling approaches, in particular, assumed a constant eddy diffusion coefficient between the major pycnoclines [8] or attempted calculating the mixing coefficients from observed double-diffusion structures [12,16].

The warming rate that is greater in the upper than deep monimolimnion (Fig. 4) suggests that the lake has been warming primarily from the surface. The effects of surface vs. deep heat sources can be illustrated with a simple model that uses assumed coefficients of vertical mixing (e.g., as suggested by previous estimates [8]) to propagate excess heat (ΔT) from the epilimnion downward. For the water of temperature T, the evolution of heat content $c\rho T$, where c is the specific heat capacity and ρ is the density of water, can be described by a reaction-diffusion equation:

$$c\rho \frac{\partial T}{\partial t} = c\rho \frac{\partial}{\partial z}\left(K_z \frac{\partial T}{\partial z}\right) - vc\rho \frac{\partial T}{\partial z} + R \qquad (1)$$

where K_z is the vertical (turbulent) diffusion coefficient, v is the vertical advection velocity, and R is the rate of heat production at depth z. Assuming that for some temperature distribution $T(z)$ the production of heat in the monimolimnion $R(z)$ can be balanced by the heat removal through diffusion and advection (a steady state), the propagation of a disturbance to that state (excess heat) can be described by replacing $T(z)$ with $T(z)+\Delta T(z,t)$ and rewriting eq. (1), taking into account that for a steady state $T(z)$ the right-hand-side of eq. (1) is zero:

$$\frac{\partial \Delta T}{\partial t} = \frac{\partial}{\partial z}\left(K_z \frac{\partial \Delta T}{\partial z}\right) - v\frac{\partial \Delta T}{\partial z} \qquad (2)$$

As the diffusion equation (1) is linear with respect to temperature T (neglecting the temperature dependencies of c and ρ), the propagation of the excess heat $c\rho\Delta T$ (eq. 2) does not depend on the direction of the temperature gradient, which simplifies the simulations. For advective velocities in Lake Kivu (inferred previously to be on the order of 1–8 meters per decade; *above* and in [8]) the advective term in eq. 2 is small in comparison to mixing by turbulent diffusion and can be neglected. Equation 2 was solved using the numerical solver of the water-column module of Aquasim, a software package designed to solve this type of equations in aquatic environments [28]. As a conservative estimate, an increase in heat fluxes of 0.5 W m^{-2} over the steady state was specified at the model's upper boundary, at the base of the mixed layer, and the resultant changes in water temperature ΔT were calculated for a forty-year time interval. The results from this simple model (Fig. 6) indicate that the observed warming below 250 m cannot be explained by the propagation of heat from the overlying waters using the currently assumed transport rates (e.g., the eddy diffusion coefficients used in [6] and [8] or the transport rates on the order of thermal diffusivity of water). Either the effective (long term) heat transport rates are higher by an order of magnitude, or the intensity of the subsurface heat sources is increasing (over a hypothetical equilibrium regime where the generated deep heat can be removed by the existing temperature gradients). An alternative way of looking at this is to calculate the characteristic length scale of diffusion $\sqrt{2K_z t}$. Using the diffusion coefficient of 1×10^{-6} m^2 s^{-1}, the maximum value used in previous modeling studies [8], suggests that after $t = 40$ years the profiles of $\Delta T(z)$ should attenuate over the characteristic length scale of 50 m, whereas observations (Fig. 5) suggest the length scale for exponential attenuation on the order of 190 m. Comparing the diffusive transport of conservative tracers, such as Na, to their removal rates from the epilimnion with the outflow [22], nevertheless, suggests that the vertical transport rates are unlikely to have been significantly underestimated in past studies (M. Schmid, pers. comm.), at least for the upper water column and assuming negligible coastal upwelling. The intensification of the subsurface heat sources thus appears to be a more likely reason for the observed deep warming.

Besides the obvious hydrothermal heat inputs, which may or may not be stable over decadal time scales considered here, warming in Lake Kivu may result from biogeochemical reactions, as originally proposed in [29]. Formation of methane from carbon dioxide and hydrogen, the process responsible for estimated 65% of methane production below 260 m [6], liberates 60 kcal/mol (the standard enthalpy of reaction $\Delta H_0 = 240.2$ kJ/mol). Acetoclastic methanogenesis releases about the same amount of energy ($\Delta H_0 = 245.2$ kJ/mol [30]). Based on comparison to 1955 measurements [31], a 1973 study [29] suggested a growth of the methane reservoir by about 1 percent per year, and a 2005 study [8] similarly claimed the increase of about 15% between 1970s and early 2000s. If methane concentrations indeed increased in the past 30 years by 3 mmol L^{-1}, formation of this amount of methane would release about 180 cal/L, enough to heat the water by 0.18 degrees, which would be consistent with the temperature

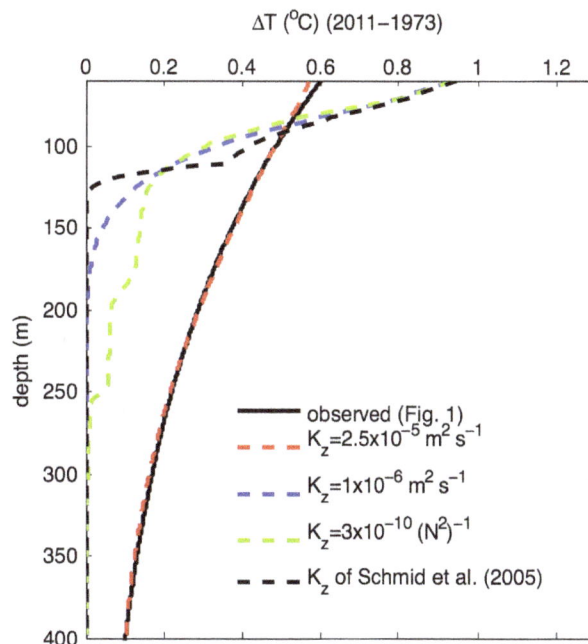

Figure 6. Simulated vs. observed change in water temperature in response to decreased epilimnetic heat removal. The change in water temperatures between 1973 and 2011 is simulated for a decrease in heat removal at the base of the mixolimnion of 0.5 W m^{-2}. The solid black line is the same as the exponential fit line in Fig. 5. Dashed lines illustrate the simulation results for different values of the mixing coefficient K_z: selected constant values, values obtained from the correlation of K_z with the stability frequency N_2 as suggested in [19], and tabulated values suggested for Lake Kivu in [8]. For comparison, the molecular thermal diffusivity in water (at 25°C) is $0.143\cdot10^{-6}$ m^2 s^{-1}.

change in Fig. 5. Some energy may go towards satisfying the energy requirements of organisms rather than converted to heat, but with the biomass likely being approximately constant over decadal time scales this fraction, which is difficult to estimate, is probably small. Similarly, if the concentrations of CO_2 also increased by ~15% (~13 mmol L^{-1}), the heat of dissolution ($\Delta H_0 = -19.3$ kJ/mol) could further increase the temperature by 0.06 degrees.

Conclusions

Lake Kivu has experienced significant warming over the course of the 20th century, with temperatures rising over the entire water column. The warming rate in the seasonally mixed epilimnion is similar to those reported in other East African lakes. This rate has likely increased over the past several decades, and is most probably linked to regional climate warming. The warming rates in the deep monimolimnion suggest that heat inputs from deep subterranean sources are not balanced by the removal of heat towards the lake surface. Unlike other deep African lakes where warming increases physical stability through increased temperature gradients, the salinity-stratified Lake Kivu is unlikely to experience strong changes in stability. Nevertheless, continued deepening of the temperature minimum will lead to rarer events of heat removal by epilimnetic mixing, which will further decrease the removal of heat from the deep waters. The warming rate is thus expected to accelerate. The warming trend, the descent of the temperature minimum, and the upward movement of thermoclines all indicate that over decadal time scales the stratification in Lake Kivu cannot be considered to be at steady state.

Acknowledgments

We thank Prof. Alfonso Mucci for help with isotopic water analyses at GEOTOP, Francois Darchambeau for collaboration in fieldwork and intellectual contributions, Sally MacIntyre for discussions and providing Onset temperature recorders, Wim Tiery for discussions and weather data compilation, and Natacha Pasche, David Krasner, Antoine Nsabimana, Elena Kovaleva and the Rwandan Ministry of Infrastructure for facilitating fieldwork and access to moored platforms. Research in Rwanda was conducted under permit MINEDUC/S&T/0011/2010.

Author Contributions

Conceived and designed the experiments: SK. Performed the experiments: SK AAA SAC REH. Analyzed the data: SK AAA. Contributed reagents/materials/analysis tools: SK SAC REH. Wrote the paper: SK SAC REH.

References

1. Verburg P, Hecky RE, Kling H (2003) Ecological consequences of a century of warming in Lake Tanganyika. Science 310: 505–507.

2. O'Reilly CM, Alin SR, Plisnier P-D, Cohen AS, McKee BA (2003) Climate change decreases aquatic ecosystem productivity of Lake Tanganyika, Africa. Nature 424: 766–768.

3. Ross KA, Smets B, De Batist M, Hilbe M, Schmid M, et al. (2014) Lake-level rise in the late Pleistocene and active subaquatic volcanism since the Holocene in Lake Kivu; East African Rift. Geomorphology, accepted.

4. Schmid M, Tietze K, Halbwachs M., Lorke A, McGinnis D, et al. (2002) How hazardous is the gas accumulation in Lake Kivu? Arguments for a risk assessment in light of the Nyiragongo volcano eruption of 2002. Acta Volcanologica 14: 115–122.

5. Tassi F, Vaselli O, Tedesco D, Montegrossi G, Darrah T, et al. (2009) Water and gas chemistry at Lake Kivu (DRC): Geochemical evidence of vertical and horizontal heterogeneities in a multibasin structure. Geochem. Geophys. Geosys. 10: Q02005.

6. Pasche N, Schmid M, Vazquez F, Schubert CJ, Wüest A, et al. (2011) Methane sources and sinks in Lake Kivu. J. Geophys. Res. 116: G03006.

7. Borges AV, Abril G, Delille B, Descy JP, Darchambeau F (2011) Diffusive methane emissions to the atmosphere from Lake Kivu (Eastern Africa). J. Geophys. Res. 116: G03032.

8. Schmid M, Halbwachs M, Wehrli B, Wüest A (2005) Weak mixing in Lake Kivu: New insights indicate increasing risk of uncontrolled gas eruption. Geochem. Geophys. Geosys. 6: Q07009.

9. Zhang Y, Kling GW (2006) Dynamics of lake eruptions and possible ocean eruptions. Annu. Rev. Earth Planet. Sci. 34: 293–324.

10. Lorke A, Tietze K, Halbwachs M, Wuest A (2004) Response of Lake Kivu stratification to lava inflow and climate warming. Limnol. Oceanogr. 49: 778–783.

11. Schmid M, Wüest A (2012) Stratification, mixing, and transport processes in Lake Kivu. In Lake Kivu: Limnology and biogeochemistry of a tropical great lake. Eds. Descy JP, Darchambeau F, Schmid M, Springer.

12. Schmid M, Busbridge M, Wüest A (2010) Double-diffusive convection in Lake Kivu. Limnol. Oceanogr. 55: 225–238.

13. Lehman JT, Litt AH, Mugidde R. Lehman DA (1998) In Environmental Change and Response in East African Lakes (ed. Lehman, J. T.) 157–172. Amsterdam, Kluwer.

14. Hirslund F (2012) An additional challenge of Lake Kivu in Central Africa - upward movement of the chemoclines. J. Limnol. 71: e4.

15. Sommer T, Carpenter JR, Schmid M, Lueck RG, Schurter M, et al. (2013) Interface structure and flux laws in a natural double-diffusive layering. J. Geophys. Res.: Oceans 118: 6092–6106.

16. Newman FC (1976) Temperature steps in Lake Kivu: a bottom heated saline lake. J. Phys. Oceanogr. 6: 157–163.

17. Aaberg AA (2013) Warming and stratification changes in Lake Kivu, East Africa. M.Sc. Thesis, University of Minnesota Duluth.

18. Schmid M, Ross K, Wüest A (2012) Comment on An additional challenge of Lake Kivu in Central Africa - upward movement of the chemoclines. J. Limnol. 71: 330–334.

19. Katsev S, Crowe SA, Mucci A, Sundby B, Nomosatryo S, et al. (2010) Mixing and its effects on biogeochemistry in the persistently stratified, deep, tropical Lake Matano, Indonesia. Limnol. Oceanogr. 55: 763–776.

20. Degens ET, von Herzen RP, Wong HK, Deuser WG, Jannasch HW (1973) Lake Kivu: structure, chemistry and biology of an East African rift lake. Geologische Rundschau, 62: 245–277.

21. Damas H (1937) Quelques caractères écologiques de trois lacs équatoriaux: Kivu, Edouard, Ndalaga. Annal. Soc. Roy. Zool. Belgique 68: 121–135.

22. Pasche N, Dinkel C, Müller B, Schmid M, Wüest A, et al. (2009) Physical and bio-geochemical limits to internal nutrient loading of meromictic Lake Kivu. Limnol. Oceanogr. 54: 1863–1873.

23. Vollmer MK, Bootsma HA, Hecky RE, Patterson G, Halfman JD, et al. (2005). Deep-water warming trend in Lake Malawi, East Africa. Limnol. Oceanogr. 50: 727–732.

24. Verburg P, Hecky RE (2009) The physics of the warming of Lake Tanganyika by climate change. Limnol. Oceanogr. 54: 2418–2430.

25. Hecky RE, Bugenyi FWB, Ochumba P, Talling JF, Mugidde R, et al. (1994) Deoxygenation of the deep-water of Lake Victoria, East-Africa. Limnol. Oceanogr. 39: 1476–1481.

26. Houghton JT, et al. (2001) In Climate Change 2001: The Scientific Basis, Contribution of Working Group 1 to the Third Assessment Report of the Intergovernmental Panel on Climate Change. Cambridge Univ. Press, Cambridge.

27. Omumbo JA, Lyon B, Waweru SM, Connor SJ, Thomson MC (2011) Raised temperatures over the Kericho tea estates: revisiting the climate in the East African highlands malaria debate. Malaria J. 10: 12.

28. Reichert P (1994) AQUASIM — a tool for simulation and data analysis of aquatic systems. Water Sci. Technol. 30: 21–30.

29. Deuser WG, Degens ET, Harvey GR, Rubin M (1973) Methane in Lake Kivu: New data bearing on its origin. Science 181: 51–54.

30. Conrad R, Wetter B (1990) Influence of temperature on energetics of hydrogen metabolism in homoacetogenic, methanogenic, and other anaerobic bacteria. Arch. Microbiol. 155: 94–98.

31. Schmitz DM, Kufferath J (1955) Problèmes posés par la présence de gaz dissous dans les eaux profondes du Lac Kivu. Acad Roy Sci Coloniales, Bull Séances 1: 326–356.

32. Tietze K (1978) Geophysical examination of Lake Kivu and its unusual methane deposit: Stratification, dynamics and gas content of the lake water. Ph.D. thesis, Christian-Albrechts- Univ. Kiel. [In German].

33. Halbwachs M, Tietze K, Lorke A, Mudaheranwa C (2002) Investigations in Lake Kivu (East Central Africa) after the Nyiragongo Eruption of January 2002, Specific study of the impact of the sub-water lava inflow on the lake stability. Technical report, European Community, ECHO; Solidarités, Aide Humanitaire d'Urgence, March 2002.

6

Effects of Ocean Acidification on the Brown Alga *Padina pavonica*: Decalcification Due to Acute and Chronic Events

Teba Gil-Díaz*, Ricardo Haroun, Fernando Tuya, Séfora Betancor, María A. Viera-Rodríguez

Centro de Biodiversidad y Gestión Ambiental, Universidad de Las Palmas de Gran Canaria, Las Palmas de Gran Canaria, Spain

Abstract

Since the industrial revolution, anthropogenic CO_2 emissions have caused ocean acidification, which particularly affects calcified organisms. Given the fan-like calcified fronds of the brown alga *Padina pavonica*, we evaluated the acute (short-term) effects of a sudden pH drop due to a submarine volcanic eruption (October 2011–early March 2012) affecting offshore waters around El Hierro Island (Canary Islands, Spain). We further studied the chronic (long-term) effects of the continuous decrease in pH in the last decades around the Canarian waters. In both the observational and retrospective studies (using herbarium collections of *P. pavonica* thalli from the overall Canarian Archipelago), the percent of surface calcium carbonate coverage of *P. pavonica* thalli were contrasted with oceanographic data collected either *in situ* (volcanic eruption event) or from the ESTOC marine observatory data series (herbarium study). Results showed that this calcified alga is sensitive to acute and chronic environmental pH changes. In both cases, pH changes predicted surface thallus calcification, including a progressive decalcification over the last three decades. This result concurs with previous studies where calcareous organisms decalcify under more acidic conditions. Hence, *Padina pavonica* can be implemented as a bio-indicator of ocean acidification (at short and long time scales) for monitoring purposes over wide geographic ranges, as this macroalga is affected and thrives (unlike strict calcifiers) under more acidic conditions.

Editor: Guillermo Diaz-Pulido, Griffith University, Australia

Funding: This study was partially funded by the Spanish MINECO "Plan Nacional" ANTROTIDAL (CGL 2011-23833) and ECOLIFE (CGL08-05407 C03). Research by all authors was supported by Canary Islands CEI: Tricontinental Atlantic Campus. Fenando Tuya was supported by the MINECO "Ramón y Cajal" program and Séfora Betancor was supported by Postgraduate Scholarship of the University of Las Palmas de Gran Canaria. The funders had no role in study design, data collection and analysis, decision to publish, or preparation of the manuscript.

Competing Interests: The authors have declared that no competing interests exist.

* Email: tebagil90@gmail.com

Introduction

Over the last centuries, atmospheric CO_2 concentrations have increased due to human activities [1]. As a global net sink, oceans have absorbed almost one third of these anthropogenic CO_2 emissions [2], causing readjustments in the carbonate chemistry and lowering the pH. This phenomenon was originally known as Ocean Acidification (OA), a term that has been broadened to include other natural events (i.e., increased volcanic activity, methane hydrate releases, long-term changes in net respiration) and anthropogenic causes (i.e., release of nitrogen and sulphur compounds into the atmosphere) [3]. This chemical process has been observed by many oceanic long-term observatories belonging to the OceanSITES programme, such as BATS (Bermuda Atlantic Time-series Study), HOT (Hawaii Ocean Time-series station) and ESTOC (European Station for Time-series in the Ocean at the Canary Islands). Such measurements provide information on the chronic, oceanic pH evolution over time. Calculations have estimated that if CO_2 emissions continue to rise, global decrease between 0.3 and 0.5 units of surface pH is expected to occur by 2100 based on anthropogenic activities [1], or between 0.06 and 0.32 units according to the latest estimations based on anthropogenic radiative forcing [3]. These values have drawn scientific attention towards their future ecological and physiological impacts, especially on calcified benthic organisms such as corals, echinoderms, gastropods and several calcareous macroalgae [4,5,6].

Recently, several studies have been carried out at naturally acidified sites [7,8,9], unveiling the effects of vent-induced acidification and the increase in total inorganic C concentrations on the community structure of brown algae, coral-reef associated macroinvertebrates and benthic foraminifera. This field-based research allows the uncovering of long-term effects [4] and combined organisms responses [9,10,11]. These studies can be also complemented with mesocosm approaches to target specific cause-effect relationships, despite some limitations (i.e., replication, realism in experiments) [12,13,14,15]. To summarize, this body of research generally points out that OA will cause ecosystem alterations, impacting calcareous but favouring fleshy organisms [6,16], with the subsequent loss of habitat complexity in the case of calcareous engineering species [9]. The Technical Report of the Secretariat of the Convention on Biological Diversity [17] highlights the main impacts of OA on marine biodiversity using evidences coming from naturally acidified locations, confirming that, although some species may take advantages, biological communities under acidified seawater conditions present less diversity and, in many cases, calcifying organisms are absent. Combined effects of elevated partial pressure of CO_2 and

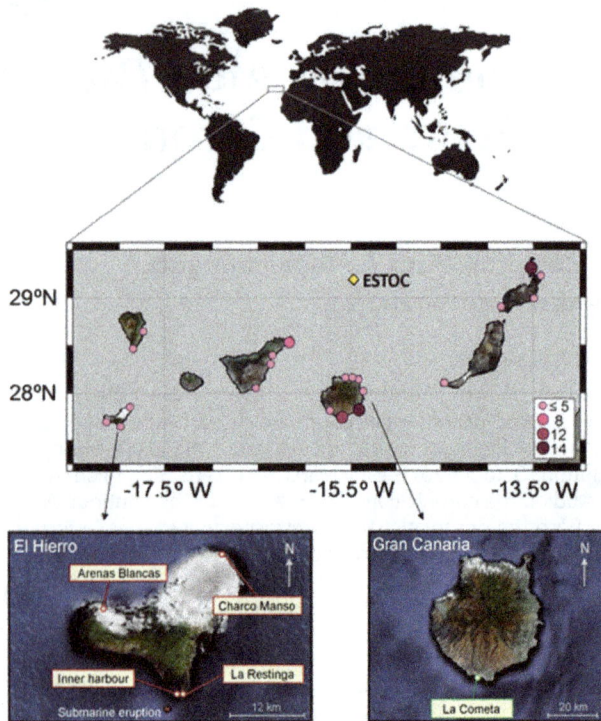

Figure 1. Collection sites of _P. pavonica_ in the Canary Islands. Top: herbarium samples denoting the frequency of available sheets (coloured dots). Bottom: sampling sites during the submarine eruption event off El Hierro Island (star) and position of the control site in Gran Canaria Island.

pCO_2, oxygen and nutrient concentrations, as well as in the redox potential by the emission of reduced sulphur and Fe(II) species (specially from November to December 2011) [25,26]. About 95% of the observed decrease in pH_T was related to the emission of CO_2, contributing to a lesser extent the emission of SO_2, H_2S/HS^- and the oxidation of dissolved reduced species during the first months [26]. The eruption produced greenish seawater plumes that occasionally extended onshore [25,27] for several weeks; November 2011 was the most intense period of volcanic activity (lowest pH_T level recorded at sea, with a mean decrease of 2.8 units within the first 100 m below the sea level, at _ca_. 2 km away from the volcano) and it officially ended in March 2012 [28]. Changes in the carbonate chemistry of the area immediately affected by the eruption are specified in [26]; i.e., in November 2011, at 5 m above the volcano, values of 7,681.5 $\mu mol\ kg^{-1}$ for total dissolved inorganic carbon, total alkalinities of 1,338.0 $\mu mol\ kg^{-1}$ and 230,316 μatm of pCO_2 were registered. This volcanic phenomenon had remarkable effects on the benthic, coastal, communities and was described as an "unprecedented episode of severe acidification and fertilization" [26]; thus, this event may be considered as an example of an acute episode promoting local acidification.

Macroalgae are often considered as indicators of the marine environment health due to their relevant roles in the structuring and functioning of coastal ecosystems [29,30,31,32]. To test the effects of OA we have selected the genus _Padina_, a brown calcified macroalga that facultatively calcifies with extracellular (on the thallus surface) aragonite needles [22,33] at a rate of 240 gm^{-2} yr^{-1} for the case of subtropical specimens [34]. Diverse members of this genus are distributed widespread in tropical to warm temperate coasts, such as the Macaronesian Islands, Mediterranean Sea, Caribbean Sea, Micronesia and Polynesia [35,36,37]. Particularly, _Padina pavonica_ (Linnaeus) Thivy plays a significant role as a dominant macrophyte in the Atlantic islands, being a conspicuous member of macroalgal communities in the sub- and intertidal rocky shore systems [38].

Here, we used _P. pavonica_ as a biological model to test the effect of natural pH alterations on algal surface thallus calcification (related to OA processes) and its potential sensitivity to acute and chronic OA events. This was accomplished through two complementary approaches: (1) by measuring the effects of the acute OA induced by the volcanic event on the proportion of calcified surface of _P. pavonica_, and (2) by ascertaining the natural calcification trend that this macroalga has followed linked to chronic OA exposure over the last decades using the longest records available for macroalgae: herbarium vouchers.

Materials and Methods

Ethics statements

This study was approved by the Canary Islands International Campus of Excellence, which is funded by the Spanish Ministry of Economy and Competitiveness and the coastal land accessed is public land under the Spanish Coast Law. In this study, permits for collections of organisms were not necessary, as seaweeds are unprotected. It was not necessary to have an Animal Care and Use permit according to the national laws.

Submarine eruption study: acute OA response

To determine the degree of impact of the submarine eruption south off El Hierro Island on the calcified surface of _Padina pavonica_, n = 17 random samples (individual and unbroken fan-shaped thalli) were collected from the intertidal at five sites (Figure 1; specific coordinates in Table S1). Two sites were

temperature levels have also been tested, revealing higher decalcification rates in coralline algae [18], loss of coral reef integrity [19] and biomass changes in macroalgal communities [20]. However, little attention has been paid to facultative calcifying organisms and their ecological performances under OA conditions [21,22], especially to the consequences of long-term (multi-decadal) exposure to rising CO_2 [9].

High uncertainty still remains regarding the potential impacts of OA on coastal systems; i.e., only few field observations have demonstrated the direct causality of anthropogenic OA on biotic responses [19]. It is suspected that a range of biological processes and functions (other than solely calcification-related issues) are likely to be affected by changes in pH [23]. In addition, OA interacts with other ocean biogeochemical processes (i.e., solubility of trace metals) and environmental changes (i.e., warming and decreasing oxygen levels at a global scale; eutrophication and pollution at local scales) [19,24]. All of this justifies the need to monitor OA at long-term scales through simultaneous measures of both chemical and biological-effects [23]. However, appropriate bio-indicators that accurately account for the biological effects of OA have not yet been established; several lists of potential organisms are available, related particularly to obligate calcifying organisms [23]. Therefore, there is need for appropriate OA-specific bio-indicators, as well as identification of the biological impacts and future ecological risks due to OA.

A submarine volcanic eruption started in mid October 2011 at _ca_. 1.8 km south offshore El Hierro Island (Canarian Archipelago, eastern Atlantic Ocean). This event caused remarkable changes in the water column chemistry such as in pH_T (total scale, at _in situ_ conditions), total dissolved inorganic carbon, total alkalinity,

Table 1. Two-way ANOVA testing the influence of "Site" (fixed factor) and "Time" (random factor) on the percentage of decalcified surface in *P. pavonica* (n = 17), pH_F (n = 10 to 15) and seawater temperature (n = 10 to 15).

	Decalcified surface (%)			Coastal pH_F			Coastal seawater temp. (°C)		
	MS	F	P	MS	F	P	MS	F	P
Ti	32895.48	184.41	0.0002	1.01	17.30	0.0002	20.49	95.51	0.0002
Si	10175.62	2.0011	0.1640	0.50	1.35	0.3474	4.51	0.79	0.5446
Ti × Si	5084.89	28.51	0.0002	0.37	6.31	0.0002	5.68	26.46	0.0002
Residual	178.39			0.06			0.21		

adjacent to the eruption (La Restinga and its inner harbour), while the other two were selected further north in the same island (Arenas Blancas and Charco Manso). Additionally, another site (La Cometa) was included - as an external control at more than 200 km from the eruption point - in Gran Canaria Island, as there were doubts as to whether El Hierro Island had any place without the influence of the submarine eruption [21]. Collections of algal material took place at three different periods corresponding to: 24–26th November 2011 (during the greatest eruption activity), 24–26th March 2012 (immediately after the volcanic activity had officially finished) and 2–5th July 2012 (*ca.*>3 months after the official cessation of the eruptive activity). At all occasions, ten to fifteen discrete *in situ* (coastal) measures of seawater temperature (°C) and pH_F data (free scale, corrected with temperature) were registered at 0.5–1 m depth by using a calibrated probe coupled to a portable multi parameter HI9829 (Hanna Instruments, USA) during low tide; the probe was dipped into open intertidal pools where the samples were collected at the lower intertidal zone. The calibration of the probe was performed following the manufacturer's instructions with the supplied reagents. Our results present accuracies in temperature and pH_F of ± 0.15°C and ± 0.01, respectively, as well as precisions of ± 0.24°C and ± 0.02.

Collected samples of *P. pavonica* were frozen at -20°C until laboratory analysis. Once there, samples were thawed, left to dry at room temperature and then digitized using a digital eyepiece camera (MVV3000) coupled to a binocular dissecting microscope. To quantify the percent of decalcified surface of this brown alga, a comparison was made between the total area of the thallus and that belonging to the decalcified zones (Figure S1) by using an image freeware tool (Image J, NIH, USA). This way, only the surface distribution of calcium carbonate, and not its total content on each thallus, was quantified, as this is a non-destructive technique (the original sample is thus conserved). Statistical analyses were carried out with univariate PERMANOVA, testing for significant differences in the mean percentage of decalcified surface (n = 17), pH_F and seawater temperature between "Site" (fixed factor) and "Time" (random factor) 2-way ANOVA; *a posteriori* tests resolved pairwise differences between sites for each level of time (months). ANOVAs based on permutations (999 in our case) were used to calculate the significance of P-values. The statistic test (pseudo-F) is a multivariate analogue of the univariate Fisher's F ratio, and in the univariate context the two are identical when using Euclidean distance as the dissimilarity measure [39]. Data satisfied homoscedasticity; thus transformations were not needed [40]. Linear regressions (simple and multiple), performed with Sigma Plot 11.0, tested if pH_F and temperature were significant predictor variables of the decalcified percentages.

Herbarium study: chronic OA response

A retrospective study was done using herbarium sheets from the main official Herbaria in the study area: TFC-Phyc from La Laguna University (http://www.gbif.es/ic_colecciones_in.php?ID_Coleccion=9767) and BCM from the University of Las Palmas de Gran Canaria (http://www.herbariobcm.org/). The surface view of 79 sheets registered as *P. pavonica]* was digitized with an Olympus 700 camera. These sheets belonged mostly to the intertidal zone (86% of the cases) of several sites (Figure 1; specific coordinates and sampled depths in Table S2), from almost a regular record between 1979 and 2012 (21 years of data out of 34). The percentage of calcified surface was quantified by comparing the number of pixels corresponding to calcium carbonate coverage to those shaping the entire thallus (Figure S2), using the Image J software (NIH, USA). We worked with binned data (12 groups of 3 years each) to overcome the uneven distribution of data sets of *P.*

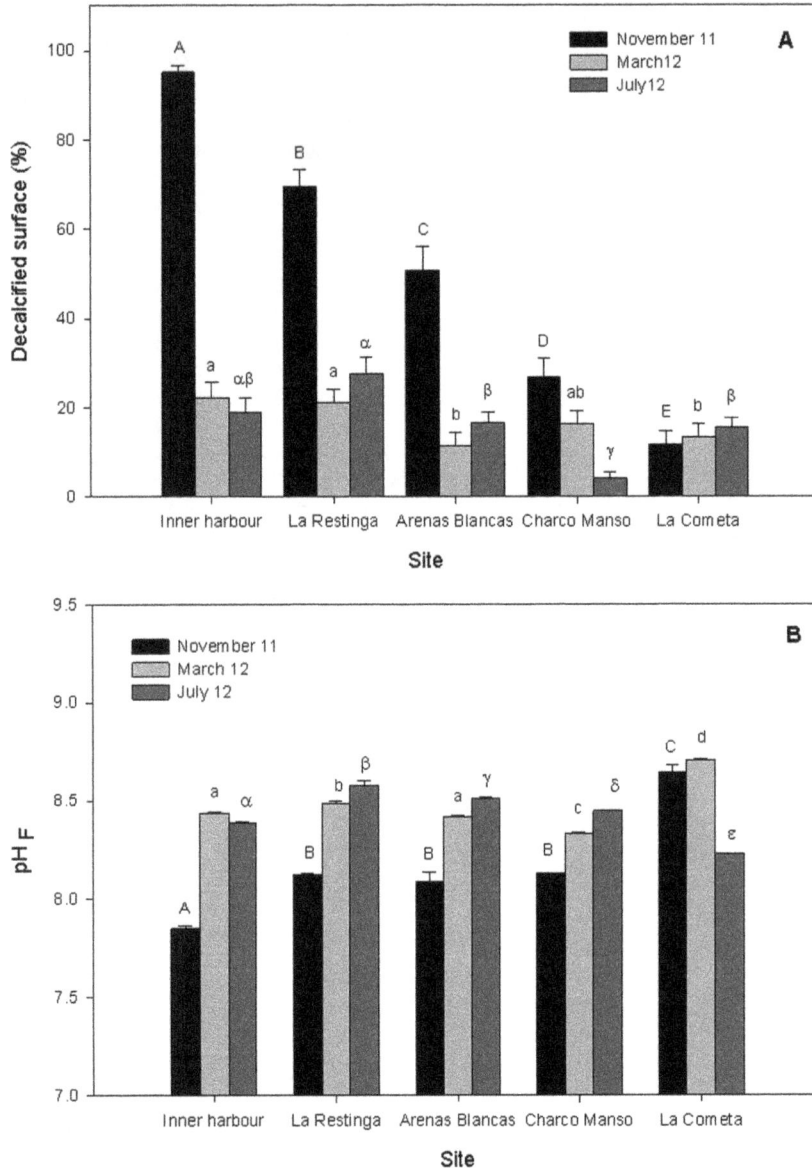

Figure 2. Mean decalcified surface percentages (n = 17) of *P. pavonica* (A) and mean pH$_F$ levels (n = 10 to 15) (B) per site and time in an acute OA response. Sites are distributed at increasing distance from the submarine eruption from left to right. The different letters above error bars refer to significant differences (P<0.05) between sites separately for each time (*post hoc* comparisons). Error bars are + SE of means.

pavonica, as some years had many more herbarium sheets than others.

Sea surface temperature (SST) and oceanic pH$_T$ values (total scale, at *in situ* conditions) were obtained from ESTOC, a European time series observatory buoy from the EuroSITES and OceanSITES network located at *ca.* 100 km north (Figure 1) from Gran Canaria Island. These data were registered at 1.5 m depth with a Sindemar Mod. SW-03 sensor (pH$_T$ values) and a SBE microcat (for SST) (EuroSITES webpage: http://www.eurosites. info/estoc.php). Data were downloaded from the Ocean CO$_2$ CDIAC webpage (http://cdiac.ornl.gov/oceans/), providing an available period from 1995 to 2009 (with irregular measurements per year). Statistical analyses were carried out with Sigma Plot 11.0 to test for linear regressions (simple and multiple) between the natural trend of the calcified percentages of the thalli, pH$_T$ and SST over time (dependent variables), as well as the relationship

between these parameters, taking pH$_F$ and temperature as the independent (predictor) variables and the calcified percentages as the dependent variable.

Results

Submarine eruption study: acute OA response

The percentage of decalcified surface differed between sites inconsistently through time (2-way ANOVA: Ti×Si, P<0.001, Table 1). In November 2011 (i.e., during the highest eruptive activity), the percentage of decalcified surfaces in *P. pavonica* showed an increasing trend (i.e., more gaps without calcium carbonate) as the distance to the submarine eruption decreased (Figure 2A; Figure S3); the highest decalcification (95.19% ± 6.32%) was detected in the inner harbour of La Restinga, in contrast to the control (11.76% ± 11.47%). However, by March

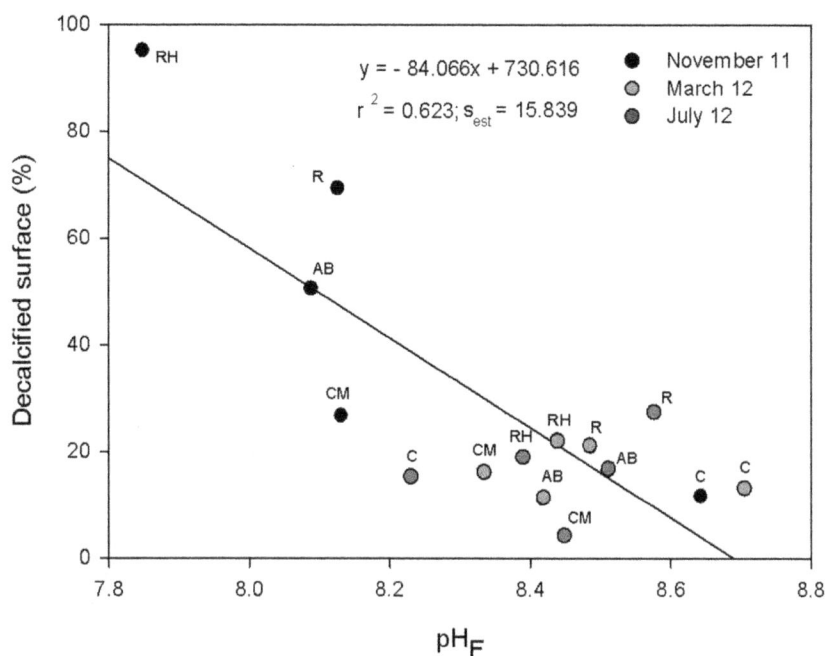

Figure 3. Relationship between decalcification and pH_F for an acute OA response. Relationship between mean decalcified surfaces (%) in *P. pavonica* and coastal pH_F values for all sites and times (n = 15): (RH) 'La Restinga harbour', (R) 'La Restinga', (AB) 'Arenas Blancas', (CM) 'Charco Manso' and (C) 'La Cometa' (control). s_{est} stands for standard error of estimate.

2012 and July 2012, significant differences in decalcified surfaces between sites were attenuated; i.e., similar values in the thalli of *P. pavonica* from El Hierro Island relative to those from the control site (Figure 2A). We have partially published some of these results in a previous study that compares the physiology and photosynthesis of decalcified and calcified thalli of *P. pavonica* [21]. Nevertheless, here we broaden the results to analyze the relationship between thalli decalcification and coastal pH_F, its variability compared to surface oceanic pH (chronic study) and to test the influence of coastal seawater temperature within this context.

Differences in the pH_F and coastal seawater temperature between sites also differed inconsistently through time (2-way ANOVA: Ti × Si, P<0.001, Table 1). In November 2011, pH_F values significantly decreased at sites located in El Hierro Island relative to the control (Figure 2B), registering 7.38±0.93 units as the lowest mean value (inner harbour of La Restinga). Coastal pH_F values at all El Hierro sites in March 2012 and July 2012 rose up to approximately 8.45 units and remained stable; these trends concur with the stabilization of the decalcified values in these months. However, the control site did not show regular pH_F values at all times. Changes in pH_F were independent relative to the pattern in temperatures at coastal sites, as the latter followed natural trends for the Canarian Archipelago (Figure S4). A multiple linear regression, testing the dependency of mean decalcified percentages to mean pH_F and seawater temperature (n = 15), showed that the former was a significant predictor of calcium carbonate coverage on *P. pavonica* thalli (P<0.05, power = 0.969). Seawater temperature had low multicollinearity with pH_F (VIF = 1.034). These results indicate the independency of seawater temperature relative to pH_F and decalcification. Hence, a simple negative linear regression (r^2 = 0.623; P<0.001, power = 0.960) was adjusted between mean decalcified percentages and

pH_F values (Figure 3), showing that a decrease in pH_F promoted decalcification in *P. pavonica*. Moreover, it can be observed how the natural pattern present in March 2012 and July 2012 is displaced in November 2011, under the volcanic influence, towards lower pH and higher decalcification percentages.

Herbarium study: chronic OA response

Herbarium-derived calcification values (n = 79) showed a decreasing trend in the calcified surface of *P. pavonica* thalli over time (Figure 4A; P<0.001, power = 1.000). A decrease of 17.17% has been quantified from 1978 to 2013, which yields a mean loss of 0.48% per year.

Oceanic pH_T values registered at ESTOC (n = 144) also showed a falling trend over time (Figure 4B; P<0.001, power = 1.000). The annual variability reflected seasonality (lowest mean pH in winter and highest in summer), a variability that was not noticed in the calcified pattern due to the applied bin. A total decrease of 0.0230 pH units was registered, declining at a mean rate of 0.0015 units per year (1995–2009). Regarding SST measurements through the 1995–2009 period (Figure S5), no specific linear trend was detected (n = 144, P>0.05), although this study had low power analysis (power = 0.03).

When the mean values of calcified percentages, pH_T and SST were adjusted for the concurring three year time periods (1993–2010) through a multiple linear regression, no significant pattern was found (n = 6, P>0.05, power = 0.638). Multicollinearity was low between pH_T and SST (VIF = 3.036). A positive linear regression (r^2 = 0.75, Figure 5) was then obtained between calcified surface percentages and pH_T (P<0.05, power = 0.628), indicating a positive relationship between the pH and the percent of surface calcification in *P. pavonica* over the last decades.

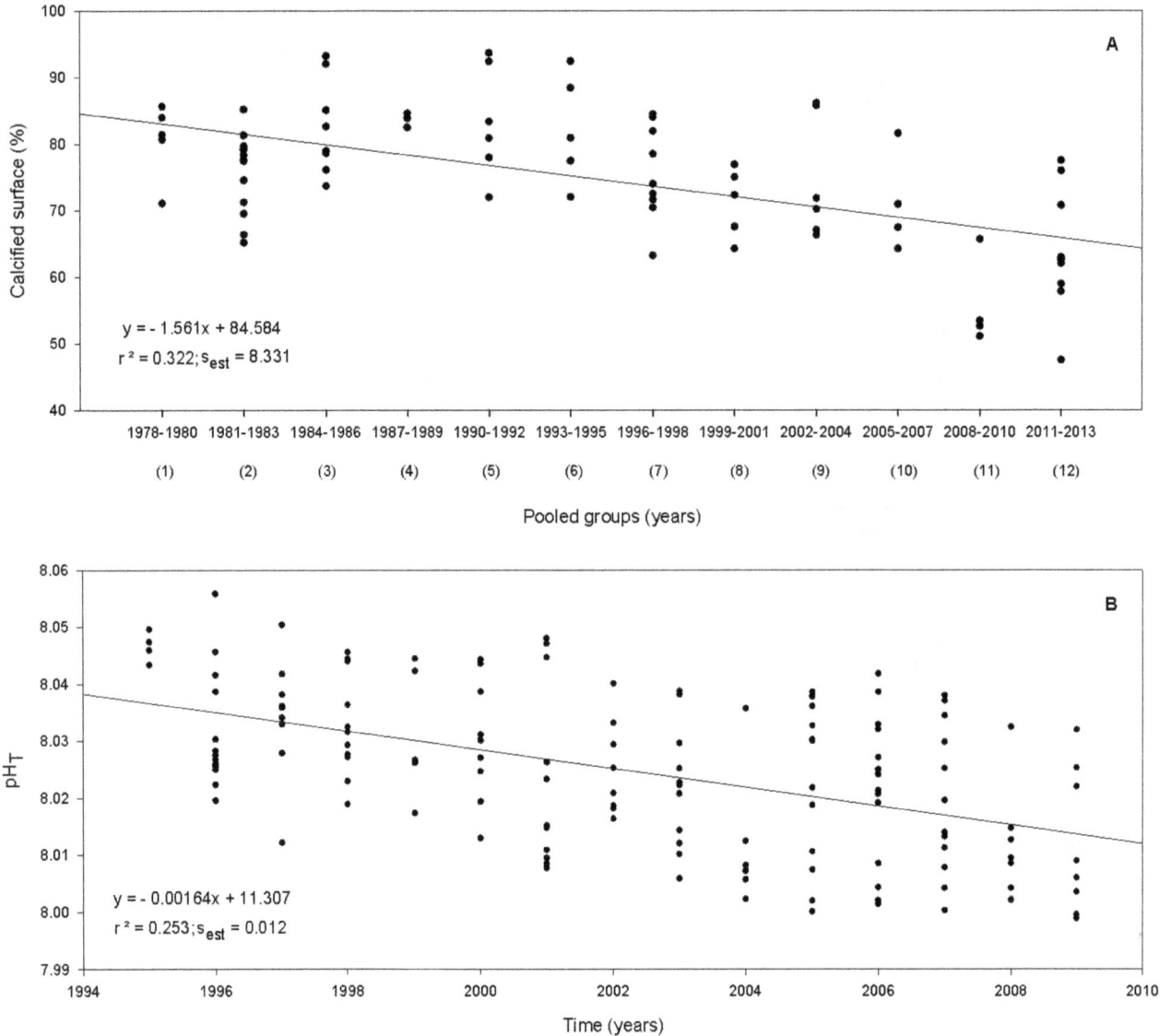

Figure 4. Calcified surface trend of *P. pavonica* thalli from 1978 to 2013 (A) and oceanic pH_T levels from ESTOC dataset between 1995 and 2009 (B). The X value in the equation represents the binned group number (1–12) in three year periods. s_{est} stands for standard error of estimate.

Discussion

The submarine volcanic eruption off El Hierro Island, already described at an oceanographic scale [25,26], affected onshore areas of the entire island in November 2011 (i.e., during the intense eruptive phase), with increasing decalcification on *P. pavonica* thalli related to decreasing pH_F values towards the submarine volcano. This result concurs with previous OA models, including field and laboratory studies, where aragonite coral reefs, calcareous algae and epibionts lose $CaCO_3$ under more acidic conditions [41,42,43]. Furthermore, our results also agree with [22], who observed that $CaCO_3$ content in *Padina* spp. is reduced with decreasing pH at volcanic CO_2 seeps in the Mediterranean Sea and Papua New Guinea.

Importantly, our results showed an apparent sensitivity of *P. pavonica* to acute pH changes as well as resilience, as the alga recovered initial calcification once the intense pH pulses had ceased. This complements the observations provided by [22], where *P.*

pavonica thrived (with reduced $CaCO_3$) at volcanic CO_2 seeps. [44] suggest that the effects of OA on calcified species causes increases in dissolution rather than a reduction in the production of calcium carbonate. Moreover, [33] observed a lack of correlation between photosynthesis and carbonate deposition in three species from the genus *Padina* (one of them was *P. pavonica*). Recent studies have already observed that this alga in more acidic environments adapts its physiological performance, behaving as a sun-adapted species instead of a shade-adapted plant as a response to low surface calcium carbonate coverage and thus increased exposure to solar radiation [21]. This could explain its thriving capacity under OA conditions. This is noteworthy, as the herbarium records also showed that *P. pavonica* is sensitive to chronic pH decrease due to OA, denoting its current decalcification trend.

Together with decalcification, *P. pavonica* under low pH conditions, as other macroalgae, decreases its content in phenolic compounds through excretion [21,45]. Both factors taking place

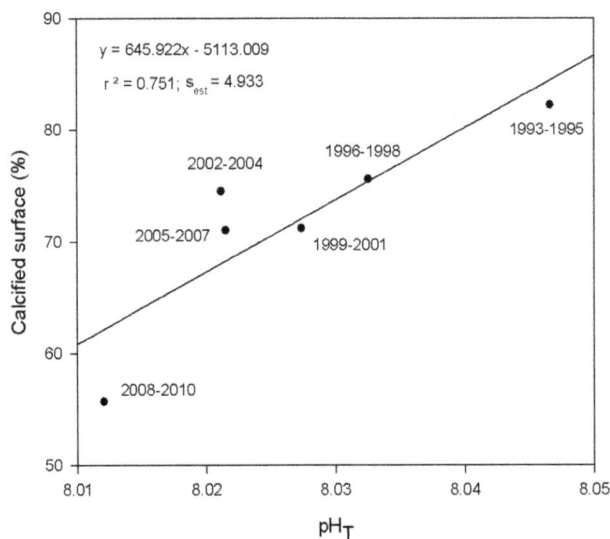

Figure 5. Linear regression between calcified surfaces (%) of *P. pavonica* and oceanic pH$_T$ (1993 to 2010). s$_{est}$ stands for standard error of estimate.

under OA conditions can influence grazing activity, as less phenolic compounds may increase thallus palatability [46] and decalcification can cause a loss in mechanical resistance to herbivory [47]. Thus, the ecological implications of the decalcification of *P. pavonica* may favour increased grazing activity of this macroalga in future scenarios. Nevertheless, according to [48], the ecological performance of *P. pavonica* also depends on its coexistence with other species and how they are affected by OA.

Natural phenomena, such as submarine eruptions or shallow submarine CO_2 seeps, are considered invaluable environments which provide the closest insight or approach to what might happen in future conditions [16,49]. In these cases, other parameters (depending on the studied site) such as seawater temperature, usually still show present-day conditions [9]. In this study, the environmental conditions around the volcanic eruption showed pH variations greater than those expected by the IS92a ("business-as-usual") scenario [1] in the inner harbour of La Restinga (0.57 units of variation) and similar values for the localities of La Restinga and Arenas Blancas (0.41 and 0.38 units of variation, respectively), including no significant variations on the coastal temperature. It is noteworthy that these values also surpass the latest estimations of global ocean surface pH changes (i.e., between 0.30 and 0.32 for the Representative Concentration Pathway 8.5) [3].

In this study, we assumed that pH$_T$ trends found in ESTOC observatory act as a baseline of pH values at an insular, onshore, scale. Superimposed on this pH variation, near shore waters may show large short-term pH variability due to events influencing primary production and/or coastal oceanographic anomalies; i.e., upwelling phenomena [19,50]. This idea cannot be implicitly contrasted here, as no overlapping periods between the open ocean and coastal pH records are available (i.e., no coastal pH data that falls within the ESTOC database period and vice versa for the years 2011–2012). Despite that coastal areas are suspected to display large and uncertain regional and local pH variations [19], our study shows that the sudden decrease in pH in a short period of time induced by the volcanic activity was superimposed on the natural variability of pH. Therefore, it is logical to think that under future pH scenarios of OA (a much slower process), the calcification of *P. pavonica* may follow a trend within a natural

range of variability, recovering in favourable conditions and decalcifying under more acidic conditions.

It is also noteworthy that for each study, pH values were measured in different scales, thus their direct comparison is not possible due to their conversion differences, being pH$_T$ in general around 0.09 units lower than pH$_F$ [51]. However, the chronic pH range seems to be comprised within the observed acute pH range. This observation, together with the positive relation of oceanic pH with calcification percentages, indicates a coincident direction trend with that of the acute values measured during the volcanic event (November 2011). This could support the idea that enriched-CO_2 volcanic events can act as windows to future pH scenarios.

Despite the evidence of exacerbated effects of combined decreased pH and increased temperatures on calcifiers [52,53,54], seawater temperature has not contributed to changes in decalcification patterns. It is possible that, during the volcanic event, the effect of seawater temperature on decalcification of *P. pavonica* was masked by the dramatic pH levels experienced. However, the results from our chronic study suggest that, most likely, there is no direct influence of coastal seawater temperature on calcification. It is possible that the link between *P. pavonica* and the temperature effect is more related to the potential of this alga to spread geographically, as indicated by [55]. For this reason, *P. pavonica* is classified as a climate change affected species for the UK, Wales, Scotland and Ireland [56,57].

In general, bio-monitors are seen as complementary tools to chemical monitoring programs, as they provide the biological context of the alterations. Monitoring programs and legislation such as the Marine Strategy Framework Directive [58] require bio-indicators to assess the status of the natural environment. In some occasions, they can act as a shortcut to monitoring all the physical-chemical parameters, as long as the potential confounding factors (biological or environmental) are taken into account.

In conclusion, *P. pavonica* is sensitive to acute and chronic environmental pH changes; this suggests that *P. pavonica* could be a suitable bio-indicator of OA on coastal habitats. New research could provide insight on how this calcified macroalga behaves seasonally and under future OA conditions by combining regular *in situ* image (non-destructive) monitoring of calcium carbonate deposition coupled to continuous physical – chemical measurements. This could supply the lack of tools in current monitoring stations that require both chemical parameters and biological effects using a suitable indicator (OA-specific) [19]. Furthermore, given the extensive distribution of this species, this research is applicable to multiple regions, resulting in a wider geographic range monitoring program that could potentially measure and show the direct effects of seawater pH changes in nature. Many questions remain regarding the real biological and biogeochemical consequences of OA for marine biodiversity and ecosystems, as well as the impacts of these changes on oceanic ecosystems and the services they provide [59]. Further direct studies with multidisciplinary approaches will allow more rigorous predictions of OA scenarios and the discovery of its effects on calcifying marine organisms.

Supporting Information

Figure S1 Example of the image treatment used to quantify the percentage of decalcified surface on *P. pavonica* thallus. (A) Total thallus area. (B) Decalcified areas.

Figure S2 Example of the pixel quantification method used for the calcified surface of herbarium *P. pavonica* thallus.

Figure S3 Representative images of sampled thalli in El Hierro Island and the control. (RH) inside 'La Restinga harbour', (R) 'La Restinga', (AB) 'Arenas Blancas', (CM) 'Charco Manso' and (C) 'La Cometa' (control), both in November 2011 (Nov) and March 2012 (Mar). Note that thalli coming from the inner harbour of La Restinga were almost totally decalcified in November 2011. Graph paper used as scale.

Figure S4 Mean seawater temperature registered on each site and time for the acute OA response study. Sites are distributed at increasing distance from the submarine eruption from left to right. The different letters above the error bars refer to significant differences ($P<0.05$) between sites for each time (*post hoc* comparisons). Error bars are + SE of means.

Figure S5 SST values registered at ESTOC between 1995 and 2009 (n = 144). These results show no specific trend over time. s_{est} stands for standard error of estimate.

Table S1 Geographical position of the sampled sites in El Hierro Island (La Restinga, La Restinga harbour,

Arenas Blancas and Charco Manso) and Gran Canaria (La Cometa).

Table S2 Herbarium specifications regarding coordinates (geographical system and UTM) and depth of collection: intertidal (Inter) and subtidal (Sub) areas.

Acknowledgments

Research by all authors was supported by Canary Islands CEI: Tricontinental Atlantic Campus. F. Tuya was supported by the MINECO 'Ramón y Cajal' program and S. Betancor was supported by Postgraduate Scholarship of the University of Las Palmas de Gran Canaria.

Author Contributions

Conceived and designed the experiments: TGD MAVR RH. Performed the experiments: TGD SB. Analyzed the data: TGD SB FT. Contributed reagents/materials/analysis tools: SB MAVR RH. Wrote the paper: TGD SB FT MAVR RH.

References

1. IPCC (2007) Climate Change 2007: The Physical Science Basis. Contribution of Working Group I to the Fourth Assessment Report of the Intergovernmental Panel on Climate Change [Solomon, S., D. Qin, M. Manning, Z. Chen, M. Marquis, K.B. Averyt, M. Tignor and H.L. Miller (eds.)]. Cambridge University Press, Cambridge, United Kingdom and New York, NY, USA, 996 pp.
2. Sabine CL, Feely RA, Gruber N, Key RM, Lee K, et al. (2004) The oceanic sink for anthropogenic CO_2. Science 305: 367–371. doi: 10.1126/science.1097403
3. IPCC (2013) Climate Change 2013: The Physical Science Basis. Contribution of Working Group I to the Fifth Assessment Report of the Intergovernmental Panel on Climate Change [Stocker, T.F., D. Qin, G.-K. Plattner, M. Tignor, S.K. Allen, J. Boschung, A. Nauels, Y. Xia, V. Ben and P.M. Midgley (eds.)]. Cambridge University Press, Cambridge, United Kingdom and New York, NY, USA, 1535 pp.
4. Porzio L, Buia M-C, Hall-Spencer JM (2011) Effects of ocean acidification on macroalgal communities. J Exp Mar Biol Ecol 400: 278–287. doi: 10.1016/j.jembe.2011.02.011
5. Fabry VJ, Seibel BA, Feely RA, Orr JC (2008) Impacts of ocean acidification on marine fauna and ecosystem processes. ICES J Mar Sci 65: 414–432. doi: 10.1093/icesjms/fsn048
6. Kroeker KJ, Kordas RL, Crim RN, Singh GG (2010) Meta-analysis reveals negative yet variable effects of ocean acidification on marine organisms. Ecol lett 13: 1419–1434. doi: 10.1111/j.1461-0248.2010.01518.x
7. Pettit LR, Hart MB, Medina-Sánchez AN, Smart CW, Rodolfo-Metalpa R, et al. (2013) Benthic foraminifera show some resilience to ocean acidification in the northern Gulf of California, Mexico. Mar Pollut Bull 73(2): 452–462. doi: 10.1016/j.marpolbul.2013.02.011
8. Bellissimo G, Rull Lluch J, Tomasello A, Calvo S (2014) The community of *Cystoseira brachycarpa* J. Agardh *emend*. Giaccone (Fucales, Phaeophyceae) in a shallow hydrothermal vent area of the Aeolian Islands (Tyrrhenian Sea, Italy). Plant Biosyst 148(1): 21–26. doi: 10.1080/11263504.2013.778350
9. Fabricius KE, De'ath G, Noonan S, Uthicke S (2014) Ecological effects of ocean acidification and habitat complexity on reef-associated macroinvertebrate communities. Proc R Soc B 281: 20132479. doi: 10.1098/rspb.2013.2479
10. Fabricius KE, Langdon C, Uthicke S, Humphrey C, Noonan S, et al. (2011) Losers and winners in coral reefs acclimatized to elevated carbon dioxide concentrations. Nature Clim Change 1: 165–169. doi: 10.1038/nclimate1122
11. Barry JP, Widdicombe S, Hall-Spencer JM (2011) Effects of ocean acidification on marine biodiversity and ecosystem function. In: *Ocean Acidification* (eds J. -L. Gattuso & L. Hansson), Chapter 10, pp. 192–209. Oxford University Press, Oxford.
12. Wernberg T, Smale DA, Thomsen MS (2012) A decade of climate change experiments on marine organisms: procedures, patterns and problems. Glob Change Biol 18: 1491–1498. doi: 10.1111/j.1365-2486.2012.02656.x
13. Price NN, Martz TR, Brainard RE, Smith JE (2012) Diel variability in seawater pH relates to calcification and benthic community structure on coral reefs. Plos One 7: 1–9.
14. Hofmann GE, Barry JP, Edmunds PJ, Gates RD, Hutchins DA, et al. (2010) The effect of ocean acidification on calcifying organisms in marine ecosystems: an organism to ecosystem perspective. Annu Rev Ecol Evol S 41: 127–147.
15. Riebesell U (2008) Acid test for marine biodiversity. Nature 454: 46–47.
16. Hall-Spencer JM, Rodolfo-Metalpa R, Martin S, Ransome E, Fine M, et al. (2008) Volcanic carbon dioxide vents show ecosystem effects of ocean acidification. Nature 454: 96–99. doi: 10.1038/nature07051
17. Secretariat of the Convention on Biological Diversity (2009) *Scientic Synthesis of the Impacts of Ocean Acidification on Marine Biodiversity*. Montreal, Technical Series No. 46, 61 pages.
18. Martin S, Gattuso JP (2009) Response of Mediterranean coralline algae to ocean acidification and elevated temperature. Glob Change Biol 15: 2089–2100. doi: 10.1111/j.1365-2486.2009.01874.x
19. IPCC (2014) Climate Change 2014: Impacts, Adaptation, and Vulnerability. Part A: Global and Sectoral Aspects. Contribution of Working Group II to the Fifth Assessment Report of the Intergovernmental Panel on Climate Change [Field, C.B., V.R. Barros, D.J. Dokken, K.J. Mach, M.D. Mastrandrea, T.E. Bilir, M. Chatterjee, K.L. Ebi, Y.O. Estrada, R.C. Genova, B. Girma, E.S. Kissel, A.N. Levy, S. MacCracken, P.R. Mastrandrea, and L.L. White (eds.)]. Cambridge University Press, Cambridge, United Kingdom and New York, NY, USA, XXX pp.
20. Olabarria C, Arenas F, Viejo RM, Gestoso I, Vaz-Pinto F, et al. (2013) Response of macroalgal assemblages from rockpools to climate change: effects of persistent increase in temperature and CO_2. Oikos 122: 1065–1079. doi: 10.1111/j.1600-0706.2012.20825.x
21. Betancor S, Tuya F, Gil-Diaz T, Figueroa FL, Haroun R (2014) Effects of a submarine eruption on the performance of two brown seaweeds. J Sea Res 87: 68–78. doi: 10.1016/j.seares.2013.09.008
22. Johnson VR, Russell BD, Fabricius KE, Brownlee C, Hall-Spencer JM (2012) Temperate and tropical brown macroalgae thrive, despite decalcification, along natural CO_2 gradients. Glob Change Biol 18: 2792–2803. doi: 10.1111/j.1365-2486.2012.02716.x
23. ICES (2013) Report of the Joint OSPAR/ICES Ocean Acidification Study Group (SGOA), 11–14 December 2012, Copenhagen, Denmark. ICES CM 2012/ACOM:83. 75 pp.
24. Hydes DJ, McGovern E, Walsham P (Eds.) (2013) Chemical aspects of ocean acidification monitoring in the ICES marine area. ICES Cooperative Research Report No. 319. 78 pp.
25. Fraile-Nuez E, González-Dávila M, Santana-Casiano JM, Arístegui J, Alonso-González J, et al. (2012) The submarine volcano eruption at the island of El Hierro: physical-chemical perturbation and biological response. Sci Rep 2(486): 1–6. doi: 10.1038/srep00486
26. Santana-Casiano JM, González-Dávila M, Fraile-Nuez E, de Armas D, González AG, et al. (2013) The natural ocean acidification and fertilization event caused by the submarine eruption of El Hierro. Sci Rep 3(1140): 1–8. doi: 10.1038/srep01140
27. Eugenio F, Matin J, Marcello J, Fraile-Nuez E (2014) Environmental monitoring of El Hierro Island submarine volcano, by combining low and high resolution satellite imagery. Int J Appl Earth Obs 29: 53–66. doi: 10.1016/j.jag.2013.12.009
28. Carracedo JC, Pérez Torrado F, Rodríguez González A, Soler V, Fernández Turiel JL, et al. (2012) The 2011 submarine volcanic eruption in El Hierro (Canary Islands). Geol Today 28(2): 53–58. doi: 10.1111/j.1365-2451.2012.00827.x

29. Collado-Vides L, Caccia VG, Boyer JN, Fourqurean JW (2007) Tropical seagrass-associated macroalgae distributions and trends relative to water quality. Estuar Coast Shelf S 73: 680–694.

30. Arévalo R, Pinedo S, Ballesteros E (2007) Changes in the composition and structure of Mediterranean rocky-shore communities following a gradient of nutrient enrichment: Descriptive study and test of proposed methods to assess water quality regarding macroalgae. Mar Pollut Bull 55: 104–113.

31. Ivesa L, Lyons DM, Devescovi M (2009) Assessment of the ecological status of north-eastern Adriatic coastal waters (Istria, Croatia) using macroalgal assemblages for the European Union Water Framework Directive. Aquat Conserv 19: 14–23.

32. Juanes JA, Guinda X, Puente A, Revilla JA (2008) Macroalgae, a suitable indicator of the ecological status of coastal rocky communities in the NE Atlantic. Ecol Indic 8: 351–359.

33. Okazaki M, Pentecost A, Tanaka Y, Miyata M (1986) A study of calcium carbonate deposition in the genus *Padina* (Phaeophyceae, Dictyotales). Br Phycol J 21: 217–224. doi: 10.1080/00071618600650251

34. Wefer G (1980) Carbonate production by algae *Halimeda, Penicillus* and *Padina*. Nature 285: 323–324.

35. N'Yeurt ADR, Payri CE (2006) Marine algal flora of French Polynesia I. Phaeophyceae (Ochrophyta, brown algae). Cryptogamie Algol 27: 111–152.

36. Abbot IA, Huisman JM (2003) New species, observations, and a list of new records of brown algae (Phaeophyceae) from the Hawaiian Islands. Phycological Research 51: 173–185. doi: 10.1046/j.1440-1835.2003.t01-1-00308.x

37. Silberfeld T, Bittner L, Fernández-García C, Cruaud C, Rousseau F, et al. (2013) Species diversity, phylogeny and large scale biogeographic patterns of the genus Padina (Phaeophyceae, Dictyotales). J Phycol 49: 130–142. doi: 10.1111/jpy.12027

38. Tuya F, Haroun RJ (2006) Spatial patterns and response to wave exposure of photophilic algal assemblages across the Canarian Archipelago: a multiscaled approach. Mar Ecol-Prog Ser 311: 15–28. doi: 10.3354/meps311015

39. Anderson MJ (2005) PERMANOVA: a FORTRAN computer program for permutational multivariate analysis of variance. Department of Statistics, University of Auckland, New Zealand.

40. Underwood AJ (1997) Experiments in Ecology: Their Logical Design and Interpretation Using Analysis of Variance. Cambridge University Press, Cambridge UK.

41. Martin S, Rodolfo-Metalpa R, Ransome E, Rowley S, Buia M-C, et al. (2008) Effects of naturally acidified seawater on seagrass calcareous epibionts. Biol Letters 4(6): 689–692. doi: 10.1098/rsbl.2008.0412

42. Orr JC, Fabry VJ, Aumont O, Bopp L, Doney SC, et al. (2005) Anthropogenic ocean acidification over the twenty-first century and its impact on calcifying organisms. Nature 437: 681–686. doi: 10.1038/nature04095

43. Kleypas JA, Buddemeier RW, Archer D, Gattuso J-P, Langdon C, et al. (1999) Geochemical consequences of increased atmospheric carbon dioxide on coral reefs. Science 284: 118–120. doi: 10.1126/science.284.5411.118

44. Roleda MY, Boyd PW, Hurd CL (2012) Before ocean acidification: calcifier chemistry lessons. J Phycol 48: 840–843. doi: 10.1111/j.1529-8817.2012.01195.x

45. Gómez I, Huovinen P (2010) Induction of phlorotannins during UV exposure mitigates inhibition of photosynthesis and DNA damage in the kelp *Lessonia nigrescens*. Photochem Photobiol 86: 1056–1063. doi: 10.1111/j.1751-1097.2010.00786.x

46. Van Alstyne KL, Paul VJ (1990) The biogeography of polyphenolic compounds in marine macroalgae: temperate brown algal defenses deter feeding by tropical herbivorous fishes. Oecologia 84: 158–163.

47. Littler MM (1976) Calcification and its role among the macroalgae. Micronesica 12: 27–41.

48. Johnson VR (2012) *A study of marine benthic algae along a natural carbon dioxide gradient*. PhD. Plymouth University, School of Biomedical and Biological Sciences, Marine Biology and Ecology Research Centre (MBERC).

49. Crook ED, Potts D, Rebolledo-Vieyra M, Hernandez L, Paytan A (2012) Calcifying coral abundance near low-pH springs: implications for future ocean acidification. Coral Reefs 31: 239–245. doi: 10.1007/s00338-011-0839-y

50. Saderne V, Fietzek P, Herman PMJ (2013) Extreme variations of pCO2 and pH in a macrophyte meadow of the Baltic Sea in summer: evidence of the effect of photosynthesis and local upwelling. Plos One 8: 1–8.

51. Lewis E, Wallace D (1998) *Program developed for CO$_2$ system calculations*. Carbon Dioxide Information Analysis Center, Oak Ridge National Laboratory, Tennessee (USA).

52. Rodolfo-Metalpa R, Houlbrèque F, Tambutté É, Boisson F, Baggini C, et al. (2011) Coral and mollusc resistance to ocean acidification adversely affected by warming. Nature Clim Change 1: 308–312. doi: 10.1038/nclimate1200

53. Diaz-Pulido G, Anthony K, Kline DI, Dove S, Hoegh-Guldberg O (2012) Interactions between ocean acidification and warming on the mortality and dissolution of coralline algae. J Phycol 48(1): 32–39.

54. Johnson MD, Carpenter RC (2012) Ocean acidification and warming decrease calcification in the crustose coralline alga *Hydrolithon onkodes* and increase susceptibility to grazing. J Exp Mar Biol Ecol 434: 94–101.

55. Hiscock K, Southward A, Tittley I, Hawkins S (2004) Effect of changing temperature on benthic marine life in Britain and Ireland. Aquat Conserv 14: 333–362. doi: 10.1002/aqc.628

56. Riley J, Kirby J, Linsley M, Gardiner G (2003) Review of UK and Scottish surveillance and monitoring schemes for the detection of climate-induced changes in biodiversity. Just Ecology, Environmental Consultation.

57. Tyler-Walters H, Hiscock K (2003) A biotope sensitivity database to underpin delivery of the Habitats Directive and Biodiversity Action Plan in the seas around England and Scotland. Report to English Nature and Scottish Natural Heritage from the Marine Life Information Network (MarLIN). Plymouth: Marine Biological Association of the UK. [Final Report].

58. Directive 2008/56/EC of the European Parliament and of the Council of 17 June 2008 establishing a framework for community action in the field of marine environmental policy [2008] OJ L164/19.

59. Koch M, Bowes G, Ross C, Xing-Haizhang (2013) Climate change and ocean acidification effects on seagrasses and marine macroalgae. Glob Change Biol 19: 103–132. doi: 10.1111/j.1365-2486.2012.02791.x

7

A Connectivity-Based Eco-Regionalization Method of the Mediterranean Sea

Léo Berline[1,3,4]*, Anna-Maria Rammou[2], Andrea Doglioli[2], Anne Molcard[1], Anne Petrenko[2]

1 Université du Sud Toulon-Var, Aix-Marseille Université, CNRS/INSU/IRD, Mediterranean Institute of Oceanography (MIO), La Garde, France, 2 Aix-Marseille Université, CNRS/INSU/IRD, Mediterranean Institute of Oceanography (MIO), Marseille, France, 3 CNRS, Laboratoire d'Océanographie de Villefranche, Villefranche-sur-Mer, France, 4 Université Pierre et Marie Curie, Paris 6, Laboratoire d'Océanographie de Villefranche, Villefranche-sur-Mer, France

Abstract

Ecoregionalization of the ocean is a necessary step for spatial management of marine resources. Previous ecoregionalization efforts were based either on the distribution of species or on the distribution of physical and biogeochemical properties. These approaches ignore the dispersal of species by oceanic circulation that can connect regions and isolates others. This dispersal effect can be quantified through connectivity that is the probability, or time of transport between distinct regions. Here a new regionalization method based on a connectivity approach is described and applied to the Mediterranean Sea. This method is based on an ensemble of Lagrangian particle numerical simulations using ocean model outputs at $1/12°$ resolution. The domain is divided into square subregions of 50 km size. Then particle trajectories are used to quantify the oceanographic distance between each subregions, here defined as the mean connection time. Finally the oceanographic distance matrix is used as a basis for a hierarchical clustering. 22 regions are retained and discussed together with a quantification of the stability of boundaries between regions. Identified regions are generally consistent with the general circulation with boundaries located along current jets or surrounding gyres patterns. Regions are discussed in the light of existing ecoregionalizations and available knowledge on plankton distributions. This objective method complements static regionalization approaches based on the environmental niche concept and can be applied to any oceanic region at any scale.

Editor: Brian R. MacKenzie, Technical University of Denmark, Denmark

Funding: The research leading to these results has received funding from the European Community's Seventh Framework Programme (FP7/2007–2013) under Grant Agreement No. 287844 for the project "Towards COast to COast NETworks of marine protected areas (from the shore to the high and deep sea), coupled with sea-based wind energy potential" (COCONET). The funders had no role in study design, data collection and analysis, decision to publish, or preparation of the manuscript.

Competing Interests: The authors have declared that no competing interests exist.

* Email: leo.berline@univ-amu.fr

Introduction

The ecoregionalization of the ocean is useful for scientific research, conservation and management of the marine environment and marine resources. For instance, ecoregionalization is needed to extrapolate punctual or transect data to broader areas and to target specific regions for interdisciplinary research (as in the Mediterranean Sea, [1]). Conservation and management goals range from selecting areas to protect [2] to defining fisheries zones or zones for monitoring and mitigating marine pollution.

To date, several approaches of ecoregionalization were used depending on the data at hand [3]. The taxonomic approach is based on species distributions and identifies areas of broadly similar assemblage of species [4–6]. The ecological approach is based on habitat characteristics; it separates areas of similar seasonal cycles of physical and biogeochemical variables [7–10]. This approach benefited from the nearly continuous coverage of satellite data. Lastly, the integrative approach is a combination of both taxonomic and ecological approaches that takes into account both the habitat and the species inhabiting it [11].

However, in the marine environment the species distribution not only results from selection by the local environment but also from dispersal of propagules and adults organisms (e.g. the metapopulation concept of Levins [12,13]). Therefore an ecoregionalization based on dispersal by ocean circulation is needed; recent studies start taking into account dispersal in defining management units [14]. However it was never achieved quantitatively at basin scale. Today this is possible, as widely available ocean circulation models provide 3 dimensional, time varying, realistic and consistent depictions of oceanic currents at basin scale. The goal of this paper is to present a regionalization method based on connectivity, assessed from ensemble Lagrangian simulations using ocean circulation model velocity outputs.

This method is applied to the Mediterranean basin, which is a target region for spatial planning owing to its high level of endemism and high biodiversity [15]. Surface circulation shows a complex pattern of larger and smaller gyres, driven by the entrance of Atlantic water at Gibraltar Strait [1], local meteorology and bathymetry. The oligotrophy increases toward the East, but productive spots also exist over shelves and deep mixing areas, thus creating a significant heterogeneity in ecosystem functioning and habitats.

Materials and Methods

The general outline of the method is as follow (Fig. 1): Lagrangian trajectories are computed from ocean circulation model velocity outputs for particles seeded over the whole model domain at three depths (0.5 m, 50 m and 100 m). The domain is divided into a regular grid (hereinafter connectivity grid) and the trajectories are used to derive the mean connection time between every pair of grid cell. In this way a mean connection time matrix is obtained and then transformed into an oceanographic distance matrix, used as input to a hierarchical clustering algorithm. Finally clustering produces a partition of the domain.

Daily outputs velocity fields for four years (2007–2010) were taken from the configuration PSY2V3 of the operational system MERCATOR OCEAN [16]. The PSY2V3 configuration covers the North Atlantic ocean and Mediterranean Sea, and is based on the NEMO-OPA primitive equations code [17] with assimilation of observed data (satellite and in situ). Here, only the domain subset covering the Mediterranean Sea was used. Daily surface forcing are provided by ECMWF [18]. The velocity components are distributed in an Arakawa C type grid [19]. The horizontal resolution is $1/12°$ (~8 km) and there are 50 fixed vertical levels with higher resolution at the surface. The vertical mixing is described by a TKE closure scheme [20] and the advection by a TVD 2nd order centered scheme [21].

The trajectories followed by numerical particles were calculated offline with the Lagrangian diagnostic tool ARIANE [22]. The trajectories only result from the horizontal advection at three depths (0.5 m hereinafter called surface, 50 m and 100 m) chosen to represent the transport in the epipelagic layer. No vertical velocity was considered to keep particles in the 0–100 m range. The one year integration time was chosen to allow particles to cover the whole basin and therefore quantify basin scale connectivity and to keep computation time reasonable. Particles were seeded every 10 km on a regular square grid covering the whole domain, totaling 25,646 initial positions for surface depth and 23,770 for depths 50 and 100 m because the domain is smaller. Particles were seeded every 3 days from the 1st to the 25th of every month, from January 1st 2007 to December 25st 2009 in order to fully sample the variability of the circulation. This represents a total of 8,309,304 particles for surface depth, respectively 7,701,480 for depths 50 and 100 m. The choice of 10 km and 3 days is a compromise between matching the horizontal resolution of the model, taking into account mesoscale processes and keeping an affordable computing time of resulting trajectories. We thus obtained three ensembles of trajectories, one per depth.

In order to quantify the connections over the model domain, the domain was divided into grid cells of 50 km×50 km on a regular square grid, the connectivity grid, with a total of 1095 cells covering only regions with depths greater than 100 m. The 50 km resolution is sufficient to keep a reasonably realistic coastline while being suitable with the seeding density chosen. Thus each connectivity grid cell contains $5*5 = 25$ particles for each initial seeding date, except grid cells including land that contains less particles.

To quantify the connectivity between each grid cell, we used the Mean Connection Time, hereinafter MCT. Defining T(i,j) as the transit time from grid cell i to grid cell j, MCT(i,j) was computed as

$$MCT(i,j) = \frac{1}{M} \sum_{n=1}^{n=M} T_n(i,j)$$

M being the number of particle transitioning from i to j. Note that for each trajectory, all intermediate transitions were used to compute the MCT. The sensitivity of MCT to the number of particles was tested. The suite of MCT matrices converged when the number of particles was greater than 6,000,000, therefore we considered that 8,309,304 particles and respectively 7,701,480 particles for depths 50 and 100 m were sufficient to obtain a robust MCT matrix. Moreover, to keep MCT robust, it was computed only when M was greater or equal to 50. Four MCT matrices of size 1095×1095 were computed: one MCT matrix from each ensemble of trajectories (MCT_0, MCT_{50}, MCT_{100} for 0.5, 50 and 100 m depths trajectories respectively) and also one MCT matrix using the three ensembles together ($MCT_{3depths}$).

Not all grid cells of the domain were connected within one year, especially remote cells (e.g. Northern Aegean and Gibraltar Strait). Thus the resulting MCT matrices had gaps (from 37% to 56%). These gaps are a problem for the steps of computing the oceanographic distance and applying hierarchical clustering on it. Therefore a gap filling procedure was introduced as follows (see Appendix S1):

○ For each unconnected pair of grid cells i-j, we looked for grid cells k so that i-k and k-j pairs are connected. There must be at least 50 grid cells k as for M.

○ Then we computed MCT(i,j) for pair i-j as the sum of the MCT(i,k) and MCT(k,j), averaged on all existing cells k, and filled the MCT(i,j) value in the matrix.

After 3 iterations of this procedure, each MCT matrix was filled. The resulting MCT values ranged from 10 days to 3000 days. This gap filling procedure avoided the very long integration time (>8 years) needed if we were to fill the whole MCT matrices from original trajectories alone.

This led to four full MCT matrices, which are asymmetric since the time to go from i to j is not equal to the time to go from j to i. Then the oceanographic distance (OD) was defined after [23] as the minimum of the two MCT values associated to each pair of grid cells i and j (travel from i to j and return travel from j to i). We chose the minimum value as it corresponds to the fastest route of transport which is also the shortest in length.

$$OD(i,j) = \min(MCT(i,j), MCT(j,i))$$

This gave four symmetric matrices, (OD_0, OD_{50}, OD_{100}, $OD_{3depths}$) where all diagonal terms (autoconnection time) were set to zero.

Finally hierarchical clustering analysis was applied on each of the oceanographic distance matrix. This method has proved to be robust in the classification of atmospheric wind data (e.g. [24]) and hydrological data (e.g. [25]). Hierarchical clustering assigns grid cells to different clusters in a way that each grid cell belongs to only one cluster [26], and each cluster belongs to a larger cluster (Fig. 2). The grid cells are grouped according to their similarity, which here is the oceanographic distance. Thus there is no distance metric applied as in usual clustering exercises. During each sequence of the clustering algorithm, the distances between the new clusters formed and the other grid cells are computed. This step requires a linkage criterion to be defined. Here we used the flexible [27] and Ward linkages [28]. WPGMA linkage was also tested ([27]) but flexible and Ward best balanced the dendrogram. For a given cut-off level of the dendrogram, we obtained a partition of the grid cells in a certain number of clusters, which is, in the spatial domain, a regionalization. Each cluster corresponded to a region on the connectivity grid whose contours were identified. Finally for each cluster, the within-cluster MCT was computed and plotted as a function of the number of clusters from 2 to 31 (Fig. 3).

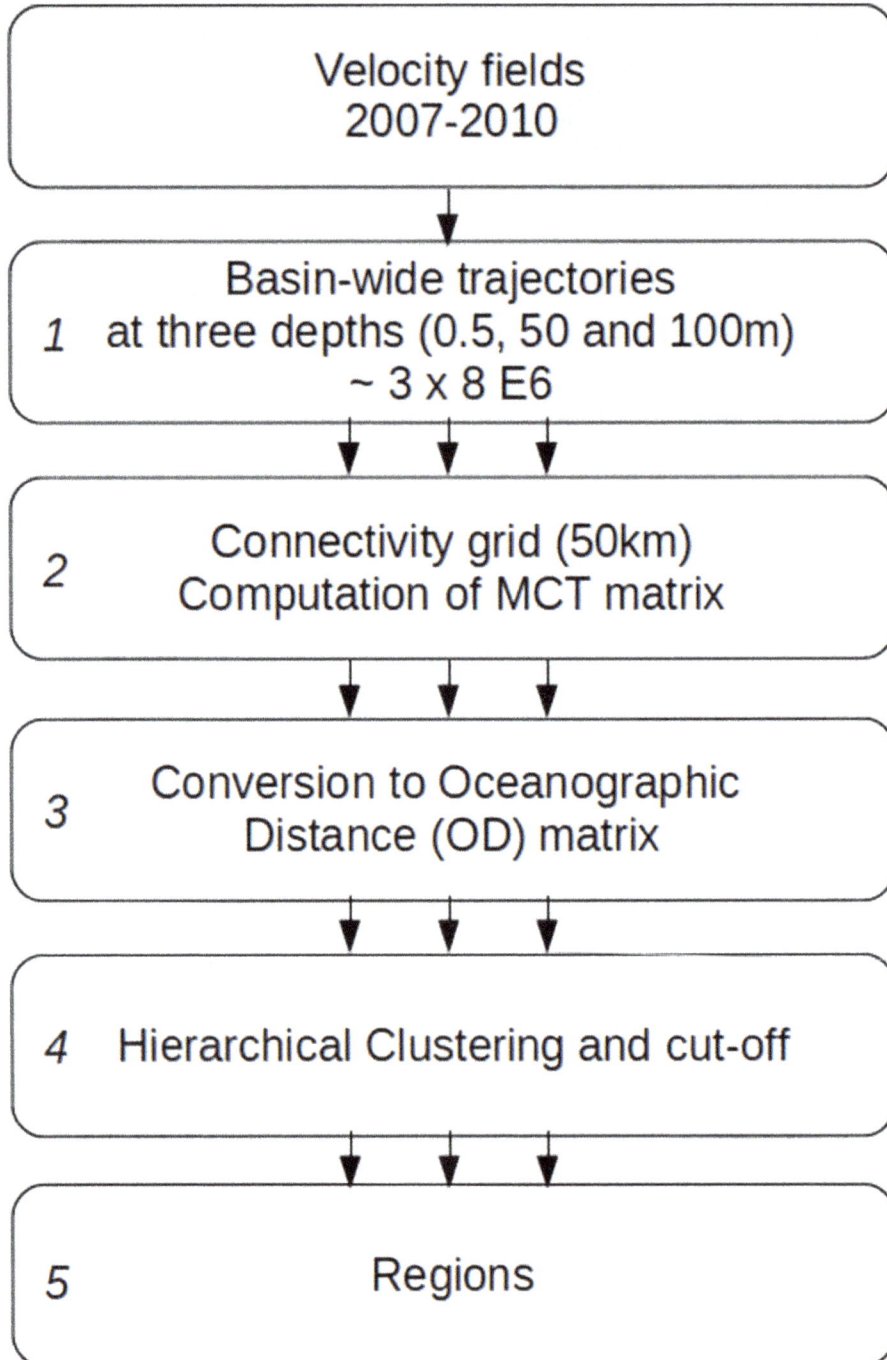

Figure 1. Schematic of the steps of the regionalization method. Note that steps 2 to 5 are repeated using trajectories at the 3 depths separately, shown with the three arrows, and then using them altogether.

Our "best estimate" regionalization was computed using flexible link and the matrix $OD_{3depths}$, built from the complete ensemble of trajectories (Fig. 4). We also computed one regionalization for each of the three depths and two linkages (6 cases). To assess the sensitivity of the regionalization results to the linkage and depth used, we computed the boundary stability, which is simply the local frequency of occurrence of a boundary in the spatial domain among the 6 cases, as defined in [29].

The choice of the optimal cut-off level and number of cluster is not straightforward here, because the distance matrix (OD) is not computed with a distance metric applied to a given dataset. Thus, usual criteria based on dataset variance within clusters cannot be used (e.g. [30]) because there is no dataset. Instead we took a simple approach comparing results from Ward and flexible linkage. For each partition into n clusters, we compute the proportion of cells classified in the same cluster with Ward and flexible (see Appendix S2). This proportion increases from 82%, to 88% from for n = 2 to n = 6 clusters, then drops to values <70% for n>6. Therefore we consider that the optimal cluster number is 6 as it gives more information while keeping consistent results

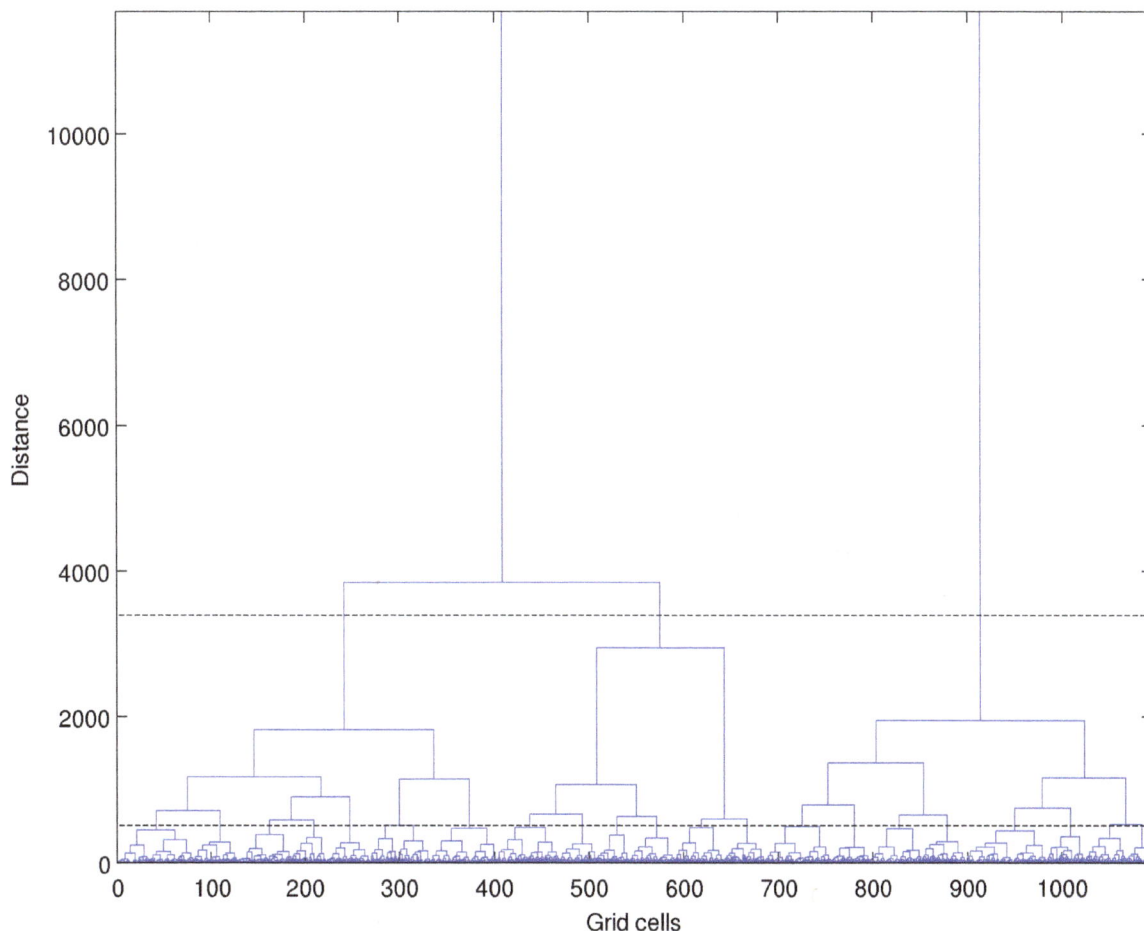

Figure 2. Cluster dendrogram of the oceanographic distance matrix OD$_{3depths}$ using the flexible linkage. Horizontal black lines show the cut-off values for 3 and 22 clusters.

among the two linkages. However, as no absolute criterion is available, we show the maximum number of clusters that we can interpret, which is 22 clusters. The clusters above 22 require detailed regional information to be interpreted, which is beyond the scope of this study.

Results

When the number of clusters increases, the within-cluster MCT diminishes, as well as the size of each region (Figs. 3 and 4). The average MCT ranges from 188 days for 2 clusters to ca. 90 days for 22 clusters.

On the basis of our interpretation of regions with respect to circulation, we retained 22 clusters (Fig. 4). The boundaries of each region were identified and colored according to the number of cluster obtained varying the cutoff distance from 10,000 (2 clusters) to 507 (22 clusters). The boundary #1 partly cuts the Sicily Strait (Fig. 4) and separates the Western and Eastern basins. The boundary #2 isolates Levantine basin from Ionian Sea and Adriatic Sea. The boundary #3 isolates the northern Ionian and Adriatic Sea from Southern Ionian. Then boundary #4 separates the Western basin into a western and an eastern part. The boundary #5 isolates the Levantine basin plus a part of AW current off Lybia from the Aegean Sea. The boundary stability map (Fig. 5) shows that some of the boundaries shown on Fig. 4 are stable (e.g. boundary #7, 11, 16) while others are variable in

position or occurrence (e.g. boundary #4). Also, some boundaries (e.g. #2, 6, 8) have only a portion that is stable.

Then considering the 22 regions, the Western basin is separated into eight regions; regions A and B in the Northern part of the basin, G, F and E in the South and C, D that contains the Tyrrhenian sub-basin, region H at the center. In the Eastern basin, the Adriatic Sea is one region I. The Ionian Sea is separated into regions J, V, T at the center and K to the east, with U and S along the coasts of Libya and Tunisia. The Aegean Sea is divided into two regions, M in the East, L in the West. The Levantine basin has four regions: two coastal regions N and O, one southern region P and one center region Q. Considering only stable boundaries, the Western basin only has 5 regions. The Eastern basin has few continuous boundaries, only 4 regions are delimited (Adriatic, South of Sicily Strait and regions U and O).

Discussion

The boundary stability shows that the majority but not all boundaries are robust to changes in linkages and depths. Often, linkages or depth changes can produce minor shifts in boundary position, hence reducing boundary stability as defined here. When a boundary is not stable, it means that either the circulation is variable, either it is located in a region where the distance (OD) among grid cells is small thus the boundary position varies according to the overall content of each cluster. Thus the

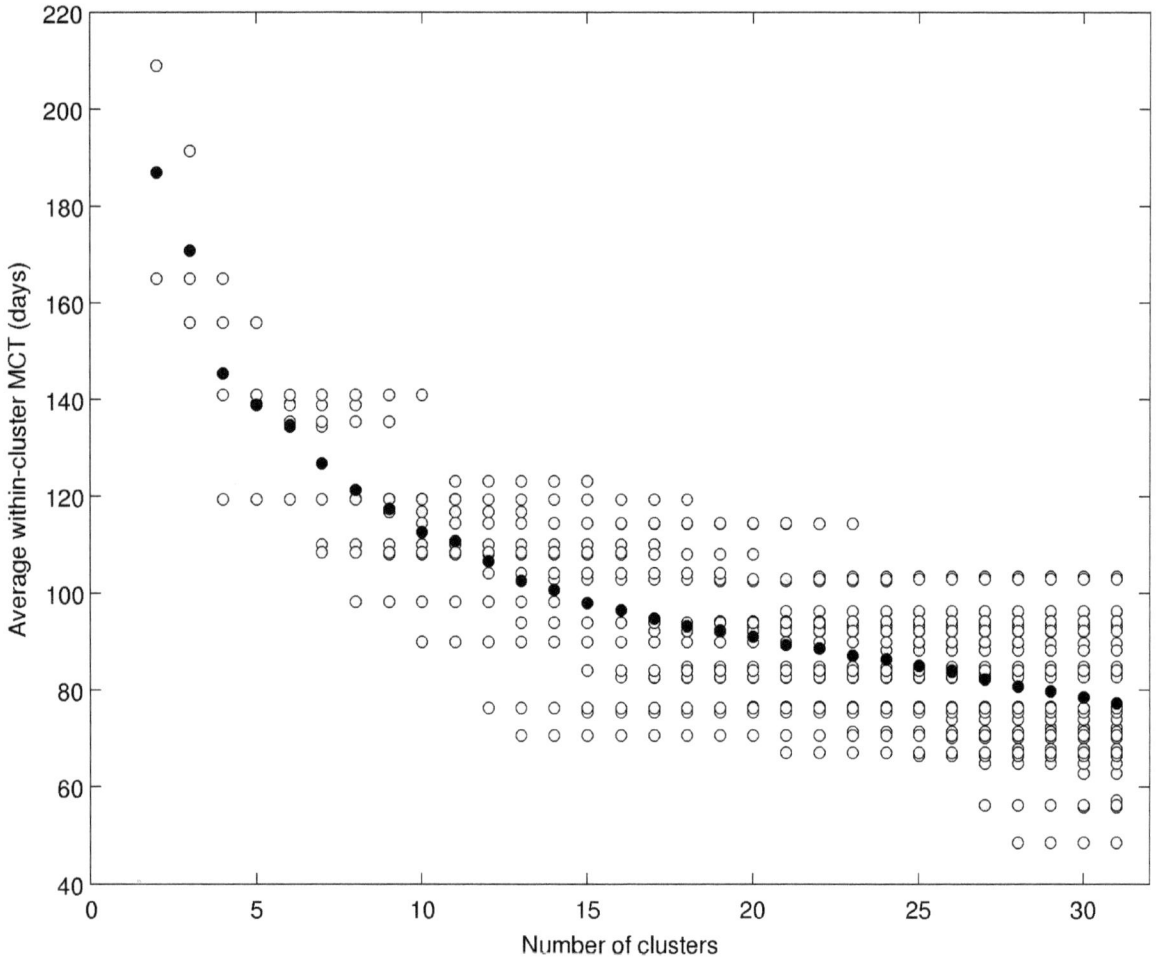

Figure 3. Within cluster mean connection time as a function of the cluster number for MCT$_{3depths}$. White dots are the mean for each cluster, black dots are the mean over all clusters.

boundary map must be analyzed jointly with the boundary stability to assess our regionalization.

1. Regions reveal circulation patterns

First the meaning of the regions obtained needs to be explained. One region contains grid cells that are connected at shorter time scale with each other than they are to the grid cells of the other regions. In the following, the relationship between the clusters boundaries, their stability and the circulation is examined in detail in comparison with the model average velocity fields (Fig. 4) and literature.

The hierarchy of cluster boundary is in good agreement with the surface general circulation scheme proposed by Millot et al. [31], their figure 2. Boundary #1 separates the Western and Eastern basins at the Sicily strait, boundary #2 isolates the Eastern Levantine, then boundary #3 the Adriatic Sea together with the northern part of the Ionian. Boundaries are often parallel to the mean velocity field. For instance boundary #16 is parallel to the Northern Current, boundary #11 parallel to the Asia Minor Current. Boundary can also separate two currents branches (the ATC along Tunisia and the AIS along Sicily, for part of boundaries #1 and 10, see [32]). This illustrates the barrier role of semi-permanent jets in the ocean. However, this is not always

the case (e.g. boundary #1 at the Sicily Strait, boundary #18 at Oranto Strait). This can occur as the MCT matrix was computed from the time varying flow field, not from the mean field shown here and because each cluster is separated according to its overall distance with other clusters.

In the Western basin, boundary #16 is associated to the path of the Northern Current [31] and is the most stable. The boundary #6 from Spain to the Baleares follows approximately the Balearic front and is also rather stable. The Tyrrhenian Sea contains regions B, C, D with partly stable boundaries. Region C east of the Strait of Bonifacio contains the wind induced cold recirculation identified by [31], which is a potential dense water formation zone [33]. The Southern region G is restricted to the Alboran Sea.

In the Eastern basin, the Ionian Sea has two Southern regions U and S. Boundary #10 follows the Sicilian current of AW and region U contains the area of accumulation of eddies of the Ionian Sea [31]. The region V can correspond to the meandering stream identified by [34] or considered as interannual variability by [31]. The South-eastern Levantine has a region O with a stable boundary #7. Region O corresponds to the eddy accumulation zone $\sum LE$ following [35]. The Asia Minor current along the Southern coasts of Turkey is captured in region N and has a stable boundary #11. Finally, the Aegean Sea is divided into an Eastern

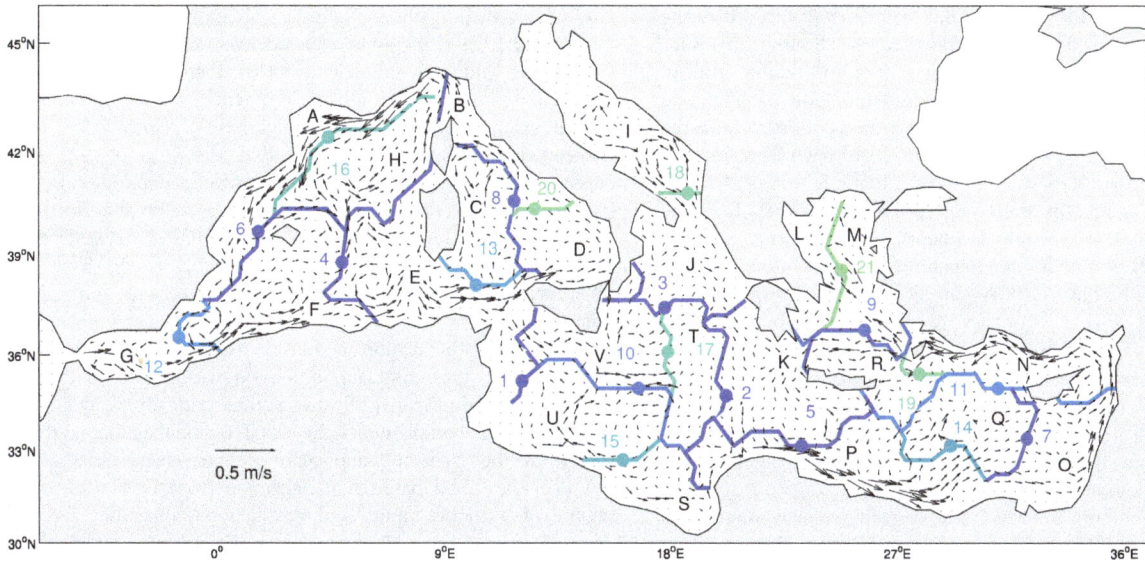

Figure 4. Map of the 21 clusters boundaries obtained from clustering of the oceanographic distance matrix $OD_{3depths}$ using the flexible link. Each boundary is colored and numbered according to the cut-off distance on the dendrogram (from blue – high distance- to green-low distance). Each region is identified by a letter from A to V. The velocity from the circulation model, averaged for the 4-year (2007–2010) and the 3 depths is overlaid as black vectors.

region M fed by AW and a North-Western region L fed by Black Sea outflow waters.

Some regions are virgin of any boundaries (Fig. 5), like the center of Gulf of Lion, the Alboran Sea, the Eastern Tyrrhenian Sea, the Northern Adriatic Sea, South of Greece, the South-East of the Levantine basin. This means that these regions are intraconnected at a time scale of less than ca 90 days (see Fig. 3).

Thus this regionalization reveals known circulation patterns and summarizes them in a way that complements the simple average velocity field analysis. It can be used to quantitatively compare the circulation patterns from contrasted periods or from different models.

2. Some boundaries coincide with major environmental boundaries and range limits of zooplankton assemblages

The identification of regions close to each other, not geographically but in terms of oceanographic connections, should help understanding the spatial distribution of properties that are passively transported by currents, such as conservative physical properties, or planktonic organisms living in the surface layer (epipelagic).

First, boundaries emerging from circulation alone often match major discontinuities in variables describing the environment. For instance a strong latitudinal salinity gradient exists near the Balearic Islands, close to our boundary #6. However, our

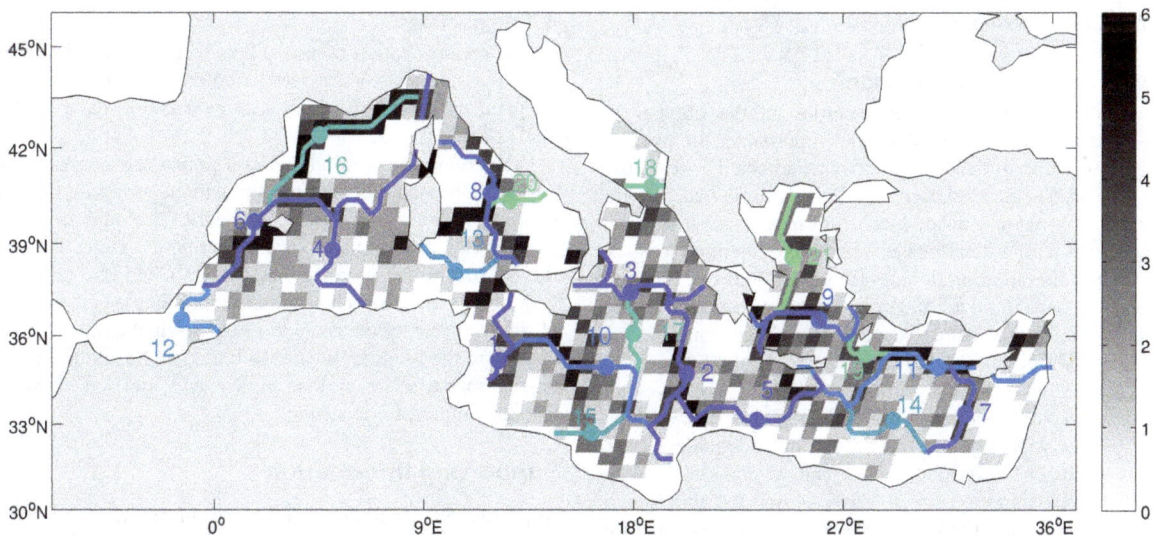

Figure 5. Map of the boundary stability (gray scale) derived from the 6 cases of clustering (3 depths $\times$ 2 linkages). Boundary stability is defined as the number of occurrence of a boundary in each grid cell among the 6 cases. Boundaries are overlaid as in figure 4.

boundary #6 coincides with the Balearic Current but not to the Balearic salinity front, located more to the South [36]. Our boundary #16 coincides with a temperature and salinity front in the Ligurian Sea, and also in phytoplankton biomass (Fig. 1 in [10]). Off the Catalan coast, boundary #16 is consistent with the alongshore distribution of fish larvae [37], although located more offshore. Also, boundary #18 south of Adriatic Sea coincides with salinity fronts as seen in MEDATLAS [38]. This results from the dynamic links between density gradients and surface currents. The boundary #21 found in the Aegean Sea parallels the front in phytoplankton biomass [10]. At the Sicily strait, corresponding to our boundary #1, a boundary was also found by [9] (their figure 2) based on a clustering of sea surface temperature and ocean color data.

Within our regions, planktonic organisms are connected at shorter time scales than between regions. Thus hydrodynamical boundaries can become faunistic boundaries as suggested by Gaylord and Gaines [39] for larvae of benthic organisms. Given the spatial resolution, the MCT can correctly resolve connections of plankton organisms with a life cycle greater than 10 days, such as most zooplankton species [40]. Indeed, consistent with boundary #6 north of the Balearic Islands, a boundary exists between Atlantic zooplankton species to the South and Mediterranean species to the North [41,42]. Also, consistent with our boundaries #1 and #2, differences in zooplankton species composition between Eastern and Western basin were reported by several authors ([43] and references therein, [44]) although the spatial resolution of zooplankton data is generally not sufficient for accurately locating boundaries.

Ecoregions drawn qualitatively from expert knowledge of species assemblages ([45] their figure 2) also distinguish Atlantic-water regions including our region G, a Northern Current region including our region A, three Adriatic regions, one Aegean Sea region including our regions L and M, and two large zonal Eastern basin regions mostly consistent with boundaries #5 and #11.

However, for living organisms such as zooplankton, circulation alone is not sufficient to explain the distribution of a given species as it is adapted to its environment, in particular to a temperature range, e.g. [46]. Thus within our connected regions environmental conditions will restrict a species distribution to its specific preferendum, i.e. its ecological niche. Moreover, we deal with particles in the 0–100 m layer, which only properly represent epipelagic zooplankters dispersal.

3. How to use this regionalization?

To use this regionalization, the question of the number of clusters to retain will arise. With our approach, no existing criterion is available to define the optimal number. However the number of clusters can be chosen based on the time scale we are interested in, as regions isolated at a given time scale become connected at a larger time scale. Therefore the time scale of interest defines the appropriate cut-off distance and the resulting cluster number and sizes (Fig. 3). For instance, one can look for the scale of dispersal of planktonic larvae and hence consider the Pelagic Larval Duration (PLD) time scale. A PLD of 120 days (e.g. a crustacean as spiny lobster *Palunirus elephas* [47]) gives an adequate cluster number of ca 8. For a PLD of ca. 70 days (e.g. a labridae fish as *Lipophrys trigloides* [48]) the adequate cluster number is ca 30. The lower bound time scale we can address with the present regionalization (~10 days) is set by the spatial resolution of our connectivity grid. Shorter time scales could be achieved with a finer connectivity grid.

Few existing studies can be compared to our regionalization because the approach is original. In the Mediterranean Sea,

Andrello et al. [49] obtained clusters of coastal marine protected areas (MPA) based on their connectivity assessed by Lagrangian simulations. Although the velocity fields, Lagrangian simulations set up and clustering method are different, we can compare the overall grouping obtained (their figure 5-A). Considering only clusters containing several MPAs (8 clusters out of 38), their clusters are mostly contained within single regions and do not spread across several regions. Exceptions occur in the Northern Ligurian Sea and Ionian Sea with MPAs located very close or even onto our regions' boundaries. This probably results from the difference in the input velocity fields and subsequent connectivity quantification.

This new regionalization method quantifies the dispersal range of organisms, This dispersal dimension was shown to explain species distribution (e.g. [50]) and is thus critically needed [51]. This approach complements the usual regionalization methods rooted in the environmental niche concept (e.g. [9,10]). For instance, the Chl-a based regionalization from [10] reflects the regime of nutrients inputs and stratification, thus they are not directly linked to surface circulation patterns. Adding our connectivity-based regionalization helps understanding the types of environment that plankton is facing, through passive horizontal transport, vertical mixing and production processes. Practically, our OD matrices could be used as a constraint during the clustering of Chl-a, as for chronological clustering [52].

Also, our regions illustrate why plankton organisms may be encountered outside their optimum range (plankton expatriates, e.g. [53]) and where transport-driven fluctuations of plankton communities are expected. Indeed fluctuations of region boundaries may produce large biogeographic fluctuations noticeable at fixed points (e.g. [54,55]). Regions can also help tracking invasions of exotic organisms, for instance the so-called lessepsian species coming from the Suez Canal [56]. Apart from living organisms, our regions could be used to quantify areas of dispersion of pollutants coming from ships or land sources [57].

Finally this regionalization is useful as a framework to interpret the genetic differentiation of a given species sampled throughout the Mediterranean (e.g. [58]). Further, our approach could be used to define a priori units for grouping existing MPA or set up new MPA (e.g. [59] for the Gulf of Lions), as envisioned in the EU Integrated Project COCONET (www.coconet-fp7.eu).

4. Perspectives

The regionalization proposed here will eventually be compared to an ongoing biogeochemistry-based regionalization [60], and to zooplankton species distribution as available in database COPE-PODS [61].

Concerning the methods, several points can be made. With a similar approach but shorter simulations, we can explore the seasonal variability of clusters boundaries that may be significant [62]. Here we used hierarchical clustering to extract clusters from the oceanographic distance, but clusters could also be computed with other methods such as graph theory that uses the asymmetry of the connectivity matrix (e.g. [49,63]). Finally, this method was applied to the Mediterranean Sea but it can be applied anywhere, at any spatial scales as long as accurate and long term model velocity outputs are available.

Supporting Information

Appendix S1   Method to fill the gap of the MCT matrix.

Appendix S2   Method for choosing the optimal cut-off distance.

Acknowledgments

Mercator-ocean (www.mercator-ocean.fr) is thanked for providing velocity outputs for configuration PSY2V3. Fabien Lombard is thanked for useful feedback on an earlier version of the manuscript. The two reviewers are thanked for their constructive comments.

Author Contributions

Conceived and designed the experiments: LB AMR AD AM. Performed the experiments: AMR LB. Analyzed the data: LB AMR. Wrote the paper: LB AMR AD AM AP.

References

1. Durrieu de Madron X, Guieu C, Sempere R, Conan P, Cossa D, et al. (2011) Marine ecosystems' responses to climatic and anthropogenic forcings in the Mediterranean. Progress in Oceanography 91: 97–166.
2. Giakoumi S, Sini M, Gerovasileiou V, Mazor T, Beher J, et al. (2013) Ecoregion-Based Conservation Planning in the Mediterranean: Dealing with Large-Scale Heterogeneity. PLoS ONE 8(10): e76449. doi:10.1371/journal.pone.0076449.
3. UNESCO (2009) Global Open Oceans and Deep Seabed (GOODS) Biogeographic Classification. Paris. UNESCO-IOC. IOC Technical Series 84 92pp.
4. Forbes E (1856) Map of the distribution of marine life. in Johnston AK, editor. The Physical Atlas of Natural Phenomena. Edinburgh (Scotland): William Blackwood and Sons. pp. 99–102.
5. Ekman S (1953) Zoogeography of the Sea. London: Sidgwick and Jackson. 417 p.
6. Briggs JC (1995) Global Biogeography. Amsterdam, Elsevier. 472 p.
7. Longhurst A (1995) Seasonal cycles of pelagic production and consumption. Progress in Oceanography 36: 77–167.
8. Longhurst A (1998) Ecological geography of the sea Academic Press, London. 560 p.
9. Oliver MJ, Irwin AJ (2008) Objective global ocean biogeographic provinces. Geophysical Research Letters, 35, L15601, doi:101029/2008GL034238.
10. D'Ortenzio F, Ribera d'Alcalà M (2009) On the trophic regimes of the Mediterranean sea: a satellite analysis. Biogeosciences 6: 139–148.
11. Koubbi P, Moteki M, Duhamel G, Goarant A, Hulley PA, et al. (2011) Ecoregionalization of myctophid fish in the Indian sector of the Southern Ocean: Results from generalized dissimilarity models. Deep Sea Research Part II: Topical Studies in Oceanography 58: 170–180.
12. Levins R (1969) Some demographic and genetic consequences of environmental heterogeneity for biological control. Bulletin of the Entomological Society of America 15: 237–240.
13. Obura D (2012) The Diversity and Biogeography of Western Indian Ocean Reef-Building Corals. PLoS ONE doi:10.1371/journal.pone.0045013.
14. Casale P, Mariani P (2014) The first 'lost year' of Mediterranean sea turtles: dispersal patterns indicate subregional management units for conservation. Marine Ecology Progress Series 498: 263–274.
15. Coll M, Piroddi C, Steenbeek J, Kaschner K, Lasram F B, et al. (2010). The Biodiversity of the Mediterranean Sea: Estimates, Patterns, and Threats. PLoS ONE, 5(8).
16. Bahurel P (2006) MERCATOR OCEAN global to regional ocean monitoring and forecasting. In: Chassignet, E P and Verron, J editors, Ocean Weather Forecasting Springer, Netherlands, pp. 381–395.
17. Madec G (2008) "NEMO ocean engine". Note du Pole de modélisation, Institut Pierre-Simon Laplace (IPSL), France, No 27 ISSN No 1288–1619, 367 p.
18. Uppala S, Kallberg PW, Simmons AJ, Andrae U, Da Costa V, et al. (2005) The ERA-40 re-analysis. Q J R Meteorol Soc, 131: 2961–3012.
19. Mesinger F, Arakawa A, Kasahara A (1976) Numerical methods used in atmospheric models, vol 1 World Meteorological Organization, International Council of Scientific Unions 37 p.
20. Blanke B, Delecluse P (1993) Variability of the tropical Atlantic ocean simulated by a general circulation model with two different mixed layer physics. Journal of Physical Oceanography 23: 1363–1388.
21. Zalesak ST (1979) Fully multidimensional flux corrected transport algorithms for fluids. J Comput Phys 31: 335–362.
22. Blanke B, Raynaud S (1997) Kinematics of the Pacific Equatorial Undercurrent: An Eulerian and Lagrangian approach from GCM results. Journal of Physical Oceanography 27: 1038–1053.
23. Alberto F, Raimondi PT, Reed DC, Watson JR, Siegel DA, et al. (2011) Isolation by oceanographic distance explains genetic structure for Macrocystis pyrifera in the Santa Barbara Channel. Molecular Ecology, 20: 2543–2554.
24. Burlando M, Antonelli M, Ratto C (2008) Mesoscale wind climate analysis: identification of anemological regions and wind regimes. Int J Climatol 28: 629–641.
25. Hjelmervik KT, Hjelmervik K (2013) Estimating temperature and salinity profiles using empirical orthogonal functions and clustering on historical measurements. Ocean Dynamics 63: 809–821.
26. Legendre P, Rogers DJ (1972) Characters and clustering in taxonomy: a synthesis of two taximetric procedures. Taxon 567–606.
27. Lance GN, Williams WT (1967). A general theory of classificatory sorting strategies 1. Hierarchical systems. The computer journal, 9: 373–380.
28. Ward JH Jr (1963) Hierarchical Grouping to Optimize an Objective Function. Journal of the American Statistical Association 58: 236–244.
29. Oliver MJ, Glenn S, Kohut JT, Irwin AJ, Schofield OM, et al. (2004), Bioinformatic approaches for objective detection of water masses on continental shelves, J Geophys Res, 109, C07S04, doi:10.1029/2003JC002072.
30. Calinski RB, Harabasz J (1974) A dendrite method for cluster analysis. Commun. Stat. 3: 1–27.
31. Millot C, Taupier-Letage I (2005) Circulation in the Mediterranean sea. In: The Mediterranean Sea, Springer, pp. 29–66.
32. Poulain PM, Zambianchi E (2007) Surface circulation in the central Mediterranean Sea as deduced from Lagrangian drifters in the 1990s. Continental Shelf Research 27(7): 981–1001.
33. Fuda JL, Etiope G, Millot C, Favali P, Calcara M, et al. (2002) Warming, salting and origin of the Tyrrhenian Deep Water (2002) Geophys. Res. Letters 29, doi:10.1029/2001GL014072.
34. Malanotte-Rizzoli P, Artale V, Borzelli-Eusebi GL, Brenner S, Crise A, et al. (2014) G.: Physical forcing and physical/biochemical variability of the Mediterranean Sea: a review of unresolved issues and directions for future research, Ocean Sciences 10: 281–322, doi:10.5194/os-10-281-2014.
35. Hamad N, Millot C, Taupier-Letage I (2005) A new hypothesis about the surface circulation in the eastern basin of the Mediterranean sea. Progress in Oceanography 66: 287–298.
36. Mariani P, MacKenzie BR, Iudicone D, Bozec A (2010) Modelling retention and dispersion mechanisms of bluefin tuna eggs and larvae in the northwest Mediterranean Sea. Progress in Oceanography 86(1): 45–58.
37. Sabatés A, Olivar MP, Salat J, Palomera I, Alemany F (2007) Physical and biological processes controlling the distribution of fish larvae in the NW Mediterranean. Progress in Oceanography, 74(2): 355–376.
38. Rixen M, Beckers JM, Brankart JM, Brasseur P (2001) A numerically efficient data analysis method with error map generation. Ocean Modelling 2: 45–60.
39. Gaylord B, Gaines SD (2000) Temperature or transport ? range limits in marine species mediated solely by flow. The American Naturalist 155: 769–789.
40. Carlotti F, Poggiale JC (2010) Towards methodological approaches to implement the zooplankton component in "end to end" food-web models. Progress in Oceanography, 84: 20–38.
41. Fernandez de Puelles ML, Pinot J-M, Valencia J (2003) Seasonal and interannual variability of zooplankton community in waters off Mallorca island (Balearic Sea, Western Mediterranean): 1994–1999. Oceanologica Acta 26: 673–686.
42. Fernandez de Puelles ML, Molinero JC (2007) North Atlantic climate control on plankton variability in the Balearic Sea, western Mediterranean. Geophysical Research Letters 34, L04608 doi:101029/2006GL028354.
43. Siokou-Frangou I, Christaki U, Mazzocchi MG, Montresor M, Ribera d'Alcalá M, et al. (2010) Plankton in the open Mediterranean Sea: a review. Biogeosciences 7: 1543–1586, doi:105194/bg-7-1543-2010.
44. Nowaczyk A, Carlotti F, Thibault-Botha D, Pagano M (2011) Distribution of epipelagic metazooplankton across the Mediterranean Sea during the summer BOUM cruise. Biogeosciences, 8: 2159–2177, doi:105194/bg-8-2159-2011.
45. Bianchi C, Morri C (2000) Marine biodiversity of the Mediterranean Sea: situation, problems and prospects for future research. Marine Pollution Bulletin 40: 367–376.
46. Beaugrand G (2005) Monitoring pelagic ecosystems using plankton indicators. ICES Journal of Marine Science 62: 333–338.
47. Queiroga H, Blanton J (2004) Interactions between behavior and physical forcing in the control of horizontal transport of decapods crustacean larvae. Advances in Marine Biology 47: 198–214.
48. McPherson E, Raventos N, (2006) Relationship between pelagic larval duration and geographic distribution of Mediterranean littoral fishes. Marine Ecology Progress Series 327: 257–265.
49. Andrello M, Mouillot D, Beuvier J, Albouy C, Thuiller W, et al. (2013) Low connectivity between mediterranean marine protected areas: a biophysical modeling approach for the dusky grouper Epinephelus marginatus. Plos One 8(7): e68564 doi:101371/journalpone0068564.
50. Wernberg T, Thomsen MS, Connell SD, Russell BD, Waters JM, et al. (2013) The Footprint of Continental-Scale Ocean Currents on the Biogeography of Seaweeds. PLoS ONE doi:10.1371/journal.pone.0080168.
51. Guisan W, Thuiller W (2005) Predicting species distribution: offering more than simple habitat models. Ecology Letters 8: 993–1009.
52. Legendre P, Dallot S, Legendre L (1985) Succession of species within a community: chronological clustering, with applications to marine and freshwater zooplankton. American Naturalist 125: 257–288.
53. Olli K, Wassmann P, Reigstad M, Ratkova TN, Arashkevich E, et al. (2007) The fate of production in the central Arctic Ocean – top-down regulation by zooplankton expatriates. Progress in Oceanography 72: 84–113.
54. Chiba S, Di Lorenzo E, Davis A, Keister JE, Taguchi B, et al. (2013) Large-scale climate control of zooplankton transport and biogeography in the Kuroshio Oyashio Extension region. Geophysical Research Letters 40: 5182–5187.

55. Berline L, Zakardjian B, Molcard A, Ourmières Y, Guihou K (2013) Modelling transport and stranding of jellyfish Pelagia noctiluca in the Mediterranean Marine Pollution Bulletin, 70: 90–99.

56. Jribi I, Bradai MN (2012) First record of the Lessepsian migrant species *Lagocephalus sceleratus* (Gmelin,1789) (Actinopterygii: Tetraodontidae) in the Central Mediterranean. BioInvasions Records 1: 49–52 doi:10.3391/bir.2012.1.1.11.

57. Olita A, Cucco A, Simeone S, Ribotti A, Fazioli L, et al. (2012) Oil spill hazard and risk assessment for the shorelines of a Mediterranean coastal archipelago. Ocean & Coastal Management 57: 44–52.

58. Serra IA, Innocenti AM, Di Maida G, Calvo S, Migliaccio M, et al. (2010) Genetic structure in the Mediterranean seagrass *Posidonia oceanica*: disentangling past vicariance events from contemporary patterns of gene flow. Molecular Ecology 19: 557–68.

59. Guizien K, Belharet M, Marsaleix P, Guarini JM (2012) Using larval dispersal simulations for marine protected area design: Application to the Gulf of Lions (northwest Mediterranean). Limnology and Oceanography, 57: 1099–1112 doi:10.4319/lo.2012.57.4.1099.

60. Reygondeau G, Irisson J-O, Guieu C, Gasparini S, Ayata S-D, et al. (2013) Toward a dynamic biogeochemical division of the Mediterranean Sea in a context of global climate change. EGU General Assembly 2013, Vienna, Austria, 07–12 April 2013. Available at: http://meetingorganizer.copernicus.org/EGU2013/EGU2013-10011.pdf.

61. O'Brien TD, Wiebe PH, Hay S (2010) ICES Zooplankton Status Report 2008/2009 ICES Cooperative Research Report No 307 152.

62. Pizzigali C, Rupolo V, Lombardi E, Blanke B (2007) Seasonal probability dispersion maps in the Mediterranean Sea obtained from the Mediterranean Forecasting System Eulerian velocity fields. Journal of Geophysical Research 112, C05012.

63. Treml EA, Halpin PN, Urban DL Pratson LF (2008) Modeling population connectivity by ocean currents, a graph-theoretic approach for marine conservation. Landscape Ecology 23: 19–36.

8

Empirical Evidence Reveals Seasonally Dependent Reduction in Nitrification in Coastal Sediments Subjected to Near Future Ocean Acidification

Ulrike Braeckman[1]*, Carl Van Colen[1], Katja Guilini[1], Dirk Van Gansbeke[1], Karline Soetaert[2], Magda Vincx[1], Jan Vanaverbeke[1]

1 Ghent University, Department of Biology, Marine Biology Research Group, Ghent, Belgium, 2 Netherlands Institute for Sea Research, Department of Ecosystem Studies, Yerseke, The Netherlands

Abstract

Research so far has provided little evidence that benthic biogeochemical cycling is affected by ocean acidification under realistic climate change scenarios. We measured nutrient exchange and sediment community oxygen consumption (SCOC) rates to estimate nitrification in natural coastal permeable and fine sandy sediments under pre-phytoplankton bloom and bloom conditions. Ocean acidification, as mimicked in the laboratory by a realistic pH decrease of 0.3, significantly reduced SCOC on average by 60% and benthic nitrification rates on average by 94% in both sediment types in February (pre-bloom period), but not in April (bloom period). No changes in macrofauna functional community (density, structural and functional diversity) were observed between ambient and acidified conditions, suggesting that changes in benthic biogeochemical cycling were predominantly mediated by changes in the activity of the microbial community during the short-term incubations (14 days), rather than by changes in engineering effects of bioturbating and bio-irrigating macrofauna. As benthic nitrification makes up the gross of ocean nitrification, a slowdown of this nitrogen cycling pathway in both permeable and fine sediments in winter, could therefore have global impacts on coupled nitrification-denitrification and hence eventually on pelagic nutrient availability.

Editor: Kay C. Vopel, Auckland University of Technology, New Zealand

Funding: U.B. was financially supported by FWO project nr G.0033.11 and the Special Research Fund of Ghent University. C.V.C. and K.G. acknowledge a postdoctoral fellowship from the Flemish Fund for Scientific Research (FWO). K.G. also received funding from the European Community's Seventh Framework Programme (FP7/2007-2013) under grant agreement no. 265847. Additional funding was provided by the Special Research Fund of Ghent University (BOF-GOA 01GA1911W). The funders had no role in study design, data collection and analysis, decision to publish, or preparation of the manuscript.

Competing Interests: The authors have declared that no competing interests exist.

* Email: Ulrike.Braeckman@UGent.be

Introduction

Over the past 250 years, the atmospheric CO_2 concentrations have increased by nearly 40% as a consequence of human activities [1]. This increase is partly counteracted by the capacity of the oceans to absorb CO_2, which occurs currently at a rate of about 10^6 metric tons of CO_2 per hour [2]. The latter process led to a decrease in surface-ocean pH by about 0.1 units since the start of the industrial revolution [3]. Climate change models predict a further decrease of 0.35 units by the end of the century for open ocean waters [3,4] and recent measurements for coastal zones even reveal acidification rates that are an order of magnitude higher [5,6].

This decrease in pH is known to have a direct or indirect negative effect on many nektonic, pelagic and benthic organisms [7–10] and thus marine food web structures in general [11]. Although knowledge on the effects of ocean acidification on the level of organisms and populations is increasing fast, the understanding of how biogeochemical processes are affected in acidified seawater is lagging behind [12]. As the ocean carbon cycle is linked with the cycles of major nutrient elements, it is to be

expected that ocean acidification will have large consequences for marine ecosystem functioning [13,14]. With respect to the N-cycle, currently available knowledge suggests that ocean acidification will lead to increased N_2 fixation while nitrification may decrease [13]. This would reduce the supply of oxidized nitrogen substrate to denitrifiers and reduce levels of nitrate-supported primary production that may instigate shifts in plankton communities [13]. Most empirical evidence illustrates that ocean acidification decreases nitrification rates in pelagic environments [15–18] but the consequences for N-cycling in soft-sediments underlying an acidified water column are far less understood. As coastal sediments are spatially and temporally heterogeneous, the available studies do not allow a proper generalisation, e.g. due to the absence of seasonal replication and the limited amount of sediment types investigated [19–25]. Part of the benthic variability in coastal soft-sediments is linked to the macrobenthic communities that vary both spatially (e.g. sediment type) and temporally (e.g. seasonal demographics). Furthermore, the impact of these communities and species on biogeochemical cycling might fluctuate seasonally [26–28] and depends on the environmental

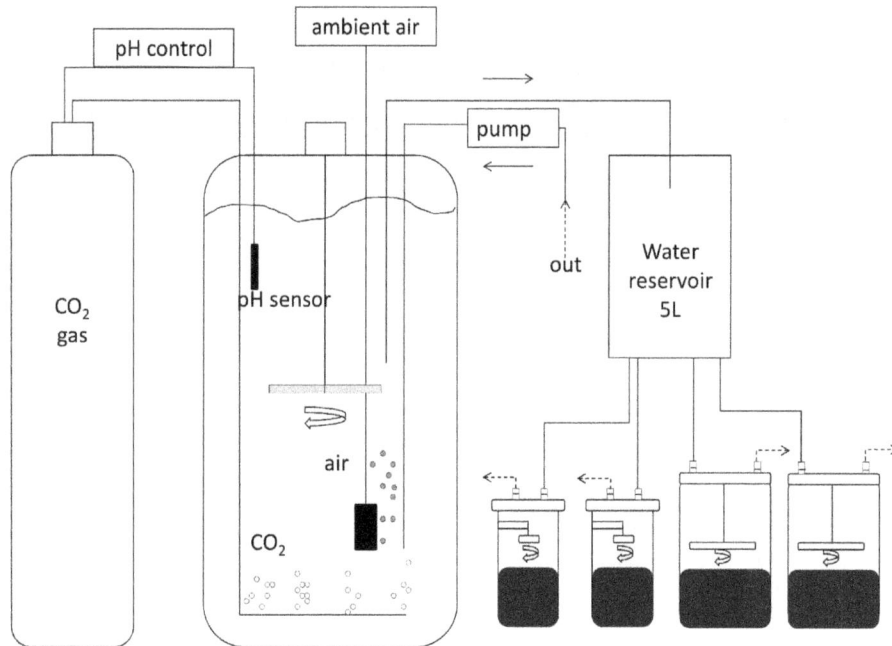

Figure 1. Scheme of experimental set-up. Dashed arrows on core outflow represent tubes that lead to the pump, directing them back into the tank. See also text for explanation. The same set-up without the CO_2-gas was used as a control.

context such as temperature or algal bloom deposition. A decrease in seawater pH has been shown to affect both macrobenthic bioturbation and bio-irrigation activities [18,22,24], which alter redox gradients and microbial communities that regulate the cycling of energy and matter.

To investigate ocean acidification effects on benthic nitrogen cycling and organic matter mineralization, we measured multiple response variables associated with benthic nitrogen cycling in closed core incubations, using two different sediment types in control and manipulated sea water carbonate chemistry. These incubations were executed during two seasons of the year representing the period prior to the annual phytoplankton bloom and the actual bloom period. Nitrification consumes NH_x ($NH_4^+ + NH_3$) and O_2, and produces NO_x ($NO_2^- + NO_3^-$) thus any effect of acidification on nitrification may be reflected in the fluxes of these substances, as well as alter the oxygen penetration depth in the sediment. We constructed integrated mass budgets of O_2, NO_x and NH_x [29] to estimate nitrification rates from the measured fluxes.

Material and Methods

Sampling and experimental set-up

We sampled a fine sandy sediment station (St. 115bis: 51° 09.2'N; 02° 37.2'E, 13 m depth) and a station with coarse, permeable sediment (St. 330: 51° 26.0'N; 02° 48.5'E, 20 m depth), before (February) and during (April) the annual phytoplankton bloom of 2012. Both stations are located in the subtidal part of the Belgian Part of the North Sea, characterized by turbid waters with a light extinction coefficient of 0.36 m^{-1} [30], excluding light penetration to the sea floor, hence also precluding microphytobenthos growth. Sampling was carried out with the *RV Simon Stevin*. At each station, a CTD cast was performed to record the water temperature.

Sediment was collected by means of a Reineck box-core at both stations. At St. 115bis with fine sandy sediment (median grain size

180 μm with 14% of mud; [31]), four separate Reineck box-corers were subsampled with a single Plexiglass tube each (internal diameter ø: 10 cm; H: 25 cm). At St. 330 with permeable sediment (median grain size 360 μm, devoid of mud; permeability: 5.3 10^{-10} m^2, [31]), we measured fluxes in centrally stirred chambers (Plexiglass, ø: 19 cm; H: 30 cm) to create a pore water flow similar to *in situ* conditions [32]. As these chambers were too large to insert in the Reineck box-core, they were filled with homogenized sediment from the upper 10 cm of the sediment from 7 Reineck box-corers. The experimental collections differed between the two sites rendering them not strictly comparable. No specific permits were required for the described field study: the location is not privately-owned or protected in any way and the field study did not involve endangered or protected species.

Sediment cores were transported within 10 h to a temperature-controlled room at *in situ* temperature (recorded from the CTD cast: February: 5°C, April: 9°C) in the lab and immediately submerged in well-aerated sea water in two set-ups (one control set-up and one set-up to be acidified; Fig. 1). Teflon coated magnets rotated by a central magnet were adjusted 5 cm above the sediment surface of the fine sandy sediment cores to ensure water mixing. The rate of water circulation was kept well below the resuspension threshold. Lids equipped with a flat stirring disc were fixed onto the permeable sediment chambers with the stirring disc rotating at 12 rpm 5.4 cm above the sediment surface. From 200 L tanks, a flow-through of seawater through inlet and outlet ports in the lids was ensured with a peristaltic pump (Watson-Marlow 520S), refreshing the core volume ca. 2× h^{-1}. The outflow of the cores was recycled and aerated in the 200 L tanks. Our set-up encompassed duplicate cores for each pH treatment – sediment type combination in each month.

Manipulation of carbonate chemistry

After 24 h of acclimatization, the acidification process was started. To create a predicted pH likely to establish within the current century in coastal waters ($\sim$pH 7.7; [33]), the carbonate chemistry of natural sea water was manipulated through controlled pumping of 100% CO_2 gas at the bottom of a 200 L seawater tank ($\sim$pH 8.0), using Dulcometer technology (ProMinent) connected to a pH electrode (Dulcotest PHE-112SE) mounted in the tank [33] (Fig. 1). This method was chosen as it best replicates ocean acidification by altering dissolved inorganic carbon (DIC) while keeping total alkalinity constant [34]. Every other day the pH electrode was calibrated using Hanna Instruments' pH NBS buffers. The control seawater unit had the same set-up except for the CO_2 gas supply and was bubbled with ambient air.

The sediments were incubated under experimental conditions for 14 days in the dark. During this acclimatization period, temperature, salinity (WTW COND 330), pH (Hanna Instruments), oxygen (Pyroscience needle-type optodes) and alkalinity (Gran titration [35], were monitored daily. Carbonate system parameters (pCO$_2$, dissolved inorganic carbon and calcite and aragonite saturation states) of the overlying water in each tank for the experiment duration were calculated from alkalinity, pH, temperature and salinity in CO2SYS [36] and can be found in Table S1.

Sediment core incubations

We quantified sediment community oxygen consumption (SCOC) and dissolved inorganic nitrogen (DIN) exchange using a direct flux approach by incubating undisturbed sediment cores. From the resulting oxygen and DIN fluxes, we modeled nitrification rates using an integrated mass budget approach [28,29].

One week after acclimatization, depth profiles of oxygen were measured in the sediment with O_2 microsensors (100 μm tip size, Unisense) in vertical increments of 250 μm (3 replicates in each core), up to 10 mm depth. pH profiles were measured in depth intervals of 1000 μm with pH microsensors (tip size 500 μm, Unisense), starting just above the sediment surface downwards, until a constant pH was reached (max. 3.5 cm). A two-point calibration was performed in pH buffers 4.01 and 7.00 before the measurements. Electrodes were connected to a picoammeter (oxygen) or pH-meter and output was displayed on an online PC using SensorTracePro (Unisense) software.

After a 14-day acclimatization period, dark incubations were initiated. Sediment cores were uncoupled from the recirculation system, closed airtight for a period long enough to measure steady changes in oxygen and nutrient concentrations while ensuring that oxygen concentration in the sediment-overlying sea water did not drop below 50% saturation. Oxygen concentration in the sediment-overlying sea water was continuously monitored with Oxygen Spot Sensors (OXSP5, Pyroscience) glued to the inner wall of the tubes. The sensors were operated with an optical oxygen meter (FireStingO2, Pyroscience), connected with a lens-spot adapter (SPADLNS) and Spot Fiber (SPFIB). Incubation time depended on the sediment type and temperature (i.e. longer incubation times for permeable sediment and measurements at low temperatures). Tubes of half the volume of a normal fine sandy sediment tube (i.e. 12.5 cm height) were incubated with tank water to estimate the influence of processes occurring only in the water column. Six (February) or five (April) times, bottom water samples were taken with a glass syringe for the determination of oxygen concentrations (in 12 mL Winkler bottles) and DIN (10 mL), the latter filtered through Whatman GF/F filters. At the same time, tank water was carefully injected into the overlying water of the tubes to compensate for the sampled volume. Tank samples were taken for oxygen and DIN analyses to correct for the

dilution in the tubes during this additional sampling. Oxygen samples were stored in the same temperature-controlled room in the dark until further analysis (within 3 days); DIN samples were stored frozen ($-20°$C).

At the end of the incubations, the tubes were opened and the sediment was subsampled for pigments and organic carbon and nitrogen (each 2 ml from upper 2 cm). The remaining sediment was sieved on a 1 mm mesh to sample the macrofauna, which was preserved in ethanol. Macrobenthos specimens were identified to the lowest possible taxonomic level (typically species level), counted and biomasses were determined (blotted wet weights). Apart from deriving the structural characteristics, density, biomass and species richness, also the functional descriptors Bioturbation Potential index (BPi) and Bioturbation Potential of the Community (BPc) were calculated [37,38].

Laboratory analyses and flux calculation

Pigments were determined by HPLC (Gilson) analysis according to Wright and Jeffrey [39]. Total organic C and N content was analyzed with an Element Analyzer N1500 (Carlo Erba). Oxygen was analyzed by automated Winkler titration [40] and DIN by automated colorimetric techniques (SKALAR). Oxygen and DIN fluxes were calculated from the significant regression slopes of concentration over time compensating for dilution by refill water. Finally, a correction for the processes occurring in the water column was made, by subtracting the rates measured in the water incubations from the total measured rate.

Mass budget modeling

The fluxes of O_2, NO_x, and NH_x across the sediment–water interface were used to estimate rates of nitrification, denitrification and total nitrogen mineralization. This was done by constructing an integrated mass balance of oxygen, nitrate and ammonium over the entire sediment column [29]. See Materials S1 for detailed methodology.

It must be noted that the DIN concentrations of the 'tank water' used in the core incubations (cf. Supra: 2.3 Sediment core incubations) differed from the field DIN concentrations [41]: while NH_x concentrations were rather similar in the field and in the lab, NO_x concentrations were one order of magnitude higher in the tank than in the field. The reported NO_x fluxes should therefore be considered as "potential fluxes".

We chose to make measurements of several aspects of the N cycle rather than analyze a large number of replicates of fewer variables using a variance based statistical approach. The latter strategy allows for detection of significant differences in the measured variables, but we aimed to obtain a more holistic view by performing measurements of different N-cycle related processes covarying with ammonium fluxes (nitrate fluxes, oxygen fluxes) in combination with modeling of individual flux terms subjected to overall mass balance constraints. This allowed us to (1) assess the robustness of single measurements and (2) understand why patterns were observed.

Statistical approach

To test differences in oxygen and pH profiles, a multivariate data matrix was constructed in which each depth horizon is considered a "variable" and each pH measurement a measure of "abundance" [24]. As not all pH profiles were measured to the same depth, including all data would make the design too imbalanced. Therefore, only the major depth horizons were considered (sediment–water interface till 5 mm below the sediment–water interface, in steps of 1 mm). Permutational ANOVAs (Permanova) were carried out to test for differences in

Table 1. Results of Generalized Least Squares for differences in chlorophyll-a content of the sediment, macrobenthic univariate descriptors and measured and estimated sediment processes among the experimental factors pH treatment, Sediment and Month.

	Factor	L-ratio	p-value
Chlorophyll-*a*	pH×Sediment×Month	7.01	0.008
Macrobenthic density	Sediment	13.12	<0.001
	Month	4.63	0.03
Macrobenthic species richness	Sediment	10.28	0.001
Macrobenthic biomass	Sediment	10.33	0.001
Bioturbation potential, BPc	Sediment	12.06	<0.001
	Month	4.09	0.043
SCOC	pH	5.44	0.02
	Sediment×Month	30.48	<0.001
NH$_x$ effluxes	Month	6.42	0.011
	Sediment	29.37	<0.001
NO$_x$ effluxes	Sediment×Month	9.48	0.002
Nitrification	pH×Month	8.07	0.004
	Sediment×Month	18.68	<0.001
Total N mineralization	Sediment×Month	17.10	<0.0001

Only significant results (p<0.05) are shown.

these sediment oxygen and pH profiles. A Euclidean distance similarity matrix was built and subsequently fully crossed 3-way Permanova's were run with factors Month, Sediment and pH Treatment. To test the effect of these factors on macrobenthic community structure, a Bray-Curtis similarity matrix was constructed. In case of significant interaction of factors, pair-wise tests were run to further investigate the observed differences. For sediment oxygen and pH profiles, SIMPER analysis was used to determine which depth was responsible for any differences identified, whereas for macrobenthic community analysis, this test indicated the species that characterized the community. Homogeneity of multivariate dispersion ('variance') was tested with PERMDISP for any of the significant terms in Permanova; if significant, this test indicates that observed patterns can be a result of both treatment and dispersion. The difference in pH in the water tanks during the acclimatization period was also tested with Permanova, using a two-factor design with pH treatment and time.

While testing the effect of pH Treatment, Sediment and Month on the univariate variables chlorophyll-*a*, SCOC, DIN fluxes, estimated nitrification, total N mineralization and macrobenthic characteristics, PERMDISP often pointed at heterogeneity of variances that complicated the interpretation of the significant Permanova's. Therefore, we adopted a linear model with a generalized least-squares extension [42–44], which allows unequal variances among treatment combinations to be modeled as a variance covariance matrix [42,43]. Following West et al. [42] and Zuur et al. [44], the most appropriate variance covariate matrix was determined using AIC scores in conjunction with plots of fitted values versus residuals with different variance covariate terms relating to the independent variables, using restricted maximum likelihood (REML, [42]). This procedure resulted in the use of a variance structure that allowed for different variances per stratum for sediment type and pH treatment (varIdent function, R package nlme). The fixed component of the model was then refined by manual backwards stepwise selection using maximum likelihood (ML) to remove insignificant variable terms. Following Under-

wood [45], the highest order significant interactions in the minimal adequate model were examined, but the nested levels within these were not. The importance of the highest order term was estimated using a likelihood ratio (L-ratio) test to compare the full minimal adequate model with a model in which the relevant variable and all the interaction terms that it was involved in, was omitted. Pair-wise tests were carried out within the R package contrast [46].

All multivariate analyses were carried out within PRIMER v6.0 with Permanova+ add-on software [47,48]. Univariate models were performed in the free statistical environment R (http://cran.r-project.org). Results are shown as mean ± standard deviation.

Results

Evaluation of experimental system

A significant difference in pH of 0.31±0.03 between control and acidified treatments was established in the water tanks in both months (Permanova, pseudo-F >36.8, p=0.001) (Table S1). Temperature and salinity remained constant and oxygen concentration stayed saturated over time. All other seawater carbonate variables remained constant over time as well (Table S1).

Pigments

Chlorophyll-*a* concentrations were affected by the interactive effects of pH Treatment, Sediment and Month (Table 1, 2). In the fine sandy sediments, chl-*a* concentrations were low in February. The bloom deposition was evidenced by higher chl-*a* concentrations at the sediment in April. Chlorophyll-*a* concentrations in permeable sediments were an order of magnitude lower in February, but also increased in April (Table 1, 2).

Macrobenthos

Macrobenthic community structure was only dependent on sediment type (Permanova, pseudo-F = 9.29 p = 0.001) and the two sediment types were strongly dissimilar in terms of macrobenthic community structure (99.91%; SIMPER analysis). The fine sandy sediment community was dominated (94% in terms of

Table 2. Average ± sd of sediment chlorophyll-a, measured and estimated sediment processes and macrobenthic parameters for each of the factors (pH treatment, Month, Sediment or their interactions) with a significant effect.

	pH	Month	Month×pH	Sediment	Sediment×Month	Sediment×Month×pH
Chlorophyll-a	–	–		–	–	Permeable · February · Control 0.05±0.07; Permeable · February · Acidified 0.02±0.03; Permeable · April · Control 0.13±0.07; Permeable · April · Acidified 0.13±0.04; Fine sandy · February · Control 0.29±0.04; Fine sandy · February · Acidified 0.68±0.33; Fine sandy · April · Control 0.81±0.47; Fine sandy · April · Acidified 11.03±4.48
SCOC (mmol O$_2$ m^{-2} d^{-1})	Control 11.73±8.09; Acidified 6.63±7.45			–	Permeable February 5.87±4.98; Permeable April 1.25±2.36; Fine sandy February 10.33±3.57; Fine sandy April 19.26±3.21	–
NHx efflux (mmol N m^{-2} d^{-1})	–	February 1.09±1.55; April 1.00±1.00			Permeable February 0.31±0.39; Permeable April 1.79±1.41	–
NOx efflux (mmol N m^{-2} d^{-1})	–				Permeable February 8.66±1.22; Permeable April 0.25±0.47; Fine sandy February 3.04±2.53	–
Nitrification (mmol N m^{-2} d^{-1})	–		February · Control 3.74±2.21; February · Acidified 0.22±1.14; April 0.87±1.24		Permeable February 3.15±2.36; Permeable April 0.02±0.87; Fine sandy February 1.17±1.72	–
N mineralization (mmol N m^{-2} d^{-1})	–				Permeable February 4.88±1.23; Permeable April 0.42±0.64; Fine sandy February 2.97±0.84	–
Density (ind. m^{-2})	–	February 725±879; April 370±412		Permeable 44±41; Fine sandy 1050±659		–
Biomass (g m^{-2})	–	–		Permeable 3±5; Fine sandy 189±137		–
Species richness (core^{-1})*	–	–		Permeable 1.13±0.99 *per 283.53 cm^2; Fine sandy 3.13±1.25 *per 78.54 cm^2		–
BPc (m^{-2})	–	February 898±1102; April 601±671		Permeable 57±51; Fine sandy 1442±785		–

numbers) by the small surface-modifying *Magelona johnstoni* and
the biodiffusing *Scoloplos armiger* polychaete species and the
surface-modifying bivalve *Macoma balthica* (SIMPER analysis; for
functional group classification see [38]). The permeable sediment
community consisted 94% of the biodiffusing polychaetes *Nephtys
cirrosa* and *Ophelia limacina* and the surface-modifying amphipod
Urothoe brevicornis (SIMPER analysis). Neither structural (density,
biomass, species richness) nor functional (BPc) univariate macrobenthic characteristics were affected by pH treatment. Density,
biomass, species richness and BPc only differed among sediments
(Table 1). The fine sandy sediment clearly displayed higher
macrobenthic density, species richness, biomass and BPc values
than the permeable sediment (Table 2). Density and BPc differed
also marginally among months (Table 1): in February, density and
BPc were higher than in April (Table 2).

pH and oxygen sediment profiles

pH profiles in the upper 5 mm of the sediment (Fig. 2) were
significantly affected by the interactive effects of Sediment and pH
(Permanova: pseudo-F = 17.31, p = 0.001) and Month and pH
(Permanova: pseudo-F = 4.23, p = 0.001). However, for both tests,
some heterogeneity of variances was detected (Permdisp:
p = 0.001), which indicates that there is also a significant dispersion
effect. Pair-wise tests revealed significant differences between pH
profiles of acidified and control treatments in permeable sediments
(p<0.05), and a marginally significant difference (p = 0.054)
between pH profiles of acidified and control treatments in April.
The surface and the first two millimeters made up 60% of the
difference between pH profiles of acidified and control permeable
sediments (Simper). In these layers, the pH was on average
0.24±0.03 units lower than in the respective control surface
sediment layers. In April, the difference between acidified and
control sediments was mainly (54%) found in the deeper 3–5 mm,
where pH values were 0.03±0.03 units lower in the acidified than
in the control sediment.

Figure 3 suggests a deeper sediment oxygen penetration in
acidified fine sediments in February than their respective controls,
but these effects were not significant. Only the interactive effects of
Month and Sediment influenced sediment oxygen penetration
marginally (Permanova: pseudo-F = 2.90, p = 0.042), where the
oxygen content of fine sandy sediments was higher in February
than in April (pair-wise tests: p<0.05) and that of permeable
sediments stayed equal (pair-wise tests: p>0.05).

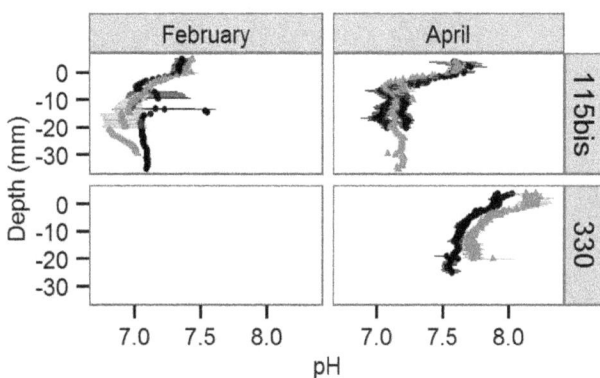

Figure 2. Sediment pH profiles. Mean ± sd pH in control (▲) and
acidified (●) treatments of the two sediment types (115bis: fine sandy;
330: permeable) and two seasons (February and April). No data for
permeable sediment in February.

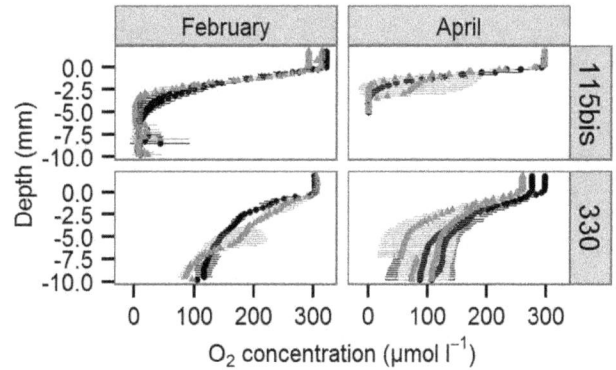

Figure 3. Sediment O₂ profiles. Mean ± sd O₂ concentration in
control (▲) and acidified (●) treatments of the two sediment types
(115bis: fine sandy; 330: permeable) and two seasons (February and
April).

Fluxes at the sediment-water interface

SCOC differed among pH treatment (Table 1, 2; Fig. 4), and
was significantly lower in acidified treatments (59±35%) than in
the respective controls. SCOC also differed among sediments and
months (Table 1, 2). Significantly higher SCOC was measured in
February compared to April in the permeable sediments (pair-wise
tests: p<0.002). In the fine sandy sediments, SCOC was higher in
April than in February (pair-wise tests: p<0.02).

All NH$_x$ fluxes were directed towards the water column, and
only differed between sediments (Table 1, 2; Fig. 4), with lower
effluxes in permeable than in fine sandy sediment, and between
months (Table 1, 2), with slightly higher effluxes in February than
in April. Also note the high variability among NH$_x$ effluxes,
especially in the acidified treatments (Fig. 4).

NO$_x$ fluxes were mainly directed into the sediment and differed
only among months and sediments (Table 1, 2; Fig. 4). In the
permeable sediments, NO$_x$ influxes were highest in February
compared to April (pair-wise tests: p<0.05), whereas they were
equal in both months in the fine sandy sediments (pair-wise tests:
p>0.05).

The nitrification rates estimated by the mass budgets were
affected by the interactive effects of pH Treatment and Month and
Sediment and Month (Table 1, 2; Fig. 4). Only in February,
estimated nitrification rates were significantly reduced by 94% in
sediments underlying an acidified water column compared to
control sediments. Nitrification estimates were significantly higher
in permeable sediments in February compared to April.

Total N mineralization estimates were only affected by the
interactive effects of Sediment and Month (Table 1, 2; Fig. 4).
Total N mineralization estimates were significantly higher in
permeable sediments in February compared to April.

Discussion

Our results indicate that ocean acidification in coastal sediments
reduces sediment community oxygen consumption and nitrification, but does not significantly affect total N mineralization. To the
best of our knowledge, this is the first observation of reduced
sediment community oxygen consumption (60% reduction) in
sediments underlying an acidified water column. SCOC is the sum
of all oxygen consuming processes in the sediment, which include
nitrification, oxidation of reduced substances other than ammonia
and nitrite, but also the oxygen uptake of the fauna present and
bacteria, whether or not stimulated by macrofaunal bio-irrigation.

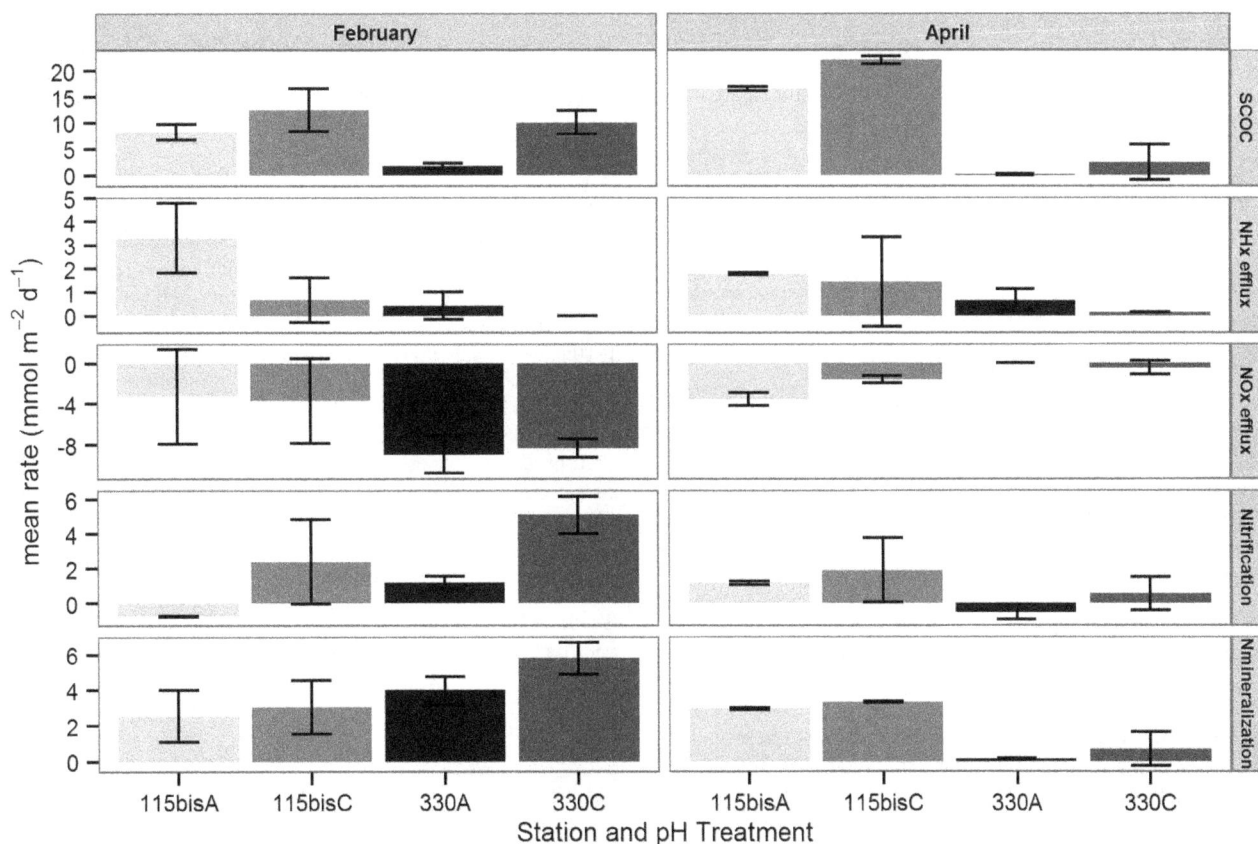

Figure 4. Sediment–water fluxes. Measured Sediment Community Oxygen Consumption (SCOC, mmol O_2 m^{-2} d^{-1}), NH$_x$ and NO$_x$ effluxes and estimated nitrification and total N mineralization (mmol N m^{-2} d^{-1}) (mean $\pm$ sd) from control and acidified treatments of the two sediment types (115bis: fine sandy; 330: permeable) and two seasons (February and April).

The effect of ocean acidification on faunal stimulation of SCOC through bio-irrigation is expected to be minor, since no real bio-irrigators were present in the macrofauna communities and such effects are only expected on the longer term (at least several weeks) [19,21,22]. As total N mineralization remained constant, the decrease in SCOC can be mainly attributed to a lower demand by nitrifying organisms. This effect was found in two biologically and biogeochemically contrasting sediments that were incubated at a realistic scenario of reduced seawater pH by 2100. A 94% reduction in nitrification rate was observed before but not during the annual phytoplankton bloom deposition.

Our results are in contrast with the two earlier studies that did not find a significant reduction in ammonia oxidation (which is the first step in nitrification) in the North Sea, in muddy sediments in January (pre-bloom) [17], nor in non-bioturbated muddy surface sediments in September (post-bloom) [18], nor in permeable sediments in August (post-bloom) [17,18]. These different seasonal and sediment settings might complicate the comparison with our study. On the other hand, other North Sea studies interpreted fluxes of reduced and oxidized N-species across the sediment–water interface, and suggested a decreased benthic nitrification mainly at much more extreme pH reductions (pH range: 5.6–7.6) [19,20,23,24] than we applied.

The potential to detect significant trends at realistic acidification scenarios depends on the accuracy of the applied methods. Measuring fluxes in closed tube incubations and correcting for processes occurring in the water column offers the advantage of a holistic approach and balanced interpretation of the on-going

processes in each sediment core, but there still exists large variability between replicate experimental units (see relatively large error bars on our figures – but nonetheless are the fluxes in this study in the same range as fluxes earlier measured in the area [41] - and the range of variation in Kitidis et al. [17]). As such, the lower nitrification rates in April, especially in permeable sediments, combined with a relatively high variability among replicates, probably prevent a statistically detectable decrease in April. In some acidified treatments (fine sandy sediment in February, permeable sediments in April), estimated nitrification rates were slightly negative, indicating a violation of the mass budget assumption that all reduced substances other than ammonia are reoxidized within the sediment [41]. This suggests that, in addition to ammonia and nitrite oxidation, other oxidation processes are affected by acidification as well. This could have included NH$_4^+$ efflux rates, which were also highly variable among replicates. More dedicated experiments, where also the fluxes of other reduced substances are measured and replication is increased, are needed to enhance our understanding of stressor effects (e.g. acidification) on benthic ecosystem functioning.

How can the observed decrease in estimated nitrification rates be explained? The lower pH imposed on the sea water reduced nitrification in both our sediments, by altering the pH within the sediment matrix. In the fine sandy sediments, we observed steep pH profiles in accordance with earlier results [20,24,49]. These profiles are driven by microbially mediated redox reactions linked to mineralization of organic matter and the dissolution and precipitation of minerals such as $CaCO_3$ (reviewed in [10]).

Hence, benthic microbial communities involved in nitrification in fine sandy sediments are likely more adapted to low pH. In addition, the dissolution of minerals acts as a buffer towards pH changes in the water column [10,50–52]. The very similar pH profiles in our fine sandy acidified and control sediments suggest that these sediments were buffered against water column acidification, which resulted in relatively small effects. In contrast, the pH in the permeable sediment was consistently lowered over the upper 1 cm of the sediment under the acidification scenario (Fig. 3). These pH changes may affect biogeochemical rates in several ways. First of all, the first step in nitrification (ammonia oxidation) requires NH_3 rather than NH_4^+ [53,54]. Since seawater buffers pH by shifting the NH_3/NH_4^+ balance towards the larger NH_4^+ ion at lower pH, a decrease in substrate for ammonia oxidation takes place during ocean acidification. In the presented experiments, a decrease in pH of 0.3 would imply a substrate decline of about 50% [55]. Secondly, large infaunal bioturbators can stimulate nitrification [56], but when these are negatively affected by pH reductions in the overlying water, the stimulating effect disappears and a negative effect on ammonia oxidising microbial organisms is observed [18]. Bioturbation is an instrumental mechanism for ammonia-oxidizing bacteria (AOB) but not for ammonia-oxidizing archaea (AOA) [57]. Consequently, bioturbation-mediated change in AOB communities has been suggested to explain the outcome of ocean acidification effects on ammonium oxidation in a pH range of 7.90–6.80 [18]. Quantification of faunal activity (bioturbation and bio-irrigation) was not specifically targeted in our experiments but we did not observe effects of acidification on the structural and functional characteristics of the macrofauna communities in our sediments, which were not inhabited by strong bio-irrigators or bioturbators. Moreover, bioturbation is not expected to decrease in short-term experiments at realistic pH declines [19,21,22].

We speculate that the observed reduction in nitrification in the present short-term study could result from rapid acidification-induced changes in microbial activity and composition, particularly in terms of the realized ratio of archaea to bacteria [58,59]. Within buffered cohesive sediments, only minor changes in the active benthic total microbial communities from surface sediments were observed in a 2-week acidification experiment [60]. However, in the same experiment, Tait et al. [61] revealed changes in the relative abundance of benthic AOB and AOA under reduced pH circumstances. In addition, under strong acidification scenarios, sedimentary AOB show drastic reductions in amoA transcription, while sedimentary AOA transcripts increase [61]. This suggests that AOB and AOA in marine sediments have different pH optima, and the impact of elevated CO_2 on N cycling may be dependent on the relative abundances of these two major microbial groups [61].

Finally, we should be aware that short-term effects of acidification do not necessarily reflect the net effect of climate forcing; long-term studies more likely include seasonal and other natural variability that bioturbators and the interaction that species undergo [22,62]. As benthic nitrification makes up most of ocean nitrification, a reduction of this nitrogen cycling in both permeable and fine sediments in winter, when coastal benthic nitrification is generally highest [63,64], could have global-wide impacts on coupled nitrification-denitrification and hence eventually on pelagic nutrient availability. High NH_x effluxes could as such precipitate a positive feedback on acidification, if eutrophication is accelerated by the NH_x flux, via oxidation of senescent phytoplankton biomass [65].

Supporting Information

Table S1 Monitoring data of 14 days prior to the incubations. Minimum, maximum, average and standard error of the monitored variables and calculated variables (pCO2, DIC and Ωaragonite and Ωcalcite) in the tanks of the different experiments.

Acknowledgments

We are grateful to the crew of RV Zeeleeuw for help with sampling at sea, to Yves Israël for the development of the experimental set-up, to Joris Bril, Bart Beuselinck, Niels Viaene and Liesbet Colson for help with sample processing.

Author Contributions

Conceived and designed the experiments: JV UB CVC. Performed the experiments: UB JV CVC KG. Analyzed the data: UB JV KS. Contributed reagents/materials/analysis tools: MV DVG KS. Wrote the paper: UB JV CVC KG KS MV.

References

1. Solomon S, Qin D, Manning M, Chen Z, Marquis M, et al. (2007) The physical science basis. Contribution of working group I to the fourth assessment report of the intergovernmental panel on climate change. Cambridge Univ. Press, Cambridge, UK, and New York.
2. Brewer PG (2009) A changing ocean seen with clarity. Proc Natl Acad Sci 106: 12213–12214.
3. Caldeira K, Wickett ME (2003) Anthropogenic carbon and ocean pH. Nature 425: 365.
4. Orr JC, Fabry VJ, Aumont O, Bopp L, Doney SC, et al. (2005) Anthropogenic ocean acidification over the twenty-first century and its impact on calcifying organisms. Nature 437: 681–686.
5. Wootton JT, Pfister CA, Forester JD (2008) Dynamic patterns and ecological impacts of declining ocean pH in a high-resolution multi-year dataset. Proc Natl Acad Sci 105: 18848–18853.
6. Provoost P, van Heuven S, Soetaert K, Laane R, Middelburg JJ (2010) Long-term record of pH in the Dutch coastal zone: a major role for eutrophication-induced changes. Biogeosciences 7: 3869–3878.
7. Andersson AJ, Mackenzie FT, Gattuso J-P (2011) Effects of ocean acidification on benthic processes, organisms, and ecosystems. Ocean acidification. Oxford University Press, Oxford. pp. 122–153.
8. Pörtner H-O, Gutowska M, Ishimatsu A, Lucassen A, Melzner F, et al. (2011) Effects of ocean acidification on nektonic organisms. Ocean acidification. Oxford University Press, Oxford. pp. 154–175.
9. Riebesell U, Tortell PD (2011) Effects of ocean acidification on pelagic organisms and ecosystems. Ocean acidification. Oxford University Press, Oxford. pp. 99–112.
10. Widdicombe S, Spicer JI, Kitidis V (2011) Effects of ocean acidification on sediment fauna. Ocean acidification. Oxford University Press, Oxford. pp. 113–122.
11. Kroeker KJ, Micheli F, Gambi MC, Martz TR (2011) Divergent ecosystem responses within a benthic marine community to ocean acidification. Proc Natl Acad Sci 108: 14515–14520.
12. Gehlen M, Gruber N, Gangsto R, Bopp L, Oschlies A (2011) Biogeochemical consequences of ocean acidification and feedbacks to the earth system. Ocean acidification. Oxford University Press, Oxford. pp. 230–248.
13. Hutchins DA, Mulholland MR, Fu F (2009) Nutrient cycles and marine microbes in a CO_2-enriched ocean. Oceanography 22.
14. Voss M, Bange HW, Dippner JW, Middelburg JJ, Montoya JP, et al. (2013) The marine nitrogen cycle: recent discoveries, uncertainties and the potential relevance of climate change. Philos Trans R Soc B Biol Sci 368: 20130121.
15. Huesemann MH, Skillman AD, Crecelius EA (2002) The inhibition of marine nitrification by ocean disposal of carbon dioxide. Mar Pollut Bull 44: 142–148.
16. Beman JM, Chow C-E, King AL, Feng Y, Fuhrman JA, et al. (2011) Global declines in oceanic nitrification rates as a consequence of ocean acidification. Proc Natl Acad Sci 108: 208–213.

17. Kitidis V, Laverock B, McNeill LC, Beesley A, Cummings D, et al. (2011) Impact of ocean acidification on benthic and water column ammonia oxidation. Geophys Res Lett 38: L21603.

18. Laverock B, Kitidis V, Tait K, Gilbert JA, Osborn AM, et al. (2013) Bioturbation determines the response of benthic ammonia-oxidizing microorganisms to ocean acidification. Philos Trans R Soc B Biol Sci 368: 20120441.

19. Widdicombe S, Needham HR (2007) Impact of CO_2-induced seawater acidification on the burrowing activity of Nereis virens and sediment nutrient flux. Mar Ecol Prog Ser 341: 111–122.

20. Widdicombe S, Dashfield SL, McNeill CL, Needham HR, Beesley A, et al. (2009) Effects of CO_2 induced seawater acidification on infaunal diversity and sediment nutrient fluxes. Mar Ecol Prog Ser 379: 59–75.

21. Wood HL, Widdicombe S, Spicer JI (2009) The influence of hypercapnia and the infaunal brittlestar Amphiura filiformis on sediment nutrient flux- will ocean acidification affect nutrient exchange? Biogeosciences 6: 2015–2024.

22. Godbold JA, Solan M (2013) Long-term effects of warming and ocean acidification are modified by seasonal variation in species responses and environmental conditions. Philos Trans R Soc B Biol Sci 368: 20130186.

23. Murray F, Widdicombe S, McNeill CL, Solan M (2013) Consequences of a simulated rapid ocean acidification event for benthic ecosystem processes and functions. Mar Pollut Bull 73: 435–442.

24. Widdicombe S, Beesley A, Berge JA, Dashfield SL, McNeill CL, et al. (2013) Impact of elevated levels of CO_2 on animal mediated ecosystem function: The modification of sediment nutrient fluxes by burrowing urchins. Mar Pollut Bull 73: 416–427.

25. Gazeau F, van Rijswijk P, Pozzato L, Middelburg JJ (2014) Impacts of Ocean Acidification on Sediment Processes in Shallow Waters of the Arctic Ocean. Plos One 9: e94068.

26. Ouellette D, Desrosiers G, Gagne JP, Gilbert F, Poggiale JC, et al. (2004) Effects of temperature on in vitro sediment reworking processes by a gallery biodiffusor, the polychaete Neanthes virens. Mar Ecol Prog Ser 266: 185–193.

27. Maire O, Duchene JC, Rosenberg R, de Mendonca J, Gremare A (2006) Effects of food availability on sediment reworking in Abra ovata and A. nitida. Mar Ecol Prog Ser 319: 135–153.

28. Braeckman U, Provoost P, Gribsholt B, Van Gansbeke D, Middelburg JJ, et al. (2010) Role of macrofauna functional traits and density in biogeochemical fluxes and bioturbation. Mar Ecol Prog Ser 399: 173–186.

29. Soetaert K, Herman PMJ, Middelburg JJ, Heip C, Smith CL, et al. (2001) Numerical modelling of the shelf break ecosystem: reproducing benthic and pelagic measurements. Deep-Sea Res Part II 48: 3141–3177.

30. Lancelot C, Spitz Y, Gypens N, Ruddick K, Becquevort S, et al. (2005) Modelling diatom and Phaeocystis blooms and nutrient cycles in the Southern Bight of the North Sea: the MIRO model. Mar Ecol Prog Ser 289: 63–78.

31. Van Oevelen D, Soetaert K, Franco MA, Moodley L, van IJzerloo L, et al. (2009) Organic matter input and processing in two contrasting North Sea sediments: insights from stable isotope and biomass data. Mar Ecol Prog Ser 380: 19–32.

32. Huettel M, Rusch A (2000) Transport and degradation of phytoplankton in permeable sediment. Limnol Oceanogr 45: 534–549.

33. Caldeira K, Wickett ME (2005) Ocean model predictions of chemistry changes from carbon dioxide emissions to the atmosphere and ocean. J Geophys Res 110: C09S04.

34. Riebesell U, Fabry VJ, Hansson L, Gattuso JP (2010) Guide to best practices for ocean acidification research and data reporting. Luxembourg: Publications Office of the European Union. 260 p.

35. Dickson AG, Sabine CL, Christian JR (2007) Guide to best practices for ocean CO_2 measurements. Available: http://aquaticcommons.org/1443/.

36. Pierrot D, Lewis E, Wallace DWR (2006) CO2SYS DOS Program developed for CO_2 system calculations. ORNL/CDIAC-105. Carbon Dioxide Information Analysis Center, Oak Ridge National Laboratory, US Department of Energy, Oak Ridge, TN.

37. Solan M, Cardinale BJ, Downing AL, Engelhardt KAM, Ruesink JL, et al. (2004) Extinction and ecosystem function in the marine benthos. Science 306: 1177–1180.

38. Queirós AM, Birchenough SNR, Bremner J, Godbold JA, Parker RE, et al. (2013) A bioturbation classification of European marine infaunal invertebrates. Ecol Evol 13: 3958–3985.

39. Wright SW, Jeffrey SW (1997) High-resolution HPLC system for chlorophylls and carotenoids of marine phytoplankton. Jeffrey Sw Mantoura Rfc Wright Sw Eds Phytoplankton Pigments Ocean Guid Mod Methods Unesco Paris Pp 327–341.

40. Parsons TR, Maita Y, Lalli CM (1984) A manual of chemical and biological methods for seawater analysis. Pergamon Press New York.

41. Braeckman U, Foshtomi MY, Van Gansbeke D, Meysman F, Soetaert K, et al. (2014) Variable importance of macrofaunal functional biodiversity for biogeochemical cycling in temperate coastal sediments. Ecosystems 17: 720–737.

42. West B, Welch KB, Galecki AT (2006) Linear Mixed Models: A Practical Guide Using Statistical Software. Taylor & Francis. 382 p.

43. Pinheiro JC, Bates D (2009) Mixed-Effects Models in S and S-PLUS. Springer. 538 p.

44. Zuur A, Ieno EN, Walker N (2009) Mixed Effects Models and Extensions in Ecology with R. Springer. 580 p.

45. Underwood AJ (1997) Experiments in ecology: their logical design and interpretation using analysis of variance. Cambridge University Press.

46. Kuhn M, Weston S, Wing J, Forester J, Thaler T (2013) Contrast: A collection of contrast methods. R package version 0.19. Available at http://CRAN.R-project.org/package=contrast.

47. Clarke KR, Gorley RN (2006) Primer v6: User Manual/Tutorial. Primer-E Plymouth.

48. Anderson MJ, Gorley RN, Clarke KR (2008) PERMANOVA+ for PRIMER: guide to software and statistical methods. Primer-E Plymouth.

49. Dashfield SL, Somerfield PJ, Widdicombe S, Austen MC, Nimmo M (2008) Impacts of ocean acidification and burrowing urchins on within-sediment pH profiles and subtidal nematode communities. J Exp Mar Biol Ecol 365: 46–52.

50. Wenzhöfer F, Adler M, Kohls O, Hensen C, Strotmann B, et al. (2001) Calcite dissolution driven by benthic mineralization in the deep-sea: in situ measurements of Ca^{2+}, pH, pCO_2 and O_2. Geochim Cosmochim Acta 65: 2677–2690.

51. Jourabchi P, Cappellen PV, Regnier P (2005) Quantitative interpretation of pH distributions in aquatic sediments: A reaction-transport modeling approach. Am J Sci 305: 919–956.

52. Morse JW, Andersson AJ, Mackenzie FT (2006) Initial responses of carbonate-rich shelf sediments to rising atmospheric pCO_2 and "ocean acidification": Role of high Mg-calcites. Geochim Cosmochim Acta 70: 5814–5830.

53. Suzuki I, Dular U, Kwok SC (1974) Ammonia or Ammonium Ion as Substrate for Oxidation by Nitrosomonas europaea Cells and Extracts. J Bacteriol 120: 556–558.

54. Ward BB (1987) Kinetic studies on ammonia and methane oxidation by Nitrosococcus oceanus. Arch Microbiol 147: 126–133.

55. Zeebe RE, Wolf-Gladrow D (2001) CO_2 in Seawater: Equilibrium, Kinetics, Isotopes: Equilibrium, Kinetics, Isotopes. Elsevier. 362 p.

56. Stief P (2013) Stimulation of microbial nitrogen cycling in aquatic ecosystems by benthic macrofauna: mechanisms and environmental implications. Biogeosciences 10: 7829–7846.

57. Laverock B, Tait K, Gilbert J, Osborn A, Widdicombe S (2013) Impacts of bioturbation on temporal variation in bacterial and archaeal nitrogen-cycling gene abundance in coastal sediments. Environ Microbiol Reports 6: 113–121.

58. Wyatt NJ, Kitidis V, Woodward EMS, Rees AP, Widdicombe S, et al. (2010) Effects of high CO_2 on the fixed nitrogen inventory of the Western English Channel. J Plankton Res 32: 631–641.

59. Gilbertson WW, Solan M, Prosser JI (2012) Differential effects of microorganism–invertebrate interactions on benthic nitrogen cycling. Fems Microbiol Ecol 82: 11–22.

60. Tait K, Laverock B, Shaw J, Somerfield PJ, Widdicombe S (2013) Minor impact of ocean acidification to the composition of the active microbial community in an Arctic sediment. Environ Microbiol Reports 5: 851–860.

61. Tait K, Laverock B, Widdicombe S (2013) Response of an Arctic Sediment Nitrogen Cycling Community to Increased CO_2. Estuaries Coasts 1–12.

62. Godbold JA, Calosi P (2013) Ocean acidification and climate change: advances in ecology and evolution. Philos Trans R Soc B Biol Sci 368: 20120448.

63. Kemp WM, Sampou P, Caffrey J, Mayer M, Henriksen K, et al. (1990) Ammonium recycling versus denitrification in Chesapeake Bay sediments. Limnol Oceanogr 35: 1545–1563.

64. Lohse L, Malschaert JFP, Slomp CP, Helder W, van Raaphorst W (1993) Nitrogen cycling in North Sea sediments: interaction of denitrification and nitrification in offshore and coastal areas. Mar Ecol Prog Ser 101: 283–283.

65. Sunda WG, Cai W-J (2012) Eutrophication Induced CO_2-Acidification of Subsurface Coastal Waters: Interactive Effects of Temperature, Salinity, and Atmospheric PCO_2. Environ Sci Technol 46: 10651–10659.

9

Characterizing Air Temperature Changes in the Tarim Basin over 1960–2012

Dongmei Peng[1,2,6]**, Xiujun Wang**[1,3]***, Chenyi Zhao**[1]**, Xingren Wu**[4]**, Fengqing Jiang**[1]**, Pengxiang Chen**[5]

1 State Key Laboratory of Desert and Oasis Ecology, Xinjiang Institute of Ecology and Geography, Chinese Academy of Sciences, Urumqi, Xinjiang, China, **2** Graduate University of Chinese Academy of Sciences, Beijing, China, **3** Earth System Science Interdisciplinary Center, University of Maryland, College Park, Maryland, United States of America, **4** IM System Group and Environmental Modeling Center, National Centers for Environmental Prediction, National Oceanic and Atmospheric Administration, College Park, Maryland, United States of America, **5** Xinjiang Meteorological Observatory, Urumqi, Xinjiang, China, **6** Information Centre of Xinjiang Xingnong-Net, Urumqi, China

Abstract

There has been evidence of warming rate varying largely over space and between seasons. However, little has been done to evaluate the spatial and temporal variability of air temperature in the Tarim Basin, northwest China. In this study, we collected daily air temperature from 19 meteorological stations for the period of 1960–2012, and analyzed annual mean temperature (AMT), the annual minimum (T_{min}) and maximum temperature (T_{max}), and mean temperatures of all twelve months and four seasons and their anomalies. Trend analyses, standard deviation of the detrended anomaly (SDDA) and correlations were carried out to characterize the spatial and temporal variability of various mean air temperatures. Our data showed that increasing trend was much greater in the T_{min} (0.55°C/10a) than in the AMT (0.25°C/10a) and T_{max} (0.12°C/10a), and the fluctuation followed the same order. There were large spatial variations in the increasing trends of both AMT (from −0.09 to 0.43 °C/10a) and T_{min} (from 0.15 to 1.12°C/10a). Correlation analyses indicated that AMT had a significantly linear relationship with T_{min} and the mean temperatures of four seasons. There were also pronounced changes in the monthly air temperature from November to March at decadal time scale. The seasonality (i.e., summer and winter difference) of air temperature was stronger during the period of 1960–1979 than over the recent three decades. Our preliminary analyses indicated that local environmental conditions (such as elevation) might be partly responsible for the spatial variability, and large scale climate phenomena might have influences on the temporal variability of air temperature in the Tarim Basin. In particular, there was a significant correlation between index of El Niño-Southern Oscillation (ENSO) and air temperature of May (P = 0.004), and between the index of Pacific Decadal Oscillation (PDO) and air temperature of July (P = 0.026) over the interannual to decadal time scales.

Editor: Inés Álvarez, University of Vigo, Spain

Funding: This research was supported by the Chinese National Basic Research Key Project (No. 2010cb951001), National Natural Science Foundation (No. 41171095), and National Science & Technology Pillar Program (No. 2013BAC10B01). The funders had no role in study design, data collection and analysis, decision to publish, or preparation of the manuscript.

Competing Interests: The authors have declared that no competing interests exist.

* Email: wwang@essic.umd.edu

Introduction

There have been numerous studies of temporal variations of air temperature at various spatial scales [1–5], which all have pointed to the fact that there has been a general warming trend in the global mean air temperature. Earlier studies indicated that the magnitude of warming in the Northern Hemisphere (0.30°C/10a) was more than double of the one (0.13°C/10a) in the Southern Hemisphere during 1977–2001 [6,7].

There has been evidence of difference in warming trend over space and time in China. For example, the warming rate of air temperature was 0.25°C/10a for the period of 1951–2004 in China [8], but 0.35°C/10a in northwest China during the period of 1961–2006 [9], which were much higher than the global average. Over the past 50 years, an increase of air temperature with a linear tendency of 0.28°C/10a was observed in Xinjiang, which was lower than that for the northwest China [10]. A couple of studies showed that the warming in the Tarim Basin was only from 0.19 to 0.22°C/10a [11,12]. These findings indicated that climate change in Xinjiang might have its own spatial and temporal characteristics due to its large extent and complex terrain.

Apart from the increasing trend in annual mean temperature (AMT), there has been evidence of differences in warming rate between seasons at various regional scales. An earlier study based on analyses of 726 stations' data (1951–2004) in China showed an increasing trend of 0.39°C/10a in winter, 0.28°C/10a in spring, 0.20°C/10a in autumn and only 0.15°C/10a in summer [8]. Some studies indicated that the increasing rate of individual season's temperature was even greater in northwest China, in particular in winter. For example, Wang et al. [13] reported a rate of 0.56°C/10a in winter, 0.35°C/10a in autumn, 0.26°C/10a in spring, and 0.22°C/10a in summer for the period of 1960–2005. A

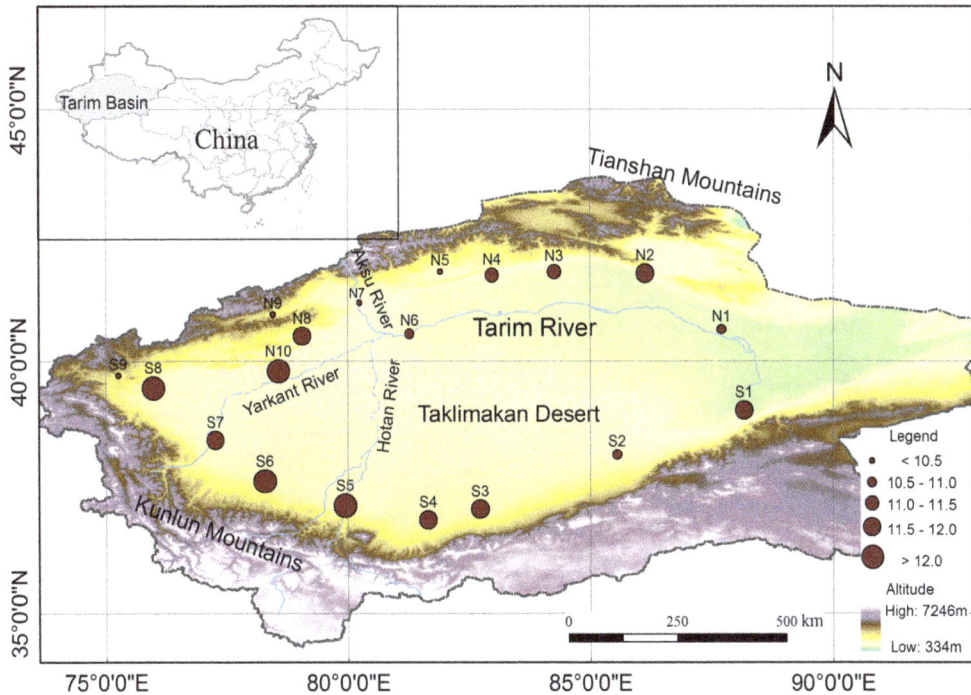

Figure 1. Topographic map with the locations of the meteorological stations in Tarim Basin and climatology AMT for the period of 1960–2012.

recent study by Li et al. [14] showed an increasing trend of 0.49, 0.36, 0.27 and 0.23°C/10a for winter, autumn, spring and summer, respectively, for the period of 1960–2010. These results imply that there are significant changes in the seasonality of air temperature in the vast arid/semi-arid regions. However, there were limited analyses for the Tarim Basin, which were based on data of 1958–2004 from six meteorological stations in the western section of the basin [15], showing an increasing trend of 0.4 and 0.3°C/10a in winter and autumn, respectively.

Large scale climate phenomena often have various impacts on local climate. For example, El Niño-Southern Oscillation (ENSO) has widespread influence on the climate of north America and east Asia. An early study indicated that ENSO had significant effects on precipitation and temperature in the basins in Xinjiang [16]. However, other studies suggested that ENSO had no significant effect on the annual temperature in the Tarim basin [15,17]. On the other hand, there was evidence of interannual to decadal variations in air temperature in the Tarim Basin [17–19]. However, little is done to evaluate the temporal variability of air temperature change and the underlying mechanisms.

While there have been some studies of the climate changes in the Tarim Basin [12,15,20], there is a lack of consistency in findings, primarily due to the difference in the studying period (thus difference in dataset). In addition, detailed analyses of the spatial and temporal variations in air temperature have been lacking for the Tarim Basin, which hampers our understanding of the climate change at local to regional scales. To address this issue, we carry out a detailed study, using the latest dataset covering the period of 1958–2012. Our approaches include anomalies, trend analyses, and fluctuation and correlation analyses. The objective of this study is to investigate the spatial and temporal variability in air temperature at seasonal to decadal time scales, and to explore the possible mechanisms responsible for these changes.

Materials and Methods

Description of study area

The Tarim Basin is located in the southern part of the Xinjiang Autonomous Region, Northwestern China, which is surrounded by high mountains (i.e., Tianshan Mountain and Kunlun Mountain) (Figure 1). The main river system, the Tarim River, consists of the Yarkant River, Aksu River, Hotan River, and Kaidu River. Rivers are primarily fed by snow and glacier melting waters from the mountains. The Tarim Basin (with a total area of 1.02×10^6 km^2) is divided into the mountains (47%, with elevation of 4000–6000 m), plains (22%, with elevation of 800–1400 m), and deserts (31%, with elevation of 1200–1500 m in the western and southern parts, and 800–1000 m in the eastern and northern parts). The climate is typical continental, with an average annual temperature of 10.6–11.5°C and precipitation of 17.4–42.0 mm [21].

Data sources and analyses

Data sources. We obtained daily mean air temperature data of 19 meteorological stations for the period of 1960–2012 from the National Climatic Center, China Meteorological Administration (http://cmdp.ncc.cma.gov.cn/cn/index.htm). These stations have the most completed dataset with acceptable quality control [22]. Figure 1 shows the spatial distribution of these stations and AMT over the period of 1960–2012. As shown in Figure 1, AMT varies from 6.57°C to 12.75°C, with higher values in the southwest stations than in the northeast stations. However, the lowest AMT was found at the stations with higher elevation in the northwest section.

We obtained three common indices of large-scale climate variability, i.e., the Southern Oscillation Index (SOI), the Pacific Decadal Oscillation (PDO), and the Indian Ocean Dipole Mode Index (DMI). The SOI data were obtained from the Climate

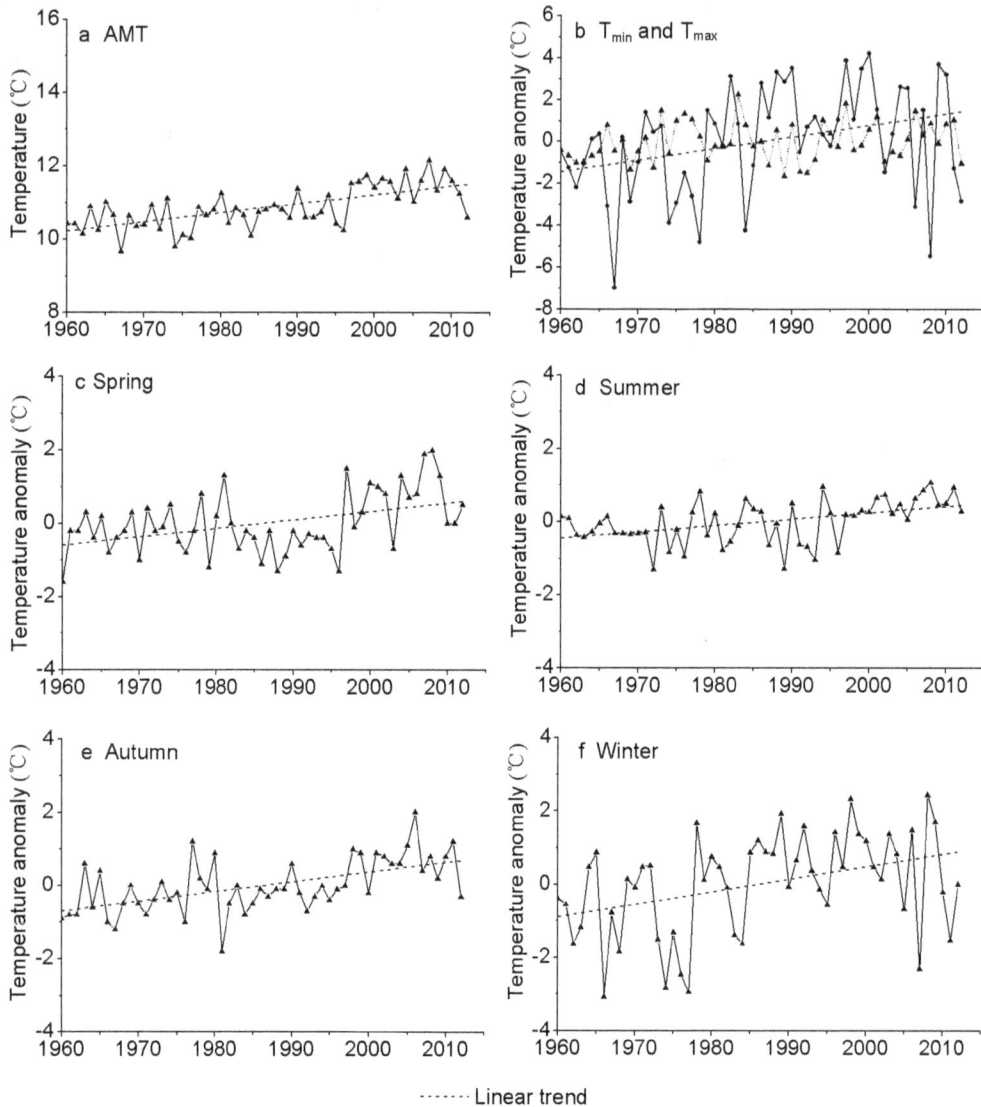

Figure 2. Time series and linear trends of AMT and temperature anomalies in the Tarim Basin. Area averaged (a) AMT, and temperature anomaly for (b) annual minimum (solid line), and maximum (dotted line), (c) spring, (d) summer, (e) autumn, and (f) winter.

Prediction Center (CPC) of the National Weather Service, U.S. (http://www.cpc.ncep.noaa.gov/), PDO index from N. Mantua (http://jisao.washington.edu/pdo/), and DMI from the web site (http://www.jamstec.go.jp/frcgc/research/d1/iod/).

Data analyses. We used daily mean temperature to derive the minimum (T_{min}) and maximum (T_{max}) temperature, and average temperatures for all twelve months, spring (March-May), summer (June-August), autumn (September-November) and winter (December-February) for each year. Our analyses included anomalies, trend analysis, fluctuation analysis, and correlation. All calculations were performed using the SPSS 19.

Temperature anomaly (TA) was determined by removing the mean value $(\overline{T})$ that was calculated for the period of 1960–2012:

$$TA(x) = T_x - \overline{T}_x \qquad (1)$$

where x represented any temperature index (e.g., T_{min} and T_{max}).

Trend analysis was carried out by linear regression that has been a common method to determine the long-term changing trend of air temperature [23]:

$$T^r = T_0^r + S(t - 1960) \qquad (2)$$

where T^r and T_0^r were the predicted temperature for year t and 1960, respectively. The slope S represented the trend of temperature increase, and the strength and significance of the linear increase were indicated by the regression coefficient (R) and P value. In general, when the P value was smaller than 0.05, the trend was significant.

Detrended fluctuation analysis (DFA) has proven useful in revealing the extent of long range correlations in time series [24,25]. We carried out a similar analysis but with a single time window to determine the extent of fluctuation in various temperature indices, which was the standard deviation of the detrended time series of anomaly (SDDA) and calculated as:

Table 1. Trend analyses with regression coefficient (R) and standard deviation of the detrended anomaly (SDDA) of various means of temperature during 1960–2012 and their correlations with AMT.

	AMT	T_{max}	T_{min}	T_{spr}	T_{sum}	T_{aut}	T_{win}
Trend (°C/10a)	0.25	0.12	0.55	0.23	0.17	0.27	0.36
R	0.668**	0.200	0.328*	0.430**	0.454**	0.565**	0.386**
SDDA (°C/10a)	0.43	0.90	2.43	0.74	0.52	0.61	1.26
R with AMT		0.259	0.533**	0.564**	0.616**	0.701**	0.335*

Significance of the regression/correlation was marked with one (P <0.05) and two (P <0.01) asterisks.

$$SDDA = \sqrt{\frac{\sum_{i=1}^{n}(T_i - T_0^r - iS)^2}{n}} \qquad (3)$$

where T_i was the ith temperature, and n the number of data points. In general, the smaller the SDDA value, the smaller the fluctuation is.

Correlation analyses were carried out to examine the relationships between AMT and other temperatures and the potential impacts of climate phenomena over the period of 1961–2013. We used annual means of PDO and DMI indices and the mean value of SOI over winter. We evaluated the relationship between various temperature anomalies with ENSO, DMI and the PDO indices. We also conducted linear and nonlinear regression analyses to examine the relationship between the change rates of AMT or T_{min} and elevation, latitude and longitude. In addition, multi-regressions were carried out to evaluate the influences of local factors and climate phenomena.

Correlation coefficient (R) was used to indicate the strength and direction of the relationship between the observation (x) and prediction (y), which was defined as the covariance of the variables divided by the product of their standard deviations:

$$R = \frac{n\sum xy - (\sum x)(\sum y)}{\sqrt{n(\sum x^2) - (\sum x)^2}\sqrt{n(\sum y^2) - (\sum y)^2}} \qquad (4)$$

where n was the number of pairs of data. The value of R ranged from -1 to $+1$. The + and – signs indicated positive and negative linear correlations, respectively. The same as the regression analysis, the correlation coefficient R was assessed by the P value. When P was smaller than 0.05, the two variables were considered to have a significant correlation.

Results

Temperature change trend and fluctuation

We first evaluated the increasing trend of air temperature averaged over the entire basin. There was a significantly linear increase in the AMT (0.25°C/10a, P <0.001) and T_{min} (0.55°C/10a, P = 0.017) (Figure 2, Table 1). While the increasing trend of T_{max} (0.12°C/10a) was not significant, and weaker than that of T_{min}, SDDA was smaller for T_{max} (0.9) than for T_{min} (2.43) (Table 1), indicating less fluctuation in T_{max} than in T_{min} relative to their change trends. Trend analyses also showed that there was a significantly (P <0.05) linear increase in the mean temperatures of all seasons, with the greatest in winter (0.36°C/10a), followed by autumn (0.27°C/10a), spring (0.23°C/10a) and summer (0.17°C/10a). However, comparisons of regressions indicated that these trends were not significantly different (P > 0.05, data not shown). SDDA was much smaller in summer than in other seasons, which indicated that temperature fluctuation relative to the trend was smaller in summer than in autumn, winter and spring in the Tarim Basin. Correlation analyses indicated that AMT had a significant relationship (P <0.001) with T_{min} and all seasons' mean temperatures, but no significant relationship with T_{max} (R = 0.26, P = 0.061) (Table 1).

There was a considerably spatial variability in the increasing rate of AMT, ranging from -0.09°C/10a to 0.43°C/10a (Figure 3a), with only one station (i.e., N4) showing a decreasing

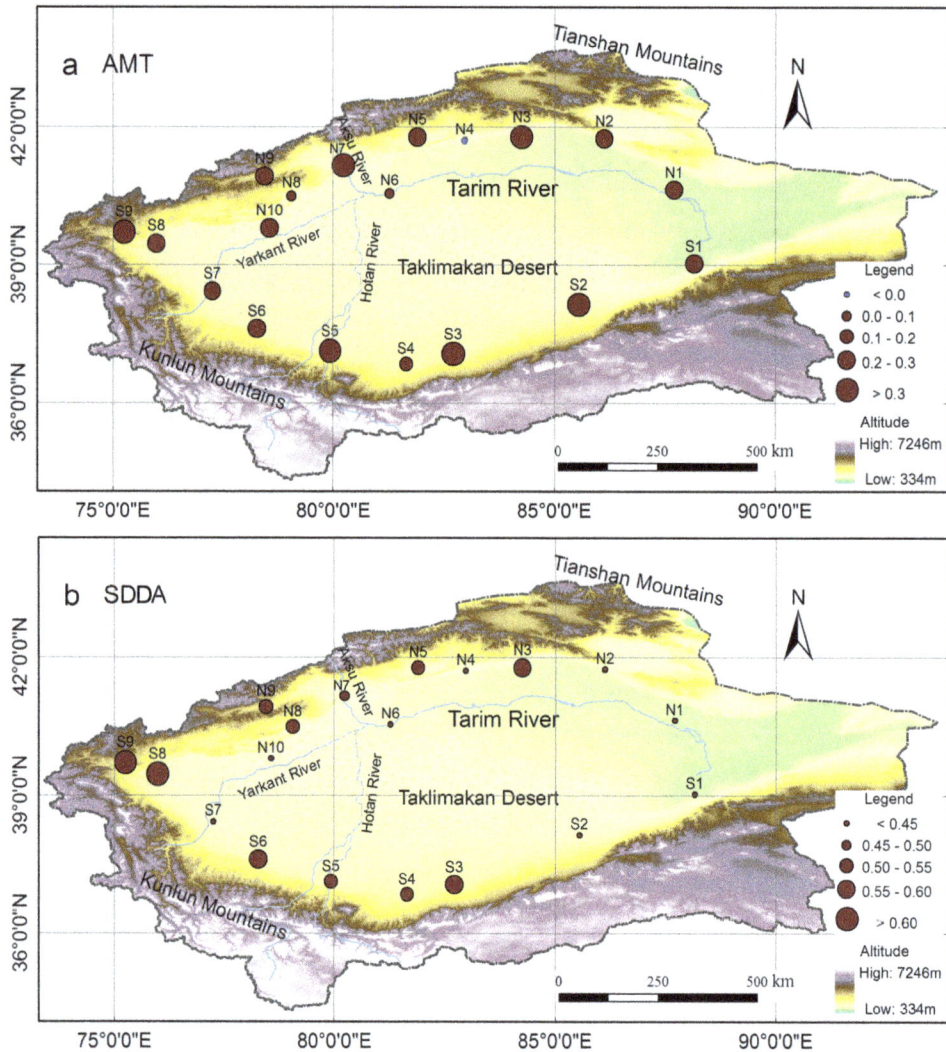

Figure 3. Spatial distributions of linear trend of AMT and its SDDA during 1960–2012 in the Tarim Basin. Spatial distributions of (a) linear trend of AMT (°C/10a) and (b) SDDA (°C/10a) of AMT.

trend (i.e., −0.09°C/10a). While there seemed no clear spatial pattern, the warming rate of AMT was similar (~0.26°C/10a) at all the stations located west of 79°E and east of the 86°E. The highest and lowest warming rates were found between 80°E and 86°E. Unlike the warming rate of AMT, the SDDA of AMT revealed a large spatial variation, with much higher values towards the west. In general, SDDA was larger at higher elevation stations than lower elevation stations, indicating that there would be less fluctuation in AMT in the regions with lower elevations.

Figure 4 shows the spatial pattern of the increasing rate of T_{min} during 1960–2012. There was a big range in the increase of T_{min}, i.e., from 0.15 to 1.12°C/10a. In most parts of the Tarim Basin, T_{min} increasing rate reached 0.55°C/10a, which was almost twice as large as the rate of AMT. Clearly, T_{min} increasing rate was significantly higher in the west than in the east. In general, T_{min} revealed a greater increasing rate to the north than to the south, with very high rates (0.80°C/10a) in the most northern edge of the basin. Overall, the SDDA of T_{min} showed a similar spatial pattern to the increasing rate of T_{min}, i.e., higher values in the northern parts than in the southern parts, and to the west than to the east of the basin. The exception was at the N4 station that revealed low

increasing rate but considerable SDDA for T_{min}. These analyses indicated that although the northwest Tarim had greater increasing rate in T_{min}, there were larger fluctuations in T_{min} of the northwest stations relative to southeast ones.

Changes in the seasonal patterns

To understand the seasonal changes, we analyzed the temperature anomalies of individual month for the past five decades (Figure 5). There were large variations between various periods except for the months from July to October. Clearly, air temperature revealed positive anomalies in all the seasons during the most recent decade, with the strongest anomaly found in March (1.54°C) and weakest one in July (0.28°C). On the other hand, there were negative anomalies in most of the months during the period of 1960–1979, with the most negative values found in the winter (−0.6 to −1.21°C). In contrast, the 1980s and 1990s had generally warm winters, with the warmest January in the 1980s and the warmest December in the 1990s. Overall, the largest decadal variability was found in early spring and late autumn.

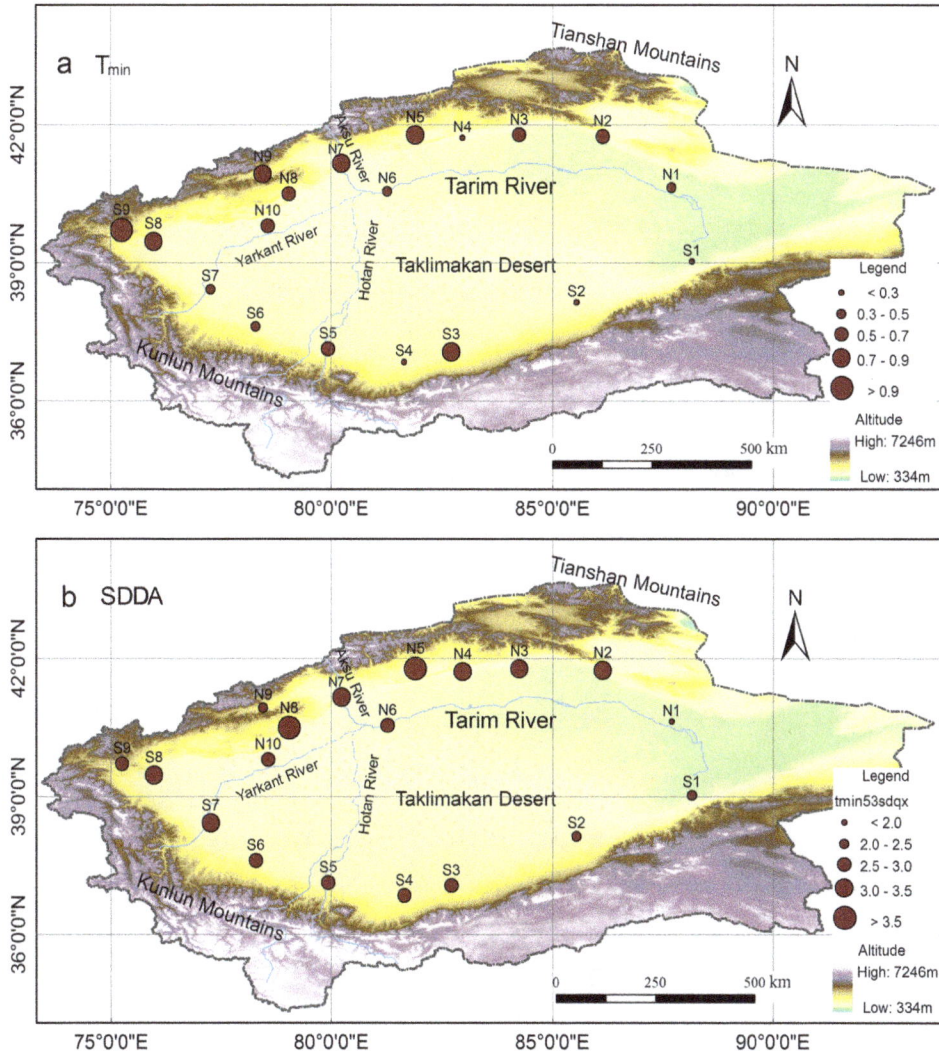

Figure 4. Spatial distributions of linear trend of T$_{min}$ and its SDDA during 1960–2012 in the Tarim Basin. Spatial distributions of (a) linear trend of T$_{min}$ (°C/10a) and (b) SDDA (°C/10a) of T$_{min}$.

We further evaluated seasonal change by looking into the temporal variations of monthly temperature anomalies for early spring and late autumn (Figure 6). The temperature anomaly showed an increasing rate of 0.20°C/10a (P = 0.107) in March and 0.32°C/10a (P = 0.004) in April. There was a similarity in temperature anomaly between March and April except during the period of 1960–1976. The temperature anomaly showed a significant increasing trend in October (0.19°C/10a, P = 0.027) and November (0.41°C/10a, P < 0.001) from 1960 to 2012. Figure 6c illustrated the temperature increase in spring (i.e., the temperature difference between March and April) and decrease in autumn (i.e., the temperature difference between October and November). Overall, the temperature increase from March to April was much greater post 1970 than prior to 1970 although there was a large year-to-year variability during the most recent five years. Our data showed a general decreasing trend in the temperature change from October to November. For example, average temperature drop was 9.48°C in the 1960s but 8.30°C in the 2000s. These results implied the possibility of shortened spring

and autumn with enhanced spring warming and delayed autumn cooling in the Tarim Basin since mid-1970s.

Relationships with climate indices and local environmental conditions

To explore the possible driving forcing for the temperature change in the Tarim Basin, we carried out correlation analyses, which showed that neither of PDO, SOI and DMI had significantly linear relationship with AMT, T$_{min}$, T$_{max}$ or mean temperature of any season in the Tarim Basin. We also conducted non-linear relationship and multi-regression analyses, but found no significant relationship (data not shown). However, Table 2 showed that the correlation coefficient value was 0.264 (P = 0.059) for the relationship between the spring temperature and the SOI, suggesting that ENSO might have influence on the basin's air temperature in spring. Similarly, there was a weak negative correlation between the PDO index and the AMT in the following year (R = −0.212, P = 0.139), implying that PDO might partly be responsible for the decadal variability of air temperature in the Tarim Basin.

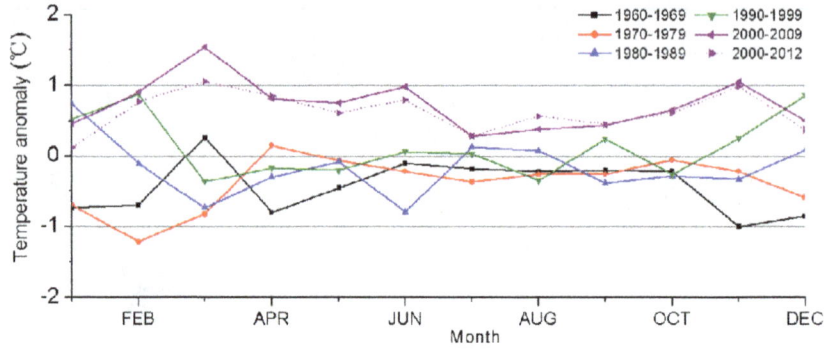

Figure 5. Monthly mean temperature anomalies for different decades in the Tarim Basin.

Table 3 illustrated that elevation, latitude or longitude had little influence on the warming rate of AMT. However, the warming rate of T_{min} showed significant relationship with elevation $(R = 0.535, \ P = 0.018)$ and longitude $(R = -0.528, \ P = 0.02)$, indicating greater warming in T_{min} at higher elevation or west stations. It seems that the significance of combined effects is about the same as that of elevation or longitude.

Discussion

Temporal and spatial variations of temperature in the Tarim Basin

In this study, a focus was placed on analyzing spatial and temporal variability of temperature change in the Tarim Basin. Our study showed that there was a large spatial variability (from

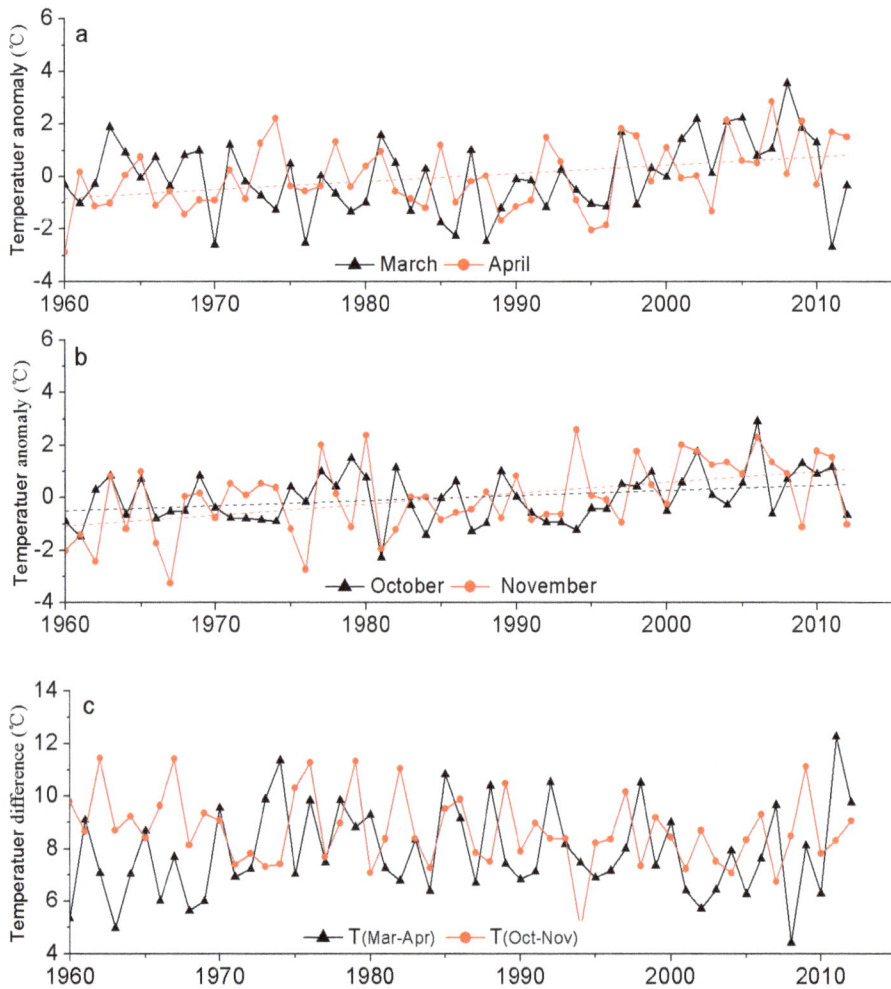

Figure 6. Time series and linear trends of monthly temperature anomalies. Trends for (a) March (black) and April (red), (b) October (black) and November (red), and (c) temperature differences between March and April (black) and between October and November (red).

Table 2. Correlations between climate indices and various temperature means[a].

	PDO		SOI[b]		DMI	
	R	p	R	p	R	p
AMT	−0.212	0.139	0.138	0.338	−0.090	0.950
T_{max}	0.003	0.984	0.218	0.129	−0.190	0.195
T_{min}	−0.066	0.647	0.068	0.640	−0.041	0.784
T_{spr}	−0.175	0.214	0.264	0.059	−0.158	0.269
T_{sum}	0.110	0.437	0.023	0.874	0.156	0.276
T_{aut}	−0.036	0.796	0.036	0.797	0.038	0.791
T_{win}	0.166	0.240	−0.144	0.310	−0.176	0.216

[a]Temperature means for 1961-2012; PDO and DMI were annual means for 1960–2011.
[b]Mean value over winter (e.g., from December 1960 to February 1961).

−0.09 to 0.43°C/10a in the AMT warming rate during the period 1960–2012 in the Tarim Basin. The spatial variability might be a result of many factors because any single factor such as elevation could not explain the variability. The average rate (0.25°C/10a) in our study was higher than those in previous studies of the Tarim Basin (Table 3), such as 0.20°C/10a during 1960–2001 [21] and 0.22°C/10a during 1960–2007 [11]. The differences in the basin scale mean rate mainly reflected the accelerated warming (0.73°C/10a) in the most recent decade, which was consistent with the analysis by Li, Chen, Shi, Chen and Li [26] who reported a greater temperature increase (0.52°C/10a) for the period of 2000–2010 in northwest China.

An earlier study based on only six stations' data from 1958–2004 in the Tarim Basin indicated that there was a significant temperature increase in autumn and winter but not in spring and summer [15]. However, our analyses using data of 1960–2012 showed a significant warming trend in all four seasons, with the highest increasing rate in winter (0.36°C/10a), and the lowest increasing rate in summer (0.17°C/10a). The difference between these two studies was probably due to the significant warming in spring in the most recent decade (Figure 5), and also reflected the large spatial variability in the Tarim Basin.

Our study also showed that the warming rate was much greater in T_{min} (0.55°C/10a) than in T_{max} (0.12°C/10a) during the period 1960–2012, and there was a significant correlation between AMT and T_{min}. These analyses indicated that air temperature increase in winter was largely responsible for the AMT increase. Overall, the seasonality had become weaker, particularly during the period of 1975–2009 in the Tarim Basin. Further analyses seemed to show an enhancement of the early spring warming and a reduction of the late autumn cooling. In addition, the increasing rate in winter (0.36°C/10a) was greater than in summer (0.17°C/10a). All these might be attributable to the weakening of the seasonality in the Tarim Basin.

Comparison between the Tarim Basin and other regions

The global mean surface air temperature had risen by 0.74±0.18°C/100y during the twentieth century and was projected to rise by 1.8–4.0°C in the twenty-first century [4,27]. For the period of 1880–2003, the linear increase of AMT over China was 0.58°C/100y, which was slightly weaker than that of the global mean [28]. However, the warming rate (0.30°C/10a) during the past two decades in China was much stronger than that of the global mean (0.19°C/10a) [29,30]. Table 4 illustrated that warming rate of the AMT was greater than 0.34°C/10a in the northwest China, but less than 0.3°C/10a in Xinjiang. Limited studies, including our study, indicated that on average, warming trend in the Tarim Basin was not as strong as that of China, which might be attributable to the smaller increase of temperature in summer.

Some studies showed an important feature associated with climate warming, i.e., the asymmetric nature over the daily cycle, with less warming observed in daily maximum temperature than in daily minimum temperature [31,32], especially in winter. A similar feature was also found over the seasonal cycle in many regions. For example, Turkes and Sumer [33] reported a significant warming of T_{min} but a weak warming and/or cooling in T_{max} in many regions of Turkey. Studies in China revealed stronger warming in T_{min} (0.32°C/10a) than in T_{max} (0.13°C/10a) [34,35]. While the increasing rate of T_{max} was comparative, the increasing rate of T_{min} was much greater in the Tarim Basin (0.55°C/10a). Similarly, based on data from 19 synoptic stations in the arid and semi-arid regions of Iran for the period of 1966–2005, Tabari [36] found that the increasing trends in the T_{min} series

Table 3. Regressions between the warming rate of temperature (AMT and T_{min}) and local variables (elevation, latitude and longitude).

	AMT		T_{min}	
Variable	**R**	**P**	**R**	**P**
Elevation	0.149	0.543	0.535	0.018
Latitude	−0.211	0.385	0.204	0.402
Longitude	−0.010	0.968	−0.528	0.020
Latitude and elevation	0.230	0.648	0.645	0.014
Latitude and longitude	0.210	0.700	0.612	0.023
Longitude and elevation	0.187	0.752	0.588	0.034
Elevation, latitude and longitude	0.255	0.791	0.689	0.019

(0.44°C/10a) were much stronger than those in the T_{max} series (0.09°C/10a). These results imply that under the global warming, arid and semi-arid regions have less extremely cold days in winter.

Impacts of local environments and large scale climate phenomena

There was evidence of warming rate increasing with the increase of latitude and/or elevation [10,37,38]. However, some studies showed a reduction in the warming rate at high elevations [39,40] or lack of clear relationship between the warming rate and elevation [41]. At the global scale, there was no simple relationship between elevation and warming rate [42]. Our analyses showed that the warming rate of AMT had little relationship with elevation (R = 0.149, P = 0.543) and latitude (R = −0.211, P = 0.385) in the Tarim Basin. However, there was a significant correlation between the warming rate of T_{min} and elevation (R = 0.535, P = 0.018), indicating greater increase of T_{min} at

higher elevation. The facts of significant relationship between AMT and T_{min} and great warming rate of T_{min} over 1960–2012 suggest that elevation may have an influence on the basin scale warming trend.

There have been limited studies that addressed the impacts of the ENSO phenomenon on the temperature in the Tarim Basin, showing inconsistent conclusions [15,17]. Using the latest dataset, our analyses indicated that ENSO events (mainly during December-February) would affect the spring temperature. Further analyses showed that there were lagged effects of both ENSO and PDO on air temperature of the Tarim Basin, i.e., ENSO on the temperature in May (R = 0.392, P = 0.004) and PDO on the temperature in July (R = 0.308, P = 0.026) of the following year.

Figure 7 illustrated that there was a large similarity in the temporal variability for the SOI, mean temperature in May and in spring, particularly post the early 1980s. In addition, the cold ENSO phases (with positive SOI, e.g., 1998–2000, and 2007)

Table 4. Increasing trends in mean air temperatures at various spatial and temporal scales.

Region	Stations	Period	Trend	References
Global		1901–2000	0.6°C/100a	[46]
Global		1906–2005	0.74°C/100a	[47]
Global		1956–2005	0.13°C/10a	[47]
China	726	1951–2004	0.25°C/10a	[8]
China	726	1905–2001	0.5–0.8°C/100a	[47]
Northern China	486	1960–2000	0.2–0.3°C/10a	[48]
Southern China	486	1960–2000	<0.1°C/10a	[49]
Northwest China	135	1960–2005	0.37°C/10a	[13]
Northwest China	138	1961–2006	0.35°C/10a	[9]
Northwest China	74	1960–2010	0.343°C/10a	[49]
Xinjiang	77	1955–2000	1°C/50a	[50]
Xinjiang	65	1961–2005	0.28°C/10a	[10]
Xinjiang	50	1961–2008	0.30°C/10a	[51]
Tarim	26	1960–2001	0.20°C/10a	[52]
Tarim	23	1959–2006	0.1–0.4°C/10a	[53]
Tarim	25	1960–2007	0.22°C/10a	[11]
Tarim	13	1957–2005	0.19°C/10a	[12]
Tarim	19	1960–2012	0.25°C/10a	This study

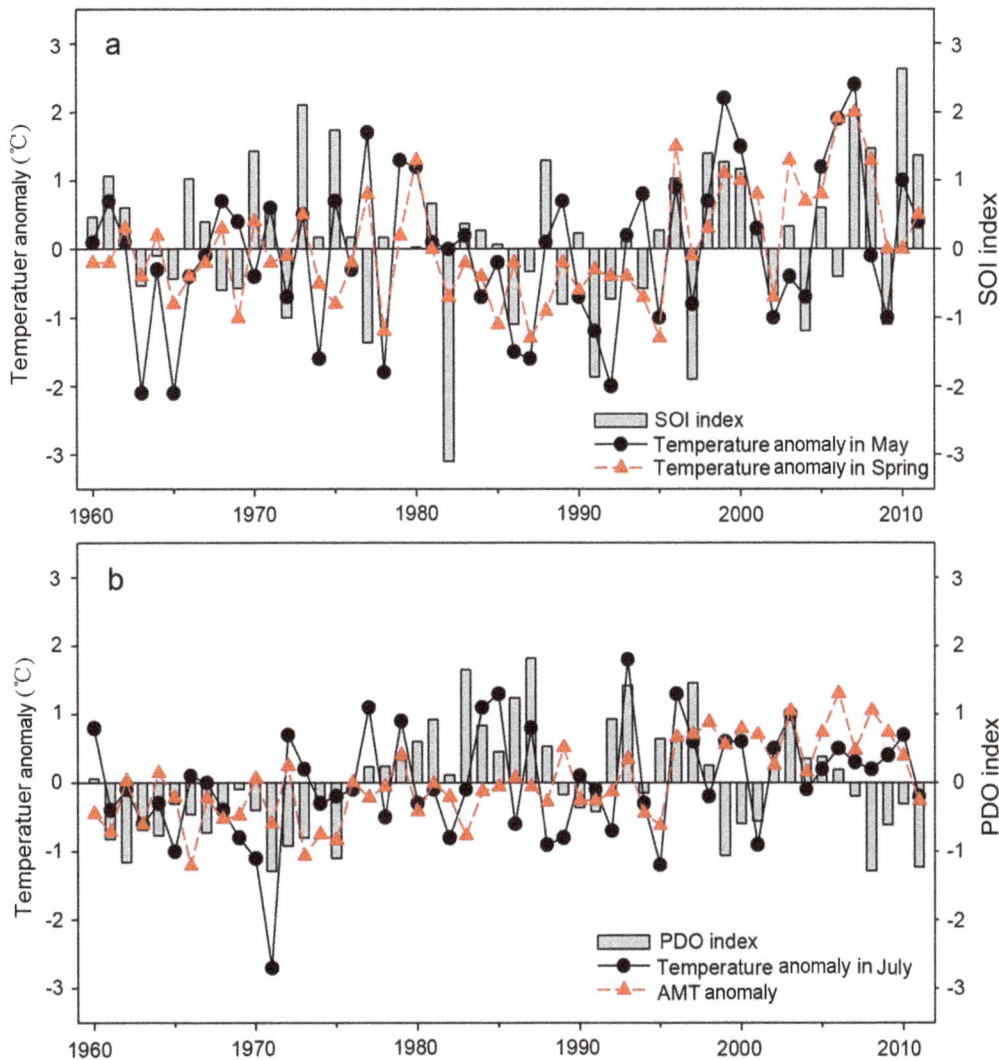

Figure 7. Time series of temperature anomalies and SOI and PDO indices in the Tarim Basin. Time series of (a) SOI and temperature anomalies in spring and May, and (b) AMT anomaly, temperature anomaly in July and PDO index.

during the most recent two decades corresponded with extremely warm temperature in May in the Tarim Basin. It seemed that there was a positive correlation between the PDO index and mean air temperature in July prior to the mid-2000s. For example, both showed an increasing trend from the early 1970s to mid-1980s, indicating a warming trend. However, for the period post late 1990s, PDO revealed mainly cold phases whereas positive anomalies (i.e., strong warming) were seen for both AMT and temperature in July.

There was evidence that the long-term climate variations in winter and summer in China might be connected to the warming trend in the sea surface temperature of the Indian Ocean [43]. Recent studies also indicated that decadal to interdecadal variability in the climate changes in Asia might be related to other climatic phenomena, such as the Arctic Oscillation [44], and Asia-Pacific Oscillation [45]. All these analyses indicated that the relationship between any of the climate indices and air temperature in the Tarim Basin might be non-linear, implying complex impacts of multi factors associated with local environmental driving and remote forcing.

Conclusions and implications

This study demonstrated a significantly increasing trend in air temperature in all four seasons during 1960–2012 in the Tarim Basin. However, there was large spatial and temporal variability in the warming rate. Temperature increase was much greater in the T_{min} (0.55°C/10a) than in AMT (0.25°C/10a) and T_{max} (0.12°C/10a), and warming rate was 0.36, 0.27, 0.23 and 0.17°C/10a in winter, autumn, spring and summer, respectively. The warming was most pronounced in the most recent decade, which might be associated with both PDO and ENSO phenomena that were in cold phases.

There was an overall weakening in the seasonality of air temperature since mid-1970s in the Tarim Basin. Apart from being less cold in winter, spring warming was another feature, which would have impacts on the hydrological cycle in the basin. Particularly, increasing temperature would lead to enhanced melting of snow and glacier in the surrounding mountains, causing extreme runoff events such as floods with a wide range of implications. Future studies are in need to better understand the climate change at various spatial scales and underlying mechanisms, and also to assess the impacts of climate change on environmental and economic aspects.

Author Contributions

Conceived and designed the experiments: XJW XRW. Performed the experiments: DMP. Analyzed the data: DMP PXC. Contributed reagents/materials/analysis tools: CYZ FQJ. Wrote the paper: DMP XJW.

References

1. Jones PD, New M, Parker DE, Martin S, Rigor IG (1999) Surface air temperature and its changes over the past 150 years. Rev Geophys 37: 173–199.
2. Hansen J, Ruedy R, Sato M, Lo K (2002) Global Warming Continues. Science 295: 275.
3. Brown SJ, Caesar J, Ferro CAT (2008) Global changes in extreme daily temperature since 1950. J Geophys Res 113: 115–125.
4. Solomon S (2007) Climate change 2007: the physical science basis: contribution of Working Group I to the Fourth Assessment Report of the Intergovernmental Panel on Climate Change. Cambridge Univ Pr.
5. You Q, Kang S, Aguilar E, Pepin N, Flügel WA, et al. (2011) Changes in daily climate extremes in China and their connection to the large scale atmospheric circulation during 1961–2003. Climate Dynamics 36: 2399–2417.
6. Jones PD, Moberg A (2003) Hemispheric and large-scale surface air temperature variations: An extensive revision and an update to 2001. Journal of Climate 16: 206–223.
7. Luterbacher J, Dietrich D, Xoplaki E, Grosjean M, Wanner H (2004) European seasonal and annual temperature variability, trends, and extremes since 1500. Science 303: 1499–1503.
8. Ren G, Xu M, Chu Z, Guo J, Li Q, et al. (2005) Changes of surface air temperature in China during 1951–2004. Climatic and Environmental Research 10: 717–727.
9. Chen SY, Shi YY, Guo YZ, Zheng YX (2010) Temporal and spatial variation of annual mean air temperature in arid and semiarid region in northwest China over a recent 46 year period. Journal of Arid Land 2: 87–97.
10. Li Q, Chen Y, Shen Y, Li X, Xu J (2011) Spatial and temporal trends of climate change in Xinjiang, China. Journal of Geographical Sciences 21: 1007–1018.
11. Xu Z, Liu Z, Fu G, Chen Y (2010) Trends of major hydroclimatic variables in the Tarim River basin during the past 50 years. Journal of Arid Environments 74: 256–267.
12. Zhou H, Zhang X, Xu H, Ling H, Yu P (2011) Influences of climate change and human activities on Tarim River runoffs in China over the past half century. Environmental Earth Sciences: 1–11.
13. Wang J, Fei X, Wei F (2008) Further study of temperature change in Northwest China in recent 50 years. Journal of Desert Research 4: 723–732.
14. Li B, Chen Y, Li W, Chen Z, Zhang B, et al. (2013) Spatial and temporal variations of temperature and precipitation in the arid region of northwest China from 1960–2010. Fresenius Environmental Bulletin 22: 362–371.
15. Chen Y, Xu C, Hao X, Weihong L, Yapeng C, et al. (2009) Fifty-year climate change and its effect on annual runoff in the Tarim River Basin, China. Quat Int 208: 53–61.
16. Zhang J, Shi Y (2002) Research on Climate Change and Short-term Forecast in Xinjiang. Beijing: China Meteorological Press.
17. Xu Z, Chen Y, Li J (2004) Impact of climate change on water resources in the Tarim River basin. Water Resources Management 18: 439–458.
18. Chen YN, Li WH, Xu CC, Hao XM (2007) Effects of climate change on water resources in Tarim River Basin, Northwest China. Journal of Environmental Sciences-China 19: 488–493.
19. Xu C, Chen Y, Li W, Chen Y, Ge H (2008) Potential impact of climate change on snow cover area in the Tarim River basin. Environmental Geology 53: 1465–1474.
20. Tao H, Gemmer M, Bai Y, Su B, Mao W (2011) Trends of streamflow in the Tarim River Basin during the past 50 years: Human impact or climate change? Journal of Hydrology 400: 1–9.
21. Chen Y, Xu C, Hao X, Li W, Chen Y, et al. (2008) Fifty-year Climate Change and Its Effect on Annual Runoff in the Tarim River Basin, China. Journal of Glaciology and Geocryology 30: 921–929.
22. Li Z, Yan Z (2009) Homogenized daily mean/maximum/minimum temperature series for China from 1960−2008. Atmospheric And Oceanic Science Letters 2: 237–243.
23. Wu S, Yin Y, Zheng D, Yang Q (2007) Climatic trends over the Tibetan Plateau during 1971–2000. Journal of Geographical Sciences 17: 141–151.
24. Király A, Jánosi IM (2005) Detrended fluctuation analysis of daily temperature records: Geographic dependence over Australia. Meteorology and Atmospheric Physics 88: 119–128.
25. Talkner P, Weber RO (2000) Power spectrum and detrended fluctuation analysis: Application to daily temperatures. Physical Review E 62: 150.
26. Li B, Chen Y, Shi X, Chen Z, Li W (2012) Temperature and precipitation changes in different environments in the arid region of northwest China. Theoretical and Applied Climatology: 1–8.
27. Brohan P, Kennedy JJ, Harris I, Tett SFB, Jones PD (2006) Uncertainty estimates in regional and global observed temperature changes: A new data set from 1850. Journal of geophysical research 111: 106–126.

28. Wang S, Gong D (2000) Enhancement of the warming trend in China. Geophysical Research Letters 27: 2581–2584.
29. Wang S, Zhu J, Cai J (2004) Interdecadal variability of temperature and precipitation in China since 1880. Advances in Atmospheric Sciences 21: 307–313.
30. Hansen J, Sato M, Ruedy R, Lo K, Lea DW, et al. (2006) Global temperature change. Proceedings of the National Academy of Sciences 103: 14288–14293.
31. Karl TR, Jones PD, Knight RW, Kukla G, Plummer N, et al. (1993) Asymmetric trends of daily maximum and minimum temperature. Papers in Natural Resources 74: 1007–1023.
32. Liu X, Cheng Z, Yan L, Yin ZY (2009) Elevation dependency of recent and future minimum surface air temperature trends in the Tibetan Plateau and its surroundings. Global and Planetary Change 68: 164–174.
33. Türkeş M, Sümer U (2004) Spatial and temporal patterns of trends and variability in diurnal temperature ranges of Turkey. Theoretical and Applied Climatology 77: 195–227.
34. Liu B, Xu M, Henderson M, Qi Y, Li Y (2004) Taking China's Temperature: Daily Range, Warming Trends, and Regional Variations, 1955–2000. Journal of climate 17: 4453–4462.
35. Wei F (2008) Probability Distribution of Minimum Temperature in Winter Half Years in China. Advances in Climate Change Research 4: 8–11.
36. Tabari H, Hosseinzadeh Talaee P (2011) Analysis of trends in temperature data in arid and semi-arid regions of Iran. Global and Planetary Change 79: 1–10.
37. Holmes JA, Cook ER, Yang B (2009) Climate change over the past 2000 years in Western China. Quat Int 194: 91–107.
38. Wang G, Shen Y, Zhang J, Wang S, Mao W (2010) The effects of human activities on oasis climate and hydrologic environment in the Aksu River Basin, Xinjiang, China. Environmental Earth Sciences 59: 1759–1769.
39. Pepin N, Losleben M (2002) Climate change in the Colorado Rocky Mountains: free air versus surface temperature trends. International Journal of Climatology 22: 311–329.
40. Vuille M, Bradley RS (2000) Mean annual temperature trends and their vertical structure in the tropical Andes. Geophys Res Lett 27: 3885–3888.
41. Pepin N, Lundquist J (2008) Temperature trends at high elevations: patterns across the globe. Geophysical Research Letters 35: L14701.
42. Pepin N, Seidel DJ (2005) A global comparison of surface and free-air temperatures at high elevations. Journal of Geophysical Research 110: D03104.
43. Hu ZZ, Yang S, Wu RG (2003) Long-term climate variations in China and global warming signals. Journal of Geophysical Research-Atmospheres 108.
44. Woo SH, Kim BM, Jeong JH, Kim SJ, Lim GH (2012) Decadal changes in surface air temperature variability and cold surge characteristics over northeast Asia and their relation with the Arctic Oscillation for the past three decades (1979–2011). Journal of Geophysical Research-Atmospheres 117.
45. Zhao P, Yang S, Wang HJ, Zhang Q (2011) Interdecadal Relationships between the Asian-Pacific Oscillation and Summer Climate Anomalies over Asia, North Pacific, and North America during a Recent 100 Years. Journal of Climate 24: 4793–4799.
46. Xu J, Li W, Ji M, Lu F, Dong S (2010) A comprehensive approach to characterization of the nonlinearity of runoff in the headwaters of the Tarim River, western China. Hydrological processes 24: 136–146.
47. Ding Y, Ren G, Zhao Z, Xu Y, Luo Y, et al. (2007) Detection, causes and projection of climate change over China: an overview of recent progress. Advances in Atmospheric Sciences 24: 954–971.
48. Qian W, Qin A (2006) Spatial-temporal characteristics of temperature variation in China. Meteorology and Atmospheric Physics 93: 1–16.
49. Li B, Chen Y, Shi X (2012) Why does the temperature rise faster in the arid region of northwest China? Journal of Geophysical Research: Atmospheres (1984–2012) 117.
50. Hao X, Chen Y, Xu C, Li W (2008) Impacts of climate change and human activities on the surface runoff in the Tarim River basin over the last fifty years. Water Resources Management 22: 1159–1171.
51. Zhang Y, Wei W, Jiang F, Liu M, Wang W, et al. (2012) Brief communication "Assessment of change in temperature and precipitation over Xinjiang, China". Natural Hazards and Earth System Sciences 12: 1327–1331.
52. Liu Z, Xu Z (2007) Spatio-temporal distribution of hydrometeorological variables and their main impact factors in the Tarim river basin. Journal of China Hydrology 5: 020.
53. Xu J, Chen Y, Li W, Ji M, Dong S, et al. (2009) Wavelet analysis and nonparametric test for climate change in Tarim River Basin of Xinjiang during 1959–2006. Chinese Geographical Science 19: 306–313.

10

The Morphometry of Lake Palmas, a Deep Natural Lake in Brazil

Gilberto F. Barroso[1]*, **Monica A. Gonçalves**[2], **Fábio da C. Garcia**[1]

1 Department of Oceanography and Ecology, Federal University of Espírito Santo, Vitória, Espírito Santo, Brazil, **2** Espírito Santo State Water Resources Agency, Vitória, Espírito Santo, Brazil

Abstract

Lake Palmas (A = 10.3 km^2) is located in the Lower Doce River Valley (LDRV), on the southeastern coast of Brazil. The Lake District of the LDRV includes 90 lakes, whose basic geomorphology is associated with the alluvial valleys of the Barreiras Formation (Cenozoic, Neogene) and with the Holocene coastal plain. This study aimed to investigate the relationship of morphometry and thermal pattern of a LDRV deep lake, Lake Palmas. A bathymetric survey carried out in 2011 and the analysis of hydrographic and wind data with a geographic information system allowed the calculation of several metrics of lake morphometry. The vertical profiling of physical and chemical variables in the water column during the wet/warm and dry/mild cold seasons of 2011 to 2013 has furnished a better understanding of the influence of the lake morphometry on its structure and function. The overdeepened basin has a subrectangular elongated shape and is aligned in a NW-SE direction in an alluvial valley with a maximum depth (Z_{max}) of 50.7 m, a volume of 2.2×10^8 m^3 (0.22 km^3) and a mean depth (Z_{mv}) of 21.4 m. These metrics suggest Lake Palmas as the deepest natural lake in Brazil. Water column profiling has indicated strong physical and chemical stratification during the wet/warm season, with a hypoxic/anoxic layer occupying one-half of the lake volume. The warm monomictic pattern of Lake Palmas, which is in an accordance to deep tropical lakes, is determined by water column mixing during the dry and mild cold season, especially under the influence of a high effective fetch associated with the incidence of cold fronts. Lake Palmas has a very long theoretical retention time, with a mean of 19.4 years. The changes observed in the hydrological flows of the tributary rivers may disturb the ecological resilience of Lake Palmas.

Editor: Andrew C. Singer, NERC Centre for Ecology & Hydrology, United Kingdom

Funding: The authors acknowledge Espírito Santo State Agencies for Environment and Water Resources (IEMA) and Agriculture, Aquaculture (SEAG) as well as Linhares Aquaculture Association (AquaLin) for field work logistics. They also would like to acknowledge Espírito Santo State and Federal research supporting agencies FAPES and CNPq, respectively, for Dr. Fábio da Cunha Garcia's researcher fellowship grant. The funders had no role in study design, data collection and analysis, decision to publish, or preparation of the manuscript.

Competing Interests: The authors have declared that no competing interests exist.

* Email: gilberto.barroso@ufes.br

Introduction

Lake morphology has been recognized as a key factor for the understanding of lacustrine structure and function. Since the late 1930s, based on Rawson's diagram [1], lake area and depth contours have been viewed as factors controlling ecosystem productivity due to light penetration, heat balance, oxygen distribution, the input of allochthonous matter, the nature of the sediments and littoral zone development. The influence of the relative shape and size of lake basins on several lake processes has been investigated. These processes include mixing dynamics [2–4], hydrology [5], sedimentation [6], dissolved organic carbon content [7], the biomass of submersed macrophytes [8], primary productivity [9] and lake metabolism [10].

Lake morphology is quantified with morphometric metrics that are descriptors of the form and size of lake basins. This analysis provides crucial knowledge in support of approaches to lake management. Geographical information systems are becoming an important tool to process and analyze morphometric metrics of areas and volumes [11–13].

Geographical location of lakes (latitude, longitude and altitude) must also be considered due to the climatic drivers of insolation, wind and precipitation. Lake typologies for many different geographical settings have been established based on lake morphology and climate [14,15].

In Brazil, with the exception of the extensive study of 61 coastal lakes in the southern portion of the country [16], lake morphometric studies are relatively scarce. Morphometric data are generally available for artificial lakes, particularly in terms of reservoir engineering and management [17]. The deepest natural lake in the country for which data have been published is Dom Helvécio, in the Middle Doce River Valley - MDRV (State of Minas Gerais, Southeastern Brazil), with maximum and mean depths of 39.2 and 11.3 m, respectively [18]. This well-known lake, whose genesis is fluvial, shows a warm monomictic pattern. The lower Doce River Valley (LDRV) (State of Espírito Santo, Southeastern Brazil) has 90 lakes, comprising a valuable water resource that needs sound environmental management.

The present study aims to improve the ecological knowledge of moderate tropical deep lakes through the determination of several morphometric factors for Lake Palmas. A geographic information system (GIS) was developed as an environment for metrics calculation. Wind climate effects on lake stratification and mixing, based on wind direction and intensity, were also integrated in the GIS approach. Vertical water column profiles were developed to explore the relationships between lake morphology and water column stratification.

Materials and Methods

No specific permissions were required to collect hydrographic data and temperature and dissolved oxygen data in Lake Palmas (19°23′S/40°17′W and 19°26′S/40°13′W). In addition, field and lab studies did not involve any biological species.

Physiography of the study area

The LDRV is the location of a district including 90 lakes with areas ranging from 0.8 ha to 62.0 km^2, a total lake area of 165.5 km^2 (Figure 1a and 1b). The LDRV Lake District and its 'lakescape' comprise lakes located in dammed alluvial valleys and lakes located on the coastal plain. According to Bozelli et al. [19], the lakes of the LDRV show both intermittent and dynamic patterns of metabolism. The intermittent pattern is found in the lakes of the alluvial valleys of the Barreiras Formation (Cenozoic, Neogene Period), which are functionally deep and can be described by a seasonal metabolism model. The dynamic metabolism pattern is characteristic of the coastal plain lakes (Cenozoic, Quaternary Period, and Holocene Epoch), which are relatively shallow and are more efficient in processing organic matter.

In the easternmost part of the LDRV, neotectonic processes with patterns of alignment from the NW to the SE control the drainage system of the major river, the Doce River, and its tributary rivers and lakes in the alluvial valleys [20]. According to Martin et al. [21], the Doce River delta and the associated Holocene sedimentation of an ancient lagoon represent a breakthrough process of regional geomorphologic evolution. This process, which started approximately 5,100 yrs B.P., is associated with sea level transgression and regression on the paleodeltaic coastal plain and the damming of the alluvial valleys of the Barreiras Formation with fluvial sediments [21].

The regional climate is characterized by relatively wet and hot summers and dry/mild cold winters. Land use is dominated by pastureland, croplands and *Eucalyptus* forestry. The major areas of urbanization are located in the southeast portions of the District, in the vicinity of Juparanã (62.0 km^2), Meio (1.3 km^2), and Aviso (0.7 km^2) Lakes [22]. In general, lakes water resources have been used for crop irrigation, recreation, fishing and, more recently, for aquaculture. There are two intensive fish farming operations in Lake Palmas, which produce tilapia in floating cages.

Hydrographic survey

The Lake Palmas (Figure 1c) shoreline was screen digitized with the geographic information system ArcGIS 10.1 ESRI (Redlands, California, USA), software licensing EFL615216336, using a digital aerial orthophotograph acquired on May 2008 at a scale of 1:15,000 and a spatial resolution of 1 m, georeferenced with Universal Transverse Mercator (UTM) projection zone 24 k and World Geodetic Datum (WGS) 1984. The polygon shapefile of the lake surface retained the projection and datum of the orthophoto. Echosounding survey lines were plotted on the polygon shapefile at a spacing of 200 m along the longitudinal axes of the lake (Figure 1d). A total of 100.7 km was selected for bathymetric sounding.

A hydrographic survey was performed in May 2011 with an Ohmex (Sway, Hampshire, UK) HydroLite XT echosounding system composed of a 210.0 kHz single beam transducer, a SonarMite V3 BT Bluetooth connection and a Trimble (Sunnyvale, California, USA) GeoXH DGPS receiver. Navigation along the survey lines was oriented with the shapefile of transects displayed with ArcPad 7.0 ESRI on a Trimble Juno GPS receiver. Spatial information was determined according to the Universal Transverse Mercator (UTM) projection and World Geodetic Datum (WGS) 1984. Lake hydrographic surveys were performed at a maximum boat speed of 5.0 km.h^{-1} (2.7 knots).

Bathymetric data processing

X, Y and Z (easting, northing and depth) data were downloaded to Ohmex SonarVista software and then exported as a *dxf* file. In ArcGIS 10.1, the *dxf* files were converted to point features in a shapefile format. The attribute table of the shapefile of depth values was edited to identify and erase depth spikes. The lake shoreline shapefile was converted from a polygon to a point file with a yield of 1,539 shoreline points, which were assigned to zero depth. This shapefile was later merged into the bathymetric survey shapefile.

The interpolation procedure for generating a surface model of the bathymetry data was conducted with Ordinary Kriging using the ESRI extension Geostatistical Analyst 10.1. The process is based on semivariogram modeling, neighborhood search and crossvalidation [23,24]. The resulting bathymetric map was presented with 5.0 m isobaths.

The intensity of the survey (L$_r$) is the ratio between the lake area in km^2 and the echosounding track length in km. The accuracy of the bathymetric map was assessed with the information value (I), which indicates a completely correct map when I = 1. I was calculated as a product of correctly identified area (I') and information number (I''). I' and I'' also vary between 0 and 1, with a value of I' = 1 indicating that all contour lines are correct and a value of I'' indicating that the number of contour lines is optimal. The equations for I, I' and I'', given by Håkanson [11,25], incorporate lake area (km^2), the distance between the sounding tracks (km), shoreline development and the number of bathymetric contour lines. The symbols for the morphometry metrics are based on Hutchinson [26].

Lake size metrics

The maximum depth (Z$_{max}$) was determined from the echosounding points after editing to remove spike data. Lake perimeters, areas, and volumes were calculated with ArcGIS 10.1 routines to determine primary morphometric parameters for lake size, such as lake surface area - A (m^2), shoreline length - L$_0$ (m), maximum length - L$_{max}$ (m), maximum breadth - B$_{max}$ (m) and volume V (m^3).

Lake form metrics

Lake form factors were calculated according to Håkanson [11] as follows: mean depth in m, $Z_{mv} = V/A$; relative depth in %, $Z_r = \left[(Z_{max} * \sqrt{\pi}) / (20 * \sqrt{A}) \right]$; shoreline development index, $L_d = \left\{ (L_0/2) * \left[\sqrt{(\pi * A)} \right] \right\}$; volume development, $V_d = [(3 * Z_{mv})/Z_{max}]$; and mean basin slope in %, $S_{mv} = \left\{ [L_0 + (2 * L_{cot})] * [Z_{max}/(20 * n * A)] \right\}$.

Where, V = lake volume in m^3, A = lake water surface area m^2 in (in km^2 for S$_{mv}$), Z$_{max}$ = maximum depth in m, L$_0$ = normalized shoreline length in km, L$_{ctot}$ = total normalized length for all contour lines in km excluding the shoreline and n = number of contour lines.

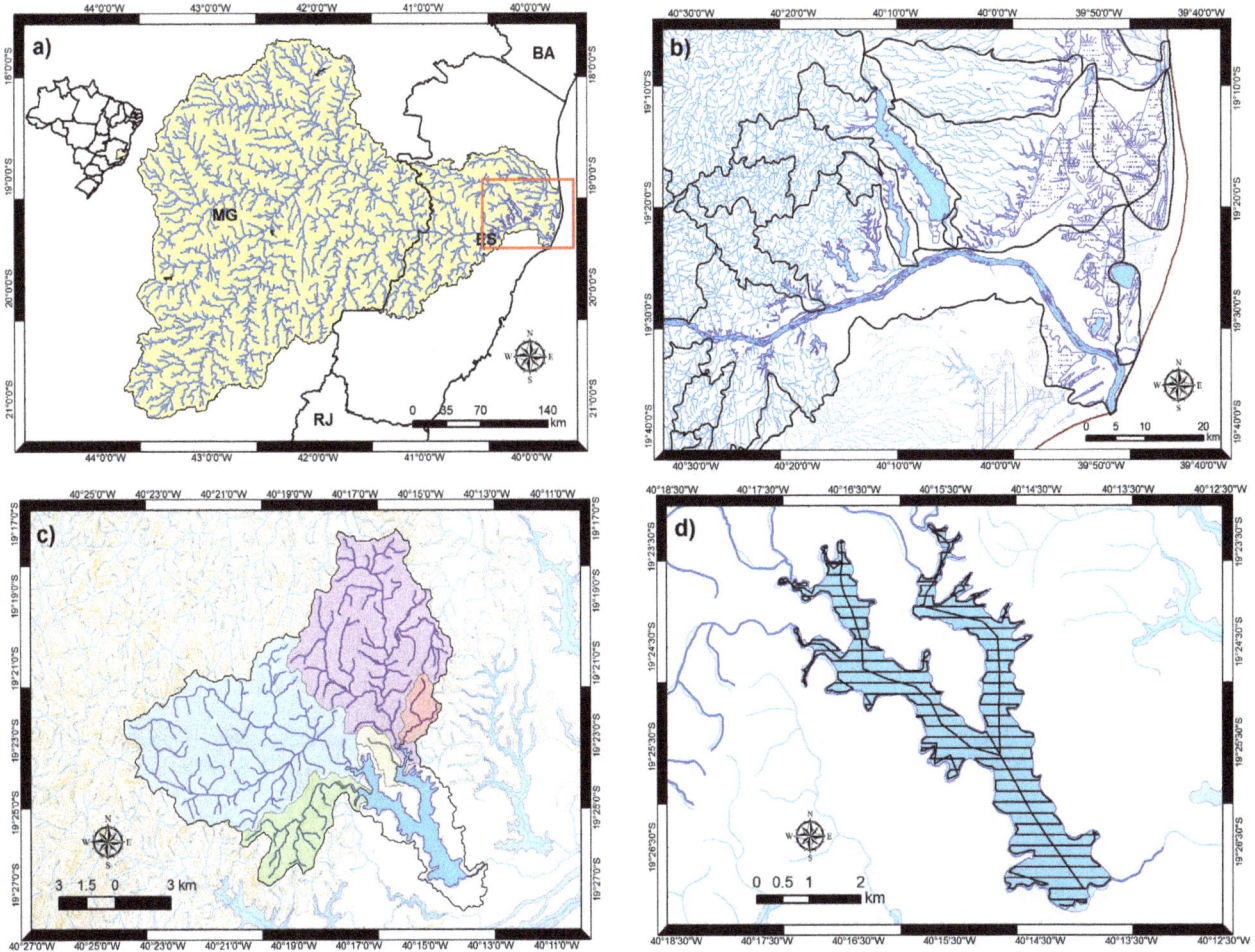

Figure 1. Study area settings: a) LDRV location on the southeastern coast of Brazil in the Doce River Basin (State of Espírito Santo); b) the LDRV Lake District, with its "lakescape"; c) Lake Palmas and its watershed and height curves; d) echosounding transects for hydrographic survey of Lake Palmas.

Water column structure

Water column profiles for temperature (°C), photosynthetically active radiation (PAR) and dissolved oxygen (mg.L^{-1}) were recorded for the field samples from 2011 to 2013 in wet/warm months and in dry/mild cold months. Based on data from 1947 to 2013 from 13 meteorological stations (National Water Agency – ANA, hidroweb.ana.gov.br), the wet months show a regional mean monthly rainfall greater than 100 mm, whereas the regional mean monthly rainfall for the dry months is less than 50 mm (Figure 2). The wet/warm season extends from October to March, with a mean monthly rainfall of 167.6±32.2 mm and a mean air temperature of 24.8±3.25°C. The dry/mild cold season extends from May to August, with a mean monthly rainfall of 46.1±2.5 mm and a mean air temperature of 21.9±3.1°C.

Water transparency was estimated with the depth of Secchi disk. A Horiba (Minami-Ku, Kyoto. Japan) U-53G multiparameter water quality meter with a 30 m cable was used for vertical profiling at 4 sampling sites along the lake axes (Figure 3). Bottom water samples were taken with a Niskin bottle. The extent of the euphotic zone (Z_{eu}) was estimated with underwater light attenuation to a depth corresponding to 1% of subsurface PAR through vertical profiling with a LiCor (Lincoln, Nebraska USA) system with a LI-250A light meter and LI-193 spherical PAR quantum sensor.

The mixing depth (Z_{mix}) in m was calculated based on the maximum discontinuity in the relative thermal resistance (RTR) [27]. The thermal resilience of the water column, based on the Effective Wedderburn number (W_e) [28,29], was calculated for each sampling site at every field sampling event:

$$W_e = \left\{ \left[\Delta\rho_\omega * g'(h_m)^2 \right] * \left[\rho_\omega * (u^*) * L \right]^{-1} \right\}$$

where, $\Delta\rho_\omega$ = the difference between the water mass density at the upper and lower limits of the thermocline (kg.m^{-3}), h_m = mixing depth (m), L = effective fetch in m, g' = the reduced gravity and u^* = the wind friction velocity, g' and u* are calculated with the following equations:

$$g' = \left[(g * \rho')/\rho_0 \right]$$

where, g' is the reduced gravity, g is the normal gravitational acceleration, ρ' is a density perturbation and ρ_0 is a standard reference density;

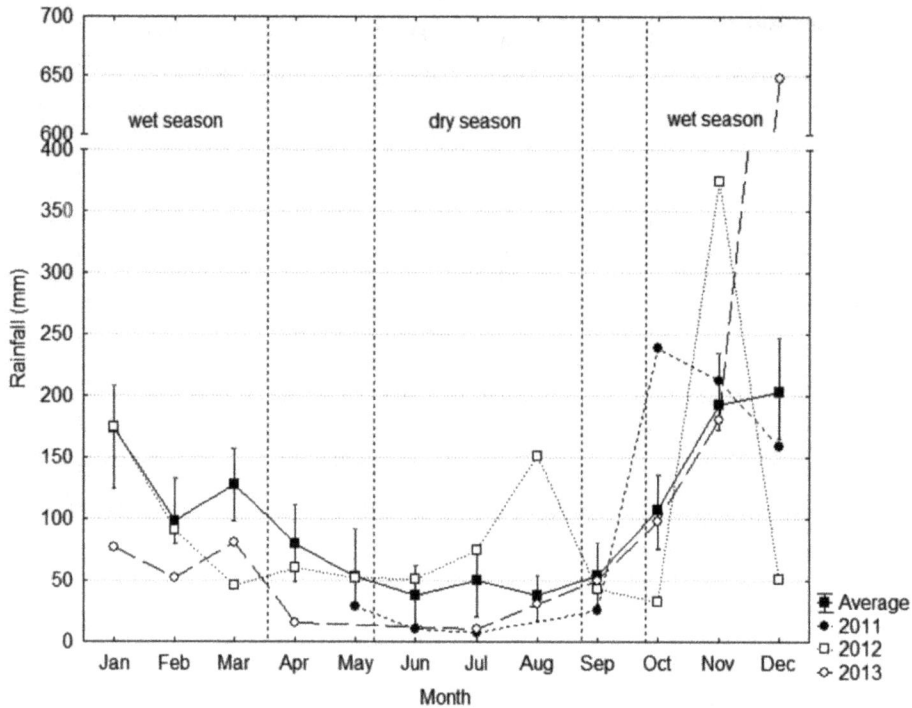

Figure 2. Regional mean monthly rainfall (1947 to 2013) and mean monthly rainfall for the study period (2011 to 2013). Data from 13 meteorological stations (National Water Agency – ANA).

$$u^* = \left[(\rho_{air}/\rho_s) * (C_d * u^2) \right]$$

Where, ρ_{air} is the air specific mass in kg.m^{-3}, ρ_s is the water specific mass at the lake surface in kg.m^{-3}; C_d is the coefficient of friction (0.0015) and u is the wind velocity in m.s^{-1}. If $W>1$, the thermal structure is stable; if $W<1$, the water column is susceptible to changes resulting from the effects of wind.

Hourly wind direction and intensity data were obtained from observations at the Linhares meteorological station (INMET - A614), approximately 18.0 km NE of Lake Palmas.

Special metrics for lake morphometry

The A/V ratio was calculated to estimate the potential evaporation rate of lake water and the resistance of the water column to mixing. The slope of the lake basin was modeled in terms of the percent rise function using ArcGIS 10.1 with the function Slope, 3D Analyst Tool. The continuous surface model for the basin slope was reclassified in terms of gentle and steep slopes. According to Duarte and Kalff [8], a slope of 5.3% is the threshold value separating gentle and steep slopes in relationship to the development of submersed rooted aquatic vegetation in lakes.

The wave base depth (Z_{wb}) in meters was calculated from the $Z_{wb} = \left[(45.7 * \sqrt{A})/(21.4 * \sqrt{A})\right]$, with A in km^2 [11]. Z_{wb} was the depth used to estimate the volume of the epilimnetic waters. The delimitation of the littoral and pelagic zones and their respective volumes was performed based on the mean depth of the euphotic zone (Z_{eu}) in meters. The area suitable for the development of rooted aquatic vegetation biomass was also based on the mean Z_{eu} and within gentle slopes (<5.3%). Volumes in m^3

for hypoxic/anoxic bottom waters were determined based on the depth corresponding to hypoxia (<2.0 mg.L^{-1}) during the stratification season.

The effective fetch, L_{ef} (km) and the wave heights (m) were estimated for 46 sites distributed along the lake surface with a grid with equal distances of 500.0 m. To estimate L_{ef}, distances from each site to the shoreline were measured according to the prevailing winds (defined as an angle of 0°) and every 6° on both sides of the 0° angle to 42° [11]. The below provides the integrated value of L_{ef}:

$$L_{ef} = \left[\Sigma xi \cos(ai)/(\Sigma xi \cos(ai) * SC') \right]$$

where, $\Sigma cos(a_i)$ is 13.5, a calculation constant, and SC' is the map scale constant of 0.35.

Wave heights (H) in m were computed for the sites for which L_{ef} was estimated, according to the Beach Erosion Board (1972) in Håkanson [11] using the following equation:

$$H = \left[0.105 * \left(\sqrt{L_{ef}} \right) \right].$$

Both L_{ef} and H were calculated based on two prevailing wind directions, one for the wet/warm season (NE) and one for the dry and wet season (SE, considering the major axis of the lake). Maps of L_{ef} and H maps were created using GIS, interpolating the point data with a spline function with tension, a neighborhood of 5 points, a weight of 0.01 and a cell resolution of 5.0 m.

The basin permanence Index (BPI) (m^3.km^{-1}), which indicates the littoral effect on basin volume, is calculated according to the ratio of lake volume ($\times 10^6$ m^3) to shoreline length (km), $BPI = V/D_L$ [30]. The dynamic ratio (DR) was calculated according to the equation: $DR = \left[\sqrt{A}/Z_{mv} \right]$ with A in km^2 [11].

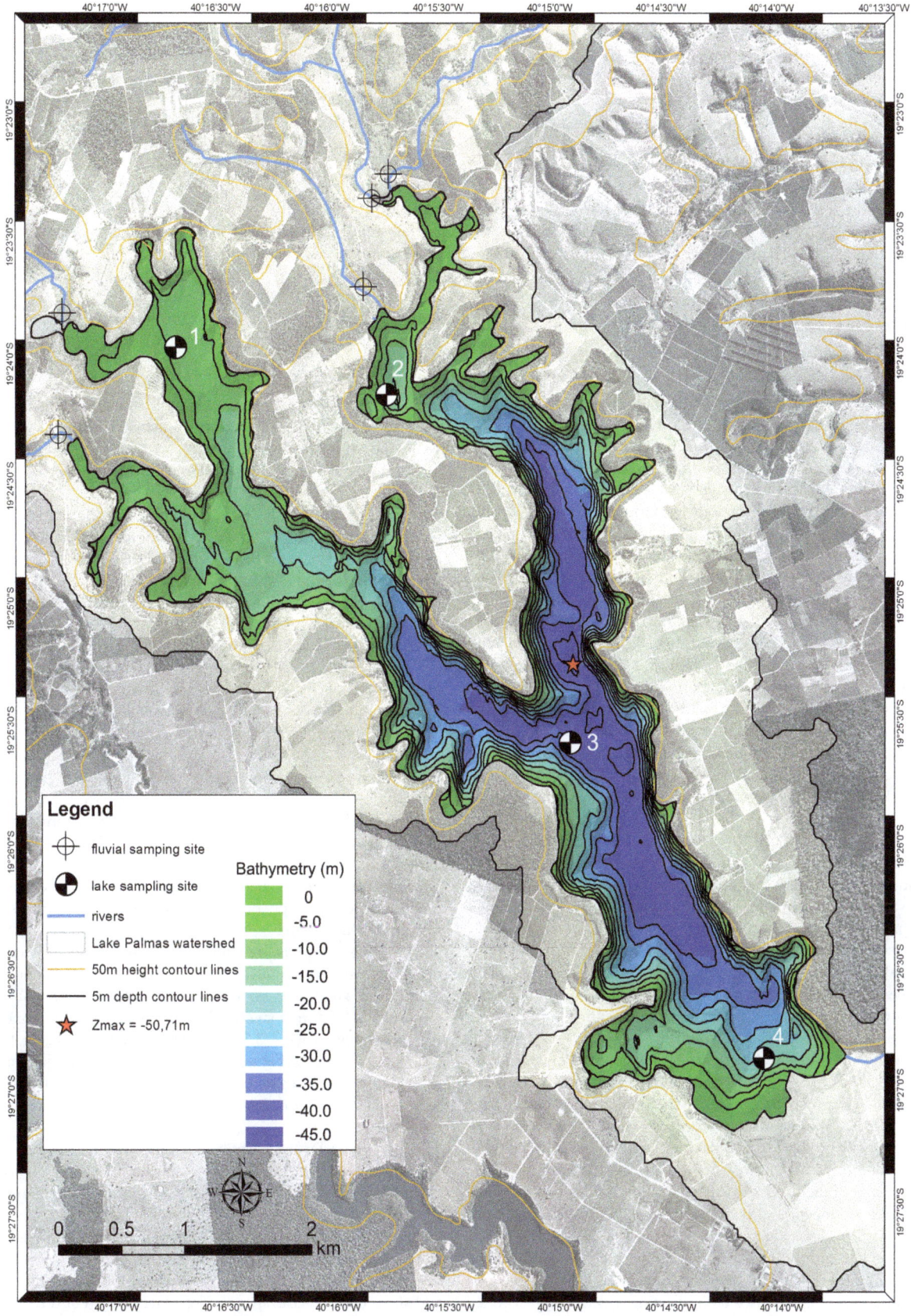

Figure 3. Lake Palmas bathymetric map.

To assess the cryptal depth (Z_c) and cryptal volume (V_c) of the lake, depth values were converted according to the altitude of the lake surface above sea level, using an altitude of 20 m as the reference value for the lake surface.

The theoretical lake water retention time was calculated according to the ratio of the lake volume (m^3) to the mean annual river tributary inflow ($m^3 \cdot s^{-1}$) $RT = V/Q_{mean}$ [41,42]. Discharge of the five tributary rivers (Figure 3) were measured during the wet/warm and dry/mild cold seasons ($n = 8$) with a SonTek (San Diego, California, USA) FlowTracker Handheld Acoustic Doppler Velocimeter (ADV). Mean annual river tributary discharges were then calculated, as well as discharge values for dry and wet seasons.

Results

The total bathymetric sounding survey track was 122.9 km, yielding a survey intensity (L_r) of 0.08. A total of 46,941 valid depth points were computed. Ordinary kriging to obtain prediction results was applied as the interpolation method to yield a continuous surface of lake depths. A neighborhood search was used, considering a smooth type within an axis range between 100 to 2,000 points. Variogram modeling was based on 9 lags with a size of 280 and a spherical model with anisotropy (a direction of 125°). The regionalized variation of point data, optimized sampling and spatial pattern determination is addressed with the Semivariogram on Figure S1. The bathymetric map, with a cell size of 10×10 m (Figure 3).

Based on depth contour intervals of 5 m, the correctly identified area (I') is 0.8571. This value means that 85.7% of the lake area was correctly identified and that 14.3% (1.5 km²) of the lake area was incorrectly estimated. The information number (I') for the 5 m contour lines was 0.9995, and the information value (I), indicating the overall map accuracy, was 0.8566.

The lake basin has a 'Y' shape aligned in a NW-SE direction, with a maximum length (L_{max}) of 7.1 km and an average breadth (B_{mv}) of 1.7 km. Other lake basin size metrics were a total surface area of 10.3 km², a shoreline length (L) of 51.9 km, a maximum depth (Z_{max}) of 50.7 m and a volume of 2.2×10^8 m³ (0.2 km³).

The lake form metrics were found to have the following values: the shoreline development index (D_L) was 4.5, the mean depth (Z_{mv}) was 21.4 m, the relative depth (Z_r) was 1.4%, the volume development (V_d) was 1.3, the mean slope (S_{mv}) was 15.8% and the $Z_{mv}:Z_{max}$ ratio was 0.42. These metrics indicate a flat-bottomed, overdeepened lake basin. Based on the value of D_L, the shoreline form is subrectangular elongate. The basin form is linear (L) according to the relative hypsographic area and volume curves (Figure S2a and S2b) as well as according to V_d.

There are 18 embayments along the lake axis, most of which are less than 15.0 m in depth. The area deeper than 40.0 m extends from the intersection of the two lake axes to the S shore and to the upper N axis. The three major deepest basins (>45.0 m) are located midway on the N-S axis next to the intersection of the two lake axes, representing an area of 1.5×10^5 m² (1.5% of the lake area) and a volume of 5.1×10^5 m³ (0.02% of the lake volume).

The lake drainage is also oriented from NW to SE, with 5 tributary streams located along the upper NE and N shores. Lake Palmas discharges into the Doce River through a drainage river located along the SW shore.

The area/volume ratio is 0.05, indicating a deep basin with a small littoral zone. The mean basin slope (S_{mv}) of 15.8% represents an overall value for the shallow areas, central basin plain and steep lateral slopes. The basin slope GIS model (Figure 4a) shows steep lateral slopes, up to 112.8%, along the central E shore as well as along the E and W shores of the promontory that separates the lake into two arms. The central basin plain is constrained with slopes lower than 20% and depths greater than 30 m. Shallow areas with a depth of less than 5 m and a slope of 5% are located at the mouths of the tributaries as well as the southernmost part of the lake. Very gentle slopes up to 2.0% are characteristic of deep basins (>40 m deep). Based on a threshold of 5.3%, in which fine sediments are retained [8,11], 9.1 km² (58.8% of the lake area) and 4.3 km² (41.3%) were classified as steep and gentle slopes, respectively (Figure 4b).

Data from field vertical profiling at 4 sampling sites (Table S1) show stratification during the warm months, with surface water temperatures reaching 31.3°C (Figure 5a), and a mixed water column during the mild cold months, with a mean water temperature of 23.2°C. The Effective Wedderburn (W_e) values, an overall indicator of the thermal stability of the water column, were 6.5 ± 7.2 during the wet/warm season and zero in the dry/mild cold season. During the dry season, the mean Z_{mix} value of 20.7 ± 8.3 m indicated the presence of a deep mixing layer but with a very weak, i.e., unstable, stratification due to the zero W value and an RTR lower than 3.1.

During the season of stratification, Z_{mix} was usually shallower or at least equal to Z_{eu} (Figure 5a), yielding higher $Z_{eu}:Z_{mix}$ ratios. In contrast, the Secchi disk depth (Z_{Sd}) was higher during the mixing season. Based on a mean Z_{eu} of 10.0 m, the volume of the euphotic layer (V_{eu}) was 9.2×10^7 m³ (41.4% of the lake volume), whereas the volume of the aphotic zone was 1.3×10^8 m³ (58.6%).

During the stratification season, hypoxic/anoxic conditions may develop below a depth of 13 m (Figure 5b). Under these conditions, the volume of anoxic waters may reach 1.1×10^8 m³ or 48.6% of the lake volume. Bottom hypoxia/anoxia was recorded during the entire wet/warm season. In contrast, DO is well distributed in the water column throughout the mixing season (Figure 5c), even showing supersaturation at the surface (Figure 5d).

Special morphometry metrics show that the lake basin has a wave base depth (Z_{wb}), an indicator of the depth of turbulent mixing, of 6.0 m, with a surface layer volume (i.e., a mixing layer) of 5.8×10^7 m³, or 26.1% of the lake volume. Thus, the volume of bottom waters was 1.6×10^7 m³, or 73.9% of the lake volume. The dynamic ratio (DR) had a value of 0.15 indicating the predominance of slope processes over wind/wave processes in sediment resuspension.

The Basin Permanence Index (BPI), 4.3 $m^3 \cdot km^{-1}$, indicates that Lake Palmas is relatively less suitable for the development of the littoral zone and rooted aquatic plants. With the same threshold of 10.0 m for Z_{eu}, the littoral and pelagic zones are represented by 8.1×10^6 (3.7% of the lake volume) and 2.1×10^8 m³ (96.3%), respectively (Figure 6a). The predicted potential areas for rooted submersed vegetation with nearshore gentle slopes comprise only 7.9×10^5 m², less than 1.0% of the bottom area of the lake (Figure 6b).

Wind pattern for the warm/wet months showed a dominance of 26% from the NE, with wind speeds up to 8.8 $m \cdot s^{-1}$ (Figure 7a). During the dry/mild cold months, the wind was predominantly from the S, with speeds up to 11.1 $m \cdot s^{-1}$ (Figure 7b).

The effective fetch (L_{ef}) model for wet/warm months with NE winds yielded values up to 0.8 km at the SW embayment (Figure 8a) and wave heights up to 0.5 m in the same embayment and at the confluence of the two axes (Figure 8c). For the dry/mild cold months, with SE winds, values up to 0.72 km were found at the lower section of the NW-SE axis and at the southern part of the land promontory (Figure 8b). Under SE winds, the wave heights were up to 0.6 m at the W shore next to the land

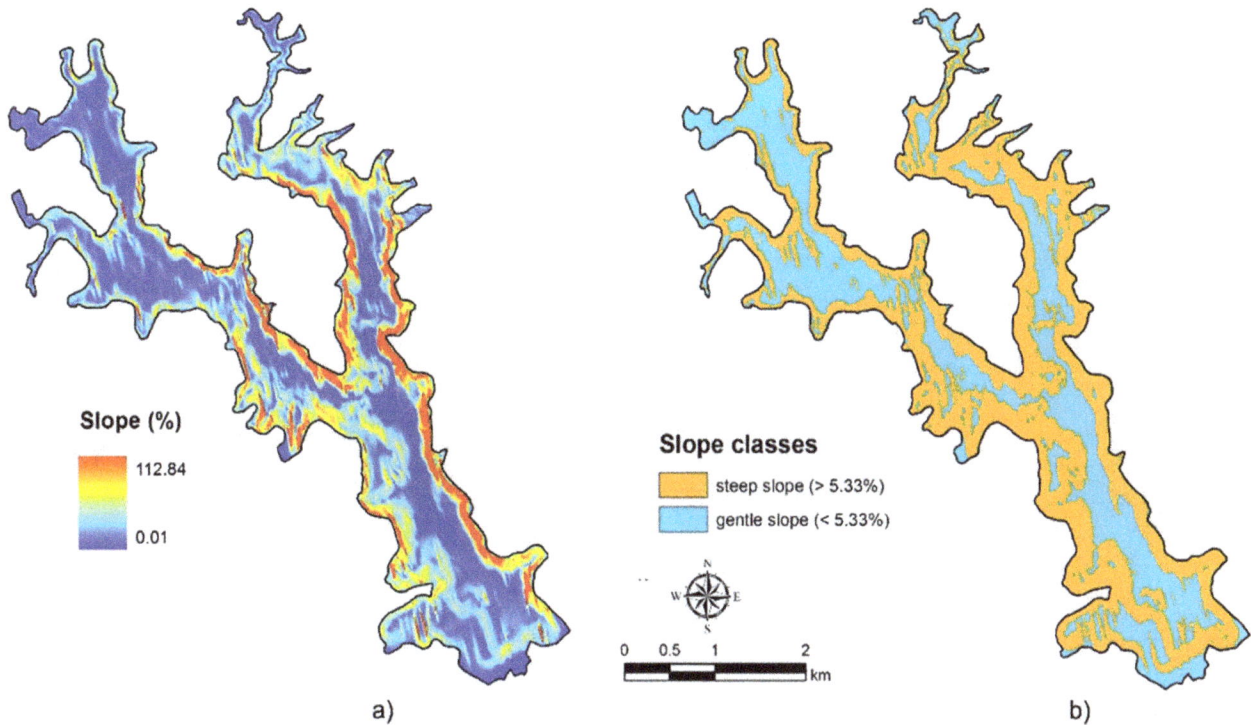

Figure 4. Lake bottom slope (%): a) slope gradient in %; b) reclassified slope: gentle (<5.33%) and steep (>5.33%) slopes.

promontory as well as in the central section of the NW-SE axis (Figure 8d).

As the lake surface is 20 m above mean sea level, the cryptal depth (Z_c) is 20.0 m, with a corresponding cryptodepression volume (V_c) of 2.8×10^7 m^3 (12.6% of the lake volume) (Figure 9). The water column is free from the influence of salt water.

The Lake Palmas watershed area (W_A) is 168.2 km^2, and the W_A/A ratio is 16.3. The mean annual, dry/mild cold and wet/warm total tributary discharge values were 0.4±0.2, 0.3±0.3 and 0.4±0.04 m^3.s^{-1}, respectively. The river discharge during the wet/warm season was 10.0% higher than the annual mean. In contrast, the dry season discharge was 14.6% lower. Zero discharge was registered three times for tributary river 1 during the dry season, but tributary 5 dried up twice during the wet/warm season.

The theoretical retention time based on the mean annual tributary discharge was 19.4 years, which may increase or decrease up to 20.7 and 17.7 years, considering the low and high discharges of the dry and wet seasons, respectively.

Discussion

According to the D_L criteria proposed by Hutchison [26], $D_L >$ 2.5 and <5.0, the shoreline form of Lake Palmas is subrectangular elongated. Although this D_L range was thought to designate lakes in overdeepened valleys associated with tectonic grabens or glaciated fjords, the geomorphology of Lake Palmas is associated with fluvial erosional and depositional processes in alluvial valleys and with Holocene sea level transgressions and regressions [21]. The relatively deep valley, from elevations up to 70 m at the Barreiras Formation plateaus down to – 30 m below sea level ($Z_{max} = -50.7$ m), may be associated with neotectonic processes, with valley alignments along the NW-SE axis [20]. The neotectonic hypothesis has also been supported by geophysical

studies at Lake Juparanã (62.0 km^2) [32], the largest lake in the LDRV.

The high V_d and Z_r values indicate that the basin form of Lake Palmas is an overdeepened valley with a relatively flat bottom. The mean lake slope is moderate despite steep areas at the SE shore of the lake and around the S section of the land promontory at the confluence of the lake axes. In addition, the relative hypsographic area and volume curves indicate a linear basin, an intermediate profile between concave and convex basins. Nevertheless, the basin linear profile represents the major component of water storage for the pelagic volume (96.3%). This characteristic is supported by the low BPI value and the low A:V ratio, reinforcing the relatively deep morphology of Lake Palmas. These metrics also emphasize a low potential for lake water evaporation and a higher potential for water column resistance to mixing.

The basin slope influences the processes of sediment erosion, transport and deposition as well as macrophyte biomass. Based on the critical value of 5.3% used to differentiate gentle slopes from steep slopes, the areas with gentle slopes (below 5.3%) are characterized by the deposition of fine sediments and the thriving stands of rooted macrophytes. Nearshore steep slopes support erosion and transport processes and decrease macrophyte biomass [8]. The dynamic ratio (DR) of 0.15 indicates the predominance of slope processes in view of the threshold value of 0.25, with higher values indicating the predominance of resuspension from wind and wave action [11].

Another feature of deep basins is the significant volume of the aphotic zone (58.0% of lake volume), which may constrain the development and distribution of the photosynthetic biota. Consequently, the limited littoral zone (3.7% of lake volume) associated with gentle slopes (<5.3%) [8] restricts the habitat of rooted submersed macrophytes to less than 1.0% of the lake bottom area. The depth of 10.0 m, corresponding to 1.0 atm of hydrostatic

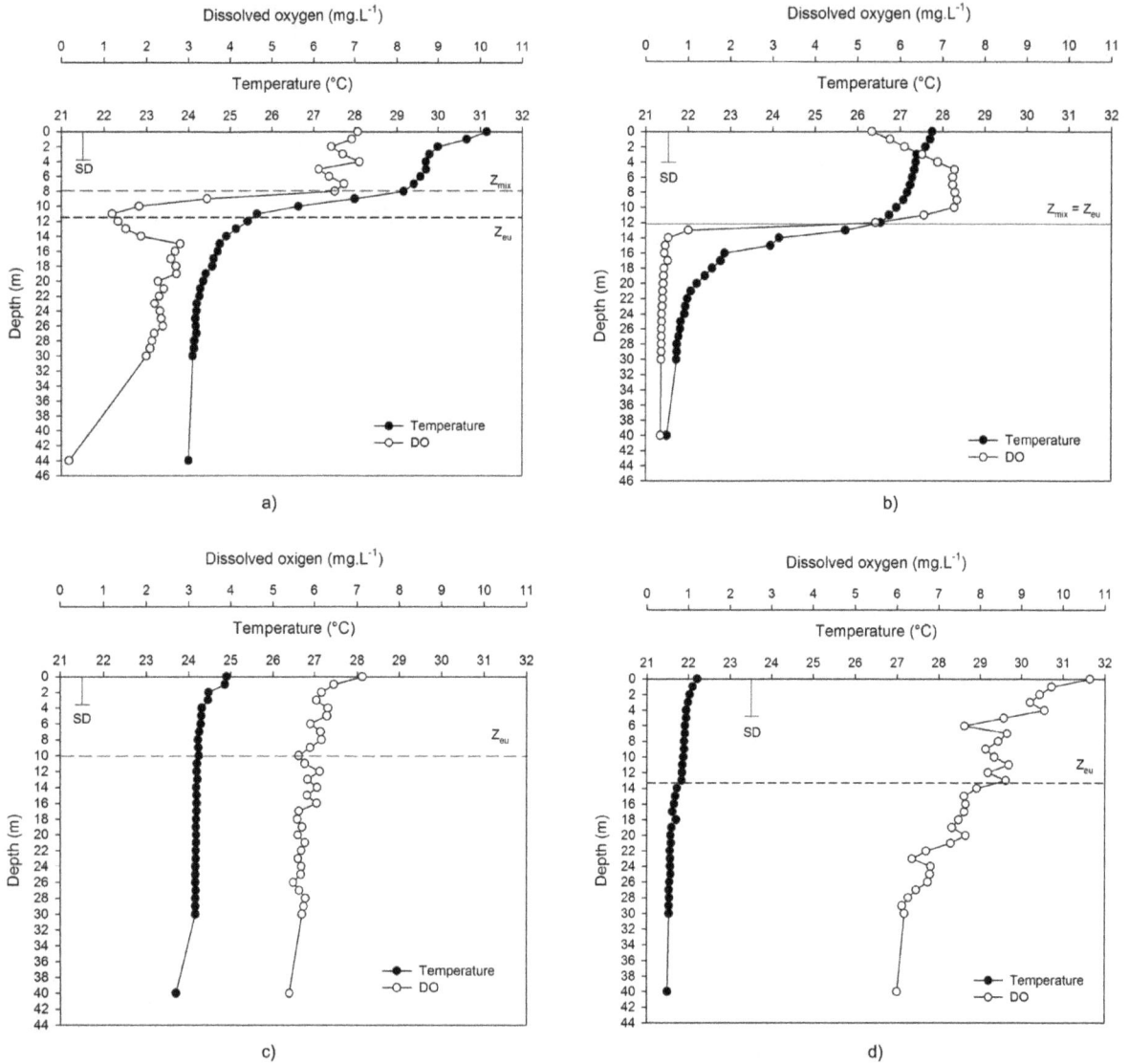

Figure 5. Typical vertical profiles of temperature and dissolved oxygen with mixing (Z_{mix}) and euphotic (Z_{eu}) zones and Secchi disk depth (SD) at the deepest Lake Palmas sampling site during the wet/warm (a and b) and dry/mild cold (c and d) seasons: a) 03/07/12, b) 03/12/13, c) 07/24/12 and d) 09/04/13.

pressure, is a threshold for the vascular system of angiosperms and defines the boundary of the lower infralittoral zone [33].

Based on published studies (Table S2), it seems that, Lake Palmas is the deepest natural lake in Brazil in terms of both Z_{max} and Z_{mv}. In the light of the predominance of the fluvial and coastal geomorphological genesis of natural lakes in Brazil, these depths are remarkable. Lake Dom Helvécio (A = 5.3 km^2), in the MDRV (State of Minas Gerais), was formerly considered the deepest natural lake in Brazil, with the following metrics: Z_{max} = 39.2 m, Z_{mv} = 11.3 m and V = 59.6×10^6 m^3 [18]. In the same Lake District of Lake Dom Helvécio, Bezerra-Neto, Briguent and Pinto-Coelho [34] determined Z_{max} and Z_{mv} values of 11.8 and 4.7 m, respectively, for Lake Carioca, with an area of 0.14 km^2 and a volume of 6.7×10^5 m^3. Schwarzbold and Schâfer [16] conducted an extensive survey of 61 coastal lakes of southern Brazil and found a lake as large as 802 km^2, with Z_{max} = 4.0 m and Z_{mv} = 2.5 m (Lake Mangueira), but Lake Figueira (7.1 km^2) was found to have the highest Z_{max} and Z_{mv} values, 11.0 and 5.7 m,

respectively. Recent hydrographic surveys conducted in other LDRV lakes determined Z_{max} values of 33.9, 31.6 and 22.1 m for Lakes Nova (A = 15.5 km^2, D_L = 4.5 and Z_{mv} = 14.7 m), Palminhas (A = 8.8 km^2, D_L = 8.1 and Z_{mv} = 14.2) and Terra Alta (A = 3.9 km^2, D_L = 3.1 and Z_{mv} = 9.0), respectively [35]. Although Lake Juparanã has a surface area 6 times greater than that of Lake Palmas, the estimated Z_{max} is approximately 20.0 m. Even when considering Amazon lakes associated with fluvial processes in the floodplain, it seems that these lakes are shallower comparing to the ones of LDRV. For instance Lake Calado (A = 8.0 km^2) [36] and Tupé (A = 0.6 km^2) [51] show during Amazon River peak flooding Z_{max} of 12.0 and 6.0 m, respectively.

Z_{wb} is a functional depth that separates areas of sediment transport occurring through resuspension via wind turbulence from areas of sediment accumulation with no resuspension. The concept is also very useful for delimiting the boundary between surface (epilimnetic) and bottom (hypolimnetic) waters [11]. Considering the variability of Z_{mix} during the stratification season,

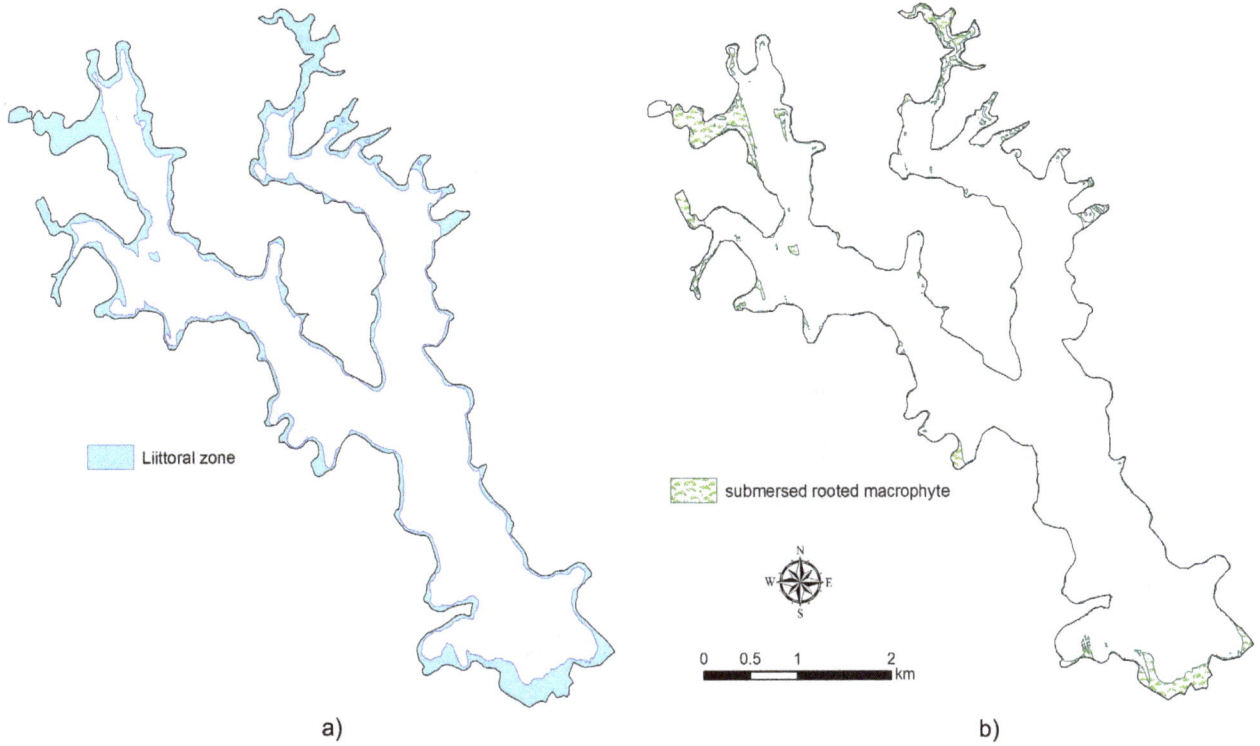

Figure 6. Lake littoral zone: a) littoral zone with a threshold of a depth of 10.0 m for the euphotic zone; b) potential submersed macrophyte area with a depth of 10.0 m or less and gentle slopes (<5.33%).

8.4 ± 2.5 m, a Z_{wb} value of 5.9 m can serve as an effective criterion to measure the significance of the physical and chemical stratification of Lake Palmas during the stratification season. Based on Z_{wb}, the epilimnion volume of Lake Palmas is only 26% of the lake volume. This value is consistent with the effects of the overdeepened basin on the resistance to mixing.

A moderate effective fetch (L_{ef}) may deepen the thermocline down to 12.0 m in the thermally stable water column of Lake Palmas. High L_{ef} values usually occur with SE winds, which are characteristic of cold fronts. According to Marchioro [37], 3 and 16 cold fronts were recorded during summer and winter/spring 2011, respectively, for Vitória (ES), which is located 90 km south of Lake Palmas. These cold fronts, characterized by S, SW and

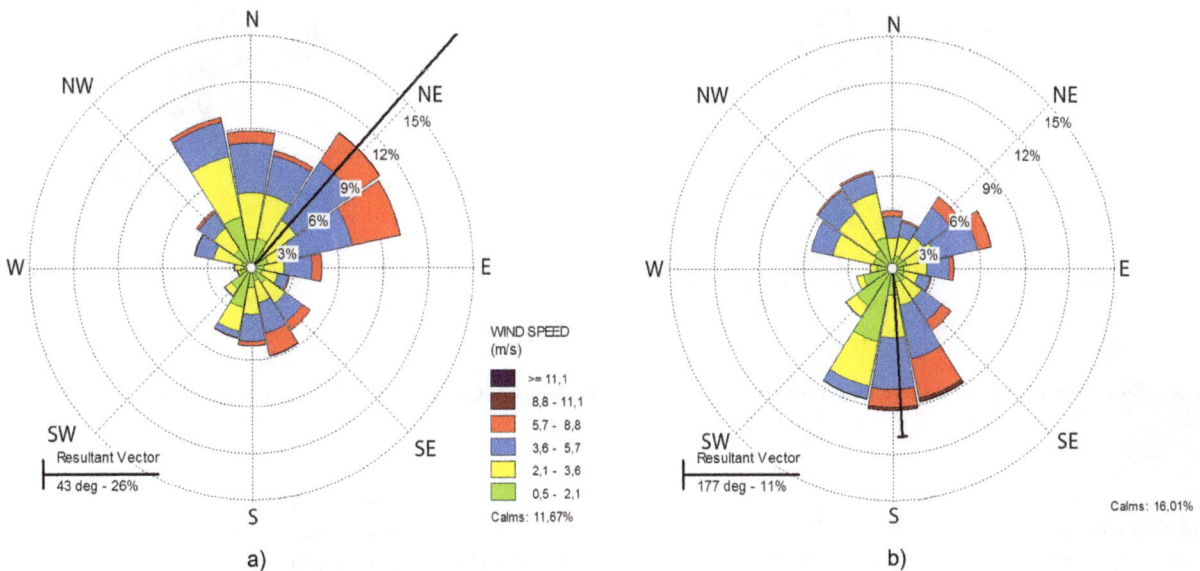

Figure 7. Wind direction frequency (%) and intensity (m.s^{-1}) from Linhares meteorological station from 2007 to 2009: a) warm and wet months; b) dry/mild cold months.

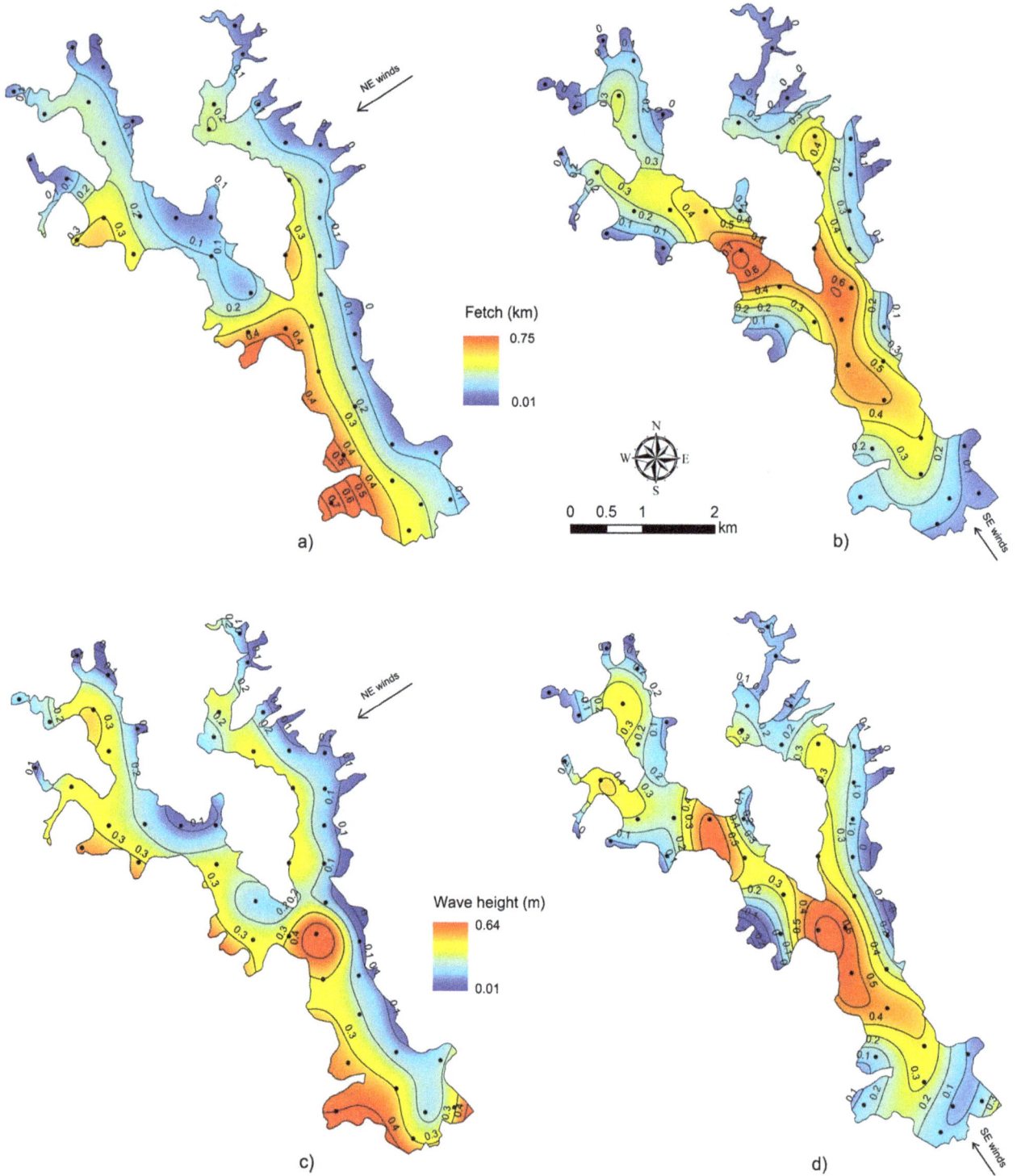

Figure 8. Effective fetch (L$_{ef}$) and wave height (H) models. a) L$_{ef}$ from NE winds; b) L$_{ef}$ from SE winds; c) wave heights from NE winds; and d) wave heights from SE winds. Contour lines show 0.1 m and km intervals.

SSE winds blowing up to 8.8 m.s^{-1}, may produce an average air temperature decrease of 7.1°C relative to the previous day's temperature. On average, cold fronts may persist up to 3.3 days. Tundisi et al. [38] have reported that the incidence of cold fronts may cause vertical mixing during the summer in reservoirs in Brazil.

The significant volume of hypoxic/anoxic bottom waters remains for the entire wet/warm season. This finding implies severe dissolved oxygen deficits and, as a consequence, the potential release of dissolved inorganic nutrients and heavy metals that are chemically bonded to the sediments. Nevertheless, Lake Palmas can be considered an oligotrophic ecosystem with a low

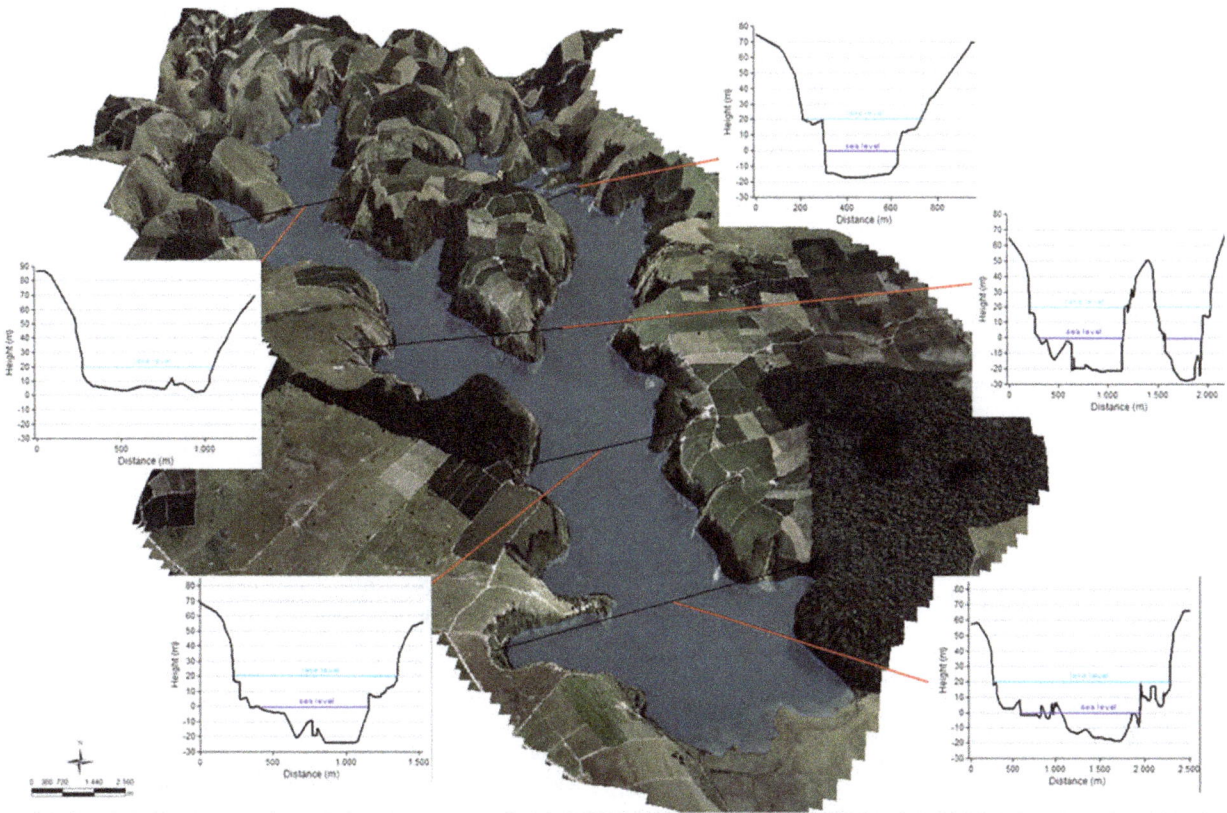

Figure 9. 3D view of the terrain model and height profiles: the light blue line refers to the lake level, the dark blue line to sea level. The cryptal volume is the volume below the dark blue line.

phytoplankton biomass, e.g., a mean chlorophyll a value of less than 1.0 $\mu g.L^{-1}$ [36]. Given that Z_{eu} reaches the metalimnion and hypolimnion layers, it may produce a suitable climate with low light and rich nutrients, supporting the cyanobacteria community. This maximum in metalimnion phytoplankton pigments has been reported for Lake Dom Helvécio [39]. Lake Palmas may also exhibit maximum metalimnetic conditions with relative low concentrations of chlorophyll a. Despite the value of $Z_{eu}:Z_{mix}$ indicates that light is not a limiting factor for phytoplankton growth during the stratification season.

These metrics describe a relatively deep basin that promotes physical and chemical seasonal stratification with strong environmental gradients and a warm monomictic pattern. Stratification may inhibit phytoplankton biomass, producing an oligotrophic state despite anoxic bottom waters. These findings agree with the concept of the intermittent metabolism of the overdeepened lakes of the LDRV [19].

Wind climate is a key driving force in the deepening of the mixing depth during the stratified season. However, this process depends on the angle of incidence of the wind on the aquatic surface, i.e., the fetch. S winds associated with the occurrence of cold fronts approaching from the S of the continent are frequent during the dry/mild cold months. These winds blow along the major axis of the lake, which is aligned with the NW-SE direction of the drainage network. This condition, associated with lower thermal radiation, is effective in breaking down the thermal stability and mixing the entire water column. Under these circumstances, the effective Wedderburn number has low values, indicating low thermal stability.

The W_A/A ratio of 16.27 indicates a relatively large drainage basin and a potentially significant discharge of river water into Lake Palmas, although the seasonality of rainfall and the potential negative effects of water and land uses in the watershed may halt the flow of tributary rivers during the dry season. The year-round unregulated use of the waters of the lake and tributary rivers for irrigation may also change the water balance in the lacustrine basin. Land use in the lake watershed is predominantly allocated to pasture, agriculture and forestry, representing 62.4% of the watershed area, whereas forested areas occupy only 32.8% of the watershed area [22]. If 30 and 100 m buffer areas are considered along the tributary river network and along the shoreline of Lake Palmas, agroecosystems represent up to 76.8 and 79.7%, respectively, of the buffer areas.

Accurate theoretical retention time (RT) estimates should be based on best knowledge of hydrological flows of tributary rivers inputs, evaporation rate, groundwater exchanges, and water consumption rates, instead of the simple ratio of the inflow to lake volume. Only 0.5% of the total number of natural lakes are known morphologically and hydrologically [40] in contrast, estimates of RT data are usually available for reservoirs construction and management. As rules of thumbs, RT is longer in deeper basin, reservoirs show shorter RT than lakes, and surface outflow from the lakes compared to deep outflow from the reservoirs. Reservoir RT in general is less than a year, with a threshold for reservoir limnology below 200 days [41,42]. The effect of flow is very significant for reservoir ecological structure and functioning, with RT >1 year show trend to stratification, eutrophication, anoxic bottom and recurrency/persistence of cyanobacteria bloom [42]. Blooms of cyanobacteria have been recorded in Funil reservoir, a tropical system in Brazil, despite the very short RT (annual mean of 41.5 days). High inputs of allochthonous nutrients from the reservoir watershed promote the eutrophic status of this artificial lake. The increase of the residence time, up to 80 days during the dry season, promotes spatial variability in the ecological structure of the reservoir along the river-dam axis [43].

The RT of 19.4 years for Lake Palmas can be considered very long, particularly for the regional wet climate (annual mean of $1,027$ mm.y^{-1}). In order to put it in perspective, and despite the lack of RT data for Brazilian natural lakes, Lakes in the Yunnan plateau, southwest of China, such as the oligotrophic Lakes Fuxianhu ($A = 211.0$ km^2, $Z_{mv} = 89.6$ m and $V = 18.9 \times 10^9$ m^3), Luguhu ($A = 48.4$ km^2, $Z_{mv} = 40.3$ m and $V = 1.9 \times 10^9$ m^3) and Cheghai ($A = 77.2$ km^2, $Z_{mv} = 25.7$ m and $V = 1.9 \times 10^9$ m^3) show RT of 35.5, 17.7 and 12.5 years, respectively [44]. These longer RTs imply in poorly flushed systems with negative correlations with total nitrogen and phosphorus and chlorophyll a. With such long RT of overdeepened lakes of Yunnan plateau are associated with oligotrophic systems, acting as a sink for inorganic and organic matter and with a delay response to additional nutrient inputs from lake watershed. In another hand, longer RT may imply in lack of resiliency after ecosystem distress from cultural eutrophication [44]. In addition to information about phosphorus and nitrogen loads input to the lake, knowledge of the theoretical residence time is a key factor for regulating the uses of Lake Palmas, such estimating lake carrying capacity for fish farming.

Hazards to water quantity and quality caused by pollution, silting and the introduction of exotic species may impair the ecological resilience of the lake. In addition, climate changes involving the intensification of extreme hydrological events (specifically, a predicted shortening of the rainfall season with fewer rainy days but with more intense and frequent storm events) [45] can be a major driver of shifts in lacustrine ecosystems in the LDRV. In December 2013, an extreme amount of regional precipitation, 650 mm of rainfall, 3 times the month mean rainfall at Linhares meteorological station, caused a major flood in the LDRV. Aditionaly, it must be considered a scenario of lake surface warming as a consequence of global warming. For deep lakes this scenario implies in an increasing loss of energy through evaporation, a deeper mixing layer, and an earlier summer stratification. These factors may lead monomitic lakes to turn into holo-oligomictic, with a complete vertical mixing occurring eventually in some years when stratification become weaker during winter [46].

Water uses in the lake watershed may also increase the stress on the theoretical retention time of Lake Palmas. Of the 5 tributary rivers, 2 showed no discharge at all at least 5 times. Rivers also became dry during the wet/warm season. These events might have resulted from river damming for irrigation purposes, given that 55 small reservoirs for irrigation purposes with an area up to 0.6 km^2 have been mapped in the Lake Palmas watershed [22].

Climate, lake morphology, and edaphic factors have been considered key drivers for trophic status of lakes, including the overdeepen basins. However, human impact factors have also been recognized, in some cases, as the leading driver for cultural eutrophication. The intensification of land and water uses in lake watersheds highlight the urgent need to regulate these uses in order to maintain a healthy lake ecosystems. In addition, climate change effects on water balance and related threats to ecosystem resilience and water security must be recognized.

Conclusions

The subrectangular elongated shape and the relatively over-deepened basin of Lake Palmas place most of the lake's volume in the pelagic compartment. Approximately one-half of the lake's volume is within the aphotic zone. The overdeepened basin promotes the physical and chemical stratification of the water column during the wet/warm months of the year. Under these conditions, only a small part of the lake volume is prone to mixing effects, and a large volume remains hypoxic/anoxic. During the dry/mild cold months, the predominance of S-SE winds, characteristic of the arrival of cold fronts, and the high effective fetch of these winds on the basin aligned along a NW-SE axis effectively promote the mixing of the lake's water column. Thus, the thermal pattern of Lake Palmas is warm monomictic. This finding is consistent with the hypothesis of a pattern of intermittent metabolism in the overdeepened lakes of the LDRV. Based on published data, Lake Palmas seems to be considered the deepest natural lake in Brazil in terms of both its maximum and mean depths. Given the very long theoretical retention time of Lake Palmas, hydrological changes in tributary rivers may increase the retention time and foster water quantity and quality problems. There are warning signs that the water balance in the basin is under pressure due to the unregulated uses of water for the year-round irrigation of croplands.

Supporting Information

Figure S1 Semivariogram for kriging interpolation of point data to generate a continuous surface describing the lake depth measurements.

Figure S2 Hypsographic curves of percent total surface (a) and total volume (b).

Table S1 Descriptive statistics of limnological variables in wet/warm and dry/mild cold seasons (2011 to 2013). Data from field vertical profiling at 4 sampling sites" from surface to the bottom.

Table S2 Morphometry of natural lakes in Brazil deeper than 6.0 m.

Author Contributions

Conceived and designed the experiments: GFB. Performed the experiments: GFB MAG FCG. Analyzed the data: GFB MAG FCG. Contributed reagents/materials/analysis tools: GFB. Contributed to the writing of the manuscript: GFB MAG FCG.

References

1. Cole GA (1994) Textbook of limnology. 4th ed., Prospect Heights: Waveland Press, Inc. 412p.
2. Kling GW (1988) Comparative transparency, depth of mixing, and stability of stratification in Lakes of Cameroon, West Africa. Limnology and Oceanography 33(1): 27–40.
3. Ambrosetti W, Barbanti L (2002a) Physical limnology of Italian lakes. 1. Relationship between morphometry and heat content. Journal of Limnology 61(2): 147–157. DOI:10.4081/jlimnol.2002.147.
4. Ambrosetti W, Barbanti L (2002b) Physical limnology of Italian lakes. 2. Relationships between morphometric parameters, stability and Birgean work. Journal of Limnology 61(2): 159–167. DOI:10.4081/jlimnol.2002.159.

5. Kvarnäs H (2001) Morphometry and hydrology of the four large lakes of Sweden. Ambio 30(8): 469–474. DOI:10.1579/0044-7447-30.8.467.

6. Blais JM, Kalff J (1995) The influence of lake morphometry on sediment focusing. Limnology and Oceanography 40(3): 582–588.

7. Rasmussen JB, Godbout L, Schallenberg M (1989) The humic content of lake water and its relationship to watershed and lake morphometry. Limnology and Oceanography 34(7): 1336–1343.

8. Duarte CM, Kalff J (1986) Littoral slope as a predictor of the maximum biomass of submersed macrophyte communities. Limnology and Oceanography 1(5): 1072–1080.

9. Fee EJA (1980) A relation between lake morphometry and primary productivity and its use in interpreting whole-lake eutrophication experiments. Limnology and Oceanography 24(3): 401–416.

10. Staehr PA, Baastrup-Spohr L, Sand-Jensen K, Stedmon C (2012) Lake metabolism scales with lake morphometry and catchment conditions. Aquatic Sciences 74(1): 155–169. DOI:10.1007/s00027-011-0207-6.

11. Håkanson L (2004) Lakes: form and function. Cladwell: The Blackburn Press. 201p.

12. Hollister J, Milstead WB (2009) Using GIS to estimate lake volume from limited data. Lake and Reservoir Research and Management 26(3): 194–199. DOI:10.1080/07438141.2010.504321.

13. Hollister JW, Milstead WB, Urrutia MA (2011) Predicting maximum lake depth from surrounding topography. PLoS ONE 6(9): e25764. DOI:10.1371/journal.pone.0025764.

14. Nõges T (2009) Relationships between morphometry, geographic location and water quality parameters of European lakes. Hydobiologia 633(1): 33–43. DOI:10.1007/s10750-009-9874-x.

15. Kosten S, Huszar V, Mazzeo N, Scheffer M, Sternberg LSL, et al. (2009) Lake and watershed characteristics rather than climate influence nutrient limitation in shallow lakes. Ecological Applications 19(7): 1791–1804. DOI:10.1890/08-0906.1.

16. Schwarzbold A, Schäfer A (1984) Genesis and morphology of Rio Grande do Sul (Brazil) costal lakes. Amzoniana. 9(1): 87–104 (In Portuguese).

17. von Sperling E (1999) Morphology of lakes and reservoirs. Belo Horizonte: Ed. Universidade Federal de Minas Gerais – UFMG, 137p (In Portuguese).

18. Bezerra-Neto JF, Pinto-Coelho RM (2008) Morphometric study of Lake Dom Helvécio, Parque Estadual do Rio Doce (PERD), Minas Gerais, Brazil: a re-evaluation. Acta Limnologica Brasiliensia 22(2): 161–167 (In Portuguese).

19. Bozelli RL, Esteves FA, Roland F, Suzuki MS (1992) Lakes of the lower Doce River: abiotic variables and chlorophyll *a* (Espírito Santo – Brazil). Acta Limnologica Brasiliensia. 4: 13–21 (In Portuguese).

20. Bricalli LL, Mello CL (2013) Lineament patterns related to lithostructural and neotectonic fracturing (State of Espírito Santo, Southeastern Brazil). Revista Brasileira de Geomorfologia, v. 14, p. 301–311 (In Portuguese).

21. Martin L, Suguio K, Flexor JM, Archanjo JL (1996) Coastal quaternary formations of the southern part of the State of Espírito Santo (Brazil). Anais da Academia Brasileira de Ciências 68(3): 389–404.

22. Barroso GF, Mello FA de O (2013) Landscape compartments and indicators of environmental pressures on fluvial and lacustrine ecosystems of the Lower Doce River Valley. Proceedings of the 15th Brazilian Symposium of Applied Physical Geography. Vitória, UFES, 158–165p. Available: http://www.xvsbgfa2013.com.br/anais/ (In Portuguese).

23. Isaaks EH, Srivastava RM (1989) Applied geostatistics. Oxford: Oxford University Press.

24. Burrough PA, MacDonnell RA (1998) Principles of geographical informations systems. Oxford: Oxford University Press.

25. Håkanson L (1978) Optimization of lake hydrographic surveys. Water Resources Research 14(4): 545–560.

26. Hutchinson GE (1957) A treatise on limnology. Volume I: Geography, physics and chemistry. New York: John Wiley & Sons, Inc.

27. Dadon JR (1995) Calor y temperatura en cuerpos lenticos. In: Lopretto EC, Tell G. Ecosistemas de aguas continentales: metodologias para su studio. Vo. 1, La Plata: Ediciones SUR, 47–56.

28. Imberger J, Hambling PF (1982) Dynamics of lakes, reservoirs and cooling ponds. Annual Review of Fluid Mechanichs 14, 153–87. DOI:10.1146/annurev.fl.14.010182.001101.

29. Reynolds CS (2006) The ecology of phytoplankton. Ecology, Biodiversity and Conservation. Cambridge: Cambridge University Press, 537 p.

30. Kerekes J (1977) The index of lake basin permanence. Internationale Revue der Gesamten Hydrobiologie und Hydrographie. 62(2): 291–293.

31. Håkanson L (1983) Principles of lakes sedimentology. Cladwell, The Blackburn Press.

32. Hatushika RS, Silva CG, Mello CL (2007) High resolution seismic stratigraphy of Lake Juparanã, Linhares (ES, Brazil) as study aid for sedimentation and quaternary tectonics. Revista Brasileira de Geofísica. 25(4): 433–442. DOI:10.1590/S0102-261X2007000400007 (In Portuguese).

33. Wetzel RG (2001) Limnology: lake and river ecosystems. 2nd ed., San Diego, Academic Press, 1006p.

34. Bezerra-Neto JF, Briguent LS, Pinto-Coelho RM (2010) A new morphometric study of Carioca Lake, Parque Estadual do Rio Doce (PERD), Minas Gerais State, Brazil. Acta Scientiarum. Biological Sciences 32(1): 49–54. DOI:10.4025/actascibiolsci.v32i1.4990 (In Portuguese).

35. Barroso GF, Garcia F da C, Gonçalves MA, Martins, FC de O, Venturini JC, et al. (2012) Integrated studies on the lacustrine system of the Lower Doce River Valley (Espírito Santo). Proceedings of the I National Seminar of Sustainable Management of Aquatic Ecosystems: Complexity, interactivity and Ecodevelopment. Arraial do Cabo, COPPE/UFRJ, 21 a 23 de março. 7p. Available: http://gestaoecossistemas.files.wordpress.com/2012/11/1-i-1-estudos-integrados-no-sistema-lacustre-do-baixo-rio-doce-espc3adrito-santo.pdf (In Portuguese).

36. Melack JM, Fisher TR (1990) Comparative limnology of tropical lakes with emphasis on the central Amazon. Acta Limnologica Brasiliensia 3: 1–48.

37. Marchioro E (2012) Incidence of cold fronts in Vitória, ES, Brazil. Acta Geografica, v. 7, p. 49–60. DOI:10.5654/actageo2012.0002.0003 (In Portuguese).

38. Tundisi JG, Matsumura-Tundisi T, Pereira KC, Luzia AP, Passerini MD, et al. (2010) Cold fronts and reservoir limnology: an integrated approach towards the ecological dynamics of freshwater ecosystems. Brazilian Journal of Biology, 70(3 Suppl): 815–824. DOI:0.1590/S1519-69842010000400012.

39. Souza MBG, Barros CFA, Barbosa F, Hajnal É, Padisák J (2008) Role of atelomixis in replacement of phytoplankton assemblages in Dom Helvécio Lake, South-East Brazil. Hydrobiologia 607: 211–224. DOI:10.1007/s10750-008-9392-2.

40. Ryanzhin SV (1999) Fundamental limnological processes and relevant indicators used in lake studies. Kondratyev, K. Y. and Filatov, N. Limnology and Remote Sensing: A Contemporary Approach, Springer: 53–78.

41. Straškraba M (1998) Coupling of hydrobiology and hydrodynamics: lakes and reservoirs. Coastal and Estuarine Science 288: 601–622.

42. Straškraba M (1999) Retention time as a key variable of reservoir limnology. In: Tundisi JG, Straskraba M. Theoretical reservoir ecology and its applications. São Carlos, Brazilian Academy of Sciences/International Institute of Ecology/Backhuys Publishers: 385–410.

43. Soares MCS, Marinho MM, Azevedo SMOF, Branco CWC, Huszar VLM (2012) Eutrophication and retention time affecting spatial heterogeneity in a tropical reservoir. Limnologica 42(3): 197–203. DOI:10.1016/j.limno.2011.11.002.

44. Liu W, Zhang Q, Liu G (2011) Lake eutrophication associated with geographic location, lake morphology and climate in China. Hydrobiologia 644(1): 289–299.

45. PBMC (2013) Contributions of the Working Group 2 to the First National Assessment Report of National Climate Change Panel. Executive Summary GT2: Impacts, vulnerability and adapting, PBMC, Rio de Janeiro (In Portuguese).

46. Ambrosetti W, Barbantti L, Sala N (2003) Residence time and physical processes in lakes. Journal of Limnology 62(1): 1–15.

47. Henry R, Tundisi JG, Calijuri M do C, Ibañez M do SR (1997) A comparative study of thermal structure, heat content and stability of stratification in three lakes. In Tundisi JG, Saijo Y (eds.) Limnological studies on the Rio Doce Lakes, Brazil. Brazilian Academy of Sciences. University of São Paulo. School of Engineering at São Carlos. Center for Water Resources and Applied Ecology. 528p.

48. Barros CF de A, dos Santos AMM, Barbosa FAR (2013) Phytoplankton diversity in the middle Rio Doce lake system of southeastern Brazil. Acta Botanica Brasilica 27(2): 327–346.

49. Brighenti LS, Pinto-Coelho RM, Bezerra-Neto JF, Gonzaga AV (2011) Morphometric features of Lake Central (Lagoa Santa, Minas Gerais State): comparison of two methodologies. Acta Scientiarum. Biological Sciences 33(3): 281–287 DOI:10.4025/actascibiolsci.v33i3.5545 (In Portuguese).

50. Nogueira F, Souza MD, Bachega I, Silva RL (2002) Seasonal and diel limnological differences in a tropical floodplain lake (Pantanal of Mato Grosso, Brazil). Acta Limnologica Brasiliensia 14(3): 17–25.

51. Aprile FM, Darwich AJ (2005) Geomorphological models for Lake Tupé. In: dos Santos-Silva E.N., Aprile FM, de Melo S, Scudeller VV (eds.). Biotupé: Physical, biodiversity and sociocultural dimensions of the Lower Negro River, Central Amazon. Manaus, AM: Editora INPA, p. 03–17. (In Protuguese).

52. Panosso R (2000) Geographical and gemorphological considerations. In: Bozelli RL, Esteves FA, Roland F (eds). Lake Batata: impacts and restoration of an Amazon ecosystem. IB-UFRJ/SBL. Rio de Janeiro. 342 p. (In Protuguese).

53. Herdendorf CE (1984) Inventory of the morphometric and limnological characteristics of the large lakes of the world. Technical Bulletin OHSU-TB-017, Columbus, The Ohio State University, 78p.

54. Niencheski LFH, Baraj E, Windom HL, França RG (2004) Natural background assessment and Its anthropogenic contamination of Cd, Pb, Cu, Cr, Zn, Al and Fe in the sediments of the southern area of Patos Lagoon. Journal of Coastal Research SI39: 1040–1043.

55. Llames ME, Zagarese HE (2010) Lakes and reservoirs of the world: South America. In: Likens GE. (ed.). Lake ecosystem ecology: a global perspective. San Diego, Academic Press: 332–342.

56. Hennemann MC, Petrucio MM (2011) Spatial and temporal dynamic of trophic relevant parameters in a subtropical coastal lagoon in Brazil. Environmental Monitoring and Assessment (Print) 181: 347–361.

57. Kjerfve B, Schettini CAF, Knoppers B, Lessa G, Ferreira HO (1996) Hydrology and salt balance in a large, hypersaline coastal lagoon: Lagoa de Araruama, Brazil. Estuarine, Coastal and Shelf Science 42: 701–725.

11

Short-Term Coral Bleaching Is Not Recorded by Skeletal Boron Isotopes

Verena Schoepf[1,2]*, Malcolm T. McCulloch[1], Mark E. Warner[3], Stephen J. Levas[2¤], Yohei Matsui[2], Matthew D. Aschaffenburg[3], Andréa G. Grottoli[2]

1 School of Earth and Environment, The University of Western Australia and ARC Centre of Excellence for Coral Reef Studies, Crawley, WA, Australia, 2 School of Earth Sciences, The Ohio State University, Columbus, Ohio, United States of America, 3 School of Marine Science and Policy, University of Delaware, Lewes, Delaware, United States of America

Abstract

Coral skeletal boron isotopes have been established as a proxy for seawater pH, yet it remains unclear if and how this proxy is affected by seawater temperature. Specifically, it has never been directly tested whether coral bleaching caused by high water temperatures influences coral boron isotopes. Here we report the results from a controlled bleaching experiment conducted on the Caribbean corals *Porites divaricata*, *Porites astreoides*, and *Orbicella faveolata*. Stable boron ($\delta^{11}B$), carbon ($\delta^{13}C$), oxygen ($\delta^{18}O$) isotopes, Sr/Ca, Mg/Ca, U/Ca, and Ba/Ca ratios, as well as chlorophyll *a* concentrations and calcification rates were measured on coral skeletal material corresponding to the period during and immediately after the elevated temperature treatment and again after 6 weeks of recovery on the reef. We show that under these conditions, coral bleaching did not affect the boron isotopic signature in any coral species tested, despite significant changes in coral physiology. This contradicts published findings from coral cores, where significant decreases in boron isotopes were interpreted as corresponding to times of known mass bleaching events. In contrast, $\delta^{13}C$ and $\delta^{18}O$ exhibited major enrichment corresponding to decreases in calcification rates associated with bleaching. Sr/Ca of bleached corals did not consistently record the 1.2°C difference in seawater temperature during the bleaching treatment, or alternatively show a consistent increase due to impaired photosynthesis and calcification. Mg/Ca, U/Ca, and Ba/Ca were affected by coral bleaching in some of the coral species, but the observed patterns could not be satisfactorily explained by temperature dependence or changes in coral physiology. This demonstrates that coral boron isotopes do not record short-term bleaching events, and therefore cannot be used as a proxy for past bleaching events. The robustness of coral boron isotopes to changes in coral physiology, however, suggests that reconstruction of seawater pH using boron isotopes should be uncompromised by short-term bleaching events.

Editor: Christian R. Voolstra, King Abdullah University of Science and Technology, Saudi Arabia

Funding: This work was funded by the U.S. National Science Foundation (OCE#0825490 to AGG and OCE#0825413 to MEW). MTM was supported by a Western Australian Premiers Fellowship and an Australian Research Council Laureate Fellowship. The funders had no role in study design, data collection and analysis, decision to publish, or preparation of the manuscript.

Competing Interests: The authors have declared that no competing interests exist.

* Email: schoepf.4@osu.edu

¤ Current address: Department of Geography and the Environment, Villanova University, Villanova, Pennsylvania, United States of America

Introduction

The world's oceans are simultaneously warming and acidifying at unprecedented pace due to rising atmospheric CO_2 concentrations [1], thereby severely threatening marine ecosystems [2,3]. Coral reefs are particularly vulnerable because they are highly sensitive to changes in both seawater temperature and pH [4–6].

Coral reefs are increasingly subject to periods with unusually warm seawater temperatures which can cause coral bleaching, defined as a significant loss of photosynthetic pigments and/or algal endosymbionts from the coral tissue [7,8–12]. Although corals can sometimes recover from bleaching events [13–18], other corals are significantly impacted by bleaching events, resulting in widespread mortality for more sensitive species [10,11,19]. As surface ocean temperatures have already increased by 0.6°C since preindustrial times and are projected to increase by at least

another 2.0°C under a business as usual scenario by the year 2100 [1], coral bleaching events are expected to increase in frequency and intensity [11,20–22], thus contributing to the worldwide decline of corals reefs [23,24].

In addition to greenhouse-induced warming, rising levels of atmospheric CO_2 concentrations have already caused a drop in surface seawater pH by approximately 0.1 unit compared to preindustrial times [25], and a further decrease of 0.3 pH units by the end of this century has been predicted [1,26]. This is of particular concern for marine calcifying organisms including corals because ocean acidification (OA) typically compromises calcification [27–33]. Although it is now recognised that scleractinian corals have a capacity to internally up-regulate pH [34–40], it has been estimated that coral calcification will decrease by up to 22% by the year 2100 [41]. Therefore, predicting how global climate change will affect ocean chemistry and marine ecosystems in the coming decades is of imminent concern and requires understanding of how factors such as ocean temperature and pH have varied naturally in the past. Since direct measurements of ocean chemistry are only available for the most recent decades, proxy records are of crucial importance.

Coral skeletal boron isotopes ($\delta^{11}B$) are promising proxies of seawater pH [31,34,40,42–45] that have been used to reconstruct paleo-pH [46–48]. Their use as a pH-proxy is based on the fact that the relative abundance of the two major aqueous boron species in seawater, boric acid ($B(OH)_3$) and borate ion ($B(OH)_4^-$), are dependent on seawater pH. Further, their isotopic composition is also pH dependent and differs by about 27‰. It is generally thought that corals mainly incorporate the charged borate ion into the skeleton as their isotopic composition reflects that of borate [34,42,49], although this has been questioned by some studies [50,51]. However, it is now established that coral skeletal $\delta^{11}B$ does not directly record seawater pH but rather reflects the pH at the site of calcification [34,35,40,42], which is typically elevated by up to 1 unit [35,37–39] or more [36] relative to ambient seawater. Therefore, species-specific calibrations have to be performed to quantify the amount of pH-upregulation in order to reconstruct seawater pH [34].

One key requirement for any proxy is that it faithfully records the environmental variable of interest without being influenced by other factors. Several experimental studies have attempted to identify potentially confounding effects of light, depth, temperature, and feeding on coral $\delta^{11}B$ but, taken together, their results are contradictory. For example, Hönisch et al. [45] showed that various light, depth, and feeding regimes did not influence $\delta^{11}B$ of *Porites compressa* and *Montipora compressa*. However, another study using branching *Acropora* sp. found that $\delta^{11}B$ decreased as light intensity increased, introducing a bias of about 0.05 pH units [43].

Regarding temperature, Reynaud et al. [44] observed no temperature effect on $\delta^{11}B$ of *Acropora* sp. cultured at 25 and 28°C and two pCO_2 levels. Interestingly, Dissard et al. [43] cultured *Acropora* sp. using the same protocol as in Reynaud et al. [44] at 22, 25 and 28°C, and found a significant temperature effect only at the lower temperatures of 22 and 25°C, corresponding to an increase in $\delta^{11}B$ of about 0.02 pH units. A significant temperature effect was not observed between 25 and 28°C, consistent with the findings of Reynaud et al. [44]. It therefore appears that other factors may have confounded this work or, at least in *Acropora*, that temperature effects on coral $\delta^{11}B$ are non-linear and only exist for lower temperature ranges. However, given that both studies only used nubbins prepared from a single parent colony of one species, it is not clear whether these results could be reproduced if nubbins were used from multiple parent colonies (thus representing genetically different individuals), and/or different species of coral such as massive *Porites* sp., which is frequently used for paleoclimate reconstruction.

Although the experimental evidence available to date suggests that temperature may only affect coral $\delta^{11}B$ at lower temperatures (i.e., <25°C), no studies have examined stressfully high temperatures that can result in coral bleaching. Evidence from a massive *Porites* core from the Great Barrier Reef suggests that significant short-term decreases in $\delta^{11}B$ could be the result of coral bleaching events [47]. They proposed that this was due to a major reduction in pH at the site of calcification associated with impaired calcification, and suggest that coral $\delta^{11}B$ may be used as a proxy for past bleaching events [47]. Although coral $\delta^{13}C$ and $\delta^{18}O$ as well as Sr/Ca have been suggested as potential bleaching proxies, they were unreliable in this regard or in the case of Sr/Ca lack

sufficient detail [52–55]. Therefore, establishing coral $\delta^{11}B$ as a bleaching proxy could present a major step forward in understanding how often bleaching events occurred in the past. Further, it is essential that temperature effects on coral $\delta^{11}B$ be better understood over a large range of temperatures and using a variety of coral species to improve the reconstruction of past seawater pH.

To address these questions, we conducted a controlled, replicated bleaching experiment using the Caribbean coral species *Porites divaricata*, *Porites astreoides* and *Orbicella faveolata*. Coral skeletal $\delta^{11}B$, $\delta^{13}C$ and $\delta^{18}O$ as well as Sr/Ca, Mg/Ca, U/Ca, and Ba/Ca were analysed in samples collected immediately after the bleaching experiment and again after 6 weeks of recovery on the reef. Several physiological measurements were concurrently performed on these corals [13,56–59], providing a rich background of physiological information within which to interpret our findings. Based on the evidence presented by Wei et al. [47], we hypothesized that $\delta^{11}B$ of bleached corals decreases compared to non-bleached controls, and approaches (i.e., increases to) the values of non-bleached controls during recovery.

Methods

Coral Bleaching Experiment

The corals used for this study were taken from an experiment where three species of Caribbean corals were experimentally bleached in two consecutive summers [13]. These corals were physiologically fully recovered (i.e., there were no significant differences between treatment and control corals in any of the measured variables) after a year on the reef prior to exposure to the elevated temperature stress for a second time [57,59]. A detailed description of the experimental design can be found in Grottoli et al. [13].

Briefly, coral fragments were collected from 9 healthy colonies of *Porites divaricata* (branching morphology), *Porites astreoides* (mounding/encrusting morphology), and mounding *Orbicella faveolata* (formerly *Montastraea faveolata*) [60] (large mounding morphology) in July 2009 from reefs near Puerto Morelos, Yucatan Peninsula, Mexico (20°50′N, 86°52′W). All collections and experiments were conducted following the rules and regulations of Mexico and imported to the USA under CITES permits held by UNAM-ICML and the Ohio State University. Half of the coral fragments from each parent colony were randomly assigned to each treatment: (1) ambient control fragments were maintained in tanks with ambient seawater temperature (30.66 ± 0.24°C), and (2) treatment fragments were placed in tanks with elevated seawater temperature (31.48 ± 0.20°C). Seawater temperature in the treatment tanks was gradually elevated over the course of a week. Corals were not fed but had access to unfiltered seawater. High-precision seawater pH measurements were not made. After a total of 15 days, temperature in all tanks was returned to ambient levels, and all coral fragments were placed back *in situ* on the reef for one year.

In July 2010, the experiment was repeated. All corals that had served as ambient control fragments the previous summer were placed in tanks with ambient seawater (30.40 ± 0.23°C), whereas all corals that had been used as treatment fragments were maintained in tanks with elevated temperature (31.60 ± 0.24°C). After 17 days, all tanks were returned to ambient temperature levels and one control and one treatment fragment per colony of each species were then frozen for geochemical analyses (= 0 weeks on the reef). All remaining fragments were placed on the back reef. All remaining corals (i.e., one control and one treatment fragment per colony of each species) were recollected from the reef after 6 weeks and frozen for geochemical analyses.

Physiological Analyses

Chlorophyll a.   Coral tissue was removed from the skeleton of a portion of each fragment with a WaterPik, homogenized, and centrifuged to separate the algal endosymbiont from the animal tissue. A subsample of the resuspended algal pellet was broken using glass beads in 100% acetone and subsequently stored at − 20°C overnight. Chlorophyll a was then determined using a Shimadzu UV-VIS spectrophotometer and the equations of Jeffrey and Humphrey [61]. Chlorophyll a content was standardized to surface area, which was determined using the aluminium foil method [62].

Calcification.   Published calcification rates determined using the buoyant weight technique [63] were reproduced from Grottoli et al. [13] with permission of the publisher (license 3424840151859).

Isotopic Analyses

Coral tissue was removed from the skeleton using a dental hygiene tool. The uppermost layer (approx. 0.25–0.5 mm) of the dried skeleton was then gently shaved with a diamond-tipped Dremel tool and ground to fine powder using agate mortar and pestles [52,64]. Approximately 2–11 mg of the skeletal powder was weighed in pre-cleaned, pre-weighed Teflon tubes. An aliquot of this powder was used for $\delta^{13}C$ and $\delta^{18}O$ analyses (see below). When there was <2 mg of powder available from individual fragments, samples from multiple fragments of the same species and time point were combined for $\delta^{11}B$ analyses.

Boron isotopes.   The following procedures were carried out in a clean lab. To remove organic matter, samples were rinsed with H_2O, incubated with 2 ml of 6.25% NaClO for 15 min including 5 min of sonication, then centrifuged for 2 min at highest speed, and the supernatant removed. Powders were then rinsed 3× with H_2O, with each rinse including 5 min sonication and 2 min centrifugation at highest speed. After cleaning, tubes were lightly capped and samples dried to constant dry weight by placing them on a 60°C hotplate with a heat-lamp next to it. Cleaned samples were dissolved in 0.58 N HNO_3. An aliquot of this solution was then diluted for trace element measurements (see below), and the remaining solution extracted for boron using ion exchange resins [35,65].

The extracted boron was analysed at the University of Western Australia using either a Neptune Plus Multi-Collector Inductively Coupled Plasma Mass Spectrometer (MC-ICP-MS; Thermo Fisher Scientific) fitted with a PFA nebulizer and a cyclonic quartz spray chamber or a NU Plasma II MC-ICP-MS (NU Instruments). The boron isotopic composition of the skeleton ($\delta^{11}B$) was reported as the per mil deviation of the stable isotopes ^{11}B:^{10}B relative to SRM-951. Sample measurements were bracketed by SRM-951 enriched to 24.7‰ and blank measurements [66], and all samples were analysed in duplicate. Repeated measurements of SRM-951 had a standard deviation (1σ) of ± 0.37‰ (n = 124) for concentrations ranging from 100–200 ppb B.

Carbon and oxygen isotopes.   A 80–100 μg aliquot of the dried and ground skeletal powder was analysed for $\delta^{13}C$ and $\delta^{18}O$ using an automated Kiel Carbonate Device coupled to a Stable Isotope Ratio Mass Spectrometer (SIRMS; Finnigan Delta IV) at The Ohio State University. Samples were not pre-treated [67]. Samples were acidified under vacuum with 100% ortho-phosphoric acid. The carbon isotopic composition of the skeleton ($\delta^{13}C$) was reported as the per mil deviation of the stable isotopes ^{13}C:^{12}C relative to Vienna-Peedee Belemnite Limestone standard (v-PDB).

Skeletal oxygen isotopes ($\delta^{18}O$) were reported as the per mil deviation of the stable isotopes ^{18}O:^{16}O relative to v-PDB. Approximately 10% of all samples were run in duplicate. The standard deviation (1σ) of repeated measurements of internal standards was ± 0.03‰ for $\delta^{13}C$ and ± 0.07‰ for $\delta^{18}O$ (n = 55).

Trace Element Analyses

From the solution used for $\delta^{11}B$ analysis, a 2–7 μL aliquot was diluted to a final concentration of 10 ppm Ca in 2% HNO_3 spiked with ~19 ppb Sc, 19 ppb Y, 0.19 ppb Pr, 0.095 ppb Bi, and 19 ppb V. Samples were then analysed for Sr/Ca, Mg/Ca, U/Ca, and Ba/Ca on an X-Series 2 Quadrupole Inductively Coupled Plasma Mass Spectrometer (Q-ICPMS; Thermo Fisher Scientific) at the University of Western Australia using the standard Xt interface and the plasma screen fitted. Between-run reproducibility (RSD) of a coral standard subjected to the same chemistry as the coral samples was 0.4, 0.4, 0.8 and 0.4% for Sr/Ca, Mg/Ca, U/ Ca and Ba/Ca, respectively (n = 6).

Statistical Analyses

Two-way analysis of variance (ANOVA) was used to test for the effects of temperature and time on coral chlorophyll a, $\delta^{11}B$, $\delta^{13}C$, $\delta^{18}O$, Sr/Ca, Mg/Ca, U/Ca, and Ba/Ca. Temperature was fixed and fully crossed with two levels (treatment, control), and time was fixed and fully crossed with two levels (0, 6 weeks on the reef). Residual values for each variable and species were calculated and tested for normality using a Shapiro-Wilk's test and homogeneity of variance was assessed with plots of expected versus residual values. Non-normal data sets were transformed. Post hoc Tukey tests were performed when main effects – but no interaction terms - were significant. A posteriori slice tests (e.g., tests of simple effects) [68] determined if the treatment and control averages significantly differed at each time interval. Calcification data was previously analysed in Grottoli et al. [13].

Corals were considered to be fully recovered once average chlorophyll a, $\delta^{11}B$, $\delta^{13}C$, $\delta^{18}O$, Sr/Ca, Mg/Ca, U/Ca, and Ba/ Ca values of treatment corals no longer significantly differed from the average values of controls. Since all fragments were exposed to identical conditions except temperature during the tank portion of the experiment, any differences in the observed responses were due to temperature effects alone and independent of seasonal variation. Bonferroni corrections were not applied because they increase the risk of false negatives [69,70]. P-values ≤0.05 were considered significant. Statistical analyses were performed using SAS software, version 9.3 of the SAS System for Windows.

Results

Physiology

A significant interaction of temperature and time was observed for area-normalized chlorophyll a concentrations of P. divaricata, with treatment corals having 42% lower and 46% higher chlorophyll a concentrations than controls after 0 and 6 weeks on the reef, respectively (Fig. 1A, Table 1). Calcification rates were the same for both treatment and control P. divaricata throughout the study (Fig. 1B) [13].

In P. astreoides, area-normalized chlorophyll a concentrations were influenced by temperature, with treatment corals having 75% and 76% lower chlorophyll a rates than the controls at both 0 and 6 weeks on the reef, respectively (Fig. 1C, Table 1). Calcification rates of treatment corals were the same as in controls initially, but were 69% lower after 6 weeks on the reef (Fig. 1D) [13].

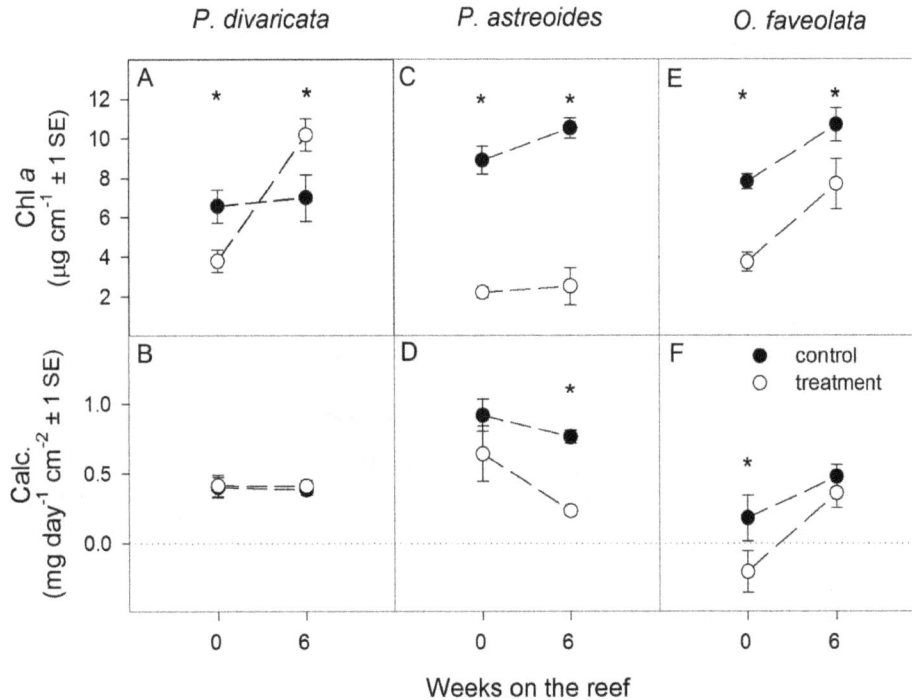

Figure 1. Average area-normalized chlorophyll a (=chl a) concentration and calcification (=calc.) rate of (A, B) Porites divaricata, (C, D) Porites astreoides, and (E, F) Orbicella faveolata after 0 and 6 weeks on the reef. Asterisks indicate significant differences between control and treatment corals at a specific time interval. Sample size ranges from 6–9. Calcification data reproduced from Grottoli et al. [13] with permission of the publisher.

In *O. faveolata*, area-normalized chlorophyll *a* concentrations were influenced by both temperature and time (Table 1). They were 52% and 28% lower in treatment corals compared to controls at 0 and 6 weeks on the reef, respectively (Fig. 1E, Table 1). Further, they were higher at 6 weeks on the reef compared to 0 weeks (Fig. 1E, Table 1). Calcification rates of treatment corals were significantly lower (−212%) than in controls initially but had fully recovered 6 weeks later (Fig. 1F) [13].

Table 1. Results of two-way ANOVA for area-normalized chlorophyll *a* content of *P. divaricata*, *P. astreoides*, and *O. faveolata*.

Variable	Effect	df	SS	F-statistic	p-value	Tukey
P. divaricata						
Chlorophyll *a*	Model	3, 28	164.4113	9.82	**0.0002**	
	Temp.	1	0.3307	0.06	0.8097	
	Time	1	82.8258	14.84	**0.0007**	
	Temp. x Time	1	63.1744	11.32	**0.0025**	
P. astreoides						
Chlorophyll *a*	Model	3, 33	469.1181	43.44	**<0.0001**	
	Temp.	1	456.8590	126.91	**<0.0001**	CO>TR
	Time	1	7.9244	2.20	0.1483	
	Temp. x Time	1	3.6588	1.02	0.3214	
O. faveolata						
Chlorophyll *a*	Model	3, 33	206.4635	12.44	**<0.0001**	
	Temp.	1	106.3292	19.22	**0.0001**	CO>TR
	Time	1	99.1921	17.93	**0.0002**	6>0
	Temp. x Time	1	2.4176	0.44	0.5136	

The effect of temperature (Temp.) was fixed with two levels (CO = control 30.4°C, TR = treatment 31.6°C), and time was fixed with 2 levels (0, 6 weeks). When main effects (but no interaction terms) were significant, Tukey post hoc results are shown. Significant p-values ($p\leq0.05$) are highlighted in bold. df = degrees of freedom, SS = sum of squares of the effect.

Isotopes

$\delta^{11}B$ was not influenced by temperature or time in any of the three species (Table 2, Fig. 2A, D, G). Similarly, $\delta^{13}C$ and $\delta^{18}O$ of *P. divaricata* were not affected by either temperature or time (Table 2, Fig. 2B, C). However, a significant interaction of temperature and time was observed for $\delta^{13}C$ and $\delta^{18}O$ of *P. astreoides* (Table 2, Fig. 2E, F). Initially, $\delta^{13}C$ and $\delta^{18}O$ did not differ between treatment and control corals, but after 6 weeks on the reef, both $\delta^{13}C$ and $\delta^{18}O$ were significantly higher in treatment compared to control corals (Fig. 2E, F). In *O. faveolata*, $\delta^{13}C$ and $\delta^{18}O$ were not influenced by temperature or time, but post hoc slice tests revealed that treatment corals had higher values than the controls at 0 weeks on the reef (Fig. 2H, I, Table 2).

Trace Elements

In *P. divaricata*, Sr/Ca, Mg/Ca, and U/Ca were not affected by either temperature or time (Table 3, Fig. 3A-C). However, Ba/Ca was affected by both temperature and time, with higher ratios found in controls than in treatment corals and increases in Ba/Ca after 6 weeks on the reef compared to 0 weeks (Table 3, Fig. 3D).

In *P. astreoides*, Sr/Ca, Mg/Ca and Ba/Ca were not affected by either temperature or time (Table 3, Fig. 3E, F, H). However, post hoc slice tests revealed that Mg/Ca ratios were higher

(+16.5%) in treatment than in control corals at 0 weeks (Table 3, Fig. 3F). A significant interaction between temperature and time was observed for U/Ca, with ratios being significantly higher (+20.3%) in treatment than in control corals after 6 weeks on the reef (Table 3, Fig. 3G).

In *O. faveolata*, Sr/Ca, Mg/Ca and U/Ca were not influenced by either temperature or time (Table 3, Fig. 3I-K). However, post hoc slice tests revealed that treatment Mg/Ca ratios were lower (−26.3%) than the controls at 0 weeks on the reef (Fig. 3J). A significant interaction of temperature and time was observed for Ba/Ca, with treatment values being higher (+5.8%) than the control initially, but no longer different after 6 weeks on the reef (Fig. 3L, Table 3).

Discussion

Isotopes

Coral skeletal boron isotopes have been established as a proxy for seawater pH, yet it remains unclear if and how this proxy is affected by seawater temperature. Specifically, it has never been tested whether coral bleaching due to elevated temperature affects coral $\delta^{11}B$. Here, we show for the first time that experimental coral bleaching does not affect $\delta^{11}B$ of three Caribbean coral species

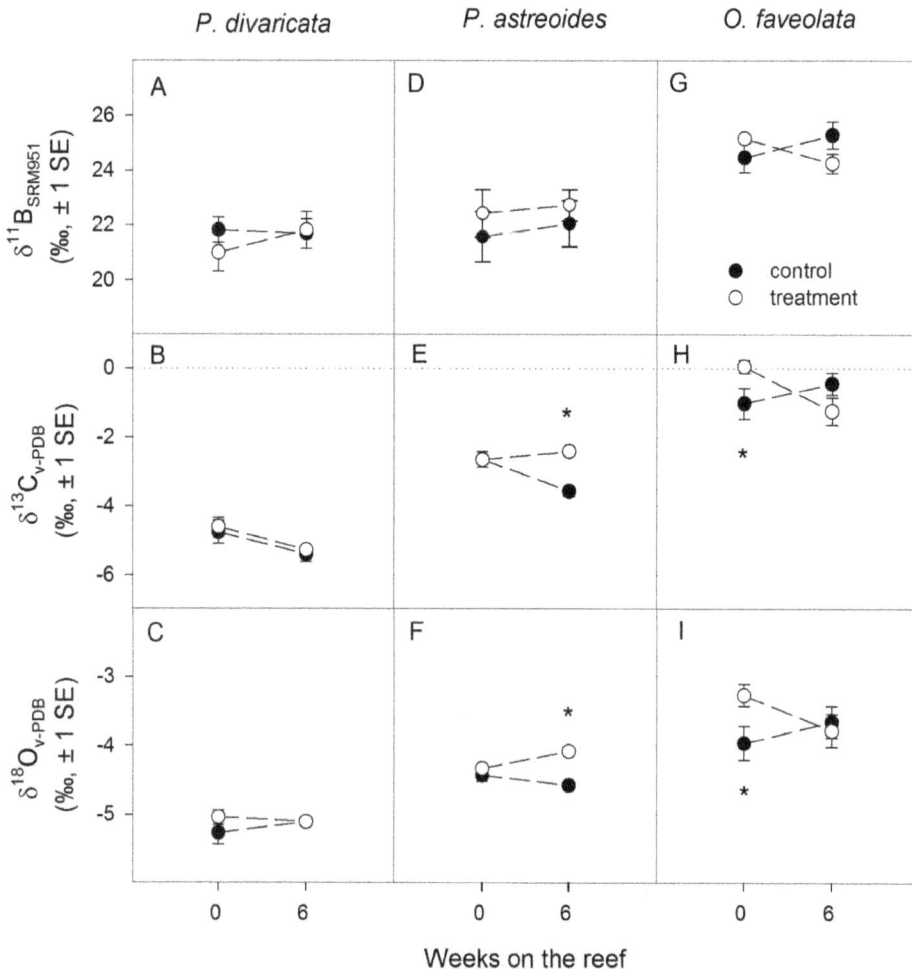

Figure 2. Average $\delta^{11}B$, $\delta^{13}C$, and $\delta^{18}O$ of (A-C) Porites divaricata, (D-F) Porites astreoides, and (G-I) Orbicella faveolata after 0 and 6 weeks on the reef. Asterisks indicate significant differences between control and treatment corals at a specific time interval. Sample size ranges from 3–9.

Table 2. Results of two-way ANOVAs for $\delta^{11}B$, $\delta^{13}C$, and $\delta^{18}O$ of *P. divaricata*, *P. astreoides*, and *O. faveolata*.

Variable	Effect	df	SS	F-statistic	p-value
P. divaricata					
$\delta^{11}B$	Model	3, 19	2.7948	0.46	0.7148
	Temp.	1	0.5602	0.28	0.6066
	Time	1	0.5602	0.28	0.6066
	Temp. x Time	1	1.0643	0.52	0.4795
$\delta^{13}C$	Model	3, 23	2.8226	2.76	0.0689
	Temp.	1	0.1219	0.36	0.5565
	Time	1	2.5603	7.51	**0.0126**
	Temp. x Time	1	0.0008	0.00	0.9610
$\delta^{18}O$	Model	3, 23	0.1570	0.89	0.4619
	Temp.	1	0.0778	1.33	0.2628
	Time	1	0.0111	0.19	0.6683
	Temp. x Time	1	0.0778	1.33	0.2628
P. astreoides					
$\delta^{11}B$	Model	3, 22	3.8106	0.32	0.8094
	Temp.	1	3.2908	0.83	0.3725
	Time	1	0.8574	0.22	0.6464
	Temp. x Time	1	0.0420	0.01	0.9189
$\delta^{13}C$	Model	3, 34	7.2196	10.56	**<0.0001**
	Temp.	1	2.9506	12.95	**0.0011**
	Time	1	1.0442	4.58	**0.0403**
	Temp. x Time	1	3.0344	13.32	**0.0010**
$\delta^{18}O$	Model	3, 34	1.1779	10.85	**<0.0001**
	Temp.	1	0.7728	21.36	**<0.0001**
	Time	1	0.0240	0.66	0.4217
	Temp. x Time	1	0.3460	9.56	**0.0042**
O. faveolata					
$\delta^{11}B$	Model	3, 13	2.6525	1.25	0.3430
	Temp.	1	0.0981	0.14	0.7173
	Time	1	0.0086	0.01	0.9145
	Temp. x Time	1	2.5358	3.59	0.0876
$\delta^{13}C$	Model	3, 33	8.3114	2.56	0.0739
	Temp.	1	0.1528	0.14	0.7100
	Time	1	1.0742	0.99	0.3275
	Temp. x Time	1	7.3876	6.82	**0.0140**
$\delta^{18}O$	Model	3, 33	2.1365	1.69	0.1903
	Temp.	1	0.6800	1.61	0.2138
	Time	1	0.0845	0.20	0.6576
	Temp. x Time	1	1.4076	3.34	0.0776

The effect of temperature (Temp.) was fixed with two levels (control 30.4°C, treatment 31.6°C), and time was fixed with 2 levels (0, 6 weeks). Significant p-values ($p \leq$ 0.05) are highlighted in bold. df = degrees of freedom, SS = sum of squares of the effect.

(Fig. 2A, D, G). This finding does not support our initial hypothesis based on evidence presented by Wei et al. [47] that coral $\delta^{11}B$ may decrease significantly during coral bleaching events. However, it is in agreement with experimental studies showing that $\delta^{11}B$ is not affected by moderate increases in temperatures from 25°C to 28°C [43,44], which do not induce coral bleaching.

Coral bleaching is caused by a significant loss of algal endosymbionts and/or photosynthetic pigments from the coral tissue [7,9,12]. In the current study, treatment corals of all species had significantly lower area-normalized chlorophyll *a* levels than controls throughout the study (Fig. 1A, C, E), primarily driven by lower endosymbiont densities [13]. Further, treatment corals of all species had a significantly lower maximum quantum yield of photosystem II (F_v/F_m) than controls for up to 6 weeks on the reef [59]. Thus, treatment corals of all three species were bleached. This demonstrates that even though corals experienced significant

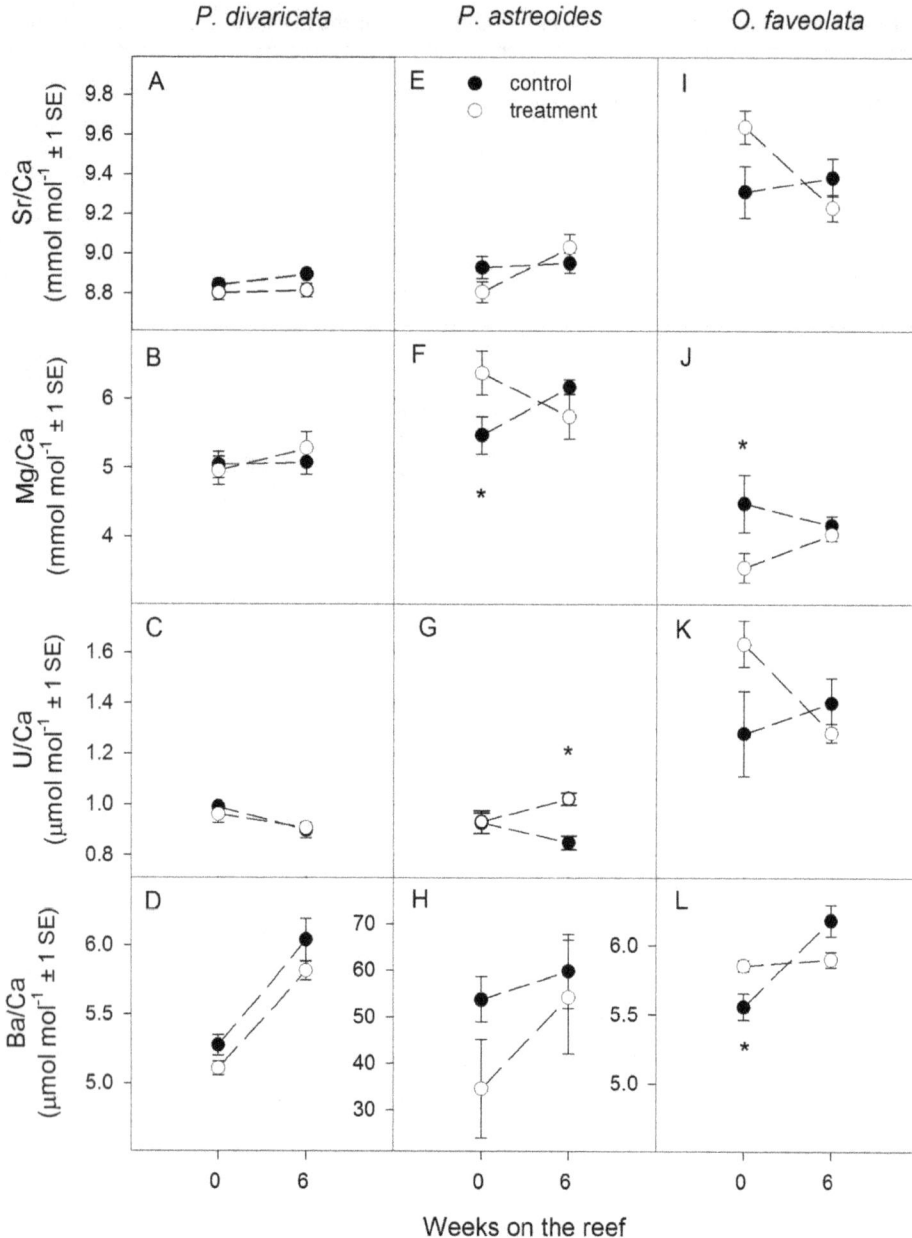

Figure 3. Average Sr/Ca, Mg/Ca, U/Ca, and Ba/Ca of (A–D) Porites divaricata, (E–H) Porites astreoides, and (I–L) Orbicella faveolata after 0 and 6 weeks on the reef. Asterisks indicate significant differences between control and treatment corals at a specific time interval. Sample size ranges from 3–7. Note that the Y-axis for Ba/Ca differs for P. astreoides and the other two species.

temperature stress and bleaching, coral skeletal $\delta^{11}B$ remained unaffected.

Calcification rates are often compromised in bleached corals [17,52,54,71], and may therefore have affected the response of skeletal isotopes and trace elements [72,73]. Indeed, calcification rates of treatment *O. faveolata* were significantly lower than in the controls at 0 weeks on the reef (Fig. 1D). In *P. astreoides*, calcification was not significantly lower than the controls until after 6 weeks on the reef (Fig. 1F) because the bleaching response of the animal host showed a lag of several weeks in this coral species [13]. Such lags in the bleaching response have been previously observed in three species of Hawaiian corals as well [17,52]. However, none of these declines in calcification resulted in significant differences

in $\delta^{11}B$ between treatment and control corals (Fig. 2D, G) despite the fact that treatment corals were significantly bleached at this time (Fig. 1C, E). Similarly, when calcification rates were maintained in bleached corals, as was the case in *P. divaricata* and in *P. astreoides* initially at the end of the heating period (Fig. 1B, D), $\delta^{11}B$ of treatment corals did not differ from the controls (Fig. 2A, D). Based on these results, coral $\delta^{11}B$ is unaffected by calcification rate.

The insensitivity of $\delta^{11}B$ to calcification rates is further corroborated by the fact that both skeletal $\delta^{13}C$ and $\delta^{18}O$ changed in response to decreases in calcification rates in two species, yet $\delta^{11}B$ did not (Figs. 1, 2). In treatment *P. astreoides* and *O. faveolata*, both $\delta^{13}C$ and $\delta^{18}O$ increased as calcification

Table 3. Results of two-way ANOVAs for Mg/Ca, Sr/Ca, Ba/Ca, and U/Ca of *P. divaricata*, *P. astreoides*, and *O. faveolata*.

Variable	Effect	df	SS	F-statistic	p-value	Tukey
P. divaricata						
Sr/Ca	Model	3, 19	0.0241	1.49	0.2557	
	Temp.	1	0.0170	3.15	0.0950	
	Time	1	0.0055	1.02	0.3273	
	Temp. x Time	1	0.0019	0.35	0.5622	
Mg/Ca	Model	3, 19	0.2820	0.42	0.7427	
	Temp.	1	0.0170	0.08	0.7868	
	Time	1	0.1575	0.70	0.4152	
	Temp. x Time	1	0.1045	0.46	0.5054	
U/Ca	Model	3, 19	0.0264	2.02	0.1520	
	Temp.	1	0.0005	0.10	0.7504	
	Time	1	0.0252	5.78	**0.0287**	
	Temp. x Time	1	0.0017	0.39	0.5435	
Ba/Ca	Model	3, 19	2.8756	27.77	**<0.0001**	
	Temp.	1	0.1797	5.21	**0.0365**	CO>TR
	Time	1	2.5673	74.38	**<0.0001**	6>0
	Temp. x Time	1	0.0038	0.11	0.7456	
P. astreoides						
Sr/Ca	Model	3, 22	0.1453	2.69	0.0751	
	Temp.	1	0.0029	0.16	0.6926	
	Time	1	0.0860	4.78	**0.0415**	
	Temp. x Time	1	0.0570	3.17	0.0911	
Mg/Ca	Model	3, 22	2.7689	2.47	0.0929	
	Temp.	1	0.3173	0.85	0.3681	
	Time	1	0.0069	0.02	0.8934	
	Temp. x Time	1	2.5033	6.71	**0.0180**	
U/Ca	Model	3, 22	0.0872	4.11	**0.0210**	
	Temp.	1	0.0442	6.24	**0.0218**	
	Time	1	0.0003	0.04	0.8529	
	Temp. x Time	1	0.0398	5.61	**0.0286**	
Ba/Ca	Model	3, 21	2242.8826	1.63	0.2181	
	Temp.	1	808.5285	1.76	0.2012	
	Time	1	871.9859	1.90	0.1851	
	Temp. x Time	1	246.0976	0.54	0.4736	
O. faveolata						
Sr/Ca	Model	3, 13	0.2863	2.48	0.1206	
	Temp.	1	0.0259	0.67	0.4308	
	Time	1	0.0962	2.50	0.1447	
	Temp. x Time	1	0.1983	5.16	**0.0464**	
Mg/Ca	Model	3, 13	1.5265	1.94	0.1879	
	Temp.	1	0.9719	3.70	0.0835	
	Time	1	0.0238	0.09	0.7696	
	Temp. x Time	1	0.5522	2.10	0.1779	
U/Ca	Model	3, 13	0.2635	1.69	0.2309	
	Temp.	1	0.0479	0.92	0.3591	
	Time	1	0.0458	0.88	0.3696	
	Temp. x Time	1	0.1922	3.71	0.0831	
Ba/Ca	Model	3, 13	0.7797	8.77	**0.0038**	
	Temp.	1	0.0001	0.00	0.9463	
	Time	1	0.3844	12.97	**0.0048**	

Table 3. Cont.

Variable	Effect	df	SS	F-statistic	p-value	Tukey
	Temp. x Time	1	0.2847	9.61	**0.0113**	

The effect of temperature (Temp.) was fixed with two levels (CO = control 30.4°C, TR = treatment 31.6°C), and time was fixed with 2 levels (0, 6 weeks). When main effects (but no interaction terms) were significant, Tukey post hoc results are shown. Significant p-values (p≤0.05) are highlighted in bold. df = degrees of freedom, SS = sum of squares of the effect.

decreased. This is unexpected given that both $\delta^{13}C$ and $\delta^{18}O$ are assumed to decrease when photosynthesis decreases and seawater temperature increases (provided there were no changes in salinity), respectively [74–79]. Instead, the enrichment in both $\delta^{13}C$ and $\delta^{18}O$ indicates that skeletal carbonate started approaching isotopic equilibrium with seawater as calcification rates slowed down [76] and that calcification rate was the primary control on $\delta^{13}C$ and $\delta^{18}O$. Similar relationships between skeletal $\delta^{13}C$, $\delta^{18}O$, and calcification were also noted in other corals when bleached [52,80]. For *O. faveolata*, this strongly suggests that some skeletal material was deposited during the bleaching treatment, even though on average net calcification rates were slightly negative. Further, in *P. divaricata*, $\delta^{13}C$ and $\delta^{18}O$ were unaffected by bleaching at any point, which is consistent with no changes in calcification rates.

Clearly, the carbon and oxygen isotopic signal was dominated by kinetic effects associated with calcification rates [76] in all three species. Kinetic effects likely overpowered any metabolic effects in $\delta^{13}C$ due to reduced chlorophyll *a* concentrations (implying reduced photosynthesis rates) as well as the expected temperature-dependent effect on $\delta^{18}O$ [74]. Although a simple data correction has been proposed to correct for kinetic isotope effects in $\delta^{13}C$ [81], it does not reliably remove kinetic isotope effects in bleached corals [82]. Unfortunately, these findings confirm that coral $\delta^{13}C$ and $\delta^{18}O$ cannot be used as a proxy for past bleaching events [17,52,54,55], and that coral $\delta^{11}B$ does not record bleaching events as suggested by Wei et al. [47], at least not in the coral species studied here.

The observed lack of a bleaching effect on coral $\delta^{11}B$ suggests that coral bleaching – and therefore the physiological status of the algal endosymbiont - does not affect pH-upregulation at the site of calcification. This is surprising given that coral calcification is enhanced in the light via photosynthesis [83–85] due to a variety of mechanisms including removal of protons from the site of calcification, supply of ATP for ion transport and/or organic matrix synthesis, production of oxygen, supply of precursors for organic matrix production, and removal of phosphates [86,87]. Thus, if the symbiotic relationship is disrupted during bleaching, any of the mechanisms listed above would likely be interrupted as well, including the impaired removal of protons from the site of calcification that affects $\delta^{11}B$ values. Clearly, this was not the case in the coral species studied here, which further suggests that decreased calcification rates of bleached corals may not be due to impaired control over the calcifying environment.

Heterotrophy and energy reserves can promote many physiological aspects including calcification in healthy corals [88], and also enhances resistance to and recovery from coral bleaching [13,18,89,90]. It is therefore possible that heterotrophic energy input may have helped bleached corals to maintain pH-upregulation at the site of calcification. However, feeding rates and total carbon acquired by feeding relative to metabolic demand (CHAR) did not differ between treatment and control corals immediately after the bleaching treatment in any of the species

[13]. It is possible that other heterotrophic sources of carbon such as dissolved and particulate organic carbon (DOC and POC, respectively) could be providing the necessary supplemental energy for calcification in bleached corals. For example, Levas [57] showed that DOC can provide 15% of daily metabolic carbon in bleached *O. faveolata* corals.

Collectively, these results demonstrate that pH-upregulation at the site of calcification is robust to significant physiological changes caused by short bleaching events. This suggests either that this process has high priority during energy allocation even under scenarios of resource limitation such as coral bleaching, or that pH-upregulation does not require that much energy and can thus be maintained during coral bleaching. However, it is possible that when coral bleaching events are prolonged, a certain threshold of resource limitation may be reached at which corals are no longer able to maintain pH-upregulation at the site of calcification. Resource limitation in this scenario would not only be related to a decrease in transferred carbon by the algal endosymbiont [18,91], but also to oxygen limitation [92,93] and potentially other mechanisms involved in light-enhanced calcification that would be interrupted due to bleaching. This would then result in decreased $\delta^{11}B$ values of bleached corals, consistent with observations by Wei et al. [47] for a *Porites* coral core, where significant decreases in boron isotopes were interpreted as corresponding to the 1998 mass bleaching event on the Great Barrier Reef. This threshold, if existent, would likely be on the order of at least several months because the physiology of *P. astreoides* in this study was affected for at least 6 weeks [13], yet showed no decrease in boron isotopes. Further, several studies have shown that coral growth rates can take several years to fully recover from bleaching [71,80,94,95].

Trace Elements

Coral bleaching did not affect trace element composition (Sr/Ca, Mg/Ca, U/Ca, and Ba/Ca) of *P. divaricata*, but influenced Mg/Ca and U/Ca ratios of *P. astreoides* as well as Mg/Ca and Ba/Ca in *O. faveolata* (Fig. 3). The lack of any temperature effect on Sr/Ca is particularly surprising, because it is generally considered to be one of the most reliable sea surface temperature (SST) proxies available [96,97,98], although it is not free of complications [99–105]. Sr/Ca sensitivity is about 0.04–0.08 mmol mol^{-1} per °C [100,101,106–109]. Thus the treatment corals in this study were expected to have Sr/Ca ratios that were 0.05–0.1 mmol mol^{-1} lower than the control corals given the inverse correlation between temperature and Sr/Ca. At 0 weeks on the reef, treatment *P. divaricata* and *P. astreoides* had indeed lower Sr/Ca ratios than the controls (0.04 and 0.12 mmol mol^{-1}, respectively), whereas treatment *O. faveolata* had higher Sr/Ca values (0.33 mmol mol^{-1}). However, none of these differences were statistically significant (Fig. 3A, E, I).

During cold and hot temperature stress, sharp increases in Sr/Ca have been observed in *Porites* coral from the Great Barrier Reef [53,96]. Marshall and McCulloch [53] proposed that temperature stress disrupts the Sr/Ca – SST relationship by inhibiting Ca transport enzymes, whereas Sr transport is unaffected. Thus, the Sr/Ca ratio increases because relatively less Ca is made available to the calcifying fluid. Similarly, Cohen et al. [99] suggested that endosymbiont photosynthesis enhances Ca but not Sr transport to the calcifying fluid, thus resulting in lower Sr/Ca ratios during times of high photosynthesis. Following this reasoning, Sr/Ca should therefore increase in bleached corals when photosynthesis is impaired and calcification slows down. However, the findings from this study do not generally support this hypothesis. Even though Sr/Ca ratios were higher in treatment than in control *O. faveolata* initially (Fig. 3I) when chlorophyll *a* concentrations and calcification rates were significantly lower (Fig. 1F), this difference was not statistically significant ($p = 0.0536$). Marshall and McCulloch [53] suggested that a particular temperature threshold needs to be reached before Sr/Ca ratios are affected by temperature stress. Potentially, this threshold differs for different coral species and therefore resulted in equivocal results here.

Changes in calcification rates due to coral bleaching may have influenced Sr/Ca [100,101,110]. Slower growing corals typically have higher Sr/Ca values than faster growing corals [100,101,110]. However, this was not generally observed, and many other studies also failed to detect growth rate effects on Sr/Ca [102,106,111–113]. It therefore seems that overall coral skeletal Sr/Ca is not influenced by the physiological changes or temperature changes occurring during short coral bleaching events.

Mg/Ca ratios were affected by coral bleaching in *P. astreoides* and *O. faveolata*, but not *P. divaricata* (Fig. 3B, F, J). Sensitivity of Mg/Ca is about 0.09–0.16 mmol mol^{-1} per °C [108,109,114], and therefore treatment corals were expected to have Mg/Ca ratios that were higher by 0.1–0.2 mmol mol^{-1} compared to control corals at 0 weeks on the reef. Although treatment *P. astreoides* had significantly higher Mg/Ca ratios than control corals at this time point (Fig. 3F), the difference was 0.9 mmol mol^{-1}, which is much higher than expected based on temperature dependence alone. Further, Mg/Ca decreased with bleaching in *O. faveolata* instead of the expected increase (Fig. 3J). However, the reliability of Mg/Ca as a SST proxy has been questioned by many studies [98,108,112]. Mg/Ca ratios showed similar patterns as calcification rates in both *P. divaricata* and *O. faveolata* and could therefore be mainly controlled by calcification rate. However, they were not significantly correlated in either species (Spearman's $r = 0.40$, $p = 0.10$ and $r = 0.63$, $p = 0.13$, respectively). They were also not significantly correlated in *P. astreoides* ($r = -0.06$, $p = 0.84$). It is therefore unlikely that Mg/Ca ratios are primarily controlled by calcification rate.

Sensitivity of U/Ca is about 0.03–0.05 µmol mol^{-1} per °C [115] and treatment corals were therefore expected to have U/Ca ratios that are 0.04–0.06 µmol mol^{-1} lower than in control corals at 0 weeks on the reef. However, either no change or the opposite (i.e., increases in U/Ca) was observed. It is therefore likely that seawater temperature is not the primary control on coral U/Ca. Potential salinity [115] and pH effects [115,116] can be excluded as treatment and control corals were kept in the same seawater (except for temperature). Coral δ^{11}B further suggests that the pH of the calcifying fluid did not significantly differ between treatment and control corals (Fig. 2).

Although Ba/Ca ratios are typically used as a proxy for upwelling of nutrient-rich water [108,117,118] and river input and flood events [119,120], in some species Ba/Ca ratios can also be influenced by SST [117,121]. If that were the case in the present study, treatment corals would be expected to have lower Ba/Ca

ratios than control corals at 0 weeks on the reef. However, this was not generally the case. As all corals were kept in the same seawater and location on the reef, both river input and temperature effects are unlikely to have caused the observed pattern.

Interestingly, Ba/Ca ratios of *P. astreoides* were an order of magnitude higher than in either *P. divaricata* or *O. faveolata*, showed high within-treatment variability, and were also much higher than typical flood or upwelling related Ba signals in corals (Fig. 3H) [118,119,120]. Considering that all three coral species were exposed to identical environmental conditions except for temperature during the bleaching treatment, this finding is difficult to explain and cannot be related to seawater Ba concentrations. Even in exceptionally high Ba seawater, coral skeletal Ba/Ca concentrations do not exceed 6.5 µmol mol^{-1} [122]. However, similar differences in Ba/Ca between different coral genera have also been observed in another study [123].

One possibility is that particulates and/or organic phases rich in Ba may have become trapped within the skeleton [124,125]. Due to its encrusting to mounding morphology, sediment settles more easily on *P. astreoides* compared to the other two species (V. Schoepf, pers. observation) and may somehow result in enriched Ba concentrations in the surrounding seawater and/or skeleton. Spawning has also been hypothesized to influence Ba/Ca ratios [125], but the seasonal peak in reproduction occurs in April for *P. astreoides* [126], whereas the corals here were collected in August and September, respectively. Overall, these findings add to the existing literature of anomalously high Ba concentrations in corals, which at present cannot be satisfactorily explained [118,124,125,127].

Overall, trace element data often showed trends that were inconsistent with either changes in seawater temperature or the observed physiological effects. This was in stark contrast to the isotopic data, which showed consistent trends across species, with calcification rate being the main influence on δ^{13}C and δ^{18}O but not δ^{11}B in bleached corals. Given that it is difficult to explain these inconsistencies based on our current understanding of trace elements, these findings demonstrate that the mechanisms of trace element incorporation in bleached corals are poorly understood and that they are likely species-specific. Species-specific differences in the bleaching response of the three coral species studied here [13] may in part be responsible for the observed inconsistencies. For example, *P. divaricata* was least affected by bleaching whereas *P. astreoides* suffered the largest declines of all three species in both symbiont and host performance (Fig. 1A–D) [13]. In contrast, *O. faveolata* showed large health declines initially but was able to recover calcification rates, energy reserves, and endosymbiont densities within 6 weeks (Fig. 1E, F) [13]. Similarly, trace elements showed no significant temperature effects and little within-treatment variability in *P. divaricata* whereas temperature effects were often evident in the other two species (Fig. 3). Although the magnitude and direction of these effects were typically inconsistent with our current knowledge about trace element incorporation and often differed between *P. astreoides* and *O. faveolata* (see above), this could nevertheless indicate that more severely bleached corals may show larger excursions in their trace elements. This study therefore highlights the need to study trace element incorporation under stressful temperatures in a variety of coral species, especially in the context of accompanying physiological information.

Implications for Paleo-Climate Reconstruction

The present study provides the first experimental evidence that short coral bleaching events do not affect boron isotopes in three Caribbean coral species. This is generally good news because it suggests that past seawater pH conditions can be reconstructed without any conflicting effects introduced by significant physiological changes due to short bleaching events. This is particularly important for massive coral species such as *O. faveolata* or *Porites* sp., which are frequently used for paleo-pH reconstruction. However, at the same time, this study indicates that coral $\delta^{11}B$ cannot be used as a novel proxy for past bleaching events as suggested by Wei et al. [47]. Similarly, our findings confirm that $\delta^{13}C$ and $\delta^{18}O$ cannot be used as bleaching proxies when calcification rates are compromised, demonstrating that there is currently no reliable proxy available to identify past bleaching events. However, future studies are needed to confirm these findings for prolonged coral bleaching events.

Regarding trace elements, Sr/Ca of bleached corals did not consistently record the 1.2°C difference in seawater temperature during the bleaching treatment, or alternatively show a consistent increase due to impaired photosynthesis and calcification. These findings suggest that the mechanisms of Sr/Ca incorporation are not well understood at temperatures that are stressful to corals. Further, the observed changes in Mg/Ca, U/Ca and Ba/Ca due to coral bleaching could not be satisfactorily explained using either temperature dependence or changes in coral physiology. There-fore, it is likely that additional factors influence these geochemical proxies and that these factors are species-specific. It is therefore not recommended to use corals with a history of bleaching for paleo-climate reconstruction using trace elements.

Acknowledgments

In Mexico, we thank R. Iglesias-Prieto, A. Banaszak, S. Enriquez, R. Smith, and the staff of the Instituto de Ciencias del Mar y Limnologia, Universidad Nacional Autonoma de Mexico in Puerto Morelos for their generous time and logisitical support. At The Ohio State University, we thank T. Huey, D. Borg, E. Zebrowski, J. Scheuermann, and M. McBride for help in the field and laboratory. At the University of Western Australia, we thank M. Holcomb, H. Oskierski, K. Rankenburg, A. Kuret, J. P. D'Olivo Cordero and J. Trotter for their help in the laboratory and general advice.

Author Contributions

Conceived and designed the experiments: AGG MEW VS. Performed the experiments: VS SJL MDA YM AGG MEW. Analyzed the data: VS. Contributed reagents/materials/analysis tools: VS MTM MEW SJL YM MDA AGG. Wrote the paper: VS MTM MEW SJL YM MDA AGG.

References

1. IPCC (2013) Climate Change 2013: The physical science basis. Summary for Policy Makers. Available: http://www.ipcc.ch. Accessed 2014 Oct 26.
2. Hoegh-Guldberg O, Bruno JF (2010) The impact of climate change on the world's marine ecosystems. Science 328: 1523–1528.
3. Harvey BP, Gwynn-Jones D, Moore PJ (2013) Meta-analysis reveals complex marine biological responses to the interactive effects of ocean acidification and warming. Ecol Evol 3: 1016–1030.
4. Wild C, Hoegh-Guldberg O, Naumann MS, Colombo-Pallotta MF, Ateweberhan M, et al. (2011) Climate change impedes scleractinian corals as primary reef ecosystem engineers. Mar Freshwater Res 62: 205–215.
5. Kleypas J, Buddemeier RW, Archer D, Gattuso J-P, Langdon C, et al. (1999) Geochemical consequences of increased atmospheric carbon dioxide on coral reefs. Science 284: 118–120.
6. Hoegh-Guldberg O, Mumby PJ, Hooten AJ, Steneck R, Greenfield P, et al. (2007) Coral reefs under rapid climate change and ocean acidification. Science 318: 1737–1742.
7. Hoegh-Guldberg O, Smith GJ (1989) The effect of sudden changes in temperature, light, and salinity on the population density and export of zooxanthellae from the reef coral *Stylophora pistillata* Esper and *Seriatopora hystrix* Dana J Exp Mar Biol Ecol 129: 279–303.
8. Fitt WK, Brown BE, Warner ME, Dunne RP (2001) Coral bleaching: interpretation of thermal tolerance limits and thermal thresholds in tropical corals. Coral Reefs 20: 51–65.
9. Jokiel PL, Coles SL (1990) Response of Hawaiian and other Indo-Pacific reef corals to elevated temperature. Coral Reefs 8: 155–162.
10. Baker AC, Glynn PW, Riegl B (2008) Climate change and coral reef bleaching: An ecological assessment of long-term impacts, recovery trends and future outlook. Estuar Coast Shelf Sci 80: 435–471.
11. Hoegh-Guldberg O (1999) Climate change, coral bleaching and the future of the world's coral reefs. Mar Freshwater Res 50: 839–866.
12. Glynn PW (1996) Coral reef bleaching: facts, hypotheses and implications. Global Change Biol 2: 495–509.
13. Grottoli AG, Warner M, Levas SJ, Aschaffenburg M, Schoepf V, et al. (2014) The cumulative impact of annual coral bleaching can turn some coral species winners into losers. Global Change Biol. doi: 10.1111/gcb.12658.
14. Nakamura T, Yamasaki H, van Woesik R (2003) Water flow facilitates recovery from bleaching in the coral *Stylophora pistillata*. Mar Ecol Prog Ser 256: 287–291.
15. Fitt WK, Spero HJ, Halas J, White MW, Porter JW (1993) Recovery of the coral *Montastraea annularis* in the Florida Keys after the 1987 Caribbean "bleaching event". Coral Reefs 12: 57–64.
16. Rodrigues LJ, Grottoli AG (2007) Energy reserves and metabolism as indicators of coral recovery from bleaching. Limnol Oceanogr 52: 1874–1882.
17. Levas SJ, Grottoli AG, Hughes AD, Osburn CL, Matsui Y (2013) Physiological and biogeochemical traits of bleaching and recovery in the mounding species of coral *Porites lobata*: Implications for resilience in mounding corals. PLoS One 8: e63267. doi:63210.61371/journal.pone.0063267.
18. Grottoli AG, Rodrigues LJ, Palardy JE (2006) Heterotrophic plasticity and resilience in bleached corals. Nature 440: 1186–1189.
19. Loya Y, Sakai K, Yamazato K, Nakan Y, Sambali H, et al. (2001) Coral bleaching: the winners and the losers. Ecol Lett 4: 122–131.
20. Donner SD (2009) Coping with commitment: Projected thermal stress on coral reefs under different future scenarios. PLoS One 4: e5712. doi:5710.1371/journal.pone.0005712.
21. Teneva L, Karnauskas M, Logan C, Bianucci L, Currie J, et al. (2012) Predicting coral bleaching hotspots: the role of regional variability in thermal stress and potential adaptation rates. Coral Reefs 31: 1–12.
22. Frieler K, Meinshausen M, Golly A, Mengel M, Lebek K, et al. (2013) Limiting global warming to 2°C is unlikely to save most coral reefs. Nat Clim Change 3: 165–170.
23. Wilkinson C (2008) Status of coral reefs of the world: 2008. Global Coral Reef Monitoring Network and Reef and Rainforest Research Centre. 296 p.
24. Carpenter KE, Abrar M, Aeby G, Aronson RB, Banks S, et al. (2008) One-third of reef-building corals face elevated extinction risk from climate change and local impacts. Science 321: 560–563.
25. Orr JC, Fabry VJ, Aumont O, Bopp L, Doney SC, et al. (2005) Anthropogenic ocean acidification over the twenty-first century and its impact on calcifying organisms. Nature 437: 681–686.
26. Caldeira K, Wickett ME (2003) Anthropogenic carbon and ocean pH. Nature 425: 365.
27. Langdon C, Atkinson MJ (2005) Effect of elevated pCO2 on photosynthesis and calcification on corals and interactions with seasonal change in temperature/irradance and nutrient enrichment. J Geophys Res 110: C09S07. doi:10.1029/2004JC002576.
28. Holcomb M, McCorkle DC, Cohen AL (2010) Long-term effects of nutrient and CO2 enrichment on the temperate coral *Astrangia poculata* (Ellis and Solander, 1786). J Exp Mar Biol Ecol 386: 27–33.
29. Schoepf V, Grottoli AG, Warner M, Cai W-J, Melman TF, et al. (2013) Coral energy reserves and calcification in a high-CO2 world at two temperatures. PLoS One 8. e75049 doi:75010.71371/journal.pone.0075049.
30. Marubini F, Ferrier-Pages C, Cuif J-P (2003) Suppression of skeletal growth in scleractinian corals by decreasing ambient carbonate-ion concentration: a cross-family comparison. Proc R Soc B 270: 179–184.
31. Krief S, Hendy EJ, Fine M, Yam R, Meibom A, et al. (2010) Physiological and isotopic responses of scleractinian corals to ocean acidification. Geochim Cosmochim Acta 74: 4988–5001.
32. Edmunds PJ, Brown D, Moriarty V (2012) Interactive effects of ocean acidification and temperature on two scleractinian corals from Moorea, French Polynesia. Global Change Biol 18: 2173–2183.
33. Comeau S, Edmunds PJ, Spindel NB, Carpenter RC (2013) The responses of eight coral reef calcifiers to increasing partial pressure of CO2 do not exhibit a tipping point. Limnol Oceanogr 58: 388–398.
34. Trotter J, Montagna P, McCulloch MT, Silenzi S, Reynaud S, et al. (2011) Quantifying the pH 'vital effect' in the temperate zooxanthellate coral *Cladocora caespitosa*: Validation of the boron seawater pH proxy. Earth Planet Sci Lett 303: 163–173.
35. Holcomb M, Venn AA, Tambutte E, Tambutte S, Allemand D, et al. (2014) Coral calcifying fluid pH dictates response to ocean acidification. Sci Rep 4. doi:10.1038/srep05207.
36. Ries JB (2011) A physicochemical framework for interpreting the biological calcification response to CO2-induced ocean acidification. Geochim Cosmochim Acta 75: 4053–4064.

37. Venn AA, Tambutte E, Holcomb M, Laurent J, Allemand D, et al. (2013) Impact of seawater acidification on pH at the tissue-skeleton interface and calcification in reef corals. Proc Nat Acad Sci USA 110: 1634–1639.

38. Venn A, Tambutte E, Holcomb M, Allemand D, Tambutte S (2011) Live tissue imaging shows reef corals elevate pH under their calcifying tissue relative to seawater. PLoS One 6: e20013. doi: 20010.21371/journal.pone.0020013.

39. Al-Horani FA, Al-Moghrabi SM, de Beer D (2003) The mechanism of calcification and its relation to photosynthesis and respiration in the scleractinian coral Galaxea fascicularis. Mar Biol 142: 419–426.

40. McCulloch MT, Falter J, Trotter J, Montagna P (2012) Coral resilience to ocean acidification and global warming through pH up-regulation. Nat Clim Change 2: 623–627.

41. Chan NCS, Connolly SR (2013) Sensitivity of coral calcification to ocean acidification: a meta-analysis. Global Change Biol 19: 282–290.

42. McCulloch MT, Trotter J, Montagna P, Falter J, Dunbar R, et al. (2012) Resilience of cold-water scleractinian corals to ocean acidification: Boron isotopic systematics of pH and saturation state up-regulation. Geochim Cosmochim Acta 87: 21–34.

43. Dissard D, Douville E, Reynaud S, Juillet-Leclerc A, Montagna P, et al. (2012) Light and temperature effect on $d^{11}B$ and B/Ca ratios of the zooxanthellate coral Acropora sp.: results from culturing experiments. Biogeosciences 9: 4589–4605.

44. Reynaud S, Hemming NG, Juillet-Leclerc A, Gattuso J-P (2004) Effect of pCO_2 and temperature on the boron isotopic composition of the zooxanthellate coral Acropora sp. Coral Reefs 23: 539–546.

45. Hönisch B, Hemming NG, Grottoli AG, Amat A, Hanson GN, et al. (2004) Assessing scleractinian corals as recorders for paleo-pH: Empirical calibration and vital effects. Geochim Cosmochim Acta 68: 3675–3685.

46. Douville E, Paterne M, Cabioch G, Louvat P, Gaillardet J, et al. (2010) Abrupt sea surface pH change at the end of the Younger Dryas in the central sub-equatorial Pacific inferred from boron isotope abundance in corals (Porites). Biogeosciences 7: 2445–2459.

47. Wei G, McCulloch MT, Mortimer G, Deng W, Xie L (2009) Evidence for ocean acidification in the Great Barrier Reef of Australia. Geochim Cosmochim Acta 73: 2332–2346.

48. Pelejero C, Calvo E, McCulloch MT, Marshall JF, Gagan MK, et al. (2005) Preindustrial to modern interdecadal variability in coral reef pH. Science 309: 2204–2207.

49. Hemming NG, Hanson GN (1992) Boron isotopic composition and concentration in modern marine carbonates. Geochim Cosmochim Acta 56: 537–543.

50. Klochko K, Cody GD, Tossell JA, Dera P, Kaufman AJ (2009) Re-evaluating boron speciation in biogenic calcite and aragonite using ^{11}B MAS NMR. Geochim Cosmochim Acta 73: 1890–1900.

51. Rollion-Bard C, Blamart D, Trebosc J, Tricot G, Mussi A, et al. (2011) Boron isotopes as pH proxy: A new look at boron speciation in deep-sea corals using ^{11}B MAS NMR and EELS. Geochim Cosmochim Acta 75: 1003–1012.

52. Rodrigues LJ, Grottoli AG (2006) Calcification rate and the stable carbon, oxygen, and nitrogen isotopes in the skeleton, host tissue, and zooxanthellae of bleached and recovering Hawaiian corals. Geochim Cosmochim Acta 70: 2781–2789.

53. Marshall JF, McCulloch MT (2002) An assessment of the Sr/Ca ratio in shallow water hermatypic corals as a proxy for sea surface temperature. Geochim Cosmochim Acta 66: 3263–3280.

54. Leder JJ, Szmant AM, Swart P (1991) The effect of prolonged "bleaching" on skeletal banding and stable isotopic composition in Montastraea annularis. Coral Reefs 10: 19–27.

55. Hartmann AC, Carilli JE, Norris RD, Charles CD, Deheyn DD (2010) Stable isotopic records of bleaching and endolithic algae blooms in the skeleton of the boulder forming coral Montastraea faveolata. Coral Reefs 29: 1079–1089.

56. Schoepf V (2013) Physiology and Biogeochemistry of Corals Subjected to Repeat Bleaching and Combined Ocean Acidification and Warming. The Ohio State University. pp. 295.

57. Levas SJ (2012) Biogeochemistry and Physiology of Bleached and Recovering Hawaiian and Caribbean Corals. The Ohio State University. pp. 238.

58. McGinley M (2012) The impact of environmental stress on Symbiodinium spp.: A molecular and community-scale investigation. University of Delaware, pp. 173.

59. Aschaffenburg M (2012) The physiological response of Symbiodinium spp. to thermal and light stress: A comparison of different phylotypes and implications for coral reef bleaching. University of Delaware. pp. 155.

60. Budd AF, Fukami H, Smith ND, Knowlton N (2012) Taxonomic classification of the reef coral family Mussidae (Cnidaria: Anthozoa: Scleractinia). Zool J Linn Soc 166: 465–529.

61. Jeffrey SW, Humphrey GF (1975) New spectrophotometric equations for determining chlorophylls a, b, c1, and c2 in higher plants, algae, and natural phytoplankton. Biochem Physiol Pflanzen 167: 191–194.

62. Marsh JA (1970) Primary productivity of reef-building calcareous red algae. Ecology 51: 255–263.

63. Jokiel PL, Maragos JE, Franzisket L (1978) Coral growth: buoyant weight technique. In: D. R Stoddart and R. E Johannes, editors. Coral Reefs: Reserach Methods. Paris: UNESCO. pp. 529–541.

64. Grottoli AG, Rodrigues LJ, Juarez C (2004) Lipids and stable carbon isotopes in two species of Hawaiian corals, Porites compressa and Montipora verrucosa, following a bleaching event. Mar Biol 145: 621–631.

65. Holcomb M, Rankenburg K, McCulloch M (2014) High-precision MC-ICP-MS measurements of $\delta^{11}B$: Matrix effects in direct injection and spray chamber sample introduction systems. In: K Grice, editor editors. Principles and Practice of Analytical Techniques in Geosciences. The Royal Society of Chemistry, pp. 254–270. In press.

66. Guerrot C, Millot R, Robert M, Négrel P (2011) Accurate and high-precision determination of boron isotopic ratios at low concentration by MC-ICP-MS (Neptune). Geostandard Newslett 35: 275–284.

67. Grottoli AG, Rodrigues LJ, Matthews KA, Palardy JE, Gibb OT (2005) Pre-treatment effects on coral skeletal $\delta^{13}C$ and $\delta^{18}O$. Chem Geol 221: 225–242.

68. Winer BJ (1971) Statistical Principles in Experimental Design. New York: McGraw-Hill.

69. Quinn GP, Keough MJ (2002) Experimental Design and Data Analysis for Biologists. New York: Cambridge University Press.

70. Moran MD (2003) Arguments for rejecting the sequential Bonferroni in ecological studies. Oikos 100: 403–405.

71. D'Olivo JP, McCulloch MT, Judd K (2013) Long-term records of coral calcification across the central Great Barrier Reef: assessing the impacts of river runoff and climate change. Coral Reefs 32: 999–1012.

72. Felis T, Paetzold J, Loya Y (2003) Mean oxygen-isotope signatures in Porites spp. corals: inter-colony variability and correction for extension-rate effects. Coral Reefs 22: 328–336.

73. Land LS, Lang JC, Barnes DJ (1975) Extension rate: A primary control on the isotopic composition of West Indian (Jamaican) scleractinian reef coral skeletons. Mar Biol 33: 221–233.

74. Weber JN, Woodhead PMJ (1972) Temperature dependence of oxygen-18 concentration in reef coral carbonates. J Geophys Res 77: 463–473.

75. Swart PK (1983) Carbon and oxygen isotope fractionation in scleractinian corals: A review. Earth Sci Rev 19: 51–80.

76. McConnaughey T (1989) ^{13}C and ^{18}O isotopic disequilibrium in biological carbonates: I. Patterns. Geochim Cosmochim Acta 53: 151–162.

77. McConnaughey TA, Burdett J, Whelan JF, Paull CK (1997) Carbon isotopes in biological carbonates: Respiration and photosynthesis. Geochim Cosmochim Acta 61: 611–622.

78. Grottoli AG (2002) Effect of light and brine shrimp levels on skeletal $d^{13}C$ values in the Hawaiian coral Porites compressa: A tank experiment. Geochim Cosmochim Acta 66: 1955–1967.

79. Grottoli AG, Wellington GM (1999) Effect of light and zooplankton on skeletal $d^{13}C$ values in the Eastern Pacific corals Pavona clavus and Pavona gigantea. Coral Reefs 18: 29–41.

80. Suzuki A, Gagan MK, Fabricius K, Isdale PJ, Yukino I, et al. (2003) Skeletal isotope microprofiles of growth perturbations in Porites corals during the 1997–1998 mass bleaching event. Coral Reefs 22: 357–369.

81. Heikoop JM, Dunn JJ, Risk MJ, Schwarcz HP, McConnaughey TA, et al. (2000) Separation of kinetic and metabolic isotope effects in carbon-13 records preserved in reef coral skeletons. Geochim Cosmochim Acta 64: 975–987.

82. Schoepf V, Levas SJ, Rodrigues LJ, McBride MO, Aschaffenburg M, et al. (in press) Kinetic and metabolic isotope effects in coral skeletal carbon isotopes: A re-evaluation using experimental coral bleaching as a case study. Geochim Cosmochim Acta.

83. Gattuso J-P, Allemand D, Frankignoulle M (1999) Photosynthesis and calcification at cellular, organismal, and community levels in coral reefs: a review on interactions and control by carbonate chemistry. Am Zool 39: 160–183.

84. Moya A, Tambutte S, Tambutte E, Zoccola D, Caminiti N, et al. (2006) Study of calcification during a daily cycle of the coral Stylophora pistillata: implications for light-enhanced calcification. J Exp Biol 17: 3413–3419.

85. Goreau TF, Goreau NI (1959) The physiology of skeleton formation in corals. II. Calcium deposition by hermatypic corals under different conditions. Biol Bull 117: 239–250.

86. Allemand D, Tambutte E, Zoccola D, Tambutte S (2011) Coral calcification, cells to reefs. In: Z Dubinsky and N Stambler, editors. Coral Reefs: An Ecosystem in Transition. Springer. pp.119–150.

87. Tambutte S, Holcomb M, Ferrier-Pages C, Reynaud S, Tambutte E, et al. (2011) Coral biomineralization: from the gene to the environment. J Exp Mar Biol Ecol. doi:10.1016/j.jembe.2011.1007.1026.

88. Houlbreque F, Ferrier-Pages C (2009) Heterotrophy in tropical scleractinian corals. Biol Rev 84: 1–17.

89. Anthony KRN, Hoogenboom MO, Maynard JF, Grottoli AG, Middlebrook R (2009) Energetics approach to predicting mortality risk from environmental stress: a case study of coral bleaching. Funct Ecol 23: 539–550.

90. Connolly SR, Lopez-Yglesias MA, Anthony KRN (2012) Food availability promotes rapid recovery from thermal stress in a scleractinian coral. Coral Reefs 31: 951–960.

91. Porter JW, Fitt WK, Spero HJ, Rogers AD, White MW (1989) Bleaching in reef corals: physiological and stable isotopic responses. Proc Natl Acad Sci USA 86: 9342–9346.

92. Holcomb M, Tambutté E, Allemand D, Tambutté S (2014) Light enhanced calcification in Stylophora pistillata: effects of glucose, glycerol and oxygen. PeerJ 2: e375.

93. Colombo-Pallotta MF, Rodriguez-Roman A, Iglesias-Prieto R (2010) Calcification in bleached and unbleached *Montastraea faveolata*: evaluating the role of oxygen and glycerol. Coral Reefs: doi:10.1007/s00338-00010-00638-x.

94. Cantin NE, Lough JM (2014) Surviving coral bleaching events: *Porites* growth anomalies on the Great Barrier Reef. PLoS One 9: e88720.

95. Carilli JE, Norris RD, Black BA, Walsh SM, McField M (2009) Local stressors reduce coral resilience to bleaching. PLoS One 4: e6324. doi: 6310.1371/journal.pone.0006324.

96. McCulloch MT, Gagan MK, Mortimer GE, Chivas AR, Isdale PJ (1994) A high-resolution Sr/Ca and δ^{18}O coral record from the Great Barrier Reef, Australia, and the 1982–1983 El Niño. Geochim Cosmochim Acta 58: 2747–2754.

97. McCulloch M, Mortimer G, Esat T, Xianhua L, Pillans B, et al. (1996) High resolution windows into early Holocene climate: Sr/Ca coral records from the Huon Peninsula. Earth Planet Sci Lett 138: 169–178.

98. Fallon S, McCulloch M, Alibert C (2003) Examining water temperature proxies in *Porites* corals from the Great Barrier Reef: a cross-shelf comparison. Coral Reefs 22: 389–404.

99. Cohen AL, Owens KE, Layne GD, Shimizu N (2002) The effect of algal symbionts on the accuracy of Sr/Ca paleotemperatures from coral. Science 296: 331–333.

100. de Villiers S, Shen GT, Nelson BK (1994) The Sr/Ca-temperature relationship in coralline aragonite: Influence of variability in (Sr/Ca)seawater and skeletal growth parameters. Geochim Cosmochim Acta 58: 197–208.

101. de Villiers S, Nelson BK, Chivas AR (1995) Biological controls on coral Sr/Ca and δ^{18}O reconstructions of sea surface temperatures. Science 269: 1247–1249.

102. Allison N, Finch AA (2004) High-resolution Sr/Ca records in modern *Porites lobata* corals: Effects of skeletal extension rate and architecture. G-cubed 5: Q05001. doi:05010.01029/02004GC000696.

103. Sun Y, Sun M, Lee T, Nie B (2005) Influence of seawater Sr content on coral Sr/Ca and Sr thermometry. Coral Reefs 24: 23–29.

104. Meibom A, Stage M, Wooden J, Constantz BR, Dunbar RB, et al. (2003) Monthly Strontium/Calcium oscillations in symbiotic coral aragonite: Biological effects limiting the precision of the paleotemperature proxy. Geophys Res Lett 30: 1418.

105. Cohen AL, Layne GD, Hart SR, Lobel PS (2001) Kinetic control of skeletal Sr/Ca in a symbiotic coral: Implications for the paleotemperature proxy. Paleoceanography 16: 20–26.

106. Alibert C, McCulloch MT (1997) Strontium/calcium ratios in modern *Porites* corals from the Great Barrier Reef as a proxy for sea surface temperature: Calibration of the thermometer and monitoring of ENSO. Paleoceanography 12: 345–363.

107. Beck JW, Edwards RL, Ito E, Taylor FW, Recy J, et al. (1992) Sea-surface temperature from coral skeletal strontium/calcium ratios. Science 257: 644–647.

108. Fallon SJ, McCulloch MT, van Woesik R, Sinclair DJ (1999) Corals at their latitudinal limits: laser ablation trace element systematics in *Porites* from Shirigai Bay, Japan. Earth Planet Sci Lett 172: 221–238.

109. Sinclair DJ, Kinsley LPJ, McCulloch MT (1998) High resolution analysis of trace elements in corals by laser ablation ICP-MS. Geochim Cosmochim Acta 62: 1889–1901.

110. Ferrier-Pages C, Boisson F, Allemand D, Tambutté E (2002) Kinetics of strontium uptake in the scleractinian coral *Stylophora pistillata*. Mar Ecol Prog Ser 245: 93–100.

111. Reynaud S, Ferrier-Pagès C, Meibom A, Mostefaoui S, Mortlock R, et al. (2007) Light and temperature effects on Sr/Ca and Mg/Ca ratios in the scleractinian coral *Acropora* sp. Geochim Cosmochim Acta 71: 354–362.

112. Mitsuguchi T, Matsumoto E, Uchida T (2003) Mg/Ca and Sr/Ca ratios of *Porites* coral skeleton: Evaluation of the effect of skeletal growth rate. Coral Reefs 22: 381–388.

113. Hayashi E, Suzuki A, Nakamura T, Iwase A, Ishimura T, et al. (2013) Growth-rate influences on coral climate proxies tested by a multiple colony culture experiment. Earth Planet Sci Lett 362: 198–206.

114. Mitsuguchi T, Matsumoto E, Abe O, Uchida T, Isdale PJ (1996) Mg/Ca thermometry in coral skeletons. Science 274: 961–963.

115. Shen GT, Dunbar RB (1995) Environmental controls on uranium in reef corals. Geochim Cosmochim Acta 59: 2009–2024.

116. Rong Min G, Lawrence Edwards R, Taylor FW, Recy J, Gallup CD, et al. (1995) Annual cycles of U/Ca in coral skeletons and U/Ca thermometry. Geochim Cosmochim Acta 59: 2025–2042.

117. Lea DW, Shen GT, Boyle EA (1989) Coralline barium records temporal variability in equatorial Pacific upwelling. Nature 340: 373–376.

118. Tudhope AW, Lea DW, Shimmield GB, Chilcott CP, Head S (1996) Monsoon climate and Arabian Sea coastal upwelling recorded in massive corals from Southern Oman. Palaios 11: 347–361.

119. McCulloch M, Fallon S, Wyndham T, Hendy E, Lough J, et al. (2003) Coral record of increased sediment flux to the inner Great Barrier Reef since European settlement. Nature 421: 727–730.

120. Alibert C, Kinsley L, Fallon SJ, McCulloch MT, Berkelmans R, et al. (2003) Source of trace element variability in Great Barrier Reef corals affected by the Burdekin flood plumes. Geochim Cosmochim Acta 67: 231–246.

121. Allison N, Finch AA (2007) High temporal resolution Mg/Ca and Ba/Ca records in modern *Porites lobata* corals. G-cubed 8: Q05001. doi:05010.01029/02006gc001477.

122. Horta-Puga G, Carriquiry JD (2012) Coral Ba/Ca molar ratios as a proxy of precipitation in the northern Yucatan Peninsula, Mexico. Appl Geochem 27: 1579–1586.

123. Pretet C, Reynaud S, Ferrier-Pagès C, Gattuso J-P, Kamber B, et al. (2014) Effect of salinity on the skeletal chemistry of cultured scleractinian zooxanthellate corals: Cd/Ca ratio as a potential proxy for salinity reconstruction. Coral Reefs 33: 169–180.

124. Hart SR, Cohen AL (1996) An ion probe study of annual cycles of Sr/Ca and other trace elements in corals. Geochim Cosmochim Acta 60: 3075–3084.

125. Sinclair DJ (2005) Non-river flood barium signals in the skeletons of corals from coastal Queensland, Australia. Earth Planet Sci Lett 237: 354–369.

126. Chornesky EA, Peters EC (1987) Sexual reproduction and colony growth in the scleractinian coral *Porites astreoides*. Biol Bull 172: 161–177.

127. Chen T, Yu K, Li S, Chen T, Shi Q (2011) Anomalous Ba/Ca signals associated with low temperature stresses in *Porites* corals from Daya Bay, northern South China Sea. Journal of Environmental Sciences 23: 1452–1459.

12

Sediment Composition Influences Spatial Variation in the Abundance of Human Pathogen Indicator Bacteria within an Estuarine Environment

Tracy L. Perkins[1], Katie Clements[2], Jaco H. Baas[2], Colin F. Jago[2], Davey L. Jones[3], Shelagh K. Malham[2,4], James E. McDonald[1]*

1 School of Biological Sciences, Bangor University, Bangor, United Kingdom, 2 School of Ocean Sciences, Bangor University, Bangor, United Kingdom, 3 School of Environment, Natural Resources and Geography, Bangor University, Bangor, United Kingdom, 4 Centre for Applied Marine Sciences, Bangor University, Bangor, United Kingdom

Abstract

Faecal contamination of estuarine and coastal waters can pose a risk to human health, particularly in areas used for shellfish production or recreation. Routine microbiological water quality testing highlights areas of faecal indicator bacteria (FIB) contamination within the water column, but fails to consider the abundance of FIB in sediments, which under certain hydrodynamic conditions can become resuspended. Sediments can enhance the survival of FIB in estuarine environments, but the influence of sediment composition on the ecology and abundance of FIB is poorly understood. To determine the relationship between sediment composition (grain size and organic matter) and the abundance of pathogen indicator bacteria (PIB), sediments were collected from four transverse transects of the Conwy estuary, UK. The abundance of culturable *Escherichia coli*, total coliforms, enterococci, *Campylobacter*, *Salmonella* and *Vibrio* spp. in sediments was determined in relation to sediment grain size, organic matter content, salinity, depth and temperature. Sediments that contained higher proportions of silt and/or clay and associated organic matter content showed significant positive correlations with the abundance of PIB. Furthermore, the abundance of each bacterial group was positively correlated with the presence of all other groups enumerated. *Campylobacter* spp. were not isolated from estuarine sediments. Comparisons of the number of culturable *E. coli*, total coliforms and *Vibrio* spp. in sediments and the water column revealed that their abundance was 281, 433 and 58-fold greater in sediments (colony forming units (CFU)/100 g) when compared with the water column (CFU/100 ml), respectively. These data provide important insights into sediment compositions that promote the abundance of PIB in estuarine environments, with important implications for the modelling and prediction of public health risk based on sediment resuspension and transport.

Editor: Newton C M. Gomes, University of Aveiro, Portugal

Funding: This research was supported by a Knowledge Economy Skills Scholarship (KESS), part-funded by the European Social Fund (ESF) through the European Union's Convergence program administered by the Welsh Government in association with Dŵr Cymru Cyf/Welsh Water Ltd. (http://www.higherskillswales.co.uk/kess/index.php.en?menu = 0& catid = 0), and a Natural Environment Research Council consortium grant under the Macronutrient Cycles Research Programme (NE/J011908/1) (http://www.nerc.ac.uk/research/funded/programmes/macronutrient/#xcollapse3). The funders had no role in study design, data collection and analysis, decision to publish, or preparation of the manuscript.

Competing Interests: The authors have declared that no competing interests exist.

* Email: j.mcdonald@bangor.ac.uk

Introduction

Estuarine environments represent some of the most biologically productive systems in the biosphere and consequently provide a wealth of economic, social and natural ecosystem services that include food, employment, recreation and habitat [1]. However, the sustainability of such systems can be severely compromised along developed and urbanised coastlines, and this is predominantly due to anthropogenic influences [2]. Almost half of the world's population are thought to live within a few hundred kilometers of the coast [3] and consequently, anthropogenic activities have a significant influence upon the health of estuarine and ocean ecosystems. Furthermore, as the global climate changes, meteorological events such as storms and floods present further impacts upon estuarine environments, and increased rainfall will

significantly impact the flow and transportation of microbial pollution from the terrestrial environment into the coastal zone [4]. Human pathogenic microorganisms are introduced into estuarine ecosystems via the release of effluent from wastewater treatment plants, ineffective septic tank systems and storm water runoff [5]. Agricultural runoff from livestock farming can also represent a major source of microbial pollution, particularly when excreted waste from poultry and livestock is re-applied to land [6]. Wildlife, especially migratory wildfowl and other birds also represent an important source of zoonotic bacterial pathogens in natural environments [7].

Human pathogenic bacteria often occur at low levels in the environment [8] and their specific detection is a laborious and costly process [9]. Consequently, the detection and enumeration of faecal indicator bacteria that are present in the gastrointestinal tract of warm-blooded animals in much higher quantities represent a more effective 'indicator' of risk [10]. Current legislation on water quality testing relies on the enumeration of faecal coliforms (including the well-characterised *E. coli*) and enterococci to assess and classify water quality [11]. However, several other groups of potentially pathogenic bacteria have been detected in estuarine and marine environments. These include *Salmonella* spp. [12], *Campylobacter* spp. [13] and *Vibrio* spp. [14]. *Salmonella* spp. are well-known foodborne pathogens and recognised as one of the main causes of gastroenteritis in humans [15], resulting in around 1.3 billion annual cases of infection worldwide [16] but their presence in bathing waters is not routinely monitored [17]. Pathogenic strains of *Campylobacter* spp. are also thought to be responsible for a large proportion of enteric illnesses in humans living in developed and industrialized countries [18]. Contamination of surface and coastal waters by *Salmonella* and *Campylobacter* spp. is thought to occur predominantly via faecal contamination from wildfowl and other birds, although other animals including domestic livestock are also reservoirs [19]. *Vibrio* spp. are ubiquitous in aquatic environments and some strains are pathogens of humans; *V. cholera* has been responsible for several previous pandemics, resulting in high human mortality rates [20], some of which originated from the consumption of contaminated seafood [21]. Consequently, filter-feeding shellfish such as mussels, scallops and oysters are especially susceptible to contamination with bacterial pathogens if grown in contaminated waters [22].

Spatial and temporal variations in bacterial abundance in an environment are controlled by the interactions of complex physical, chemical and biological parameters, such as available nutrients [23], organic matter [24], sediment grain size [25], clay content [26], heavy metal content [27], predation by protozoa [28], competition [29], temperature [30], salinity [31], sunlight intensity [32] and seasonal variations [33]. Consequently, association with particulate material and sediments offers several advantages in terms of the survival and persistence of bacterial pathogens, such as protection from UV light [34,35], protection from phage attack in saline conditions [36], shelter from predation [37] and a greater organic matter content compared to the water column [38]. However, hydrodynamic processes (e.g. tides and storms), recreational activities and mechanical disturbances such as commercial dredging all have the potential to re-suspend sediment particles and their associated pathogens back into the water column, resulting in periodic elevated levels of pollution [39,40]. Currently, classification of both bathing and shellfish water quality relies solely upon the enumeration of FIB in water samples, despite the well-established paradigm that allochthonous bacterial pathogens in the water column become preferentially attached to particulate material, which promotes their downward flux to the bottom sediments where they are typically found in greater abundance [41,42].

The Conwy catchment, North Wales, UK, has a population of approximately 112,000, 80% of which live in coastal resorts that represent the main economic and tourist areas, with over 5.4 million visitors per annum. The Conwy estuary directly impacts commercially important shellfish beds, bathing waters and beaches; hence deterioration of the water quality could have major socioeconomic consequences. The catchment covers an area of ~300 km^2 which consists of large areas of land utilised for agriculture [43], some of the highest mountains in the UK [44], in addition to residential and commercial areas. Consequently, the Conwy river and it's tributaries (such as the river Gyffin and river Ganol) receive wastewater effluent from several waste water treatment plants in addition to septic tank discharges. Microbio-

logical inputs from wildlife, agriculture, wastewater effluent, septic tank discharges and run-off from storm events therefore represent potential point and diffuse sources of FIB in the Conwy estuary.

Previous analysis of FIB concentrations in the water column across four transverse transects of the Conwy estuary revealed significant spatial variations in waterborne *E. coli* numbers on contrasting sides of the estuary (45). These data suggested that significantly different levels of microbial pollution were present in the east and west sides of the river (45) and consequently, the choice of sampling points for routine water quality monitoring could have a significant impact upon the classification of microbiological water quality in this area. The Conwy estuary has a dynamic tidal cycle, with resuspension and deposition of finer sediment particles observed on the shallower banks of the river, resulting in the formation of mud flats. Consequently, the localised resuspension of sediment associated FIB into the water column was one proposed explanation for such contrasting FIB counts in the water column of the Conwy estuary (45), yet the spatial relationships between sediments composition (grain size and organic matter content) and FIB abundance in the Conwy and beyond has received little attention.

Whilst most studies focus on enumeration of *E. coli* and enterococci in water samples for the assessment of public health risk, the aim of our study was to address the paucity of information regarding the ecology of FIB along with other potentially human-pathogenic bacteria in estuarine sediments, and to identify sediment characteristics that promote pathogen abundance in estuarine environments. A second aim was to determine if the enumeration of *E. coli*, coliforms and enterococci, that represent the current standard 'indicator' organisms for faecal contamination and water quality monitoring, are a suitable indicator for the co-occurence of other potentially pathogenic bacterial groups. This was achieved by determining spatial variations in the abundance of culturable *E. coli*, total coliforms, enterococci, *Salmonella*, *Campylobacter* and *Vibrio* spp. (to be used as a proxy for potential human bacterial pathogens in the estuary and referred to as 'pathogen indicator bacteria' (PIB) herein) in estuarine sediments of the Conwy Estuary, UK, in relation to sediment grain size, organic content and other physico-chemical factors.

Materials and Methods

Study site

Sample locations were selected to correspond to those tested in a previous study that observed spatial variations of waterborne *E. coli* within the study area [45], in addition to possible point and diffuse sources of pollution that include mudflats used by wading birds and agricultural land (transect 1), input from the river Ganol (transect 2), the river Gyffin (transect 3) and Conwy marina (transect 4), (Figure 1).

Estuarine sampling

Estuarine sediment and water samples were collected from four transverse transects of the estuary over two consecutive sampling days (transects 1 and 2 collected on the 14/05/12, transects 3 and 4 on the 15/05/12). The Bangor University research vessel "Macoma", a 7.9 m Cheetah catamaran survey boat was utilised for sample collection, which was initiated at slack water. For each sampling point, three replicate sediment samples were collected using a manually operated Van Veen sediment grab, approximately 50 g of bottom sediment from the centre of each grab was transferred into sterile 50 ml polypropylene tubes (VWR International Ltd., Leicestershire, UK), water samples were collected in

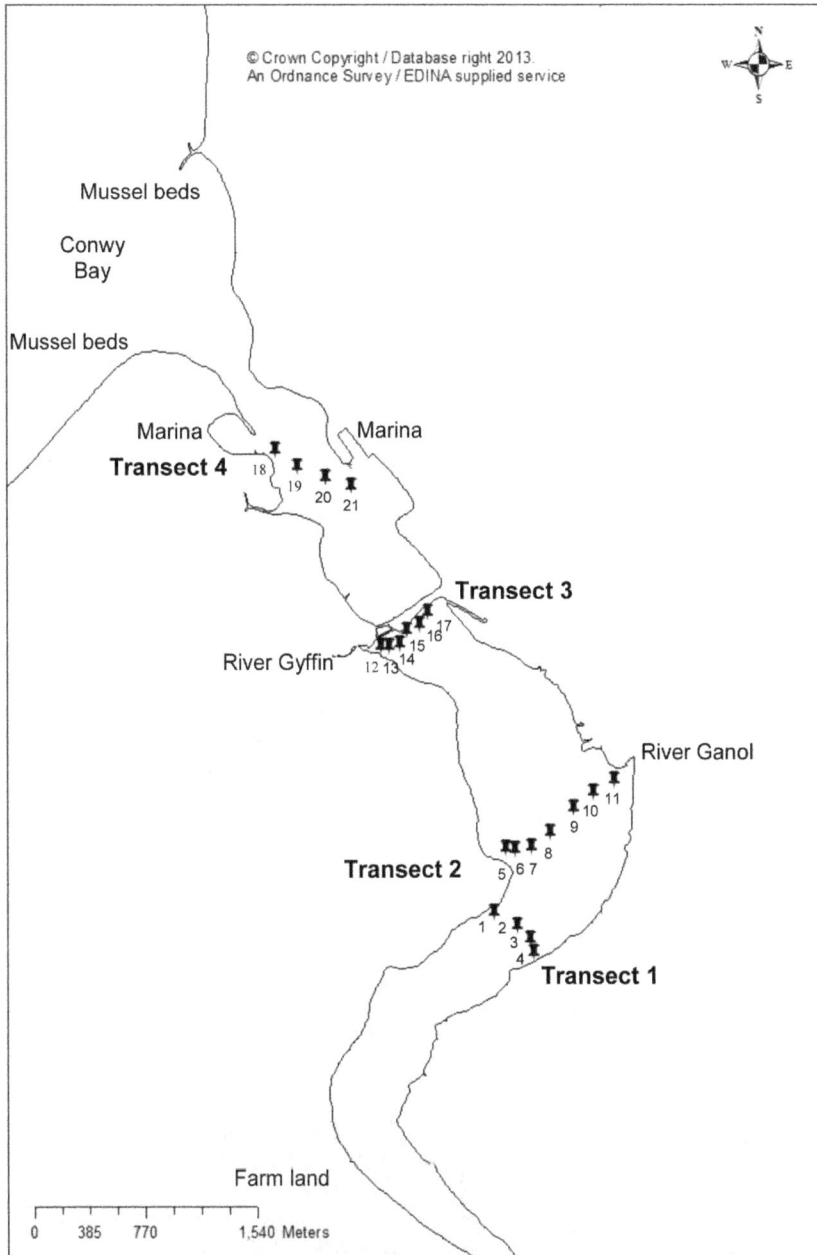

Figure 1. A map of the study site; the Conwy Estuary, North Wales, UK. Water and sediment samples were collected in triplicate from four transverse transects of the Conwy estuary (twenty one sampling sites).

triplicate from 0.2 m below the water surface using a sterile 1 L polypropylene container, then transported to the laboratory where water samples were processed within 4 h and sediment samples within 6 h.

Isolation and cultivation of target bacterial groups from sediment

One gram of each sediment sample was transferred to a 7 ml sterile bijou tube (Starlab UK Ltd., Milton Keynes, UK) and suspended in 5 ml Ringers solution (Oxoid Ltd., Basingstoke, UK) to obtain a 1:5 (w/v) dilution. Each sample was vortexed for 90 s to disassociate and resuspend bacteria from the sediment. Aliquots of the resulting supernatant for each sample were transferred

aseptically onto agar plates containing a selective medium for *E. coli*, total coliforms, enterococci, *Vibrio* spp., *Campylobacter* spp., *Salmonella* spp. and total heterotrophs. The optimum volume of supernatant used to inoculate each selective medium was determined in a previous study (data not shown). All selective medium plates were inverted and incubated according to manufacturer's recommendations. Resulting colony forming units (CFUs) provided enumeration of bacterial groups. Details describing the selective media used and incubation times are described in Table S1 in the supplemental material. The CFU data for sample point 13 represents the average of only two sediment samples due to one of the replicates being lost as a result of the presence of gravel clasts. All other CFU data represent the average of three independent replicate samples.

Isolation and cultivation of target bacterial groups from water

Water samples were processed within 4 h in accordance with the Revised Bathing Water Directive (2006/7/EC). Enumeration of bacteria in water samples was achieved by using vacuum–filtration as described in [45]. Briefly, water samples were homogenised by shaking and 50 ml of water filtered under vacuum through a 0.2 μm cellulose acetate membrane (Sartorius Stedim Biotech., Gottingen, Germany). Subsequently, the membranes were aseptically transferred onto sterile agar plates containing selective medium for the enumeration of *E. coli*, total coliforms, *Vibrio* spp. and heterotrophic bacteria. Agar plates were inverted and incubated according to manufacturer's recommendations (Table S1). Resulting Colony Forming Units (CFUs) were enumerated 24 h post incubation.

Sediment particle size analysis

Sediment grain size was determined by laser diffraction after 1 min sonication to separate particles, using a Malvern Hydro 2000 MU particle size analyser in conjunction with the Mastersizer 2000 software. Three replicate sediment samples from each site were pooled and homogenised. Approximately 1 g of sediment was added to the particle size analyser and 3 independent size determinations were made. This was repeated 3 times using the same pooled sample to determine an overall average.

Determination of sediment organic matter content

The loss on ignition method (LOI) was used to determine organic matter content of sediment samples. Three replicate sediment samples from each site were pooled and homogenised. Approximately 20 g of fresh sediment from each sample was placed in a pre-weighed crucible and dried at 95°C for 24 h. Approximately 4 g of the resultant dried sediment was pre-weighed, transferred to another crucible and placed into a muffle furnace at 550°C for 4 h. Organic matter content was calculated as the difference between the weight of the dry sediment and weight of the residue post-combustion. This was repeated 3 times using the same pooled sample to determine an overall average. Moisture content per g of fresh sediment was determined by calculating the percentage difference between wet weight and dry weight after 24 h at 95°C (Table S2).

In situ physico-chemical measurements of estuarine transect sample sites

Conductivity (calculated to practical salinity units (PSU)), temperature and depth measurements were recorded *in situ* using a YSI 600LS CTD scanner attached to a YSI 650 MDS data logger. This was deployed at each sample site in parallel with collection of water and sediment samples. Salinity and temperature measurements were recorded from the water immediately above the sediment bed and 0.2 m below the water surface. (Conductivity readings were calculated to determine salinity) (Table S3).

DNA extraction, 16S rRNA gene PCR and sequencing of bacterial isolates

To validate the identity of bacterial colonies on selective microbiological medium, DNA was extracted from a total of 30 isolates that matched or were similar to the expected colony morphology/phenotype for *E. coli*, enterococci, *Campylobacter* and *Vibrio spp.* for 16S rRNA gene PCR amplification and sequencing. For *E. coli* and enterococci, little variation in colony morphology and phenotype was observed across all of the colonies counted, and so only a small number of isolates were sequenced for confirmation.

DNA was extracted from 30 isolated bacterial colonies using the ISOLATE genomic DNA mini kit (Bioline Reagents Ltd., London, UK) following the manufacturer's protocol. Agarose gel electrophoresis (1%) was used to visualise the DNA extracted from each bacterial isolate. Subsequently, the 16S rRNA gene of each isolate was amplified via PCR using the primer pair pA (5′-AGAGTTTGATCCTGGCTCAG-3′) and pH (5′-AAGGAG-GTGATCCAGCCGCA-3′) (45). PCR reactions consisted of 10 pmol of each primer (pA and pH), 1x MyTaq red mix (Bioline), approximately 100 ng of template DNA and ddH$_2$O to a total reaction volume of 50 μl. The PCR conditions were as follows: an initial denaturation step at 95°C for 1 min, followed by 35 cycles of 95°C for 15 s, 55°C for 15 s, 72°C for 10 s and a final hold at 4°C. PCR amplicons were visualised using 1% agarose gel. The expected 16S rRNA gene amplicon size was approximately 1500 bp and amplicons of the expected size were subsequently excised from the agarose gel and purified using the Isolate PCR and Gel Kit (Bioline Reagents Ltd., London, UK) following the manufacturer's instructions. Purified 16S rRNA gene amplification products for each bacterial isolate were sequenced in both the forward and reverse orientation by Macrogen Europe (Netherlands). The Geneious Pro software package, Geneious 6.1 version created by Biomatters, (Available from http://www.geneious.com/) was used to quality clip each sequence and assemble the forward and reverse reads of each strain into a contiguous sequence. The sequence identity of each contig was determined using NCBI BLASTn. Identification of sequenced strains is given in Table S4 of the supplemental material.

Statistical analysis

Using the Statistical Package for Social Sciences SPSS v20, (IBM Corp., Armonk. NY), basic correlations were performed using the average data calculated for each site to determine the relationships between cultured bacterial abundance in sediments and water with different tested parameters. The non-parametric Spearman Rank Correlation Coefficient (r_s) was used due to the data being not normally distributed.

Results

The relationship between PIB abundance, sediment grain size and organic content

There were marked spatial differences in the abundance of sediment-associated PIB across all of the 21 sample sites within the estuary as determined by culture counts on selective medium (Table 1). Mean densities of *E. coli* and total coliforms in sediments ranged from 0 to 2.4×10^4 CFU/100 g and 0 to 5.4×10^5 CFU/100 g wet weight, respectively (for all mean bacterial densities from all sampling sites see Table S5 in the supplemental material). Enterococci and *Salmonella* abundance varied from 0 to 1×10^4 CFU/100 g and 0 to 2.5×10^4 CFU/100 g wet weight, respectively. *Vibrio* spp. were detected at all 21 sample sites (6.7×10^3 to 1.2×10^6 CFU/100 g wet weight) and this reflects their status as indigenous members of marine and aquatic environments. In addition, direct colony counts on *Campylobacter* selective media indicated the presence of 0 to 3.5×10^4 CFU/100 g wet weight. However, 16S rRNA gene sequencing of putative *Campylobacter* spp. isolated from a subsequent sampling survey on the same selective medium revealed that none of the sequenced isolates were *Campylobacter* spp. (Table S4). Culturable hetero-trophic bacteria were enumerated as a proxy for the abundance of the indigenous estuarine microbial community and their abundance ranged from 6.6×10^4 to 8.7×10^6 CFU/100 g wet weight. The enumeration of culturable heterotrophic bacteria alongside our target PIB groups enabled analysis of the relationship between total culturable heterotrophic bacteria counts and PIB counts in sediments, these data demonstrate that the abundance of heterotrophic bacteria within the sediments showed significant positive correlations with the abundance of all PIB groups measured (Table S6). Sediment grain size composition within sediment samples ranged from clay (0 to 18%), silt (0 to 65%), very fine sand (0 to 15%), fine sand (3 to 67%) medium sand (0 to 59%), coarse sand (0 to 17%) and very coarse sand (0 to 16%). The organic matter content of the sediment samples varied from 0.3 to 6% across the sample sites (Figure 2).

The abundance of each cultured bacterial group within the sediments showed significant positive correlations with the abundance of all other measured bacterial groups (Table S6). *Vibrio* spp. were also more abundant in sediments that had higher densities of FIB. In addition, the abundance of all isolated PIB groups showed a significant positive correlation with both sediment clay content (grain size <4 μm) (*E. coli*, enterococci, total coliforms, and *Vibrio* spp. $r_s = 0.543$ (p<0.011) $r_s = 0.664$ (p< 0.001), $r_s = 0.495$ (p<0.023), $r_s = 0.663$ (p<0.001) respectively) and silt content (grain size 4 μm–63 μm) (*E. coli*, enterococci, total coliforms and *Vibrio* spp., $r_s = 0.570$ (p<0.007) $r_s = 0.687$ (p< 0.001), $r_s = 0.547$ (p<0.010), $r_s = 0.688$ (p<0.001) respectively). Significant negative correlations were observed between the abundance of all PIB groups with fine sand (125 μm–250 μm) and medium sand (250 μm–500 μm) (Table S7). Sediments with high amounts of clay and silt along with very fine sand contained the greatest proportion of organic material (significant positive correlation; organic matter content and clay, $r_s = 0.917$ (p<0.001), organic matter content and silt, $r_s = 0.926$ (p<0.001), organic matter content and very fine sand, $r_s = 0.810$ (p<0.001). Significant negative correlations were evident between organic matter content and fine sand and also organic matter content and medium sand (Table S8). Sediments with high organic matter therefore also showed significant positive correlations with the abundance of all isolated PIB (Table S9).

Comparison of PIB abundance between sediment and water samples

The average abundance data for culturable PIB (*E. coli*, total coliforms and *Vibrio* spp.) in sediment and water samples across all 21 sample sites revealed that *E. coli*, coliforms and *Vibrio* spp. were 281, 433 and 58-fold more abundant in the sediment (CFU/ 100 g) than the water column (CFU/100 ml), respectively (Table 1).

The effect of depth, salinity and temperature on PIB abundance

Collection of water samples was initiated at slack water, with a range of water column depths from 0.5 to 12.3 m across all sites. Salinity and temperature measurements taken from above the bottom sediment varied between 9.3 to 30.0 psu and 10.8 to 11.2°C, respectively. At 0.2 m below the water surface the temperature varied from 10.8 to 11.2°C and salinity ranged from 10.1 to 27.8 psu. Bacterial abundance within the sediments revealed no significant correlations with any of the physico-chemical parameters measured (Table S10). *Vibrio* spp. enumerated from the water samples were the only cultured species to show significant correlations with depth, temperature or salinity. Temperature had a significant negative correlation with the abundance of *Vibrio* spp. ($r_s = 0.614$ p<0.003), while salinity had a significant positive correlation with abundance ($r_s = 0.544$ p< 0.011) (Table S11).

Taxonomic identification of bacterial isolates using 16S rRNA gene sequencing

None of the isolates derived from the selective medium for *Campylobacter* (n = 7) were positively identified as the target group. However, 91% of *E. coli* (n = 11), 100% of enterococci (n = 3) and 67% of *Vibrio* (n = 9) isolates were positively identified as the target group (Table S4). No sequence data were obtained for *Salmonella* isolates and they should therefore be considered as presumptive *Salmonella* spp.

Discussion

Results from this study demonstrate that sediments represent a significant reservoir for PIB and that sediment characteristics (grain size and organic matter content) influence spatial variations in PIB abundance in estuarine environments. Here, clay and silt fractions comprised a higher organic matter content in comparison to medium and coarse sand and harboured significantly higher densities of PIB. Previous studies have also concluded that the presence of finer sediment particles can have a positive impact on FIB abundance [25,46] and that organic material is a contributing factor to the prolonged persistence and survival of FIB in sediments [24,47,48]. However, almost all of the previous studies focus specifically on 'classic' faecal indicators such as *E. coli*, coliforms and enterococci, and here we have expanded knowledge on comparative PIB–sediment interactions with sediments for a greater repertoire of bacterial groups.

Results revealed there was not only a large spatial variation in PIB abundance that correlated to sediment grain size and organic matter content, but PIB abundance was also found to be significantly higher in the sediments when compared to the overlying water column, with levels of *E. coli* reaching over 3 orders of magnitude higher in the sediment. It is well documented that bacterial association with sediments circumvents the negative

Table 1. Average abundance of culturable pathogen indicator bacteria in sediment (CFU/100 g wet weight) and water (CFU/ 100 ml) across 21 estuarine sample sites.

Bacterial group	Sediment	Water	Fold - difference
E. coli	5.9×10^3	2.1×10^1	281
Total coliforms	1.3×10^5	3.0×10^2	433
Vibrio spp.	4.5×10^5	7.8×10^3	58

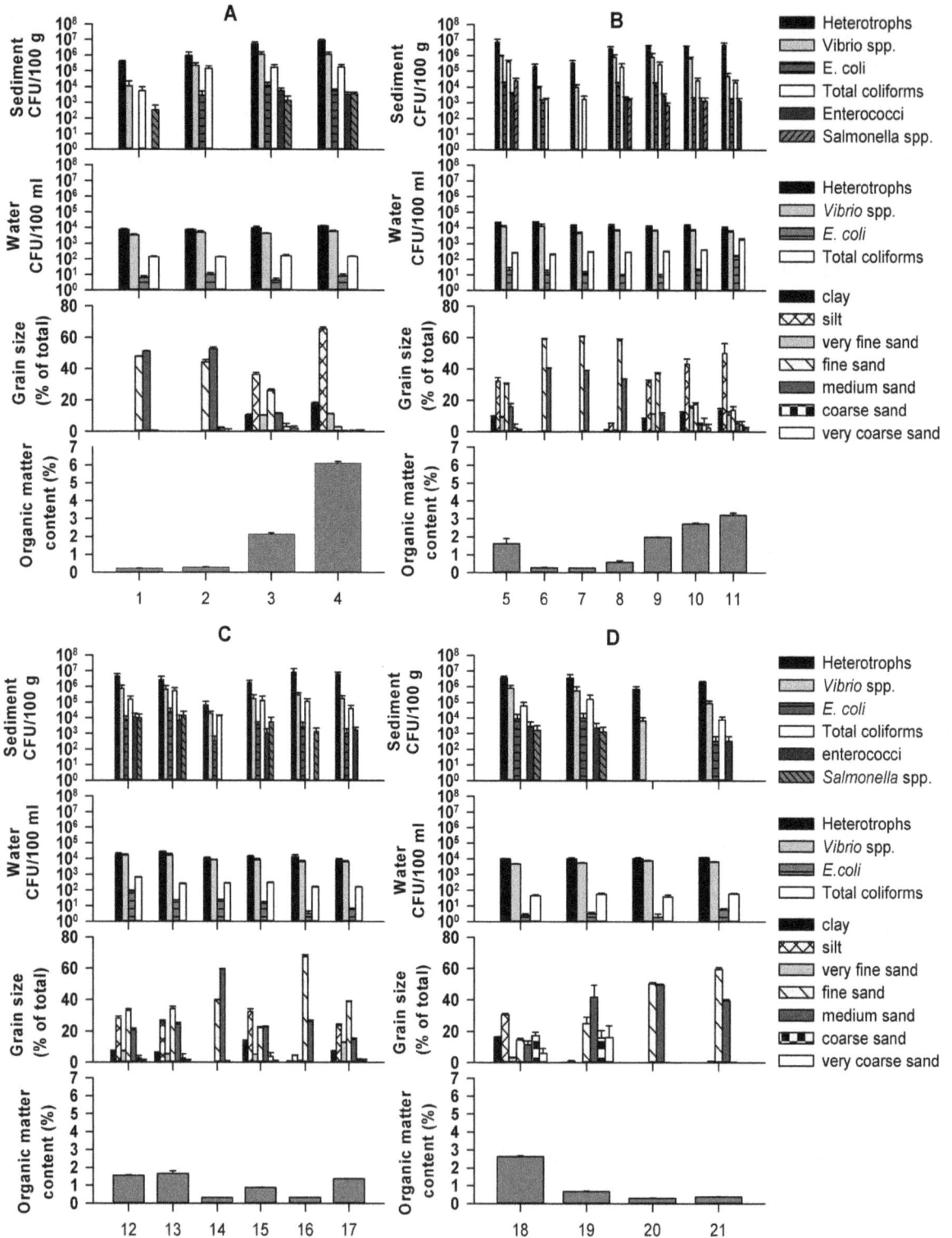

Figure 2. Bacterial abundance (CFU/100 g wet weight) compared to sediment grain size, organic matter content in sediments and bacterial abundance in the water column (CFU/100 ml), across four transverse transects. (A) Transect 1, (B) Transect 2, (C) Transect 3, (D) Transect 4. The X-axis represents sample points (n = 3 replicate samples for A, B, C and D except for sediment samples site 13, n = 2), mean values are plotted and error bars represent the SEM).

effects of environmental stresses, such as providing protection from UV light (34), which in turn can augment the survival and even growth of FIB. For example, DNA fingerprinting analyses

performed on *E. coli* populations in beach sand and sediment revealed the possibility that some strains may have become naturalized to this environment [52] and growing evidence also suggests that there may be free-living strains of FIB surviving and multiplying within the water column of environmental waters, independently of a host [53].

The abundance of *E. coli* and enterococci in sediments had a significant positive correlation with the abundance of other PIB. Consequently, the enumeration of FIB as the 'classic' bacterial indicators currently conducted as part of routine water quality monitoring represent suitable indicators for the presence of other PIB groups. *E. coli* and enterococci are the predominant indicator species for monitoring faecal pollution of aquatic environments within the European Union (EU) despite numerous studies highlighting the differential environmental survival of different FIB groups [48,51], in addition to different strains of the same group [54]. In this study, the presence and abundance of enterococci had a stronger significant positive correlation with silt and clay when compared with other FIB, which may suggest greater survival times of enterococci under favourable conditions that certain sediment types may provide. However, it is well established that the survival times of different species, and even strains of the same species, varies considerably in aquatic environments, both between and within species, and this must be taken into consideration. For example, in comparison to other FIB, enterococci survive for longer periods in sediments [48] and within estuarine environments [31], and it has been proposed that the increase in survival under harsh conditions may be due to their membrane composition [55,56]. Such evidence suggests that enterococci may be a more robust indicator of the persistence of faecal contamination rather than more recent contamination within coastal and estuarine environments.

The enumeration of total heterotrophs revealed significant positive correlations with all other cultured PIB groups and therefore the same trends were seen within the indigenous microbial community (Table S6). It should be noted that despite obtaining high colony counts on selective microbiological media for *Campylobacter* spp., 16S rRNA gene sequence analysis of some of these isolates suggested that none were *Campylobacter* spp. These data highlight the potential pitfalls of culture-based analyses of microbial taxa using a selective microbiological medium, and care must be taken when interpreting microbial culture counts. Despite this, the sequenced isolates of enterococci, *Vibrio* spp. and *E. coli* indicate that these media were selective for the desired bacterial target groups (Table S4). Little variation in temperature and salinity could be explained by the small geographical variation along each transect. There was no significant correlation between temperature and salinity and the abundance of PIB, with the exception of *Vibrio* spp. enumerated from the water column. Here, *Vibrio* spp. had a significant positive correlation with salinity and a significant negative correlation with temperature, supporting previous proposals that temperature and salinity have important implications for *Vibrio* spp. population dynamics [30]. The depth at which sediment samples were taken had no impact on the abundance of PIB enumerated from the sediments.

This study highlights the risk of periodic elevated FIB and other PIB in the water column due to the resuspension of microbial contaminated sediments. Furthermore, the risk of microbial pollution from the resuspension of sediments is not taken into consideration when assessing microbial pollution of recreational and shellfish harvesting waters. The time and place of sampling in addition to tidal and hydrodynamic conditions may impact upon FIB concentrations in the water column. Despite several reports on the importance of sediments as a reservoir for FIB [28,29], the enumeration of waterborne FIB has received much more attention. Quilliam *et al.*, (2011) revealed significant spatial variations in waterborne *E. coli* numbers on contrasting sides of

the same four transverse transects of the Conwy estuary studied here, indicating that significantly different levels of microbial pollution were present in the east and west sides of the river, and the localised re-suspension of sediment associated FIB into the water column was one proposed explanation for such contrasting FIB counts in the water column [45]. Despite this, the dynamics of sediment transport and re-suspension in relation to PIB concentrations is poorly understood. The attachment of PIB to particulate matter in the water column provides a platform for transportation and downward flux to the bed sediment, cyclical changes in tidal flow and the salinity of the water will also impact the downward flow of particle-associated PIB to the bed sediments. Conversely, under certain hydrodynamic conditions (e.g. tidal cycles and storm flow events), bed stress and turbulence can also impart the re-suspension of sediments and subsequent transportation and deposition of particle associated FIB to other areas of the estuary. Due to the hydrodynamics of an estuarine system, fine sediments are usually deposited around the banks of the basin [49] and our data support this trend, with finer particles detected in greater abundance in sediments on the east and west sides of the estuary and coarser sand deposited in the central channel. FIB levels in water can be affected by other factors such as temperature [33], exposure to sunlight and salinity [31], nutrient concentrations [5], predation by protozoan [50] and competition [51]. However, here we demonstrate that sediments also contribute to the distribution of FIB and other PIB in an estuarine environment.

Experimental results confirm that estuarine sediments harbour PIB, which may potentiate their prolonged persistence and survival in this environment. This study also identifies areas of high microbial contamination within the Conwy estuary, which highlights the risk of sediment and PIB resuspension back into the water column under turbulent hydrodynamic conditions that result in sediment bed stress and promote the erosion of sediments. To our knowledge, this is the first comprehensive study of the co-occurrence of *E. coli*, coliforms, enterococci, *Salmonella* and *Vibrio* spp. (PIB) in relation to sediment composition. These data show that all PIB groups studied are strongly correlated with the presence of clay, silt and organic matter content in sediments. In addition, the presence of *E. coli* and total coliforms strongly correlates with the abundance of the other PIB tested, both allochthonous and autochthonous, suggesting that culture-based determinations of *E. coli* and coliform abundance in sediments represent a useful surrogate for the presence of other PIB groups.

Conclusion

Faecal contamination in aquatic environments is currently assessed by measuring culturable *E. coli* and enterococci counts in water samples only. Here, we demonstrate that sediment composition, specifically clay, silt and organic matter content, are linked with greater PIB abundance in sediments. The enhanced abundance of viable PIB in sediments therefore has implications for water quality and public health when considering the potential for resuspension of bed sediments, particularly finer particles such as clay and silt that we have demonstrated to contain higher concentrations of PIB. It may therefore be necessary to incorporate PIB loadings in bottom sediments into routine monitoring protocols and hydrodynamic models to adequately assess their risk to human health. The detection of spatial variations of PIB within sediments also highlights the necessity for further research on the interactions of pathogens with sediments and their role in the survival, persistence and transportation of PIB within environmental waters.

Supporting Information

Table S1 Selective media used to enumerate target bacterial groups.

Table S2 Sediment dry weight determined from $1\,g^{-1}$ wet weight.

Table S3 Details the depth at which the sediment samples were collected, salinity measurements taken for both water samples (0.2 m from the surface) and directly above the sediment samples, calculated as Practical Salinity Units (PSU). Temperature was recorded at a depth of 0.2 m from the surface and directly above the sediment.

Table S4 Identification of sequenced isolates.

Table S5 Bacteria counts, sediment (CFU/100 g) versus water (CFU/100 ml). Data shown as mean (n = 3, sample point 13 n = 2 for sediment).

Table S6 Correlation coefficient (r_s) matrix demonstrating the relationship between the abundance of each cultured bacterial group within estuarine sediments (n = 21).

Table S7 Correlation coefficient (r_s) matrix demonstrating the relationship between the abundance of each cultured bacterial group within estuarine sediments and sediment grain size (n = 21).

Table S8 Correlation coefficient (r_s) matrix demonstrating the relationship between estuarine sediment grain size (%) and organic matter content (%) (n = 21).

Table S9 Correlation coefficient (r_s) matrix demonstrating the relationship between the abundance of each cultured bacterial group within estuarine sediments and sediment organic matter content (n = 21).

Table S10 Correlation coefficient (r_s) matrix demonstrating the relationship between the abundance of each cultured bacterial group within estuarine sediments and physico-chemical parameters measured directly above the bottom sediments (n = 2).

Table S11 Correlation coefficient (r_s) matrix demonstrating the relationship between the abundance of each cultured bacterial group within estuarine water and physico-chemical parameters measured at 0.2 m depth (n = 21).

Acknowledgments

We are grateful to Ben Winterbourn, Gwynn Parry Jones, Peter Hughes and Ben Powell for their assistance with estuarine sampling.

Author Contributions

Conceived and designed the experiments: TLP JHB CFJ DLJ SKM JEM. Performed the experiments: TLP KC JHB JEM. Analyzed the data: TLP KC JHB JEM. Contributed reagents/materials/analysis tools: JHB CFJ SKM DLJ JEM. Contributed to the writing of the manuscript: TLP KC JHB DLJ SKM JEM.

References

1. Costanza R, d'Arge R, de Groot R, Farber S, Grasso M, et al. (1997) The value of the world's ecosystem services and natural capital. Nature 387: 253–260.
2. Vitousek PM, Mooney HA, Lubchenco J, Melillo JM (1997) Human domination of Earth's ecosystems. Science 277: 494–499.
3. Shuval H (2003) Estimating the global burden of thalassogenic diseases: human infectious diseases caused by wastewater pollution of the marine environment. J Water Health 1: 53–64.
4. GESAMP (2001) The sea of troubles. GESAMP Rep. Stud. 70: 19–20.
5. Hong H, Qiu J, Liang Y (2010) Environmental factors influencing the distribution of total and fecal coliform bacteria in six water storage reservoirs in the Pearl River Delta Region, China. J Environ Sci (China) 22: 663–668.
6. Topp E, Scott A, Lapen DR, Lyautey E, Duriez P (2009) Livestock waste treatment systems for reducing environmental exposure to hazardous enteric pathogens: some considerations. Bioresour Technol 100: 5395–5398.
7. Obiri-Danso K, Jones K (2000) Intertidal sediments as reservoirs for hippurate negative Campylobacters, Salmonellae and faecal indicators in three EU recognised bathing waters in North West England. Wat Res 34: 519–527.
8. Straub TM, Chandler DP (2003) Towards a unified system for detecting waterborne pathogens. J Microbiol Meth 53: 185–197.
9. Edge TA, Boehm AB (2011) Classical and molecular methods to mearusre fecal bacteria. In: Sadowsky MJ, Whitman RL, editors. The fecal bacteria. Washington DC: ASM Press: pp. 241–273.
10. Carrero-Colon M, Wickman GS, Turco RF (2011) Taxonomy, phylogeny and physiology of fecal indicator bacteria. In: Sadowsky MJ, Whitman RL, editors. The fecal bacteria. Washington DC: ASM Press: pp. 23–38.
11. 2006 EC European Parliament, Directive 2006/7/EC of the European Parliament, concerning the management of bathing water quality and repealing directive 76/160/EEC. Off J Eur Communities 64: 37–51.
12. Berthe T, Touron A, Leloup J, Deloffre J, Petit F (2008) Faecal-indicator bacteria and sedimentary processes in estuarine mudflats (Seine, France). Mar Pollut Bull 57: 59–67.
13. Alonso JL, Alonso MA (1993) Presence of Campylobacter in marine waters of Valencia, Spain. Wat Res 27: 1559–1562.
14. Cox AM, Gomez-Chiarri M (2012) Vibrio parahaemolyticus in Rhode Island coastal ponds and the estuarine environment of Narragansett bay. Appl Environ Microb 78: 2996–2999.
15. Polo F, Figueras MJ, Inza I, Sala J, Fleisher JM, et al. (1998) Relationship between presence of Salmonella and indicators of faecal pollution in aquatic habitats. Fems Microbiol Lett 160: 253–256.
16. Coburn B, Grassl GA, Finlay BB (2007) Salmonella, the host and disease: a brief review. Immunol Cell Biol 85: 112–118.
17. Figueras MJ, Polo F, Inza I, Guarro J (1997) Past, present and future perspectives of the EU bathing water directive. Mar Pollut Bull 34: 148–156.
18. Levin RE (2007) Campylobacter jejuni: A review of its characteristics, pathogenicity, ecology, distribution, subspecies characterization and molecular methods of detection. Food Biotechnol 21: 271–374.
19. Thorns CJ (2000) Bacterial food-borne zoonoses. Rev Sci Tech 19: 226–239.
20. Dziejman M, Balon E, Boyd D, Fraser CM, Heidelberg JF, et al. (2002) Comparative genomic analysis of Vibrio cholerae: genes that correlate with cholera endemic and pandemic disease. Proc Natl Acad Sci U S A 99: 1556–1561.
21. Cabral JP (2010) Water microbiology. Bacterial pathogens and water. Int J Environ Res Public Health 7: 3657–3703.
22. Bakr WMK, Hazzah WA, Abaza AF (2011) Detection of Salmonella and Vibrio species in some seafood in Alexandria. J Amer Sci 7: 663–668.
23. Pommepuy M, Guillaud JF, Dupray E, Derrien A, Le Guyader F, et al. (1992) Enteric bacteria survival factors Wat Sci Tech 25: 93–103.
24. Pote J, Haller L, Kottelat R, Sastre V, Arpagaus P, et al. (2009) Persistence and growth of faecal culturable bacterial indicators in water column and sediments of Vidy Bay, Lake Geneva, Switzerland. J Environ Sci (China) 21: 62–69.
25. Garzio-Hadzick A, Shelton DR, Hill RL, Pachepsky YA, Guber AK, et al. (2010) Survival of manure-borne E. coli in streambed sediment: effects of temperature and sediment properties. Water Res 44: 2753–2762.
26. Burton GA, Jr., Gunnison D, Lanza GR (1987) Survival of pathogenic bacteria in various freshwater sediments. Appl Environ Microbiol 53: 633–638.
27. Jones GE (1964) Effect of chelating agents on the growth of Escherichia coli in seawater. J Bacteriol 87: 483–499.

28. Davies CM, Long JA, Donald M, Ashbolt NJ (1995) Survival of fecal microorganisms in marine and freshwater sediments. Appl Environ Microbiol 61: 1888–1896.

29. Marino RP, Gannon JJ (1991) Survival of fecal-coliforms and fecal streptococci in storm-drain sediment. Water Res 25: 1089–1098.

30. Singleton FL, Attwell R, Jangi S, Colwell RR (1982) Effects of temperature and salinity on *Vibrio-cholerae* growth. Appl Environ Microbiol 44: 1047–1058.

31. Bordalo AA, Onrassami R, Dechsakulwatana C (2002) Survival of faecal indicator bacteria in tropical estuarine waters (Bangpakong River, Thailand). J Appl Microbiol 93: 864–871.

32. Davies CM, Evison LM (1991) Sunlight and the survival of enteric bacteria in natural waters. J Appl Bacteriol 70: 265–274.

33. Faust MA, Aotaky AE, Hargadon MT (1975) Effect of physical parameters on the in situ survival of *Escherichia coli* MC-6 in an estuarine environment. Appl Microbiol 30: 800–806.

34. Davies-Colley RJ, Donnison AM, Speed DJ, Ross CM, Nagels JW (1999) Inactivation of faecal indicator microorganisms in waste stabilisation ponds: Interactions of environmental factors with sunlight. Water Res 33: 1220–1230.

35. Sinton LW, Davies-colley RJ, Bell RG (1994) Inactivation of enterococci and fecal-coliforms from sewage and meatworks effluents in seawater chambers. Appl Environ Microbiol 60: 2040–2048.

36. Roper MM, Marshall KC (1974) Modification of the interaction between *Escherichia coli* and bacteriophage in saline sediments. Microbial Ecol 1: 1–13.

37. Davies CM, Bavor HJ (2000) The fate of stormwater-associated bacteria in constructed wetland and water pollution control pond systems. J Appl Microbiol 89: 349–360.

38. Gerba CP, McLeod JS (1976) Effect of sediments on the survival of *Escherichia coli* in marine waters. Appl Environ Microbiol 32: 114–120.

39. An YJ, Kampbell DH, Breidenbach GP (2002) Escherichia coli and total coliforms in water and sediments at lake marinas. Environ Pollut 120: 771–778.

40. Grimes DJ (1975) Release of sediment-bound fecal coliforms by dredging. Appl Microbiol 29: 109–111.

41. Danovaro R, Fonda Umani S, Pusceddu A (2009) Climate change and the potential spreading of marine mucilage and microbial pathogens in the Mediterranean sea. PLoS One 4: e7006: doi:10.1371/journal.pone.0007006.

42. Droppo IG, Liss SN, Williams D, Nelson T, Jaskot C, et al. (2009) Dynamic existence of waterborne pathogens within river sediment compartments. Implications for water quality regulatory affairs. Environ Sci Technol 43: 1737–1743.

43. Thorn CE, Quilliam RS, Williams AP, Malham SK, Cooper D, et al. (2011) Grazing intensity is a poor indicator of waterborne *Escherichia coli* O157 activity. Anaerobe 17: 330–333.

44. Oliver LR, Seed R, Reynolds B (2008) The effect of high flow events on mussels (*Mytilus edulis*) in the Conwy estuary, North Wales, UK. Hydrobiologia 606: 117–127.

45. Quilliam RS, Clements K, Duce C, Cottrill SB, Malham SK, et al. (2011) Spatial variation of waterborne *Escherichia coli* - implications for routine water quality monitoring. J Water Health 9: 734–737.

46. Howell JM, Coyne MS, Cornelius PL (1996) Effect of sediment particle size and temperature on fecal bacteria mortality rates and the fecal coliform/fecal streptococci ratio. J Environ Qual 25: 1216–1220.

47. Craig DL, Fallowfield HJ, Cromar NJ (2004) Use of microcosms to determine persistence of *Escherichia coli* in recreational coastal water and sediment and validation with in situ measurements. J Appl Microbiol 96: 922–930.

48. Haller L, Amedegnato E, Pote J, Wildi W (2009) Influence of freshwater sediment characteristics on persistence of fecal indicator bacteria. Water Air Soil Pollut 203: 217–227.

49. Malham SK, Rajko-Nenow P, Howlett E, Tuson KE, Perkins TL, et al. (2014) The interaction of human microbial pathogens, particulate material and nutrients in estuarine environments and their impacts on recreational and shellfish waters. Environ Sci Process Impact 16: 2145–2155.

50. Barcina I, Lebaron P, Vives-Rego J (1997) Survival of allochthonous bacteria in aquatic systems: a biological approach. FEMS Microbiol Ecol 23: 1–9.

51. Korhonen LK, Martikainen PJ (1991) Survival of *Escherichia coli* and *Campylobacter jejuni* in untreated and filtered lake water. J Appl Bacteriol 71: 379–382.

52. Ishii S, Hansen DL, Hicks RE, Sadowsky MJ (2007) Beach sand and sediments are temporal sinks and sources of *Escherichia coli* in Lake Superior. Environ Sci Technol 41: 2203–2209.

53. Power ML, Littlefield-Wyer J, Gordon DM, Veal DA, Slade MB (2005) Phenotypic and genotypic characterization of encapsulated *Escherichia coli* isolated from blooms in two Australian lakes. Environ Microbiol 7: 631–640.

54. Anderson KL, Whitlock JE, Harwood VJ (2005) Persistence and differential survival of fecal indicator bacteria in subtropical waters and sediments. Appl Environ Microbiol 71: 3041–3048.

55. Evison LM (1989) Comparative studies on the survival of indicator organisms and pathogens in fresh and seawater. Wat Res 20: 309–352.

56. Okpookwasili GC, Akujobi TC (1996) Bacteriological indicators of tropical water quality. Environ Toxic Water 11: 77–81.

13

Aggregation and Sedimentation of *Thalassiosira weissflogii* (diatom) in a Warmer and More Acidified Future Ocean

Shalin Seebah[1], Caitlin Fairfield[2], Matthias S. Ullrich[1], Uta Passow[2]*

1 Molecular Life Science Research Center, Jacobs University Bremen, Bremen, Germany, 2 Marine Science Institute, University of California Santa Barbara, Santa Barbara, California, United States of America

Abstract

Increasing Transparent Exopolymer Particle (TEP) formation during diatom blooms as a result of elevated temperature and pCO_2 have been suggested to result in enhanced aggregation and carbon flux, therewith potentially increasing the sequestration of carbon by the ocean. We present experimental results on TEP and aggregate formation by *Thalassiosira weissflogii* (diatom) in the presence or absence of bacteria under two temperature and three pCO_2 scenarios. During the aggregation phase of the experiment TEP formation was elevated at the higher temperature (20°C vs. 15°C), as predicted. However, in contrast to expectations based on the established relationship between TEP and aggregation, aggregation rates and sinking velocity of aggregates were depressed in warmer treatments, especially under ocean acidification conditions. If our experimental findings can be extrapolated to natural conditions, they would imply a reduction in carbon flux and potentially reduced carbon sequestration after diatom blooms in the future ocean.

Editor: David William Pond, Scottish Association for Marine Science, United Kingdom

Funding: This work was supported by National Science Foundation grants OCE-0926711 & OCE-1041038 to UP and Helmholtz Graduate School for Polar and Marine Research and Jacobs University Bremen to SS. The funders had no role in study design, data collection and analysis, decision to publish, or preparation of the manuscript

Competing Interests: The authors have declared that no competing interests exist.

* Email: uta.passow@lifesci.ucsb.edu

Introduction

Globally, gravitational sinking of marine snow (>0.5 mm) contributes significantly to the biological carbon pump, leading to carbon sequestration into the deep ocean. Large sedimentation events are frequently associated with diatom blooms, because most bloom- forming diatoms form marine snow-sized aggregates. Coagulation of diatoms is impacted by diatom species [1], bacteria species and activities [2–4], and extrapolymeric substances (EPS) [5,6], especially transparent exopolymer particles (TEP) [7–9] and may be described using aggregation theory [10–12].

Atmospheric pCO_2 values are expected to rise to a global average of 750 ppm (IPCC scenario IS92a, IPCC 2007) and perhaps even beyond 1000 ppm by the end of this century [13]. Increasing atmospheric CO_2 concentrations do not solely result in higher sea-surface temperatures due to intensified radiative forcing, but also lead to ocean acidification [14]. The term ocean acidification describes the increase in dissolved inorganic carbon (DIC) and the concomitant decrease in pH in surface waters [15]. Changing oceanic conditions due to globally rising temperatures and ocean acidification may influence the functioning of the biological pump and its specific responses to these changes are currently under intense investigation [16].

The potential increase in TEP concentration as a result of elevated temperature and pCO_2 has been suggested to result in

enhanced aggregation and flux [17,18], although this finding has also been challenged [19]. Reduced production of coccoliths under ocean acidification conditions is thought to reduce sedimentation of carbon due to a reduction in ballasting [20–24]. Coagulation of organic matter with lithogenic minerals has not been found to be impacted by ocean acidification [25].

Here, the combined effects of changed carbonate chemistry and temperature on the coagulation of axenic and xenic cultures of the diatom *Thalassiosira weissflogii* were investigated. Xenic cultures contained the marine gammaproteobacterium *Marinobacter adhaerens* HP15, which has been shown to promote TEP production and induce aggregation of the diatom *Thalassiosira weissflogii* [2]. Aggregation, TEP concentration and sinking velocities of aggregates were monitored during the four-day aggregation experiments.

Materials and Methods

Experimental design and set up

The combined impact of temperature, ocean acidification and bacterial activity on the formation of TEP and aggregates by the diatom *Thalassiosira weissflogii* was tested in a full factorial design. The established bilateral model system between the diatom *T. weissflogii* and the marine bacterium *M. adhaerens* HP15 [26,27] was used to investigate the bacterial contribution to TEP formation and aggregation under the different environmental conditions. Three different carbonate chemistry regimes were selected to reflect: (i) the present-day conditions, with the partial pressure of CO_2 (pCO_2) ranging between 300–350 µatm (termed Ambient) and (ii) two future ocean scenarios with pCO_2 ranging from 750–850 µatm (designated Future 1) and 1000–1250 µatm (referred to as Future 2). For each carbonate chemistry regime, two temperatures were chosen, 15°C and 20°C (Table 1).

An axenic diatom culture of *Thalassiosira weissflogii* (CCMP 1336) and the bacterium *M. adhaerens* HP15 were used for experiments. *M. adhaerens* HP15 was isolated from marine particles collected from the surface waters of the German Bight (Grossart et al. 2004). *M. adhaerens* HP15 attaches preferentially to *T.weissflogii* cells and impacts TEP production and aggregation [2].

Due to logistical reasons, the experiment was run in two sections. First the six treatments at 15°C, and a few days later the six treatments at 20°C were incubated in duplicate rolling tanks and in darkness at three rotations per minute (rpm). Solid body rotation was established in roller tanks (1.15-L Plexiglas cylinders with a diameter of 14 cm and a depth of 7.47 cm) within 2–3 hours, which assured that aggregates remained suspended, never contacting container walls [28]. Incubation in roller tanks in the dark mimicked sinking of aggregates through the water column to depth. The experiment was terminated after 96 hrs. Prior to the experiment, the diatom and bacterial cultures were separately acclimatized to the respective temperature and carbonate chemistry regimes for more than 8 generations to avoid a stress reaction to changed environmental conditions.

During the acclimatization phase diatoms were grown at 50 µE s^{-1} for a 12-hr light period, in a semi-continuous batch approach, to ensure continuous exponential growth and restrict changes in the carbonate system. Diatom numbers and total alkalinity (TA), pH and DIC were monitored daily during this phase and cultures diluted (factor 2–6) before a cell concentration of 60, 000 cells mL^{-1} was reached or the pH changed by more than 0.25 units. *M. adhaerens* HP15 was acclimatized to the respective temperature and pCO_2 conditions overnight in sterile culture flasks with aeration of approximately 250 rpm.

After the acclimatization phase, triplicate roller tanks were filled bubble-free under sterile conditions with diatom cells at a final concentration of 3×10^3 cells mL^{-1} and bacterial cells at a final concentration of 3×10^5 cells mL^{-1}. Diatom blooms in coastal and upwelling areas regularly reach cell concentrations of 10^4 cells mL^{-1}, when small diatoms dominate and may reach 10^5 cells mL^{-1}, for example in the upwelling area of the Benguela current or off the California coast [29,30]. The chosen diatom concentration is thus still ecologically relevant while providing enough cells to allow rapid aggregation and provide enough aggregates for the required measurements. Prior to inoculation with the diatom culture, the bacterial cells were washed twice in sterile seawater to minimize carry-over of nutrients or bacterial growth-derived matter into the ASW media. One replicate roller tank per treatment was sacrificed at the beginning and two replicates per treatment at the end of the experiment. TEP concentration, and aggregate size, number, and sinking velocity, as well as the carbonate system parameters (TA, pH and DIC) were analyzed at both time points. Values are given as averages ± standard deviation of the duplicate tanks, with standard deviations calculated using error propagation, where appropriate.

Aggregates were defined as particles ≥0.5 mm. During sampling, aggregates, when present, were first removed using a cut-off pipette [31], their sinking velocities determined and all aggregates of one tank combined in a known volume of artificial seawater, creating aggregate slurry. The surrounding seawater (SSW), which remained in the tank after the manual removal of the aggregates, was sampled thereafter. TEP was measured in aggregate slurries and in the SSW; the carbonate system parameters were determined in the SSW.

Cultures and media

Autoclaving of natural seawater (collected off Santa Barbara 34° 23′ N 119° 50′ W) for media preparation was not an option since the carbonate chemistry of seawater is severely impacted by degassing. The pH of freshly collected natural seawater from the Santa Barbara Channel increased from 7.58 to 8.67 during autoclaving. Stirring the autoclaved seawater while leaving the beaker open to the atmosphere only reduced the pH to 7.89

Table 1. Design of multifactorial experiment with 12 treatments testing aggregation of the diatom *T. weissflogii* in the presence or absence of bacteria at two temperatures and three pCO_2 scenarios.

Treat. #	Temp. °C	pCO_2	Bact.
1	15	Am	Ax
2	15	Am	HP
3	15	F1	Ax
4	15	F1	HP
5	15	F2	Ax
6	15	F2	HP
7	20	Am	Ax
8	20	Am	HP
9	20	F1	Ax
10	20	F1	HP
11	20	F2	Ax
12	20	F2	HP

Xenic treatments contained the bacterium *M. adhaerens* HP15. Each treatment was prepared in triplicate; one replicate was harvested initially (t = 0) and two after 96 hr. incubation on roller tables in the dark. See text for specifics on pCO_2 treatments. Ax = axenic, HP = *M. adhaerens* HP15 added, Am = Ambient, F1 = Future 1, F2 = Future 2.

(Table 2). The repeated filtration of natural seawater through 0.2 μm pore-sized filters (Millipore, MA, USA) did not satisfactorily remove all bacterial contaminants. We therefore opted for the use of artificial seawater (ASW) [32] for media preparation. Using ASW imparts the added benefit of easily and precisely manipulating DIC concentrations. ASW was prepared with a DIC concentration of 2,050 μmol kg^{-1} for ambient treatments and supplemented with vitamins and trace metal solutions as in F/2 medium [33]. Macronutrients were added to a final concentration of 59 μM nitrate, 3.6 μM phosphate and 53.5 μM silicic acid to create ASW-media. The carbonate chemistry of future treatments was adjusted as described below.

Diatom cells were counted in a Sedgwick-Rafter Cell S50 (SPI Supplies, West Chester, PA, USA) using an inverted Axiovert 200 microscope (Zeiss, Jena, Germany). The axenicity of the diatom culture was checked by epifluorescence microscopy [34] after staining with the dye 4′, 6-diamidino-2-phenylindol (DAPI) [35].

Carbonate chemistry perturbations and analysis

Since TEP production has been shown to be impacted by bubbling [36–38], the carbonate system was chemically perturbed as described in Passow [25,39]. Briefly, to mimic future ocean conditions, appropriate amounts of 0.1 M HCl (mL kg^{-1}), 0.1 M NaHCO$_3$ (mL kg^{-1}) and 0.001 M Na$_2$CO$_3$ (mL kg^{-1}) were added to change DIC and pH while keeping TA constant. Measurements of pH and TA confirmed that our perturbations changed the system as expected and reflected those anticipated in a future ocean.

The carbonate system was monitored by measuring pH, TA and DIC. Samples for pH were collected bubble-free in 20-mL scintillation vials and the pH (total scale, pH$_T$) was measured with a spectrophotometer using the indicator dye m-cresol purple (Sigma-Aldrich) within 2 hours of sampling. The measurement temperature was held at 25°C and the absorbance measured at 730 nm, 578 nm and 434 nm before and after dye addition [40,41]. The pH was calculated following the standard operating procedure (SOP 7) [42]. Samples for TA and DIC measurements were collected following SOP 1 [42]. TA and DIC samples were poisoned with 0.02% saturated HgCl$_2$ by volume and sent for analysis to the Dickson Laboratory at the Scripps Institution of Oceanography, UCSD.

The program CO$_2$ Sys [43] was used to calculate the carbonate system. The dissociation constants K1 and K2 from Roy [44] were used since these have been described as the most appropriate for ASW [15]. Any two of the main carbonate parameters (pH, TA, DIC, pCO$_2$) describe the carbonate system sufficiently and the other parameters can be calculated from the measured ones. In 50 different samples, we measured three carbonate parameters (pH, TA and DIC) to over-determine the carbonate system.

Quantification of TEP

TEP concentrations were measured colorimetrically by filtration of samples onto 0.4-μm pore size polycarbonate filters (Millipore, MA, USA) and subsequent staining with Alcian blue [45]. TEP concentration was determined in quadruplicate and expressed as Gum Xanthan equivalents per liter (GXeq L^{-1}).

Aggregate number, size and sinking velocity

The number of aggregates >0.5 mm was counted and the sinking velocity of 10 to 20 aggregates per tank per treatment measured by gently transferring individual aggregates from the roller tanks to a tall 1 liter cylinder containing sterile ASW media [28]. Prior to measuring sinking velocity of aggregates, the ASW medium was incubated overnight at either 15°C or 20°C to ensure that aggregates experienced no change in environmental conditions during the sinking velocity determinations. The time taken for each aggregate to sink 25 cm was recorded. The dimensions of the aggregate axes (x, y, and z direction) were measured under a dissecting microscope, using grid paper and the aggregated volume was calculated by assuming an ellipsoid shape. The equivalent spherical diameter (ESD) was calculated. Sinking velocity could only be measured on 5–8 aggregates for some treatments, because of the lack of aggregates. The slopes of the sinking velocity vs. aggregate size relationships at 15°C and 20°C were compared by calculating both pooled and unpooled error variance and the appropriate t and p values.

No specific permissions were required to collect seawater samples off Santa Barbara and no endangered or protected species were used for our experiments. The data of this study is deposited at the Oceanographic Data Repository BCO-DMO (Biological and Chemical Oceanography Data Management Office; http://www.bco-dmo.org/) in accordance with NSF guidelines. Doi: 10.1575/1912/6845; http://hdl.handle.net/1912/6845.

Results

Over-determination of the carbonate chemistry

The pCO$_2$ concentrations in samples where the carbonate system was over-determined were calculated using all three possible carbonate parameter combinations. The slight but consistent discrepancy in the pCO$_2$ concentrations depending on the parameter combination used for calculation (Fig. 1) is well known and has been described [39,46]. Results from over-determined carbonate system parameters confirm that our measurements were consistent and all treatments exposed to targeted conditions.

Table 2. Effect of autoclaving on the carbonate chemistry of seawater.

Sample	pH (total)
Fresh seawater before autoclaving	7.58
After autoclaving (with or without nutrients)	8.66±0.01
after stirring for 24 hrs.	8.39
after stirring for 72 hrs.	7.94
after stirring for 96 hrs.	7.90
after stirring for 120 hrs.	7.89

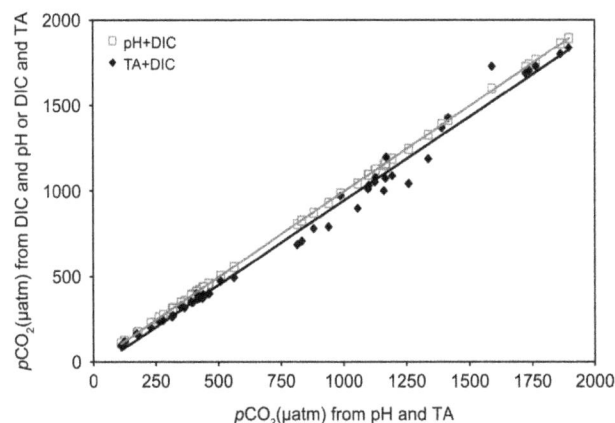

Figure 1. Relationship between pCO_2 determined from DIC and pH or TA vs. pCO_2 determined from pH and TA. Data stems from 50 random samples from the experiment, where the carbonate chemistry was over-determined.

Acclimatization phase

By frequent dilutions of the diatom cultures with media adjusted to the appropriate pH and TA, it was possible to maintain the carbonate system of the cultures within a narrow range during the acclimatization phase. The average TA was 2332 ± 32 µmol kg^{-1} and did not significantly vary statistically between any of the treatments (p>0.05). The maximal temporal shift in pH experienced in each pCO_2 treatment due to growth of phytoplankton was kept to <0.3 pH units, usually <0.2 units (Table 3). In coastal upwelling systems or during phytoplankton blooms, *in situ* variations in pH may easily be that large. For example, off California daily ranges of pH are frequently 0.2 to 0.3 units [47,48]. *T. weissflogii* were acclimatized 8 and 11 days, depending on growth rate. Exponential growth rates were a significant function of temperature, but not pCO_2, although growth rates were slightly decreased under Future 2 conditions at both temperatures (Table 3).

Aggregation experiment

Carbonate system. The average TA was 2351 ± 7 µmol kg^{-1} in all treatments. Initial Ambient pH$_T$ at *in vitro* temperature was 8.15 ± 0.00 and 8.14 ± 0.00 in 15°C and 20°C treatments, respectively. The initial pH$_T$ in Future 1 treatments were 0.34 ± 0.01 (15°C treatments) and 0.35 ± 0.01 (20°C treatments) units lower than the respective Ambient treatments, while those of Future 2 were 0.47 ± 0.05 (15°C treatments) and 0.50 ± 0.02 (20°C treatments) units lower. The associated initial pCO_2 values for

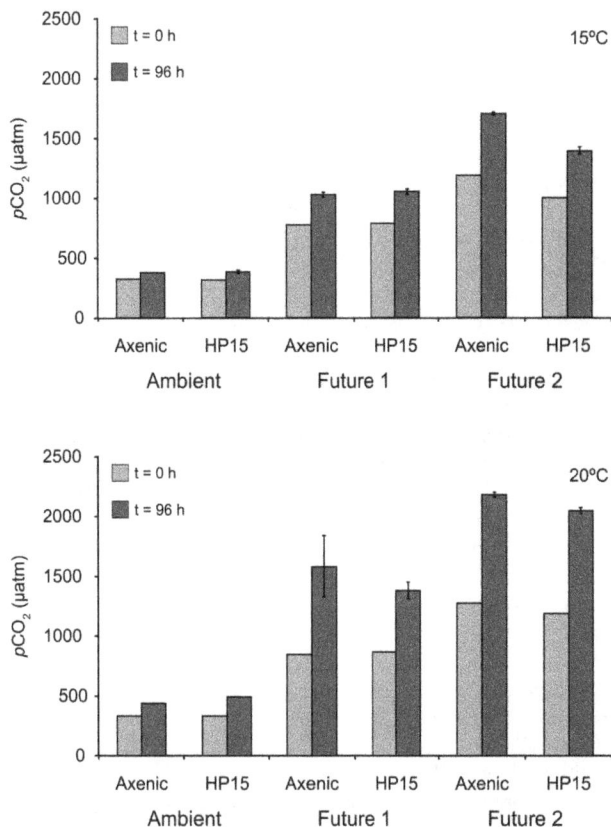

Figure 2. Initial and final pCO_2 during the incubations at 15°C and 20°C.

Ambient, Future 1 and Future 2 were 294 ± 1 µatm; 722 ± 7 µatm and 1021 ± 123 µatm in 15°C treatments and 304 ± 2 µatm, 782 ± 14 µatm and 1139 ± 55 µatm in 20°C treatments (Fig. 2).

During the 96 hr. incubation in rolling tanks, the pH$_T$ dropped between 0.06 and 0.22 units, with the largest temporal change in the two Future treatments at 20°C, reflecting the combination of higher respiration rates at higher temperatures and a weaker buffering system under future carbonate chemistry conditions. The simultaneous temporal change in pCO_2 ranged from 54 to 840 µatm, with the final pCO_2 in the 20°C Future 2 treatment reaching almost 2000 µatm (Fig. 2).

TEP formation. TEP were inadvertently added to each treatment with the diatom inoculum, and differences in initial concentrations reflect differences in TEP concentrations after the acclimatization phase. Initial TEP concentrations in the 12

Table 3. Exponential growth of *T. weissflogii* and pH range during the acclimatization phase.

	Treatment	μ (d^{-1})	pH$_T$	No. of days acclimatized
15°C	Ambient	0.51	7.93–8.21	11
	Future 1	0.52	7.57–7.76	11
	Future 2	0.49	7.45–7.66	11
20°C	Ambient	0.86	8.04–8.23	8
	Future 1	0.86	7.61–7.84	8
	Future 2	0.82	7.46–7.67	8

treatments ranged between 356 µg GXeq. L^{-1} and 1148 µg GXeq. L^{-1}, with significantly higher initial TEP concentrations in the six treatments incubated at 15°C compared to those at 20°C treatments (Mann-Whitney U-test, p<0.05).

During the 96 hr. incubation TEP remained about constant or increased moderately in treatments at 15°C, whereas the increase was appreciably higher in all treatments that were incubated at 20°C (Fig. 3). As a result, average TEP production of all 20°C treatments was significantly higher than that in 15°C treatments (Table 4), irrespective of $p CO_2$ conditions or the presence of bacteria (Mann-Whitney U-test, p<0.05). The presence of *M. adhaerens* HP15 had no effect on the amount of TEP generated during the experiment; TEP production averaged 565 µg GXeq L^{-1} with or without bacteria (Table 4). The $p CO_2$ conditions also did not influence TEP production significantly (Table 4; (Mann-Whitney U-test, p>0.05), but variability was high: Within each temperature the highest production of TEP was observed under Future 1 conditions, suggesting the possibility of some, albeit non-linear and complex impact of $p CO_2$, on TEP production. However, this could not be resolved with our experimental set-up.

Between 23 and 50% of all TEP was incorporated in aggregates (Table 5). Although the fraction of TEP in aggregates was always higher in the 20°C treatments compared to the 15°C treatments (34±9% vs. 26±3%), this trend was too small compared to the high variability to be statistically significant. The fraction of TEP enclosed in aggregates was largest under Ambient conditions (36±10%) and smallest under Future 2 conditions (25±3%). The partitioning of TEP between the aggregated and un-aggregated phase was independent of TEP concentration and the presence of bacteria.

Aggregation

Between 2 and 36 aggregates formed in the tanks, with significantly fewer aggregates in 20°C treatments compared to the 15°C treatments (Mann-Whitney U-test, p<0.05). Total aggregate volume, which is a better indicator of aggregate formation because it combines size and abundance of aggregates, was also significantly smaller in 20°C treatments compared to 15°C treatments. Total aggregate volume was high (>1 cm^3) in all treatments at 15°C, as well as in both ambient treatments at 20°C (Fig. 4). In contrast, total aggregate volume was small (<0.7 cm^3) in all four future treatments at 20°C, independent of the presence

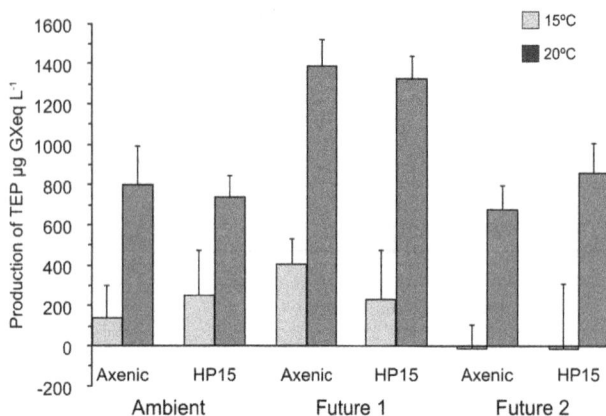

Figure 3. Production of Transparent Exopolymer Particles (TEP) during the incubations in all treatments, calculated as net change during the 96 hrs. experiment, and errors calculated using error propagation.

or absence of bacteria. Total aggregate volume was not a significant function of total TEP concentration. Not only was the relationship not significant, but higher TEP concentration tended to result in a smaller total aggregate volume.

Sinking velocity

Measured sinking velocities of aggregates >0.5 mm ranged between 8 and 110 m d^{-1}. Sinking velocity increased with the size of aggregates (equivalent spherical diameter, ESD), but was independent of $p CO_2$ or bacterial presence (Fig. 5). The slope of the velocity-size relationship depended on treatment (Table 6) and the slope of the sinking velocity vs. size relationship was significantly smaller (p<0.001, df=111, slope-t-test) for the 20°C treatments $(y = 2.4(\pm 0.3)x+16.1(\pm 1.8)$, df=44, $r^2 = 0.59$) compared to 15°C treatments $(y = 6.3(\pm 0.4)x+19.7(\pm 2.8)$, df=67, $r^2 = 0.80$) (Fig. 5). This resulted in relatively low sinking velocities, especially of larger aggregates, in 20°C treatments: For example, an aggregate with an ESD of 7 mm sank with 60 m d^{-1} in 15°C treatments and with 30 m d^{-1} in 20°C treatments.

Discussion

TEP play a key role for aggregation and flux [49] and thus may drive future changes in the functioning of the biological carbon pump [16]. Specifically, it has been hypothesised that increased TEP production under ocean acidification conditions would result in increased aggregation and carbon flux, strengthening the pump [17,18], although this has also been contested [19]. Under current conditions TEP provide the matrix of marine snow [50], and drive the aggregation of diatoms: Total aggregate volume has been found to be a positive function of total TEP concentration [2,9,51], except in the presence of certain minerals (e.g. illite) which promote aggregation [25]. Different scenarios are possible under future conditions of the ocean. Increased TEP production in a future ocean does not necessarily result in increased aggregation, because the stickiness of TEP may also change [19]. Moreover, characteristics of TEP that may change under high $p CO_2$ and elevated temperatures would impact the packaging of aggregates, and thus their sinking velocities; and consequently flux attenuation and the biological pump. Our experiment, which investigated the response of high temperature and high $p CO_2$ on TEP formation, aggregation and aggregate sinking velocity allowed us to investigate if the relationships between TEP and aggregation, and that between aggregate size and sinking velocity are likely to change in a future ocean.

TEP formation

TEP production during the aggregation experiment was a positive function of incubation temperature. Earlier work has also found increased TEP production by diatoms at elevated temperatures [52,53], but a careful study investigating a larger range of temperatures in eurythermal diatoms found a subsequent decrease when temperatures were increased further [54], suggesting an optimal type response curve. Increased production of TEP at higher temperature is due to increased release of TEP precursors by diatoms [55] and bacteria [56,57] at higher temperature.

The influence of $p CO_2$ on TEP production was less clear: At 20°C TEP production peaked under Future 1 conditions, but at 15°C TEP production at Ambient and Future 1 was similar, and no TEP was generated under the Future 2 scenario. As in our experiment, the impact of $p CO_2$ on TEP production by diatoms has been found to be non-linear in other studies ([25] and references within), with ambiguous results suggesting a complex relationship between TEP-production and $p CO_2$, possibly dependent on other

Table 4. Comparison of average TEP production (µg GXeq. L^{-1}) and aggregation, as measured by total aggregate volume (Agg. Vol.), combining treatments with the same temperature, carbonate conditions, or state of axenicity, respectively.

Treatment	TEP production µg GXeq. L^{-1}	Total Agg. Vol. cm3	n
15°C	166±164*	2.04±0.60*	6
20°C	965±311*	0.82±0.73*	6
Am	483±335	2.01±0.74	4
F1	837±605	1.21±1.15	4
F2	376±458	1.07±0.66	4
Ax	565±508	1.56±0.95	6
HP	565±498	1.30±0.92	6

N = number of treatments, each in duplicate.
*: averages significantly (p<0.05) different from each other, paired t-test.

environmental conditions (light, temperature) or nutrient availability. Significant differences in bacterial catabolism of TEP as a function of environmental conditions would have resulted in differences between treatments with and without *M. adhaerens*, which we did not observe.

The addition of *M. adhaerens* HP15 had no systematic effect on initial TEP concentration, likely because the added volume of acclimatized *M. adhaerens* HP15 cultures was very small. Net TEP production during the aggregation experiment was also not influenced by *M. adhaerens* HP15. Heterotrophic bacteria are known to generate TEP and other EPS, but also utilize it, and their influence on TEP concentrations in the presence of diatoms varies [4]. For example; TEP production by *Thalassiosira rotula* was enhanced by bacteria during exponential phase but reduced during stationary growth of the diatom [3]. *Skeletonema costatum* exhibited a different lifestyle pattern with higher TEP production in the axenic culture compared to the non-axenic one [3]. In an earlier experiment, the presence of *M. adhaerens* HP15 resulted in increased TEP production after 4 and 7 days at 18°C, but TEP was only measured in the surrounding (aggregate-free) seawater [2]. In the experiment presented here TEP concentration in the surrounding seawater was also higher (~12%) in the presence of *M. adhaerens* HP15, but total TEP concentration was not affected, implying that TEP incorporated in aggregates decreased in the presence of bacteria. Possibly the bacteria impact the fraction of TEP included in aggregates, rather than TEP production *per se*. Conceivably bacterial modification of TEP decreased its propensity to aggregate, or bacterial activity dissociated TEP from aggregates.

Aggregation

Ranges in size and total volume of aggregates in this experiment were within the ranges found under similar conditions [58,59,60]. Aggregation, measured as total aggregated volume, was appreciably higher in all 15°C treatments compared to 20°C treatments. This appears to contradict a study, which found increased aggregation at higher temperatures using a natural diatom population incubated 2.5°C and 8.5°C [53]. The formation of micro-aggregates by the diatom *Skeletonema* sp. in a high turbulence environment was also significantly reduced at 10°C compared to 20°C, but a further increase to 30°C had no effect [61]. However, the carbonate system was not perturbed in either of these studies; and in our experiment the change in temperature had no significant effect on aggregation if only the Ambient treatments are considered. Our study emphasizes that the simultaneous change in temperature and the carbonate system influenced aggregation differently than that of temperature alone: Total aggregate volume was greatly reduced at 20°C under both future pCO_2 scenarios, suggesting synergistic effects between pCO_2 and temperature.

Contrary to expectations, total aggregate volume was not a function of TEP concentration. The lower aggregation in Future 20°C treatments suggests a decreased probability that colliding particles remained attached, called stickiness, in Future 20°C treatments. Aggregation rate is a function of collision rate and stickiness. Collision rate in rolling tanks, where solid body rotation is established, depends largely on particle abundance, size and differential settling [11,62]. Because cell concentrations, sizes, turbulence and shear, all of which promote collisions, were near identical in all treatments, differences must be a function of TEP

Table 5. TEP in Aggregates, absolute amount and fraction.

Treatment	15°C		20°C	
	GXeq. L^{-1}	%	GXeq. L^{-1}	%
Am HP	410	31	432	36
Am Ax	280	26	659	50
F1 HP	319	26	471	28
F1 Ax	299	24	641	35
F2 HP	260	23	349	24
F2 Ax	260	25	381	29

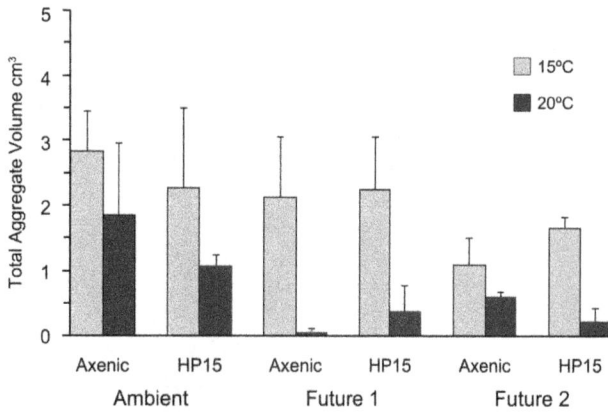

Figure 4. Total aggregate volume after the incubations in all treatments; error bars represent the range of replicates.

or stickiness. As the four treatments with the highest total TEP concentrations resulted in the lowest aggregation, it may be deduced that the average stickiness of particles was lower in Future 20°C treatments compared to the others.

Aggregation was not impacted by the presence or absence of *M. adhaerens* HP15. The impact of heterotrophic bacteria on aggregation can be extremely varied: Bacteria may increase aggregation and stability of aggregates [63], or diatom aggregation may be reduced due to hydrolysis of diatom surface mucus by attached bacteria [3,64]. Moreover, the role of heterotrophic bacteria for diatom aggregation varies appreciably between algae species and environmental conditions: The presence of bacteria was demonstrated to be a prerequisite for aggregate formation for *T. weissflogii*, but not for *Navicula* sp. [65]. A different study revealed that whether coagulation of *T. rotula* was promoted or reduced in the presence of bacteria depended on light conditions [3]. Earlier work has revealed that the presence of *M. adhaerens* HP15 greatly increased total aggregate volume of *T. weissflogii*

[2]. In those experiments, total aggregate volume in the presence of bacteria was 2–3 times higher (10 cm^3) than in our study, whereas almost no aggregates formed in the axenic cultures. In the present experiment total aggregate volume was independent of the presence of *M. adhaerens* HP15, and total aggregate volume was comparably small. The observed differences in TEP production and aggregation between both experiments were possibly caused by differences in EPS due to experimental (start conditions, time in rolling tank) or cell physiological differences. Composition of extracellular substances released by diatoms varies with growth stage and environmental conditions [66]. Variation in EPS chemistry and tertiary structure between diatom and bacteria EPS are known to result in differences in TEP formation [67,68], and EPS other than TEP can lead to the formation of aggregates [69]. Moreover, *T.weissflogii* may also aggregate by direct cell to cell attachment [1], and the factors that determine which aggregation process dominates await exploration. The observed discrepancies between similar experiments highlight the complexity (non-linearity) of the bacteria- diatom interactions and aggregation processes under different environmental conditions.

Sinking velocities of aggregates

The sinking velocity to size relationship of aggregates differed appreciably between treatments, implying differences in aggregate content (excess density) and packaging (porosity) [70]. Sinking velocities of large aggregates formed in 20°C treatments were significantly smaller than those of comparable size formed in 15°C treatments. Sinking velocity is a function of the viscosity of seawater, but this effect should increase sinking velocity at higher temperatures, rather than decrease it [71]. TEP content of aggregates was smaller in the 15°C treatments compared to the 20°C future treatments, as both total TEP concentration and fraction of TEP in aggregates were smaller in 15°C treatments. TEP are positively buoyant and a high proportion of TEP in aggregates reduces their sinking velocity [19,72,73]. Differences in chemical composition or quantity of EPS may easily explain differences in packaging and thus sinking velocities. Additionally, differences in the cellular silica content, the biochemical composition of organic matter, or the size

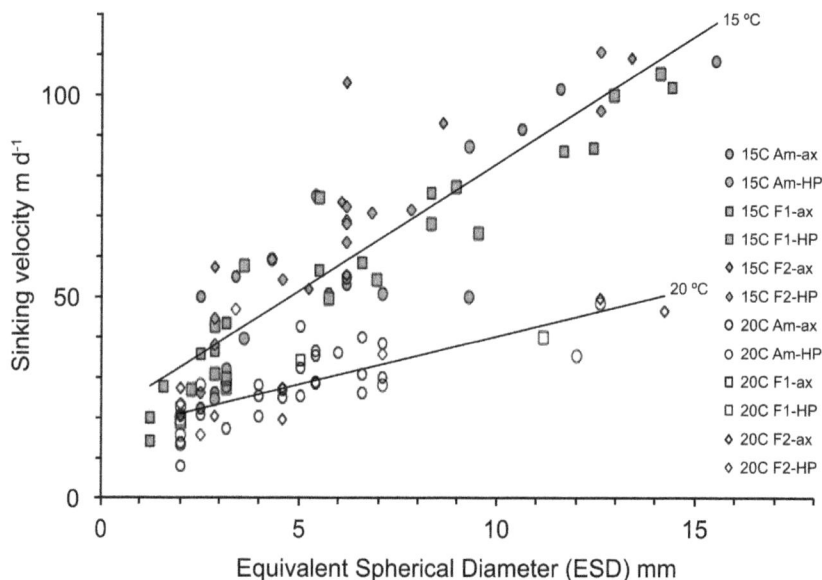

Figure 5. Sinking velocity vs. size (equivalent spherical diameter) of aggregates >0.5 mm that formed in the different treatments.
Lines represent the regression of aggregates incubated at 15°C and 20°C, respectively.

Table 6. Slopes of sinking velocity vs. equivalent spherical diameter (ESD) regressions in each treatment (Amb = Ambient, F1and F2 = Future 1 and Future 2, Ax = axenic, HP = *M. adhaerens* HP15).

Treatment	slope	n	r^2
15 Amb-Ax	6.5	13	0.88
15 Amb-HP	6.3	11	0.54
15 F1-Ax	6.0	12	0.96
15 F1-HP	6.0	13	0.85
15 F2-Ax	5.5	7	0.57
15 F2-HP	6.6	13	0.91
20 Amb-Ax	4.4	22	0.61
20 Amb-HP	2.2	7	0.78
20 F1-Ax	NA	1	NA
20 F1-HP	2.2	4	0.93
20 F2-Ax	2.3	6	0.86
20 F2-HP	3.3	6	0.32

of cells [74–78] may explain reduced sinking velocity of aggregates consisting of cells grown at higher temperature. Future experiments will need to address these factors.

Conclusions

The respective roles of TEP, bacteria and diatoms for the formation of diatom aggregates, and the influence of temperature and pCO_2 on coagulation, are complex and not well understood. Differences in environmental conditions and in physiological growth stage of the organisms as well as species specific life strategy differences all contribute to a high variability in the characteristics of the EPS that is produced by diatoms and bacteria or hydrolyzed by bacteria. Our results clearly indicate that the known relationship between TEP and aggregation, and between aggregate size and sinking velocity do not hold under different temperature and pCO_2 scenarios. In contrast to theoretical predictions [17,18], higher pCO_2 combined with elevated

temperatures resulted in increased TEP production, but decreased aggregation and decreased sinking velocity of aggregates, suggesting decreased carbon flux at 1000 m (sequestration flux). Our results thus refute, at least in a general sense, the hypothesis that elevated temperature and ocean acidification as expected in the future ocean will result in increased carbon flux and thus in a negative feed-back to the biological carbon pump [17,18]. More generally, these results provide an example that established relationships, like that between TEP concentration, aggregation and flux, may not extend to future conditions, and care must be taken, when basing predictions on such empirical relationships.

Author Contributions

Conceived and designed the experiments: UP SS MU CF. Performed the experiments: SS CF UP. Analyzed the data: SS CF UP. Contributed reagents/materials/analysis tools: UP MU. Wrote the paper: UP SS MU.

References

1. Crocker KM, Passow U (1995) Differential aggregation of diatoms. Marine Ecology Progress Series 117: 249–257.
2. Gaerdes A, Iversen MH, Grossart H-P, Passow U, Ullrich M (2011) Diatom associated bacteria are required for aggregation of *Thalassiosira weissflogii*. ISME Journal 5: 436–445.
3. Grossart HP, Czub G, Simon M (2006) Algae-bacteria interactions and their effects on aggregation and organic matter flux in the sea. Environmental Microbiology 8: 1074–1084.
4. Simon M, Grossart HP, Schweitzer B, Ploug H (2002) Microbial ecology of organic aggregates in aquatic ecosystems. Aquatic Microbial Ecology 28: 175–211.
5. Decho AW (1990) Microbial exopolymer secretions in ocean environments: Their role(s) in food web and marine processes. Oceanography and Marine Biology Annual Review 28: 73–153.
6. Decho AW, Herndl GJ (1995) Microbial activities and the transformation of organic matter within mucilaginous material. Science of the Total Environment 165: 33–42.
7. Jackson GA (1995) TEP and coagulation during a mesocosm experiment. Deep-Sea Research, II 42: 215–222.
8. Passow U, Alldredge AL (1995) Aggregation of a diatom bloom in a mesocosm: The role of transparent exopolymer particles (TEP). Deep-Sea Research II 42: 99–109.
9. Passow U, Alldredge AL, Logan BE (1994) The role of particulate carbohydrate exudates in the flocculation of diatom blooms. Deep-Sea Research I 41: 335–357.
10. Burd AB, Jackson GA (2009) Particle Aggregation. Annual Review of Marine Science 1: 65–90.

11. Jackson GA (1990) A model of the formation of marine algal flocks by physical coagulation processes. Deep-Sea Research 37: 1197–1211.
12. Kiørboe T, Lundsgaard C, Olesen M, Hansen JLS (1994) Aggregation and sedimentation processes during a spring phytoplankton bloom: A field experiment to test coagulation theory. Journal of Marine Research 52: 297–323.
13. Raupach MR, Marland G, Ciais P, Le Quere C, Canadell JG, et al. (2007) Global and regional drivers of accelerating CO2 emissions. Proceedings of the National Academy of Sciences of the United States of America 104: 10288–10293.
14. Houghton J (1995) Comment on "The roles of carbon dioxide and water vapour in warming and cooling the Earth's troposphere" J. Barrett, Spectrochim. Acta Part A, 51 (3) (1995) 415. Spectrochimica Acta Part A: Molecular and Biomolecular Spectroscopy 51: 1391–1392.
15. Zeebe RE, Wolf-Gladrow A (2001) CO2 in Seawater: Equilibrium, Kinetics, Isotopes. Amsterdam: Elsevier Science. 346 p.
16. Passow U, Carlson C (2012) The Biological Pump in a High CO2 World. Marine Ecology Progress Series 470: 249–271.
17. Riebesell U, Schulz KG, Bellerby RGJ, Botros M, Fritsche P, et al. (2007) Enhanced biological carbon consumption in a high CO_2 ocean. Nature 450: 545–548.
18. Arrigo KR (2007) Carbon cycle: Marine manipulations. Nature 450: 491–492.
19. Mari X (2008) Does ocean acidification induce an upward flux of marine aggregates? Biogeosciences 5: 1023–1031.
20. Biermann A, Engel A (2010) Effect of CO2 on the properties and sinking velocity of aggregates of the coccolithophore *Emiliania huxleyi*. Biogeosciences 7: 1017–1029.

21. Engel A, Abramson L, Szlosek J, Liu Z, Stewart G, et al. (2009) Investigating the effect of ballasting by CaCO3 in *Emiliania huxleyi*, II: Decomposition of particulate organic matter. Deep Sea Research II 56: 1408–1419.

22. Engel A, Szlosek J, Abramson L, Liu Z, Lee C (2009) Investigating the effect of ballasting by CaCO$_3$ in *Emiliania huxleyi*: I. Formation, settling velocities and physical properties of aggregates. Deep Sea Research II 56: 1396–1407.

23. Gehlen M, Alvain S, Bopp L, Moulin C (2006) Modeling the potential response of ocean biogeochemistry to climate change. Geophysical Research Abstracts 8: 3492.

24. Gehlen M, Gangsto R, Schneider B, Bopp L, Aumont O, et al. (2007) The fate of pelagic CaCO$_3$ production in a high CO$_2$ ocean: a model study. Biogeosciences 4: 505–519.

25. Passow U, Rocha CLDL, Fairfield C, Schmidt K (2014) Aggregation as a function of pCO2 and mineral particles. Limnology and Oceanography 59: 532–547.

26. Kaeppel EC, Gaerdes A, Seebah S, Grossart H-P, Ullrich MS (2012) Marinobacter adhaerens sp nov., isolated from marine aggregates formed with the diatom *Thalassiosira weissflogii*. International Journal of Systematic and Evolutionary Microbiology 62: 124–128.

27. Gaerdes A, Ramaye Y, Grossart H-P, Passow U, Ullrich MS (2012) Effects of *Marinobacter adhaerens* HP15 on polymer exudation by *Thalassiosira weissflogii* at different N: P ratios. Marine Ecology Progress Series 461: 1–14.

28. Ploug H, Terbrüggen A, Kaufmann A, Wolf-Gladrow D, Passow U (2010) A novel method to measure particle sinking velocity in vitro, and its comparison to three other in vitro methods. Limnology and Oceanography: Methods 8: 386–393.

29. Venrick EL (1998). Spring in the California Current: The distribution of phytoplankton species, April 1993 and April 1995. Marine Ecology Progress Series 167, 73–88.

30. Hart 1934 in Raymont JEG (1980), second edition. Plankton and Productivity in the Oceans: Phytoplankton. Pergamon International Library, New York.

31. Passow U, De La Rocha CL (2006) Accumulation of mineral ballast on organic aggregates. Global Biogeochemical Cycles 20: 7.

32. Kester D, Duedall I, Connors D, Pytkowicz R (1967) Preparation of artificial seawater. Limnology and Oceanography 12: 176–179.

33. Guillard RRL (1975) Culture of phytoplankton for feeding marine invertebrates. In: Smith WL, Chanley MH, editors. Culture of marine invertebrate animals: Plenum Press. pp. 108–132.

34. Kirchman D (1993) Statistical analysis of direct counts of microbial abundance. In: Kemp P, Sherr B, Sherr E, Cole J, editors. Handbook of methods in aquatic microbial ecology. London: Lewis Publishers. pp. 117–119.

35. Porter KG, Feig YS (1980) The use of DAPI for identifying and counting aquatic microflora. Limnology and Oceanography 25: 943–948.

36. Mopper K, Zhou J, Ramana KS, Passow U, Dam HG, et al. (1995) The role of surface-active carbohydrates in the flocculation of a diatom bloom in a mesocosm. Deep-Sea Research II 42: 47–73.

37. Schuster S, Herndl GJ (1995) Formation and significance of transparent exopolymeric particles in the northern Adriatic Sea. Marine Ecology Progress Series 124: 227–236.

38. Zhou J, Mopper K, Passow U (1998) The role of surface-active carbohydrates in the formation of transparent exopolymer particles by bubble adsorption of seawater. Limnology and Oceanography 43: 1860–1871.

39. Passow U (2012) The Abiotic Formation of TEP under different ocean acidification scenarios. Marine Chemistry 128–129: 72–80.

40. Clayton TD, Byrne RH (1993) Spectrophotometric seawater pH measurements: total hydrogen ion concentration scale calibration of m-cresol purple and at sea results. Deep Sea Research I 40 (10), 2115–2129.

41. Fangue N, O'Donnell M, Sewell M, Matson P, MacPherson A, et al. (2010) A laboratory-based experimental system for the study of ocean acidification effects on marine invertebrate larvae. Limnology and Oceanography Methods 8, 441-452.

42. Dickson AG, Sabine CL, Christian JR (2007) Guide to best practices for CO$_2$ measurments. PICES Special Publication 3: 1–191.

43. Lewis E, Wallace DWR (1998) Program developed for CO$_2$ System calculations. ORNL/CDIAC-105 Carbon dioxide informations Analysis center, Oak Ridge National Laboratory, U.S. Department of Energy.

44. Roy RN, Roy LN, Vogel KM, Portermoore C, Pearson T, et al. (1993) The dissociation constants of carbonic-acid in seawater at salinities 5 to 45 and temperatures 0 degrees to 45°C. Marine Chemistry 44: 249–267.

45. Passow U, Alldredge AL (1995) A dye-binding assay for the spectrophotometric measurement of transparent exopolymer particles (TEP). Limnology and Oceanography 40: 1326–1335.

46. Hoppe CJM, Langer G, Rokitta SD, Wolf-Gladrow D, Rost B (2012) Implications of observed inconsistencies in carbonate chemistry measurements for ocean acidification studies. Biogeoscience 9: 2401–2405.

47. Hofmann G, Smith JE, Johnson K, Send U, Levin LA, et al. (2011) High Frequency Dynamics of ocean pH: A Multi-Ecosystem Comparison. PLoS ONE 6 (12), e28983.

48. Frieder C, Nam S, Martz TR, Levin LA (2012). High temporal and spatial variability of dissolved oxygen and pH in a near shore California kelp forest. Biogeosciences 9, 3917–3930.

49. Logan BE, Passow U, Alldredge AL, Grossart H-P, Simon M (1995) Rapid formation and sedimentation of large aggregates is predictable from coagulation

50. Alldredge AL, Passow U, Logan BE (1993) The abundance and significance of a class of large, transparent organic particles in the ocean. Deep-Sea Research I 40: 1131–1140.

51. Engel A, Thoms S, Riebesell U, Rochelle-Newall E, Zondervan I (2004) Polysaccharide aggregation as a potential sink of marine dissolved organic carbon. Nature 428: 929–932.

52. Fukao T, Kimoto K, Kotani Y (2012) Effect of temperature on cell growth and production of transparent exopolymer particles by the diatom Coscinodiscus granii isolated from marine mucilage. Journal of Applied Phycology 24: 181–186.

53. Piontek J, Haendel N, Langer G, Wohlers J, Riebesell U, et al. (2009) Effects of rising temperature on the formation and microbial degradation of marine diatom aggregates. Aquatic Microbial Ecology 54: 305–318.

54. Claquin P, Probert I, Lefebvre S, Veron B (2008) Effects of temperature on photosynthetic parameters and TEP production in eight species of marine microalgae. Aquatic Microbial Ecology 51: 1–11.

55. Passow U (2002) Production of transparent exopolymer particles (TEP) by phytoplankton and bacterioplankton. Marine Ecology Progress Series 236: 1–12.

56. Passow U (2000) Formation of Transparent Exopolymer Particles, TEP, from dissolved precursor material. Marine Ecology Progress Series 192: 1–11.

57. Yong-Xue D, Chin-Chang H, Santschi PH, Verdugo P, Wei-Chun C (2009) Spontaneous Assembly of Exopolymers from Phytoplankton. Terrestrial, Atmospheric & Oceanic Sciences 20: 741–747.

58. Gaerdes A, Iversen MH, Grossart H-P, Passow U, Ullrich M (2011) Diatom associated bacteria are required for aggregation of *Thalassiosira weissflogii*. ISME Journal 5 (3), 436–445.

59. Passow U, Rocha CLDL, Fairfield C, Schmidt K (2014) Aggregation as a function of pCO2 and mineral particles. Limnology and Oceanography 59 (2), 532–547.

60. Moriceau B, Ragueneau O, Garvey M, Passow U (2007) Evidence for reduced biogenic silica dissolution rates in diatom aggregates. Marine Ecology Progress Series 333, 129–142.

61. Thornton DCO, Thake B (1998) Effect of temperature on the aggregation of *Skeletonema costatum* (Bacillariophyceae) and the implication for carbon flux in coastal waters. Marine Ecology Progress Series 174: 223–231.

62. Jackson GA, Burd AB (1998) Aggregation in the marine environment. Environmental Science & Technology 32: 2805–2814.

63. Heissenberger A, Herndl GJ (1994) Formation of high molecular weight material by free-living marine bacteria. Marine Ecology Progress Series 111: 129–135.

64. Smith DC, Steward GF, Long RA, Azam F (1995) Bacterial mediation of carbon fluxes during a diatom bloom in a mesocosm. Deep-Sea Research II 42: 75–97.

65. Grossart HP, Kiorboe T, Tang KW, Allgaier M, Yam EM, et al. (2006) Interactions between marine snow and heterotrophic bacteria: aggregate formation and microbial dynamics. Aquatic Microbial Ecology 42: 19–26.

66. Myklestad SM (1995) Release of extracellular products by phytoplankton with special emphasis on polysaccharides. Science of the Total Environment 165, 155–164.

67. Verdugo P (2012) Marine Microgels. Annual Review of Marine Science 4 (1), 375–400.

68. Verdugo P, Santschi P (2010) Polymer dynamics of DOC networks and gel formation in seawater. Deep Sea Research II 57, 1486–1493.

69. Bhaskar PV, Grossart H-P, Bhosle NB, Simon M (2005) Production of macroaggregates from dissolved exopolymeric substances (EPS) of bacterial and diatom origin. FEMS Microbiology Ecology 53: 255–264.

70. Lam PJ, Bishop JKB (2007) High biomass, low export regimes in the Southern Ocean. Deep Sea Research Part II: Topical Studies in Oceanography 54: 601–638.

71. Taucher J, Bach LT, Riebesell U, Oschlies A (2014) The viscosity effect on marine particle flux: A climate relevant feedback mechanism. Global Biogeochemical Cycles 28 (4), 2013GB004728.

72. Azetsu-Scott K, Passow U (2004) Ascending marine particles: Significance of transparent exopolymer particles (TEP) in the upper ocean. Limnology and Oceanography 49: 741–748.

73. Engel A, Schartau M (1999) Influence of transparent exopolymer particles (TEP) on sinking velocity of *Nitzschia closterium* aggregates. Marine Ecology-Progress Series 182: 69–76.

74. Durbin EG (1977) Studies on the autecology of the marine diatom *Thalassiosira nordenskoeldii*: II The influence of cell size on growth rate, and carbon, nitrogen, chlorophyll and silica content. Journal of Phycology 13: 150–155.

75. Mejia LM, Isensee K, Mendez-Vicente A, Pisonero J, Shimizu N, et al. (2013) BSi content and Si/C ratios from cultured diatoms (*Thalassiosira pseudonana* and *Thalassiosira weissflogii*): Relationship to seawater pH and diatom carbon acquisition. Geochimica Et Cosmochimica Acta 123: 322–337.

76. Montagnes DJS, Franklin DJ (2001) Effect of temperature on diatom volume, growth rate, and carbon and nitrogen content: Reconsidering some paradigms. Limnology and Oceanography 46: 2008–2018.

77. Ishida Y, Hiragushi N, Kitaguchi H, Mitsutani A, Nagai S, et al. (2000) A highly CO$_2$-tolerant diatom, *Thalassiosira weissflogii* H1, enriched from coastal sea, and its fatty acid composition. Fisheries Science 66: 655–659.

78. Richardson TL, Cullen JJ (1995) Changes in buoyancy and chemical composition during growth of a coastal marine diatom: Ecological and biogeochemical consequences. Marine Ecology Progress Series 128: 77–90.

14

Dissolution Dominating Calcification Process in Polar Pteropods Close to the Point of Aragonite Undersaturation

Nina Bednaršek[1,2,3]*, Geraint A. Tarling[2], Dorothee C. E. Bakker[3], Sophie Fielding[2], Richard A. Feely[1]

1 NOAA Pacific Marine Environmental Laboratory, Seattle, Washington, United States of America, 2 British Antarctic Survey, High Cross, Cambridge, United Kingdom, 3 Centre for Ocean and Atmospheric Sciences, School of Environmental Sciences, University of East Anglia, Norwich Research Park, Norwich, United Kingdom

Abstract

Thecosome pteropods are abundant upper-ocean zooplankton that build aragonite shells. Ocean acidification results in the lowering of aragonite saturation levels in the surface layers, and several incubation studies have shown that rates of calcification in these organisms decrease as a result. This study provides a weight-specific net calcification rate function for thecosome pteropods that includes both rates of dissolution and calcification over a range of plausible future aragonite saturation states (Ω_{ar}). We measured gross dissolution in the pteropod *Limacina helicina antarctica* in the Scotia Sea (Southern Ocean) by incubating living specimens across a range of aragonite saturation states for a maximum of 14 days. Specimens started dissolving almost immediately upon exposure to undersaturated conditions ($\Omega_{ar} \sim 0.8$), losing 1.4% of shell mass per day. The observed rate of gross dissolution was different from that predicted by rate law kinetics of aragonite dissolution, in being higher at Ω_{ar} levels slightly above 1 and lower at Ω_{ar} levels of between 1 and 0.8. This indicates that shell mass is affected by even transitional levels of saturation, but there is, nevertheless, some partial means of protection for shells when in undersaturated conditions. A function for gross dissolution against Ω_{ar} derived from the present observations was compared to a function for gross calcification derived by a different study, and showed that dissolution became the dominating process even at Ω_{ar} levels close to 1, with net shell growth ceasing at an Ω_{ar} of 1.03. Gross dissolution increasingly dominated net change in shell mass as saturation levels decreased below 1. As well as influencing their viability, such dissolution of pteropod shells in the surface layers will result in slower sinking velocities and decreased carbon and carbonate fluxes to the deep ocean.

Editor: Pauline Ross, University of Western Sydney, Australia

Funding: NB was supported by the FAASIS (Fellowships in Antarctic Air-Sea-Ice Science, MEST-CT-2004-514159), a European Union Marie Curie Early Stage Training Network. GT and SOF were supported by the Ecosystems Programme at the British Antarctic Survey. GT and DCEB received additional support from the UK Ocean Acidification Research Programme, funded by the Natural Environment Research Council, the Department for Environment, Food and Rural Affairs and the Department for Energy and Climate Change (grant no. NE/H017267/1). RF was supported by the NOAA Ocean Acidification Program. The funders had no role in study design, data collection and analysis, decision to publish, or preparation of the manuscript.

Competing Interests: The authors have declared that no competing interests exist.

* Email: nina.bednarsek@noaa.gov

Introduction

Formation and dissolution of calcium carbonate ($CaCO_3$), and carbon export from the surface to the deep ocean are important mechanisms in the global carbon cycle, immediately related to the control of atmospheric CO_2 (carbon dioxide) and regulation of the dissolved CO_2 concentration and pH [1–4]. $CaCO_3$ can occur in calcite, aragonite, or high-magnesium calcite form, and different planktonic species produce shells or skeletons of one of these mineral forms. Aragonite is the more soluble form of $CaCO_3$, and its formation and dissolution is determined by the $CaCO_3$ saturation state (Ω_{ar}), which is the product of the concentrations of calcium (Ca^{2+}) and carbonate ions (CO_3^{2-}) at the in situ temperature, salinity, and pressure, divided by the apparent stoichiometric solubility product for the structural form of $CaCO_3$ (K^*_{sp}):

$$\Omega_{ar} = \frac{[Ca^{2+}][CO_3^{2-}]}{K^*_{sp}} \qquad (1)$$

Surface waters are generally supersaturated with CO_3^{2-} but their concentration tends to decrease with depth. Below the saturation horizon ($\Omega_{ar} < 1$), the concentration of these ions decreases to the point where aragonite starts to dissolve [5–9]. Carbonate ions in the surface ocean buffer the increased uptake of atmospheric CO_2, leading to a decrease in their concentration and a shallowing of the saturation horizon [2,10–11]; this process is referred to as ocean acidification. The greater solubility of aragonite in colder waters means that ocean acidification effects will be most evident in polar regions [11–13]. For instance, the Southern Ocean is predicted to reach surface undersaturation seasonally by about 2038 [14].

Among the most vulnerable organisms to ocean acidification are pteropods, which are thin-shelled aragonite producers particularly abundant at high latitudes. Pteropods are a major component of polar ocean food webs, and they play a key role in energy transfer and carbon fluxes in these regions by exerting a high grazing pressure with large feeding webs, faeces, and pseudofaeces sinking rapidly and transferring carbon to the ocean interior [15]. Furthermore, as a ballast for other particulate organic matter [16], the rapid sinking and dissolution of the pteropod shells at depth is an important contributor of carbon and alkalinity in the deep ocean [5].

As pteropods exert little control over the pH and carbonate chemistry of their calcifying fluid, they are more sensitive to the effects of ocean acidification than other calcifying marine organisms [17]. Corals, for example, can elevate the pH of their calcifying fluid relative to the surrounding sea water, buffering ocean acidification effects [18–19]. Nevertheless, pteropods could potentially counteract the loss of shell through calcification processes within the shell interior, as occurs in aragonite-based corals [20]. Lischka et al. [21], for example, presented evidence of "repair" calcification in pteropods, with shell thickening at sites of previous shell dissolution.

The response of thecosome pteropods to aragonite saturation state in terms of rates of calcification was considered by Comeau et al. [22], who carried out incubations with the Arctic pteropod, *Limacina helicina*. They noted that calcification did not cease at Ω_{ar} levels below 1, but in fact was still evident at Ω_{ar} of 0.6. However, the rate of calcification was sensitive to saturation state, and demonstrated a logarithmic decrease from Ω_{ar} levels of 2.0 to 0.6 such that, by Ω_{ar} equal to 1, the calcification rate was less than half of that observed at Ω_{ar} of 2.0. The study only determined gross calcification rates but did not also assess whether such rates would be sufficient to counteract dissolution.

One approach to accounting for rates of dissolution is to apply dissolution kinetic algorithms as follows:

$$R = k \cdot (1 - \Omega_{ar})^n \; for \; \Omega_{ar} < 1 \qquad (2)$$

where R is the rate of dissolution (%); k, the dissolution rate constant (d^{-1}); and n, the dimensionless reaction order [6].

The dissolution rate constant for aragonite has been principally derived through studies on non-living biogenic material and used to estimate dissolution rates as part of the global aragonite cycle [6,23]. Gangstø et al. [23] considered the dissolution of abiogenic aragonite to be a first order reaction (n = 1) with a dissolution rate constant (k) of 10.9 d^{-1}. Such an approach is supported by observations of Bednaršek et al. [24], who showed that, in the natural environment as well as in laboratory experiments, dissolution of the shell of the Southern Ocean pteropod *L. helicina* can be rapid and substantial when exposed to Ω_{ar} levels near or below 1.0. Nevertheless, the rate of dissolution in living pteropods may not simply be a function of abiogenic dissolution kinetics since living individuals have specific shell structures and mechanisms that can slow down damage from dissolution processes [25]. Thus, it is important to verify that dissolution rates of abiogenic material agree with dissolution rates of living organisms.

In this study, we consider the consequences to the shells of living specimens of *Limacina helicina ant.* of exposure to seawater undersaturated for aragonite. We examine the shells in two ways to derive an overall level of gross dissolution: firstly, by examining the shell aperture to determine the penetration thickness of dissolution and/or any calcification, and secondly, by estimating the proportion of the surface shell undergoing dissolution. As the

dissolution rate of aragonite is an important biogeochemical parameter, this study will determine whether it can be equally applied to living aragonite producers as it can to abiogenic aragonite. Finally, pteropods are a major source of $CaCO_3$ to the deeper layers of the Southern Ocean, and any processes acting to decrease shell weight through dissolution will impact this flux. This study will examine how measured dissolution levels will impact the amount of pteropod-derived $CaCO_3$ flux leaving the surface layers for eventual export to the ocean interior.

Materials and Methods

Sampling methods

Sampling and incubation were carried out aboard the RRS *James Clark Ross* during cruise JR177 to the Scotia Sea in January and February 2008 along a transect from 62.21°S 44.4°W to 50.6°S 35.1°W. Samples were collected in accordance with a 5-year permit for science operations and sampling (No. S3-3/2005) issued to the British Antarctic Survey by the British Foreign and Commonwealth Office under Article VII of the Antarctic Treaty and Article 17 of the Protocol on Environmental Protection to the Antarctic Treaty.

The methods for animal collection, the perturbation experiments, and pteropod shell analysis have been described in Bednarsek et al. [24,26–27] and are only briefly summarized here. Juvenile and adult pteropods were collected with a variety of nets including a vertically hauled Bongo net with 100 and 200 μm meshed nets, a MOCNESS with 330 μm nets, and a towed Bongo with 300 μm and 600 μm meshed nets. Ship speed was between 1 and 2 knots during towing operations. The towed nets were generally more effective at capturing adult specimens, while juveniles were caught in both the vertical and towed nets. All deployments were carried out between 0 and 400 m. Captured pteropods were counted and examined under a light microscope to look for any instances of damage to the shell. Pteropods with evidence of cracks, pits, etchings, or perforations were excluded from the incubations.

Control populations

Shelled pteropods are prone to mechanical damage, the two main causes being: (1) net sampling, and (2) incubation in the experimental chambers. This was accounted for by including two types of control in the experimental design: (1) natural environment control samples, caught in the upper 125 m, where $\Omega_{ar} = 1.82 \pm 0.60$, made up from individuals without any sign of net-induced damage that were preserved in 70% ethanol immediately upon collection; (2) experimental control samples, which were individuals incubated in the same experimental setup as those in perturbed conditions but in which pCO_2 (partial pressure of CO_2) was maintained at present day levels. This was achieved through bubbling air with a 375 ppm (μmol/mol) CO_2 mixing ratio through filtered seawater such that Ω_{ar} was 1.70 ± 0.08. Undamaged pteropods were incubated in these conditions for eight days.

Perturbation experiments

The effects of dissolution on shells of pteropods was determined at three different Ω_{ar} saturation states: (1) supersaturation, simulating present day Ω_{ar} levels; (2) transitional state, where Ω_{ar} was close to 1; and (3) undersaturation, where Ω_{ar} was less than 1. The incubation conditions were achieved by bubbling synthetic air containing CO_2 mixing ratios (BOC Special Products) of 375, 500, 750, and 1200 ppm through filtered seawater from the ship's surface seawater supply that was previously filtered using a

0.22 µm GF/F filter. The bubbling was conducted either in 15 L carboys (for incubation of the adults) or in 2 L flasks (for incubation of the juveniles) and was carried out on deck at 3–4°C with the gas mixtures being introduced via micro-porous airstones. The pCO_2 in the headspace was measured using a LI-COR CO_2/H_2O analyser 6262. The bubbling was stopped and the porous stone apparatus was removed once the water had reached the target pCO_2 before adding the pteropods. The flask was then sealed, taking care to reduce the headspace to a minimum. The flasks were kept at a constant temperature of 4°C and blacked-out for the duration of the experiments.

To ascertain the exact chemical perturbations achieved by the bubbling procedure, water was taken at the start and end of each experiment (250 ml+50 µl $HgCl_2$ solution). These samples were analysed for total alkalinity (TA) and total dissolved inorganic carbon (DIC) using a VINDTA model 3C, calibrated with Certified Reference Materials (batch 76 from Andrew Dickson, Scripps Institute of Oceanography). The analytical precision and accuracy for TA was $±2$ µmol kg^{-1} and $±4$ µmol kg^{-1} respectively and, for DIC, $±1$ µmol kg^{-1} and $±2$ µmol kg^{-1} respectively. The remaining carbonate parameters were determined through application of the CO_2SYS software (http://cdiac.ornl.gov/ftp/cp2sys) [28] using equilibrium constants from Mehrbach et al. [29] (as refitted by Dickson and Millero [30]) and the total pH scale. The error in pH_T was $±0.0062$ and, in pCO_2, $±5.7$ µatm [31]. The start and end values of TA and DIC were used to determine an average Ω_{ar} value ($±$ min/max value) for each perturbation experiment (Table 1).

Pteropod shell analysis

Pteropod samples were examined with scanning electron microscopy (SEM) to determine the level of shell dissolution, following Bednaršek et al. [26]. Abiogenic crystals were removed from the shells using 6% diluted hydrogen peroxide H_2O_2 followed by dehydration and drying using 2,2-dimethoxypropane and 1,1,1,3,3,3-hexamethyldisilazane. After the samples were mounted on an aluminium stub, oxygen plasma etching was carried out to expose microstructural elements of the shell and make them visible on the shell surface under SEM. This procedure was demonstrated to be non-destructive and very efficient. The SEM analysis was done with a JEOL JSM 5900LV fitted with a tungsten filament at an acceleration voltage of 15 kV and a working distance of about 10 mm [26]. The samples were gold coated in vacuo with a Polaron SC7640 sputter coater. Typical coating thicknesses ranged from 7 to 21 nm. The shell surface was examined by moving in small incremental steps around and across the shell and taking SEM micrographs (10–15 per animal). The first SEM micrograph was taken at the first whorl and the last at the growing edge (Fig. 1). A total of 38 animals were analysed equating to 60–90 micrographs per perturbation experiment. The SEM magnification was calibrated against known surface areas prior to any image analysis, upon which a calibration curve was produced [26]. This calibration curve was used to estimate the surface area of the shell.

Dissolution of shell carbonate

To quantify the level of dissolution from SEM images, a number of further analytical steps were necessary, which we outline briefly. The calcified layers of the pteropod shell are made up of two layers: the outer prismatic layer and the inner crossed-lamellar layer. Previous work has shown that the level of dissolution of the prismatic and crossed-lamellar layers varied according to the type of dissolution observed (i.e., Type I, II, and III) [26]. We initially evaluated the extent of dissolution expressed as percentage of

dissolved shell area of the three dissolution types, separately. This was then converted to the respective percentage of shell area over which the prismatic and crossed-lamellar layers were affected. The level of penetration into these layers was estimated through comparative measurements of size-normalized thickness in pristine and incubated shells. The percentage of shell lost to dissolution was expressed in terms of $CaCO_3$ loss by applying previously reported relationships for length to shell weight and inorganic to organic carbon [27].

In pristine shells, the prismatic layer constituted 20% of the shell thickness and the crossed-lamellar layer, 80%. The level of dissolution of the calcified layers was only slight in Type I dissolution and those areas were not considered when calculating the level of $CaCO_3$ lost to dissolution. For Type II dissolution, it was assumed that the prismatic layer (20% of shell mass) was completely dissolved but that the dissolution of the crossed-lamellar layer was only minor and constituted a negligible loss of $CaCO_3$. In areas with Type III dissolution, it was implicit that the prismatic layer was completely dissolved, and that there was partial dissolution of the crossed-lamellar layer. The scoring of shells into areas of no dissolution, Type I, Type II or Type III dissolution was carried out "blind" (i.e., without knowledge of the treatment) and only related back to the treatment once the scores had been established.

The next analytical stage was to determine the mean level of dissolution of the crossed lamellar layer in Type III dissolution. This was carried out at the shell aperture, where a natural cross-section of the shell could be resolved. All measurements were expressed as thickness-to-length (T/L) ratios to normalize for the shell size. Shell length measurements (expressed as shell diameter) were made at 90° to the direction of the shell aperture using a light microscope with a calibrated graticule (Fig. 2). Measurements of shell thickness were made at the same position of each shell aperture using SEM. Comparisons of T/L ratios were made between three sample sets: natural environment controls, experimental controls, and 14 day incubations of live juveniles held at undersaturation conditions. It is to be noted that this represented a subset of all experimental incubations given the time-intensive nature of these measurements. The proportional difference in T/L ratio values between the controls and the undersaturation incubation specimens was denoted as u. A mean value for u was determined across all measurements made on aperture cross-sections and was assumed to be the mean depth to which the prismatic plus the crossed-lamellar layer was dissolved in all areas exhibiting Type III dissolution ($D_{TypeIII}$). Consequently, the proportion of shell dissolved as a result of Type II and Type III dissolution (D_{total}) was calculated as follows:

$$D_{TypeII} = sa_{TypeII} \cdot 0.2 \qquad (3)$$

$$D_{TypeIII} = sa_{TypeIII} \cdot u \qquad (4)$$

$$D_{Total} = D_{TypeII} + D_{TypeIII} \qquad (5)$$

where D_{TypeII} and $D_{TypeIII}$ is the proportion of shell-loss resulting from Type II and Type III dissolution respectively, sa is the relative surface area of the shell affected by the different types of dissolution, and u is the relative thickness of the shell lost in areas of Type III dissolution and 0.2 represents the proportion of the shell mass within the prismatic layer. D_{Total} represents the total proportion of shell lost per specimen. D_{Total} actually represents

Table 1. Physical and chemical parameters for the experiments; the values indicate the average and range ($\pm$) of two analyses, one of which was determined prior to the experiment and one after the experiment.

Treatment	Temperature (°C)	Salinity	Phosphate (µmol kg⁻¹)	Silicate (µmol kg⁻¹)	TA (µmol kg⁻¹)	DIC (µmol kg⁻¹)	pH$_T$	pCO$_2$ (µatm)	HCO$_3^-$ (µmol kg⁻¹)	CO$_3^{2-}$ (µmol kg⁻¹)	Ω$_{ar}$
Natural control, live juveniles	2.9±2.8	33.86±0.13	1.33±0.61	12.47±31.70	2290.6±3.7	2123.0±61.9	8.13±0.1	318±111	1983.1±91.9	121.6±39.6	1.82±0.6
Exp. control (8 d) live juveniles	4.0±2.0	33.83±0.02	1.55±0.25	10.41±2.11	2360.3±3.4	2211.5±12.7	8.07±0.0	387±24	2077.8±16.6	112.7±5.4	1.70±0.1
Supersaturation (14 d) live juveniles	4.0±2.0	33.83±0.02	1.55±0.25	10.41±2.11	2345.0±16.3	2220.0±1.4	8.00±0.1	455±50	2097.1±8.9	98.4±10.1	1.49±0.1
Transitional state (14 d) live juveniles	4.0±2.0	33.83±0.02	1.55±0.25	10.41±2.11	2353.2±15.8	2285.3±6.7	7.83±0.1	704±117	2177.5±11.3	69.7±10.9	1.06±0.2
Transitional state (14 d) live adults	4.0±2.0	33.83±0.02	1.55±0.35	10.41±.27	2298.0±30.0	2239.2±45.8	7.81±0.1	727±105	2135.4±46.8	64.2±6.5	0.97±0.1
Undersaturation (4 d) live juveniles	4.0±2.0	33.83±0.02	1.55+0.35	10.41±.27	2323.1±8.9	2295.5±5.4	7.70±0.0	940±27	2192.7±5.0	51.6±1.8	0.78±0.0
Undersaturation (14 d) live juveniles	4.0±2.0	33.83±0.02	1.55±0.35	10.41±.27	2330.3±6.4	2298.3±0.0	7.73±0.0	883±12	2191.0±4.4	54.8±0.9	0.83±0.0
Undersaturation (14 d) dead juveniles	4.0±2.0	33.83±0.02	1.55+0.35	10.41±.27	2317.9±11.9	2283.45±18.0	7.72±0.0	894±49	2181.3±17.4	53.9±1.9	0.81±0.0

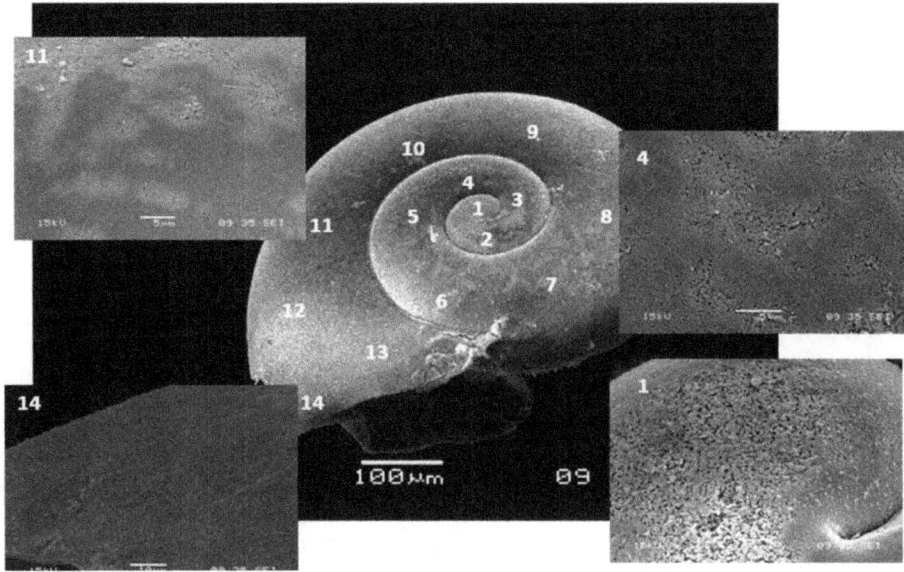

Figure 1. SEM micrograph of a *Limacina helicina antarctica* specimen with signs of surface shell dissolution. SEM micrograph exhibiting signs of dissolution at sites 1, 4, and 11 on the shell surface.

minimum gross dissolution since this estimation method is also influenced by any ongoing calcification that would act to counter the gross amount of dissolution. As discussed later, this source of bias was likely to be minimal over the course of the incubations.

The next step was to convert proportional shell loss to absolute loss of shell mass. This required estimating initial shell mass (M_0, mg) from measurements of shell length, achieved by applying the following conversion by Bednaršek et al. [27] for populations extracted from the study region:

$$DW = 0.137 L^{1.5005} \tag{6}$$

where DW is dry weight in mg and L is shell diameter in mm.

DW was converted to total carbon mass by multiplying by a factor of 0.25 [32] and then partitioned into particulate inorganic carbon (PIC) and particulate organic carbon (POC) fractions by applying a ratio of 0.27:0.73, as derived by Bednaršek et al. [27]. The mass of $CaCO_3$ (i.e., M_0 in mg) was determined by multiplying PIC by 8.33, which is the molecular mass ratio of

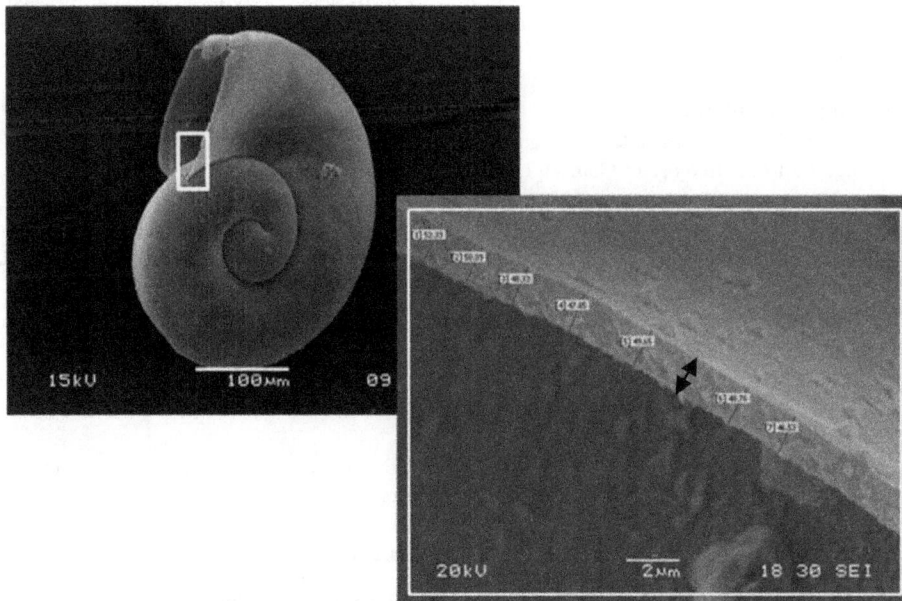

Figure 2. SEM micrograph of *Limacina helicina antarctica* showing the shell thickness measurements. SEM micrograph showing the shell thickness measurements with indication of the position where shell thickness measurements were made.

carbon to $CaCO_3$. The daily mass loss of $CaCO_3$ (V, mg d^{-1}) from the shell during incubation was then calculated as:

$$V = M_0 \cdot \frac{D_{Total}}{100} \cdot \frac{1}{t} \qquad (7)$$

where t is the duration of the incubation in days. The above term was alternatively expressed as the percentage of $CaCO_3$ mass loss d^{-1} (v), which is $V \cdot 100$ divided by M_0.

Changes to mineral structure

To determine whether there were any changes in the mineral structure at the growing edge of juvenile pteropod shells, Raman analysis was used to distinguish the occurrence of $CaCO_3$ minerals other than aragonite. The samples were studied directly with no sample preparation. Instead, the growing edges of 25–30 animals from various experimental setups were examined in small incremental steps. The Raman bands for $CaCO_3$ occur at ~1085 Δcm^{-1} and a minor mode band at ~155 Δcm^{-1}. Aragonite has two additional bands at ~207 and 704 Δcm^{-1}. Laboratory measurements were performed with a laser Raman spectrometer, manufactured by Kaiser Optical Systems, Inc. The spectrometer and laser were connected to an optical probe head which was integrated into a Leica microscope with a 10 times magnifying objective (f/2.0. at 5.8 mm working distance, ~50 μm spot size). Measurements were made on specimens exposed to manipulated conditions with high pCO_2 levels to compare to those that were only exposed to ambient conditions.

Statistical treatment

The data generated by the image segmentation analysis of shell surface dissolution were non-normally distributed and therefore were subjected to a square root transformation followed by an arcsine transformation [33]. Data on shell length and thickness were already converted to a ratio and so were not subjected to any further transformations. Datasets were analysed with t-tests provided they passed tests for normality (Kolmogorov-Smirnov test) and equal variance (variability about the group means). Non-normal datasets were otherwise analysed non-parametrically using a Mann-Whitney rank-sum test.

Dissolution and calcification simulations

Our estimates of $CaCO_3$ mass loss (V) under different aragonite saturation levels were compared to the findings of Comeau et al. [22], who estimated the rate of gross calcification in *L. helicina* exposed to varying conditions of aragonite saturation state and temperature. Comeau et al. [22] found a logarithmic relationship between aragonite saturation state and the amount of $CaCO_3$ (Q, μmol (g wet weight)$^{-1}$ h^{-1}) precipitated, as follows:

$$Q = A\ln(\Omega_{ar}) + B \qquad (8)$$

where A is 0.57 ± 0.4 and B is 0.25 ± 0.02. We converted the calcification rate Q from Eq (8) for a range of Ω_{ar} values into mg $CaCO_3$ per ind d^{-1} by firstly assuming an average shell diameter (L) of 0.31 mm [27] and determining the equivalent DW (mg) using Eq. (6). Wet weight (WW) was estimated by dividing DW by 0.28, following Davis and Wiebe [34]. The average WW of an individual was entered in Eq. (8) to derive the rate in terms of μmol $CaCO_3$ ind^{-1} h^{-1}. This was converted to mg $CaCO_3$ ind^{-1} d^{-1} by applying a molar mass of 1 mole per 100.09 g and multiplying by 24 hours per day. Although there are known genetic differences between the Arctic and Antarctic populations of *L. helicina* [35],

we assumed these to have a negligible effect on calcification rates in the present analysis.

Trajectories of mean shell weight were derived for two scenarios: first, for supersaturation levels ($\Omega_{ar} = 1.8$), where only calcification would be performed; and second, for an undersaturation level of 0.8, where there would be both calcification and dissolution occurring. This level of undersaturation corresponds to the mean level achieved in the undersaturation incubations performed in the present study. It further represents the level of undersaturation that would prevail in the Southern Ocean surface waters by 2050 [14]. A period of 100 days for the trajectories was set based on the average productivity period at mid-latitudes in the Southern Ocean [36], when the majority of growth and development occurs in the pteropod population [27].

To estimate the rate of calcification using Eq. (8), it was necessary to estimate the growth in WW over the 100 day scenario. Hence, it was assumed that WW grew in direction proportion to shell weight, M, when in supersaturated conditions. For every daily increment in shell mass due to calcification, m_{calc}, an increment of growth in WW was also added at a ratio of 1:6.35, ($m_{calc}:WW$). The next daily increment of m_{calc} was then determined from the new WW and the process repeated, as follows:

$$m_{calc,t+1} = m_{calc,t} + \left(\frac{C \cdot 100.09 \cdot 24}{1000}\right) \; for \; t = 1 \; to \; 99 \qquad (9)$$

where

$$C = \frac{Q \cdot WW_t}{1000} \qquad (10)$$

and

$$WW_t = m_{calc,t} \cdot 6.35 \qquad (11)$$

For the undersaturation scenario, it was assumed that growth in the non-shell fraction of WW remained the same but that growth in shell mass progressed at a different rate, as determined by Eq. (5).

The effect of gross dissolution (m_{diss}) was determined as follows:

$$m_{diss,t+1} = m_{diss,t} - (m_{diss,t} \cdot D_{Total}) \; for \; t = 1 \; to \; 99 \qquad (12)$$

A trajectory of net shell mass (M_t, mg $CaCO_3$ ind^{-1}) over the 100 day simulation period was determined as:

$$M_t = \sum_{t=1}^{t=99} M_{calc,t} - \sum_{t=1}^{t=99} M_{diss,t} \qquad (13)$$

It is to be noted that m_{diss} was assumed to be negligible in supersaturation conditions.

Results

Shell aperture analysis

We found that exposure to undersaturated conditions thinned the shell at the shell-aperture (Fig. 3). On average, specimens incubated for 14 days in undersaturated conditions were thinner by 39%±9% compared to specimens from the natural environ-

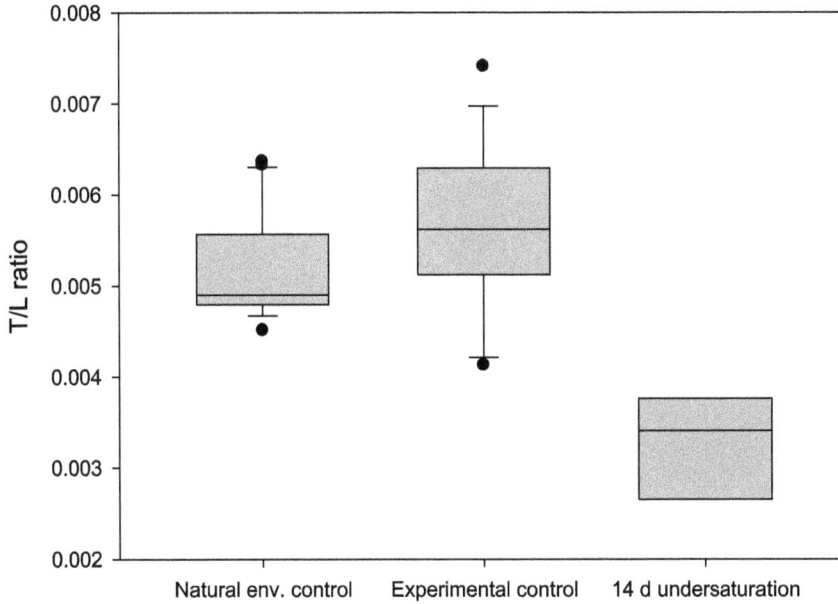

Figure 3. Thickness-to-length ratio of juvenile *Limacina helicina antarctica*. Thickness-to-length (T/L) ratio of juveniles from natural environment and experimental control populations, and from specimens incubated for 14 days in undersaturation conditions. Horizontal line represents the median; box limits, the 25th and 75th percentiles; whisker limits, 10th and 90th percentiles; and dots, the outliers.

ment controls and the experimental controls. This value (expressed as a proportion) was assumed to be the value of u in all subsequent calculations of shell dissolution rates.

Shell-surface analysis

Saturation state effects for juvenile pteropods. Levels of dissolution over the entire shell were quantified with image analysis. Pteropods in the experimental control displayed higher levels of Type I dissolution (56%±7 in the 8 day at $\Omega_{ar} = 1.70\pm0.1$), compared to the natural control samples (0.3%,

Fig. 4). This is likely to be a consequence of rearing pteropods in enclosed vessels where biological (respiration, calcification), chemical (dissolution) and physical (headspace-water CO_2 exchange) processes altered the pCO_2 level. Nevertheless, Type I dissolution represented only a minor loss of $CaCO_3$ from the shell and was considered negligible in the development of dissolution regressions. Specimens kept at supersaturated conditions for 14 days ($\Omega_{ar} = 1.49\pm0.15$) exhibited up to 55%±11% Type I and 9%±7% Type II dissolution, but there was no evidence of any Type III dissolution. Areas of Type III dissolution were present on

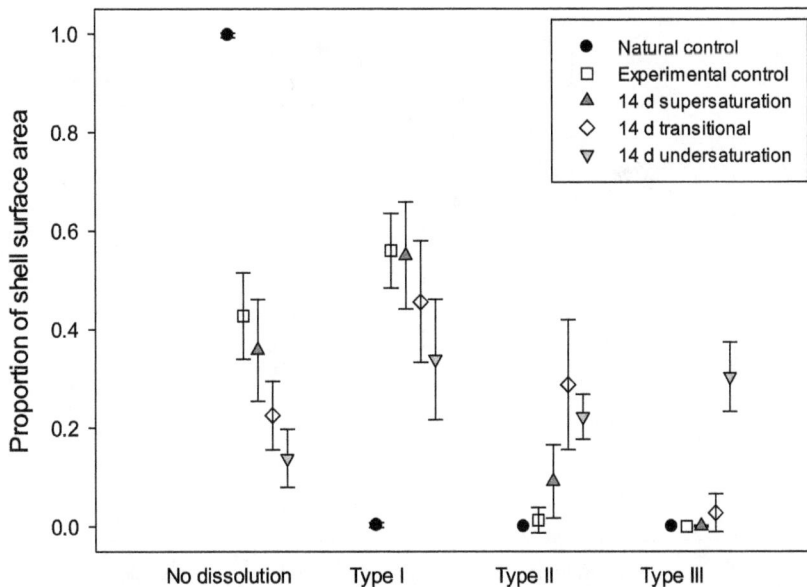

Figure 4. Proportion of shell surface dissolution in live juveniles incubated in different saturation conditions. Proportion of shell surface dissolution in live juveniles incubated in supersaturation conditions for 8 days and 14 days in different saturation conditions.

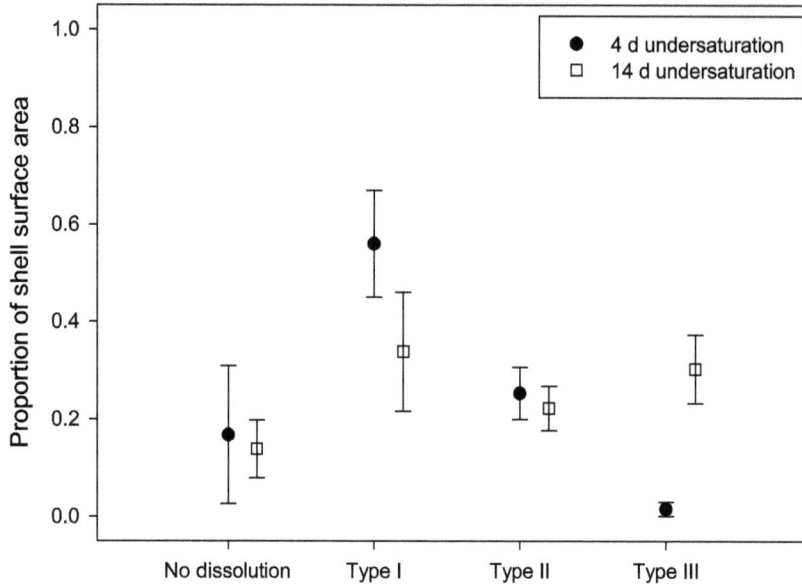

Figure 5. Proportion of shell surface dissolution in live juveniles incubated in undersaturation conditions for 4 and 14 days. Proportion of shell surface dissolution in live juveniles incubated in Ω_{ar} undersaturation conditions for 4 days or 14 days.

specimens incubated at either $\Omega_{ar} \sim 1$ or in undersaturated conditions ($\Omega_{ar} \sim 0.8$), with more extensive dissolution in the latter.

Temporal effects. Type II and Type III dissolution was evident in juveniles incubated in undersaturated conditions for 4 days ($\Omega_{ar} = 0.78 \pm 0.03$) and 14 days ($\Omega_{ar} = 0.83 \pm 0.02$) (Fig. 5). The level of Type II dissolution was similar, at $27\% \pm 6\%$ after 4 days and $25\% \pm 7\%$ after 14 days. However, whereas only around $3\% \pm 3\%$ of the shell was covered with Type III dissolution after 4 days, surface dissolution extended to $31\% \pm 6\%$ after 14 days. Therefore, Type II dissolution occurred almost immediately on

exposure to undersaturation conditions, whereas Type III dissolution mainly became apparent between 4 and 14 days.

Size and maturity effects. We found levels of Type I dissolution to be significantly greater in adults ($59\% \pm 11\%$) than in juveniles ($23\% \pm 7\%$) (Fig. 6; Normality and equal variance passed, $t = -5.59$, $df = 5$, $P = 0.003$). However, there was no significant difference in the levels of Type II and Type III dissolution between adults and juveniles (Type II: adult $21\% \pm 15\%$ vs. juvenile $29\% \pm 13\%$, Type III: adult $1\% \pm 1\%$ vs. juvenile $3\% \pm 4\%$; Normality and equal variance passed, $t = -0.71$, $df = 5$, $P = 0.51$ for Type II; $t = 0.84$, $df = 5$, $P = 0.44$ for Type III). Therefore,

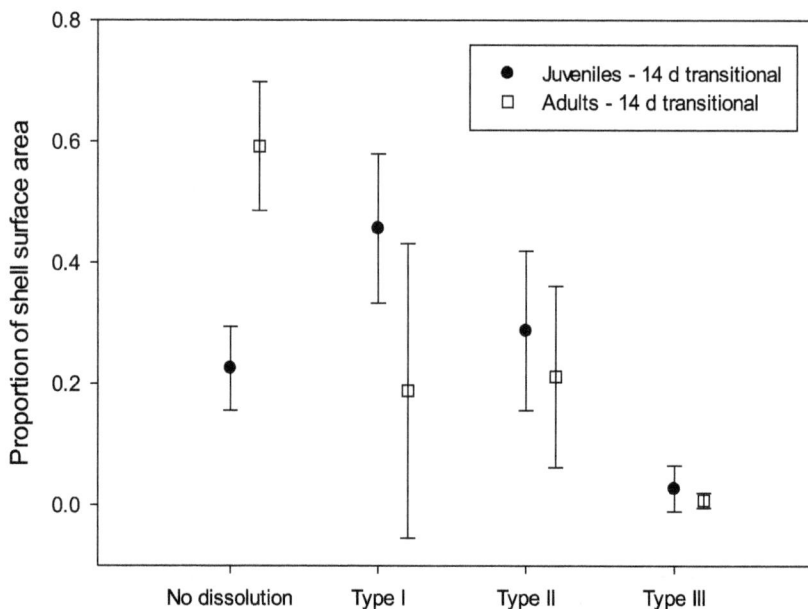

Figure 6. Proportion of shell surface dissolution in live pteropods incubated in transitional conditions for 14 days. Proportion of shell surface dissolution in live specimens that were either juvenile or adult, incubated in Ω_A transitional conditions for 14 days.

Table 2. Calculation of the CaCO$_3$ mass loss for various aragonite saturation states for G2 (Proportions were converted to percentages for clarity).

Pteropod length (mm)	0.320					
Pteropod weight (mg)	0.025					
Pteropod carbon (Larson, 1978)	0.006					
PIC (0.27 of carbon) (mg)	0.002					
POC (0.73 of carbon) (mg)	0.005					
CaCO$_3$ mass (mg)	0.014					
Dissolution in 4 days	Ω~1.5	SD at Ω~1.5	Ω~1	SD at Ω~1	Ω~0.8	SD at Ω~0.8
Type II surface area ($_{TypeII}$)					27.0%	(±6.0%)
Type III surface area ($_{TypeII}$)					3.0%	(±3.0%)
Type II% shell loss (D_{TypeII})					5.4%	(±1.2%)
Type III% shell loss ($D_{TypeIII}$)					1.2%	(±1.2%)
Total %shell loss (D_{Total})					6.6%	(±2.4%)
Total % shell loss d^{-1}					1.6%	(±0.6%)
CaCO$_3$ loss ind^{-1} d^{-1} (mg)					2.30E-04	(±8.31E-05)
Dissolution in 14 days	Ω~1.5	SD at Ω~1.5	Ω~1	SD at Ω~1	Ω~0.8	SD at Ω~0.8
Type II surface area (sa_{TypeII})	9.0%	(±6.0%)	29.0%	(±7.0%)	25.0%	(±13.0%)
Type III surface area (sa_{TypeII})	0.0%	(±3.0%)	3.0%	(±6.0%)	31.0%	(±1.0%)
Type II% shell loss (D_{TypeII})	1.8%	(±1.2%)	5.8%	(±1.4%)	5.0%	(±2.6%)
Type III% shell loss ($D_{TypeIII}$)	0.0%	(±1.9%)	1.2%	(±2.3%)	12.1%	(±0.4%)
Total % shell loss (D_{Total})	1.8%	(±3.1%)	7.0%	(±3.7%)	17.1%	(±3.0%)
Total % shell loss d^{-1}	0.1%	(±0.2%)	0.5%	(±0.3%)	1.2%	(±0.2%)
CaCO$_3$ loss ind^{-1} d^{-1} (mg)	1.80E-05	(±2.37E-05)	6.97E-05	(±3.75E-05)	1.71E-04	(±2.99E-05)

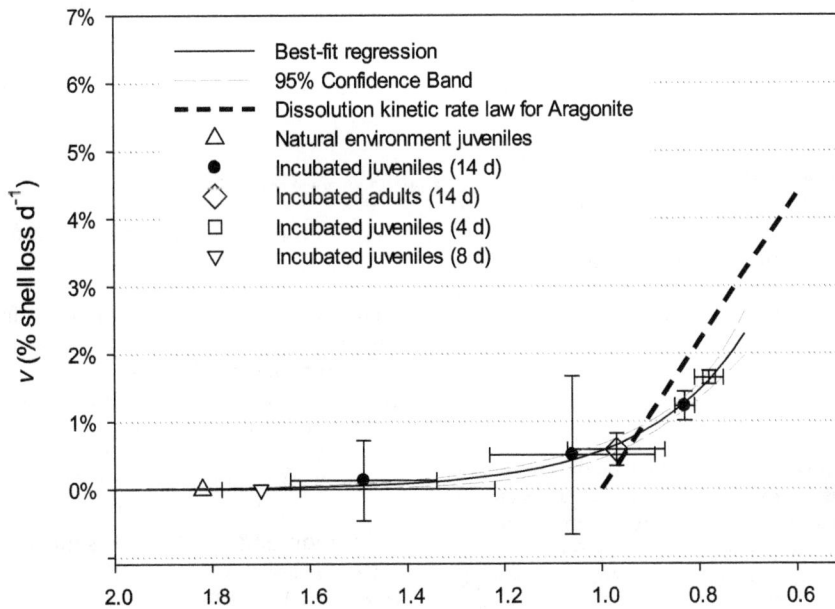

Figure 7. Percentage of shell mass loss across a range of aragonite saturation states. Shell mass loss d^{-1} (v) in live and dead specimens incubated between 4 and 14 days. Solid line represents a 2-parameter exponential function (±95% confidence intervals) fitted to all live specimens' data points. The bold hashed line represents the dissolution kinetic rate law for aragonite. Error bars show ±1 SD on Ω$_A$ values in the incubations and v respectively.

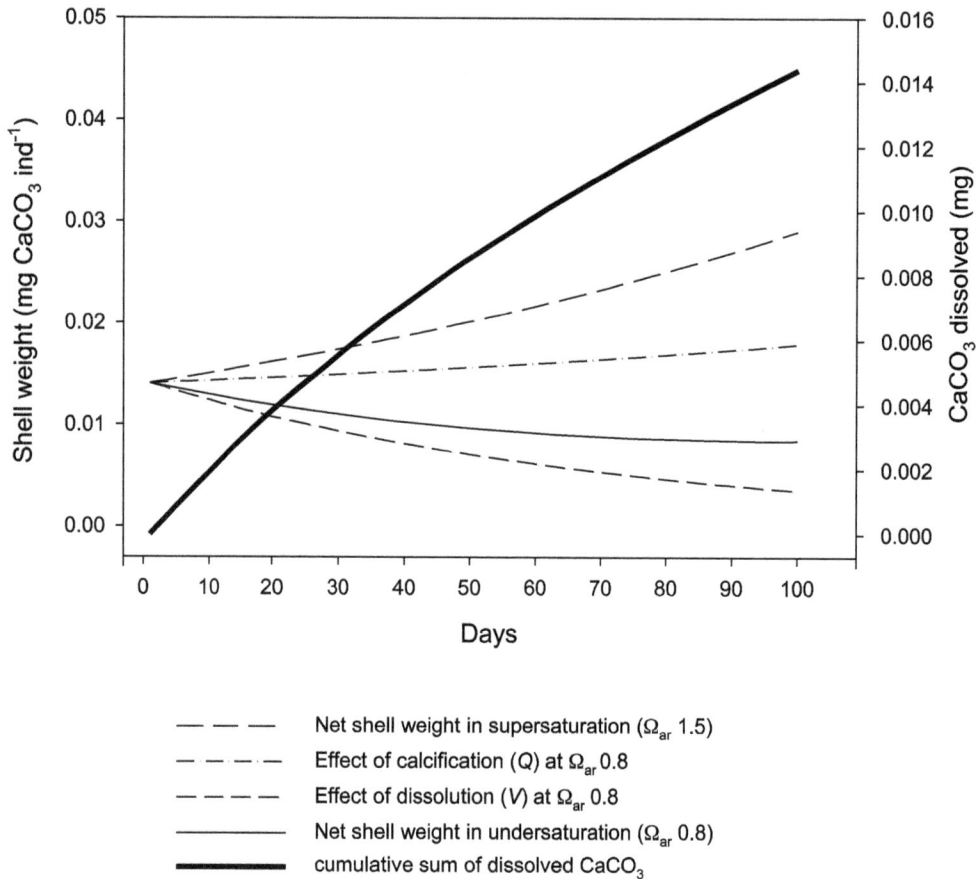

Figure 8. Simulation of dissolution and calcification on the shell weight exposed to supersaturation and undersaturation conditions. Simulation of the effect of dissolution and calcification on the shell weight when exposed to supersaturation or undersaturation conditions for 100 days. Bold line indicates the amount of $CaCO_3$ dissolved over the course of the simulation where Ω_{ar} is 0.8.

juveniles and adults showed the same Type II and Type III dissolution response at transitional saturation levels ($\Omega_{ar} \sim 1$).

Changes in mineral structure

Aragonite was consistently observed in all the spectra made by Raman spectroscopy regardless of the Ω_{ar} saturation state. No mineral structure other than aragonite was found anywhere on the growing edge of any juvenile or adult pteropod. Therefore, the animals did not change their mineralization process in response to perturbations in the saturation state.

Shell mass loss due to dissolution as a function of saturation state

The percentage shell mass loss due to dissolution over the course of the incubations was a minimum of $1.8\% \pm 3.1\%$ under supersaturated conditions for 14 days to a maximum of $17.1\% \pm 3.0\%$ in undersaturated conditions for 14 days (Table 2). In terms of shell mass loss d^{-1} (v), this equated to $0.1\% \pm 2\%$ in supersaturated conditions, $0.5\% \pm 0.3\%$ at $\Omega_{ar} \sim 1$ and between $1.2\% \pm 0.2\%$ in undersaturated conditions ($\Omega_{ar} \sim 0.8$). The decrease in aragonite saturation levels from 1 to 0.8 therefore resulted in a two- to threefold increase in the rate of dissolution. When expressed in terms of equivalent loss in $CaCO_3$ per individual, this is an increase from 0.07 µg d^{-1} to a maximum of 0.23 µg d^{-1} over the range of these saturation states.

This rate of % shell loss d^{-1} was most adequately represented by a 2-parameter exponential growth function, as follows:

$$v\% = 65.76e^{-4.7606\Omega_{ar}}$$

$$\left(Adj.\ R^2 = 0.99,\ F = 609.27,\ 7\ df,\ p < 0.0001\right) \tag{14}$$

Both our observations and the fitted function show levels of dissolved shell loss at Ω_{ar} levels greater than 1, which was not predicted by the dissolution rate algorithm (Fig. 7). At Ω_{ar} levels below 0.9 however, the fitted function shows a slower increase in v shell loss d^{-1} than the rate kinetics. This reflects our observations that, for Ω_{ar} levels of around 0.8, v was between 0.7% and 1.7%, and not 2.2% as predicted by Eq. (2).

Dissolution and calcification simulations

To compare growth trajectories between saturation conditions, we simulated the effect of exposure to undersaturated ($\Omega_{ar} \sim 0.8$) compared to supersaturated conditions ($\Omega_{ar} \sim 1.8$) for 100 days (Fig. 8). It was assumed that any dissolution would be negligible when in supersaturation conditions, and shells grew through calcification according to the rate derived by Comeau et al. ([37]; Eq. (8)). v was set at 1.4% d^{-1}, representing the average of dissolution observed on live specimens incubated for either 4 or 14 days at $\Omega_{ar} \sim 0.8$ (Table 2).

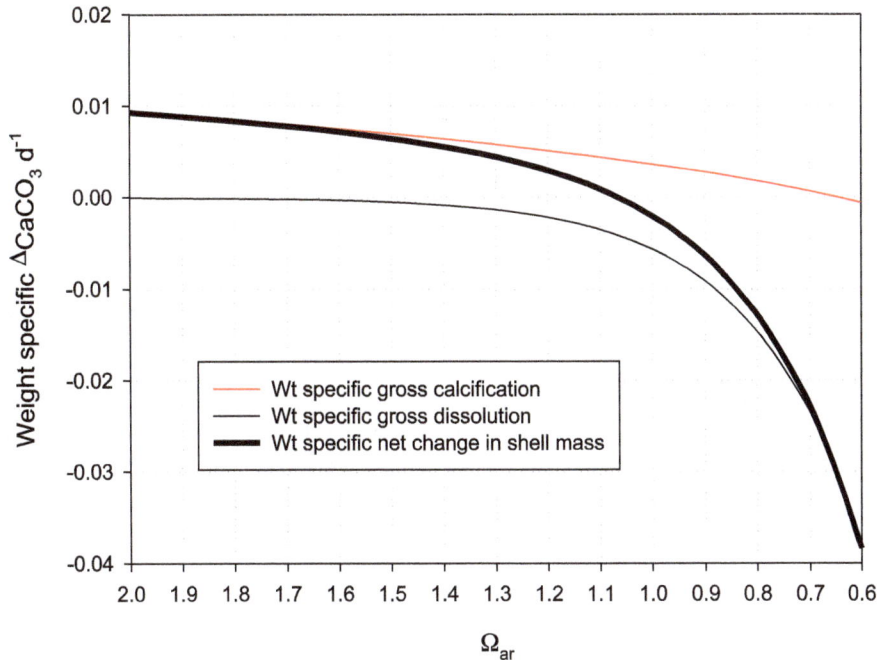

Figure 9. Weight-specific rates of net change processes as a function of aragonite saturation state. Weight-specific rates (d^{-1}) of net change in shell mass as a function of aragonite saturation state.

We found that the rate of gross calcification within undersaturated conditions compensated little for the loss of shell mass in these conditions. The total amount of $CaCO_3$ lost by this juvenile pteropod through dissolution over 100 days would have been 0.014 mg.

Weight-specific function for net calcification

The combined effects of gross dissolution and gross calcification on shell growth was examined through deriving trajectories for weight-specific net calcification against Ω_{ar} for *L. helicina* (Fig. 9). The trajectory was fitted closely by a three parameter exponential function as follows:

$$\tau = -0.5705 + 0.5783 \cdot \left(1 - e^{-4.1752\Omega_{ar}}\right)$$

$$(Adj.\ R^2 = 0.99,\ F = 8412.94,\ 17df, p < 0.0001) \tag{15}$$

where τ is weight-specific net calcification (d^{-1}).

Whereas gross calcification ceases at an Ω_{ar} level of around 0.6, the additional influence of gross dissolution means that net calcification (τ) will become 0 at an Ω_{ar} level of approximately 1.

Calculation of potential sinking flux

Declines in shell weight may in turn influence the level sinking flux. We estimated this through firstly assuming a standing stock of *L. helicina ant* of 32 mg C m^{-2}, containing 9 mg m^{-2} of PIC, following Bednaršek et al. [24]. Applying a PIC:$CaCO_3$ mass ratio of 8:33 gives a mean population shell mass of 74.97 mg $CaCO_3$ m^{-2}. Bednaršek et al. [24], also derived a Production:Biomass ratio of 0.06 d^{-1}, which, when applied to the population shell mass, gives a population shell mass production rate of 4.50 mg $CaCO_3$ m^{-2} d^{-1}. Based on observations made during this study, a juvenile exposed to $\Omega_{ar} \sim 0.8$ for 100 days would reduce in individual shell mass by 50%, which in turn would reduce the potential sinking flux to 2.25 mg $CaCO_3$ m^{-2} d^{-1}.

Discussion

Gross dissolution

In this study, we directly estimated the amount of $CaCO_3$ shell lost to dissolution and found Type I dissolution was common in all incubations, although its absence from the natural control specimens indicated that at least some of this dissolution was an experimental artefact. Quantitatively, Type I dissolution represents a very minor loss of $CaCO_3$ from the shell and can be ignored in terms of gross dissolution, while Type II and III dissolution represent a much greater amount of $CaCO_3$ loss. The amount of shell surface covered by the latter two dissolution types increased with decreasing Ω_{ar} levels and longer periods of exposure. Nevertheless, when converted into a rate of shell mass loss due to dissolution, all undersaturated incubations ($\Omega_{ar} \sim 0.8$) resulted in a loss-rate of around 0.2 µg $CaCO_3$ ind^{-1} d^{-1}, which equates to approximately 1.4% of total shell mass d^{-1} regardless of the duration of the incubation.

Shell loss mitigation processes

We found the dissolution response to Ω_{ar} undersaturated conditions to be relatively rapid; however, it was lower than predicted for the dissolution rate of abiogenic aragonite [6]. Like many molluscs, pteropods maintain an outer organic layer [38–39] that is analogous to the periostracum in molluscan groups such as the bivalves. In bivalves, the periostracum is comprised of chemically robust proteins and is believed to protect the shell from dissolution [40]. Partial chemical resistance and mechanical degradation could also be rendered through the multiple shell layers that provide both elasticity and hardness [39]. In addition, a microstructure with a higher organic content provides higher dissolution resistance through the shrouding of the crystals [41]. The full function of the outer organic layer in pteropods remains to be revealed but it appears to be able to offer some protection to the shell when faced with undersaturated conditions.

One possible means of dissolution mitigation is through 'repair calcification' [21,25] where pteropods with shell dissolution were found to have affected areas repaired with new crystals. Although we found no evidence of repair, our methods were not ideal for resolving it. Furthermore, dissolution damage could be repaired over longer timescales on return to saturation conditions [21]. It is to be noted that, if repair calcification was taking place during the present incubations, this would result in an underestimation of the true value of gross dissolution since we assumed that specimens did not add any further shell mass over the course of the incubation.

Functional response to undersaturated conditions

There have been a number investigations focused on the dissolution process, particularly targeting abiogenic aragonite (e.g., [6,23,42–46]). A study by Gangstø et al. [23] proposed an aragonite dissolution rate constant of 10.9 d^{-1}. We compared the predictions of this rate to our own fitted relationship and found that abiogenic dissolution rate law only applies to $\Omega_{ar} \leq 1$. However, we also found evidence of dissolution at transitional saturation levels; in live juveniles, daily shell mass loss amounted to 0.5% in specimens incubated in $\Omega_{ar} = 1.03$. The fitted function in turn reflected this transitional level mass loss, with the upward inflection starting at around Ω_{ar} of 1.3, reaching 0.6% d^{-1} at Ω_{ar} equal to 1. Such dissolution at transitional saturation levels has previously been reported by Betzer et al. [5] in the North Pacific, who found marked reductions in aragonite fluxes between two supersaturated depth horizons (100 m and 400 m), implying a loss in pteropod shell mass through shell dissolution. Although Betzer et al. [5] had no direct explanation for the pattern, they did refer to the findings of Morse et al. [45] and McGowan and Hayward [47], who proposed that freshly calcified aragonite surfaces in young pteropods (1–3 days old) are significantly more soluble than aged aragonite surfaces (30–70 days old). Greater solubility may therefore be expected in younger specimens or where new shell growth is occurring in older specimens.

There is a reasonable agreement to the predictions of the abiogenic rate of dissolution and our fitted function at Ω_{ar} levels just below 1. However, at increasing levels of undersaturation, we found that the abiogenic rate of dissolution overestimated the rate of shell mass loss in live organisms compared to observations. For instance, at Ω_{ar} of 0.8, juveniles showed a shell mass loss rate of between 0.8% and 1.5% d^{-1}, whereas the rate law predicted 2.2% d^{-1}. We advocate that it is unsafe to apply abiogenic dissolution rates when predicting the dissolution of aragonite in live organisms in biogeochemical models without also taking into account biological protection mechanisms.

Net calcification

In incubations of L. helicina carried out in the Arctic, Comeau et al. [22] found that calcification continued even in undersaturated conditions down to $\Omega_{ar} \sim 0.6$. We demonstrated that the loss from dissolution would be twice the amount contributed by net calcification, leading to a net decrease in the mass of the shell of 0.007 mg (50% of original shell mass) over the 100-day simulation period.

Comeau et al. ([48], their Table 3) made projections under the IPCC (Intergovernmental Panel on Climate Change) SRES (Special Report on Emission Scenarios) A2 scenario for anthropogenic CO_2 emissions on the rates of gross calcification by pteropods at a number of oceanic sites where L. helicina has been caught, including sites in the Arctic and Antarctic. At one site in the Arctic (83.58° N, 98.58° W), the projection was for Ω_{ar} to drop to 0.4 by 2095, by which point gross calcification in pteropods would cease. At another Arctic site (Svalbard, 79.8° N, 11.8° E)

and in the Southern Ocean (62.8° S, 60.8° E), the prognosis for 2095 was for Ω_{ar} to drop to 1.1 and for gross calcification to continue at a rate of between 50 and 60% of the preindustrial rate. According to the net calcification function derived by the present study, such Ω_{ar} levels would result in L. helicina being incapable of calcifying enough to offset dissolution. They would be unable to grow in shell mass in any of these polar oceanic regions that they typically inhabit.

Influence of net calcification on net aragonite flux

We estimated potential sinking fluxes of 4.50 mg $CaCO_3$ m^{-2} d^{-1} in supersaturated conditions ($\Omega_{ar} \sim 1$), and 2.25 mg $CaCO_3$ m^{-2} d^{-1} in undersaturated conditions ($\Omega_{ar} \sim 0.8$), assuming that juveniles are exposed to undersaturated conditions over a 100 day productivity period. Attempts at measuring pteropod sinking flux have been made by determining accumulation rates of bottom sediments [49–50] or vertical fluxes measured with sediment traps [5,49] but these approaches have been criticised because of the combined effects of dissolution in deeper layers and predation [5,51–54]. As an alternative, sinking fluxes can be determined based on productivity rates or instantaneous growth rates [55–56]. At Ocean Station PAPA, aragonite production was measured at 4.4 mg $CaCO_3$ m^{-2} d^{-1}, split between L. helicina (2.6±0.3) and Clio pyramidata (1.8±0.2) [55]. Similar levels were found in the Bahamas (2.8±0.3 mg $CaCO_3$ m^{-2} d^{-1}), the equatorial Pacific (6.6±1.2 mg $CaCO_3$ m^{-2} d^{-1}), and the Central Pacific (1.4±0.6 mg $CaCO_3$ m^{-2} d^{-1}) [56]. On average, productivity values were around 0.5 mg $CaCO_3$ m^{-2} d^{-1} greater than estimates made from sediment traps in the same regions [5,56]. 50% decrease in the sinking flux that we predict would occur under undersaturated conditions would have a much greater significance to the overall carbonate cycle in the Southern Ocean, as well as other high-latitude regions, where pteropods are found in high abundances.

Accompanying the decrease in overall shell mass is also the decrease in shell weight in terms of how fast it will sink through the water column. Byrne et al. [6] proposed that the decrease in sinking rate scales with loss of mass, as follows:

$$s = s_o \left(\frac{M}{M_0} \right) \qquad (16)$$

where s is the revised sinking velocity (cm s^{-1}), s_o, the original sinking velocity (cm s^{-1}), M, the remaining shell mass (mg $CaCO_3$), and M_0, the original shell mass (mg $CaCO_3$).

Byrne et al. [6] measured the sinking speed of Limacina inflata, of the same size as juvenile L. helicina ant. in the present study (~0.3 mm shell diameter, ~0.014 mg $CaCO_3$) to be 1.4 cm s^{-1}. Exposure to Ω_{ar} levels of 0.8 for 100 days would reduce shell mass to 0.007 mg $CaCO_3$, resulting in the sinking speed being reduced to 0.7 cm s^{-1}. As a consequence, the partially dissolved shell would take twice the amount of time to sink to the bottom of a 3000 m water column (5.7 days versus 2.5 days). In undersaturated conditions, the level of dissolution in these lighter shells will be even greater, making their sinking rates even slower. With respect to the carbonate cycle, slower sinking speed will result in a longer retention time in the upper water column, which may have a mitigating effect in neutralising anthropogenically induced acidification of mid-water depths [8–9,49,57]. Nevertheless, the lighter, slower-sinking pteropods would have a diminished impact in their role as ballast to sinking particulate organic matter [16]. This will result in greater subsurface water column remineralization of this particulate organic material and, ultimately, a less effective carbon pump.

Concluding remarks

In modelling the sensitivity of pelagic calcification to ocean acidification, Gangstø et al. [23,58] determined that anthropogenic CO_2 emissions may lead to irreversible changes in Ω_{ar} for several centuries. Even under optimistic emission scenarios, the ratio of open-water $CaCO_3$ dissolution will continue until 2500 where it will be 30–50% higher than at pre-industrial times. The consequence is a severe loss of suitable habitat for aragonite calcifiers. This in turn will result in a depletion of the rate of carbon and carbonate flux to the deep ocean. As confidence intervals of future projections increasingly narrow, the argument is progressing beyond whether suitable habitat will be lost to when and to what extent. The application of results obtained in this study will now enable regions of imminent habitat loss to be identified and monitored and the consequences to the sinking flux to be estimated.

Acknowledgments

The authors thank the officers and crew of the RRS *James Clark Ross* for their support during cruise JR177. We would also like to thank Dr. Sheri White (WHOI), for help on Raman spectroscopy and Bertrand Lézé (UEA) on technical support for SEM and dissolution analysis. We are grateful to Sandra Bigley from Pacific Marine Environmental Laboratory, National Oceanic and Atmospheric Administration for help on editing.

Author Contributions

Conceived and designed the experiments: NB GT SF DB. Performed the experiments: NB. Analyzed the data: NB GT DB. Contributed reagents/materials/analysis tools: NB GT DB. Wrote the paper: NB GT SF DB RAF.

References

1. Feely RA, Sabine CL, Lee K, Millero FJ, Lamb MF, et al. (2002) In situ calcium carbonate dissolution in the Pacific Ocean. Global Biogeochem Cycles 16(4): 91-1-91-12.
2. Feely RA, Sabine CL, Lee K, Berelson W, Kleypas J, et al. (2004) Impact of anthropogenic CO_2 on the $CaCO_3$ system in the oceans. Science 305(5682): 362–366.
3. Iglesias-Rodriguez MD, Armstrong R, Feely RA, Hood R, Kleypas J, et al. (2002) Progress made in study of ocean's calcium carbonate budget. Eos Trans AGU 83(34): 365, 374–375.
4. Zeebe RE, Wolf-Gladrow DA (2001) CO_2 in seawater: Equilibrium, kinetics, isotopes. Amsterdam: Elsevier Science Ltd.
5. Betzer PR, Byrne RH, Acker JG, Lewis CS, Jolley RR, et al. (1984) The oceanic carbonate system — A reassessment of biogenic control. Science 226(4678): 1074–1077.
6. Byrne RH, Acker JG, Betzer PR, Feely RA, Cates MH (1984) Water column dissolution of aragonite in the Pacific Ocean. Nature 312(5992): 321–326.
7. Feely RA, Chen CTA (1982) The effect of excess CO_2 on the calculated calcite and aragonite saturation horizons in the northeast Pacific. Geophys Res Lett 9(11): 1294–1297.
8. Feely RA, Byrne RH, Betzer PR, Gendron JF, Acker JG (1984) Factors influencing the degree of saturation of the surface and intermediate waters of the North Pacific Ocean with respect to aragonite. J Geophys Res Oceans 89(NC6): 631–640.
9. Feely RA, Byrne RH, Acker JG, Betzer PR, Chen CTA, et al. (1988) Winter summer variations of calcite and aragonite saturation in the northeast Pacific. Mar Chem 25(3): 227–241.
10. Feely RA, Sabine CL, Byrne RH, Millero FJ, Dickson AG, et al. (2012) Decadal changes in the aragonite and calcite saturation state of the Pacific Ocean. Global Biogeochem Cycles 26: GB3001.
11. Orr JC, Fabry VJ, Aumont O, Bopp L, Doney SC, et al. (2005) Anthropogenic ocean acidification over the twenty-first century and its impact on calcifying organisms. Nature 437(7059): 681–686.
12. Feely RA, Alin SR, Newton J, Sabine CL, Warner M, et al. (2010) The combined effects of ocean acidification, mixing, and respiration on pH and carbonate saturation in an urbanized estuary. Estuar Coast Shelf Sci 88(4): 442–449. doi:10.1016/j.ecss.2010.05.004.
13. Steinacher M, Joos F, Frölicher TL, Plattner G-K, Doney SC (2009) Imminent ocean acidification in the Arctic projected with the NCAR global coupled carbon cycle-climate model. Biogeosciences 6(4): 515–533.
14. McNeil BI, Matear RJ (2008) Southern Ocean acidification: A tipping point at 450-ppm atmospheric CO_2. Proc Natl Acad Sci U S A 105(48): 18860–18864.
15. Le Fèvre J, Legendre L, Rivkin RB (1998) Fluxes of biogenic carbon in the Southern Ocean: Roles of large microphagous zooplankton. J Mar Syst 17(1–4): 325–345.
16. Klaas C, Archer DE (2002) Association of sinking organic matter with various types of mineral ballast in the deep sea: Implications for the rain ratio. Global Biogeochem Cycles 16(4): 1116.
17. Ries JB (2012) Oceanography: A sea butterfly flaps its wings. Nat Geosci 5(12): 845–846.
18. Ries JB, Cohen AL, McCorkle DC (2009) Marine calcifiers exhibit mixed responses to CO_2-induced ocean acidification. Geology 37(12): 1131–1134.
19. Trotter J, Montagna P, McCulloch M, Silenzi S, Reynaud S, et al. (2011) Quantifying the pH 'vital effect' in the temperate zooxanthellate coral *Cladocora caespitosa*: Validation of the boron seawater pH proxy. Earth Planet Sci Lett 303(3–4): 163–173.
20. Rodolfo-Metalpa R, Houlbrèque F, Tambutté E, Boisson F, Baggini C, et al. (2011) Coral and mollusc resistance to ocean acidification adversely affected by warming. Nat Clim Change 1(6): 308–312.
21. Lischka S, Budenbender J, Boxhammer T, Riebesell U (2011) Impact of ocean acidification and elevated temperatures on early juveniles of the polar shelled pteropod *Limacina helicina*: Mortality, shell degradation, and shell growth. Biogeosciences 8(4): 919–932.
22. Comeau S, Jeffree R, Teyssié J-L, Gattuso J-P (2010) Response of the Arctic pteropod *Limacina helicina* to projected future environmental conditions. PLoS ONE 5(6): e11362. doi:10.1371/journal.pone.0011362.
23. Gangstø R, Gehlen M, Schneider B, Bopp L, Aumont O, et al. (2008) Modeling the marine aragonite cycle: Changes under rising carbon dioxide and its role in shallow water $CaCO_3$ dissolution. Biogeosciences 5(4): 1057–1072.
24. Bednaršek N, Tarling GA, Bakker DCE, Fielding S, Jones EM, et al. (2012c) Extensive dissolution of live pteropods in the Southern Ocean. Nat Geosci 5(12): 881–885.
25. Lischka S, Riebesell U (2012) Synergistic effects of ocean acidification and warming on overwintering pteropods in the Arctic. Glob Change Biol 18(12): 3517–3528.
26. Bednaršek N, Tarling GA, Bakker DCE, Fielding S, Cohen A, et al. (2012a) Description and quantification of pteropod shell dissolution: A sensitive bioindicator of ocean acidification. Glob Change Biol 18(7): 2378–2388.
27. Bednaršek N, Tarling GA, Fielding S, Bakker DCE (2012b) Population dynamics and biogeochemical significance of *Limacina helicina antarctica* in the Scotia Sea (Southern Ocean). Deep Sea Res Part 2 59: 105–116.
28. Lewis E, Wallace DWR (1998) Program developed for CO_2 system calculations. ORNL/CDIAC-105. Carbon Dioxide Information Analysis Center, Oak Ridge National Laboratory. Oak Ridge, Tennessee: US Department of Energy.
29. Mehrbach C, Culberson CH, Hawley JE, Pytkowicz RM (1973) Measurement of the apparent dissociation constants of carbonic acid in seawater at atmospheric pressure. Limnol Oceanogr 18(6): 897–907.
30. Dickson AG, Millero FJ (1987) A comparison of the equilibrium constants for the dissociation of carbonic acid in seawater media. Deep Sea Res A 34(10): 1733–1743.
31. Millero FJ (1995) Thermodynamics of the carbon dioxide system in the oceans. Geochim Cosmochim Acta 59(4): 661–677.
32. Larson RJ (1986) Water content, organic content, and carbon and nitrogen composition of medusae from the northeast Pacific. J Exp Mar Bio Ecol 99(2): 107–120.
33. Zar JH (1999) Biostatistical analysis, 4th edition. Upper Saddle River, New Jersey: Prentice Hall.
34. Davis CS, Wiebe PH (1985) Macrozooplankton biomass in a warm-core Gulf Stream ring: Time series changes in size, structure, taxonomic composition, and vertical distribution. J Geophys Res Oceans 90(NC5): 8871–8884.
35. Hunt B, Strugnell J, Bednaršek N, Linse K, Nelson RJ, et al. (2010) Poles apart: The "bipolar" pteropod species *Limacina helicina* is genetically distinct between the Arctic and Antarctic oceans. PLoS ONE 5(3): e9835. doi:10.1371/journal.pone.0009835.
36. Tarling GA, Shreeve RS, Ward P, Atkinson A, Hirst AG (2004) Life-cycle phenotypic composition and mortality of *Calanoides acutus* (Copepoda: Calanoida) in the Scotia Sea: A modelling approach. Mar Ecol Prog Ser 272: 165–181.
37. Comeau S, Gorsky G, Jeffree R, Teyssié J-L, Gattuso J-P (2009) Impact of ocean acidification on a key Arctic pelagic mollusc (*Limacina helicina*). Biogeosciences 6(9): 1877–1882.
38. Bé AWH, Gilmer RW (1977) A zoogeographic and taxonomic review of Euthecosomatous pteropoda. In: Ramsey ATS, editor.Oceanic micropaleontology, Vol. 1.London: Academic Press. pp. 733–808.
39. Sato-Okoshi W, Okoshi K, Sasaki H, Akiha F (2010) Shell structure of two polar pelagic molluscs, Arctic *Limacina helicina* and Antarctic *Limacina helicina antarctica* forma antarctica. Polar Biol 33(11): 1577–1583.

40. Waite JH, Saleuddin ASM, Andersen SO (1979) Periostracin—Soluble precursor of sclerotized periostracum in *Mytilus edulis L.* J Comp Physiol 130(4): 301–307.

41. Harper EM (2000) Are calcitic layers an effective adaptation against shell dissolution in the Bivalvia? Journal of Zoology 251(02): 179–186.

42. Busenberg E, Plummer LN (1986) A comparative study of the dissolution and crystal growth kinetics of calcite and aragonite. In: Mumpton FA, editor. Studies in diagenesis.US Geological Survey Bulletin 1578. pp. 139–168.

43. Gehlen M, Gangstø R, Schneider B, Bopp L, Aumont O, et al. (2007) The fate of pelagic $CaCO_3$ production in a high CO_2 ocean: A model study. Biogeosciences 4(4): 505–519.

44. Keir RS (1980) The dissolution kinetics of biogenic calcium carbonates in seawater. Geochim Cosmochim Acta 44(2): 241–252.

45. Morse JW, Mucci A, Millero FJ (1980) Solubility of calcite and aragonite in seawater of 35‰, salinity at 25°C and atmospheric pressure. Geochim Cosmochim Acta 44(1): 85–94.

46. Morse JW, Arvidson RS, Lüttge A (2007) Calcium carbonate formation and dissolution. Chem Rev 107: 342–381.

47. McGowan JA, Hayward TL (1978) Mixing and oceanic productivity. Deep Sea Res 25(9): 771–793.

48. Comeau S, Gattuso J-P, Nisumaa AM, Orr J (2012) Impact of aragonite saturation state changes on migratory pteropods. Proc R Soc Lond B Biol Sci 279(1729): 732–738.

49. Berner RA (1977) Stoichiometric models for nutrient regeneration in anoxic sediments. Limnol Oceanogr 22(5): 781–786.

50. Berner RA, Honjo S (1981) Pelagic sedimentation of aragonite — Its geochemical significance. Science 211(4485): 940–942.

51. Emerson S, Jahnke R, Bender M, Froelich P, Klinkhammer G, et al. (1980) Early diagnesis in sediments from the eastern equatorial Pacific.1. Pore water nutrient and carbonate results. Earth Planet Sci Lett 49(1): 57–80.

52. Gardner WD, Hinga KR, Marra J (1983) Observations on the degradation of biogenic material in the deep ocean with implications on accuracy of sediment trap fluxes. Journal of Marine Research 41(2): 195–214.

53. Harbison GR, Gilmer RW (1986) Effects of animal behavior on sediment trap collections: implications for the calculation of aragonite fluxes. Deep Sea Res A 33(8): 1017–1024.

54. Price BA, Killingley JS, Berger WH (1985) On the pteropod pavement of the eastern Rio Grande Rise. Mar Geol 64(3–4): 217–235.

55. Fabry VJ (1989) Aragonite production by pteropod molluscs in the subarctic Pacific. Deep Sea Res A 36(11): 1735–1751.

56. Fabry VJ (1990) Shell growth rates of pteropod and heteropod mollusks and aragonite production in the open ocean: Implications for the marine carbonate system. Journal of Marine Research 48(1): 209–222.

57. Whitfield M (1974) Accumulation of fossil CO_2 in atmosphere and in sea. Nature 247: 523–525.

58. Gangstø R, Joos F, Gehlen M (2011) Sensitivity of pelagic calcification to ocean acidification. Biogeosciences 8: 433–458. doi:10.5194/bg-8-433-2011.

15

Timing of the Departure of Ocean Biogeochemical Cycles from the Preindustrial State

James R. Christian[1,2]*

1 Canadian Centre for Climate Modelling and Analysis, Victoria, B.C., Canada, **2** Fisheries and Oceans Canada, Institute of Ocean Sciences, Sidney, BC, Canada

Abstract

Changes in ocean chemistry and climate induced by anthropogenic CO_2 affect a broad range of ocean biological and biogeochemical processes; these changes are already well underway. Direct effects of CO_2 (e.g. on pH) are prominent among these, but climate model simulations with historical greenhouse gas forcing suggest that physical and biological processes only indirectly forced by CO_2 (via the effect of atmospheric CO_2 on climate) begin to show anthropogenically-induced trends as early as the 1920s. Dates of emergence of a number of representative ocean fields from the envelope of natural variability are calculated for global means and for spatial 'fingerprints' over a number of geographic regions. Emergence dates are consistent among these methods and insensitive to the exact choice of regions, but are generally earlier with more spatial information included. Emergence dates calculated for individual sampling stations are more variable and generally later, but means across stations are generally consistent with global emergence dates. The last sign reversal of linear trends calculated for periods of 20 or 30 years also functions as a diagnostic of emergence, and is generally consistent with other measures. The last sign reversal among 20 year trends is found to be a conservative measure (biased towards later emergence), while for 30 year trends it is found to have an early emergence bias, relative to emergence dates calculated by departure from the preindustrial mean. These results are largely independent of emission scenario, but the latest-emerging fields show a response to mitigation. A significant anthropogenic component of ocean variability has been present throughout the modern era of ocean observation.

Editor: Vanesa Magar, Centro de Investigacion Cientifica y Educacion Superior de Ensenada, Mexico

Funding: The authors have no support or funding to report.

Competing Interests: The author has declared that no competing interests exist.

* Email: jim.christian@ec.gc.ca

Introduction

The direct effect of anthropogenic CO_2 on ocean chemistry already exceeds the range of natural variability in many locations [18]. However, many aspects of ocean biogeochemistry are forced only indirectly by CO_2, via the effect of atmospheric CO_2 on climate. Detecting an anthropogenic climate change signal in ocean biogeochemical data is difficult due to short data records and high natural variability [10,13,24]. Trends are not monotonic, and even strong anthropogenic forcing is subject to modulation by a variety of physical and biogeochemical processes [14]. In addition, the effects of climate warming are complex and competing processes can offset each other. For example, primary production will tend to increase with increasing temperature, but the same increases in temperature cause increasing stratification that limits the supply of nutrients to the surface ocean [37]. In addition, some fields have opposing trends in different regions [23].

Variability in climate can be divided into forced and unforced components. In model simulations the unforced component is referred to as the model's internal variability, or natural variability. Models never reproduce the timing of natural variability exactly, but a good model will reproduce the statistical characteristics [38]. A "white noise" spectrum implies approximately equal variability across the spectrum of frequencies resolved, while a "red noise"

spectrum implies greater variance at lower frequencies (i.e., significant autocorrelation in time). Climate variability is generally considered to have a "red" spectrum [19,25], but modern data records are too short for complete characterization and it is therefore difficult to know with confidence whether the models' 'natural' variability is too weak (or too strong). Nonetheless we know that climate models can accurately reproduce important aspects of observed climate variability [17,40].

There is a relatively well-established, although certainly not uniform, methodology for detection of anthropogenic climate change and attribution of those changes to specific forcing factors [6,9,20,34]. In this literature only a handful of papers to date have dealt with ocean biogeochemistry [1,8,18,24]. Detection normally refers to a demonstration that observed variability in the climate system exceeds the range expected from natural variability at some specified (e.g., 5%) level of significance [5,21]. Detection period refers to the length of a data record required to unequivocally detect an anthropogenic signal, while detection time indicates the point in time at which the signal becomes detectable [35]. The latter is related to the "time of emergence", or the point in time at which the anthropogenic signal emerges from the "historical envelope" of natural variability, but there are subtle differences between the two. In model simulations, detection of even extremely small signals is possible given a sufficiently large ensemble; so one could argue that the term should not be used at all for studies using only model simulations. Emergence implies that thereafter the signal remains consistently outside the envelope of natural variability as defined e.g., by some multiple of the standard deviation of the unforced control simulation [29,31]. In this study I define emergence as the point at which the anthropogenic signal exceeds natural variability (estimated from a preindustrial control simulation) at some specified threshold, and remains continuously in excess of that threshold (except when reversals are clearly attributable to mitigation). Note that this definition differs from that of [29], who include both natural and anthropogenic forcing up to 2005 as part of the envelope of historical variability from which emergence is estimated. It is also important to note that as observing systems inevitably have incomplete coverage, the point of potential detection of an anthropogenic signal is always later than the point of emergence as defined here.

While this sort of analysis does not offer any immediate hope of unambiguously detecting an anthropogenic signal in the real world, it represents a useful measure of the 'true' point of departure from the preindustrial climate, which is surprisingly early in many cases. Having an estimate of the emergence time is useful when examining historical data records, few if any of which are entirely free of anthropogenic influence. It can also be useful to impacts research in that it provides information about the magnitude of the anthropogenic signal relative to the natural variability [9,27].

Modern coupled climate/carbon cycle models provide a homogeneous data set with which to conduct such an experiment, which despite its shortcomings overcomes some of the problems of earlier studies. Some early detection studies were limited by the lack of extended control simulations with models identical to their forced runs, or their control runs were done with ocean-only models for which stochastic forcing had to be employed to generate internal variability [35]. The current data set includes multiple realizations of the forced simulations, and an extended control run (with an identical model) from which the preindustrial variability can be estimated. It can not be known with certainty that the model does not underestimate natural variability, but if such biases exist they are probably small [12].

Methods

Model description

The model used is the Canadian Earth System Model, which has been previously described by [3,4,12]. In the version used here (CanESM2.0), the atmosphere model is run at T63 spectral resolution, a 128×64 horizontal grid. The 256×192 ocean model, with six grid cells (2x, 3y) to each atmosphere grid cell, has a longitude resolution of approximately $1.4°$ and latitude resolution of approximately $0.94°$. The ocean model has 40 vertical levels (increased from 29 in CanESM1, with all of the additional levels in the upper 300 m), with a resolution of 10–15 m in the upper 100 m. The ocean carbon cycle model is based on the Ocean Carbon Model Intercomparison Project II protocols [32] and couples the carbon cycle to an NPZD ecosystem model via a fixed C/N Redfield Ratio and a temperature-dependent rain ratio (ratio of inorganic to organic carbon in vertical flux at the base of the euphotic zone) [12,41].

Simulations used here have specified atmospheric CO_2 concentrations; runs with freely-varying atmospheric CO_2 give similar results. Historical (1850–2005) forcing includes volcanic eruptions and solar variability. Greenhouse gas forcing after 2005 is provided by the Representative Concentration Pathways (RCPs)

[30]. RCPs 8.5, 4.5, and 2.6 are often referred to as the "no mitigation", "moderate mitigation", and "strong mitigation" scenarios respectively; in RCP8.5 emissions continue to increase and atmospheric CO_2 concentration exceeds 900 ppm by 2100 [4].

Nine data fields were considered: sea surface temperature and salinity, mixed layer depth (MLD), surface ocean pCO_2 and air-sea CO_2 flux, surface nitrate concentration, total primary production, organic export flux at 100 m, and dinitrogen fixation. All of these are standard (2D, monthly) data fields of the 5th Coupled Model Intercomparison Project [39]; all data are in the public domain. The choices are somewhat arbitrary, but cover a range of fields commonly measured by oceanographers and used to diagnose the performance of ocean biogeochemical models. MLD was excluded from some analyses because in the historical simulation it was available only for a single realization (see below Figure 1).

Statistics

For global and station means, emergence from the envelope of natural variability was defined as the first year the mean of an ensemble of five realizations of the forced (historical, RCP) run differed from the mean of the unforced preindustrial (1850 greenhouse gas concentrations) control run by n standard deviations ($n\sigma$) of the interannual variability of the control run, and remained in excess of $n\sigma$ thereafter, with $n = 3$ for global metrics [31] and $n = 2$ for individual locations. The different criteria for global means and specific locations are arbitrary but yield consistent results across the range of stations considered; for individual locations a 2σ threshold was applied, because in many cases emergence would not occur by 2100 with the more conservative 3σ threshold. The ensemble mean is used to isolate the anthropogenic signal by averaging out the effect of internal variability. Three hundred years of control run were used to calculate the standard deviation. Annual means were used in all cases. Trends were not corrected for drift in the control run, but drift is small as only the surface and euphotic zone are considered, and is also present in the forced runs. Drift will bias the method slightly towards later emergence because it increases the preindustrial variability. In some cases the fields emerged but later fell back within the $n\sigma$ window under the "strong mitigation" scenario (RCP2.6); these were considered emergent if they remained outside the window for 50 years (in practice, 20–25 years would be sufficient as the longest any fields remained outside the window in an unforced control run was ~15 years).

Spatial 'fingerprints' were calculated for the 14 ocean regions defined by Sarmiento et al. [36] (hereafter S02) and for zonal means of $10°$ latitude bands as in [31]; results are largely insensitive to which of these was used (see below Results). S02 divided the ocean up by both latitude (regions are bounded by $15°$ and $45°$ north and south and exclude areas north of $60°N$) and basin, even in the Southern Ocean which was divided into several 'sectors'. So if there are interbasin differences in trends there will be a small amount of information in this pattern that is not present in simple zonal means.

Prior to spatial averaging, data were normalized as

$$x_n = \left(\frac{x - \overline{x}}{\sigma_x} \right) \qquad (1)$$

Where $\overline{x}$ and σ_x are the mean and standard deviation of all valid data from the preindustrial control run, except for dinitrogen fixation (DNF) where only data equatorward of $40°$ latitude were

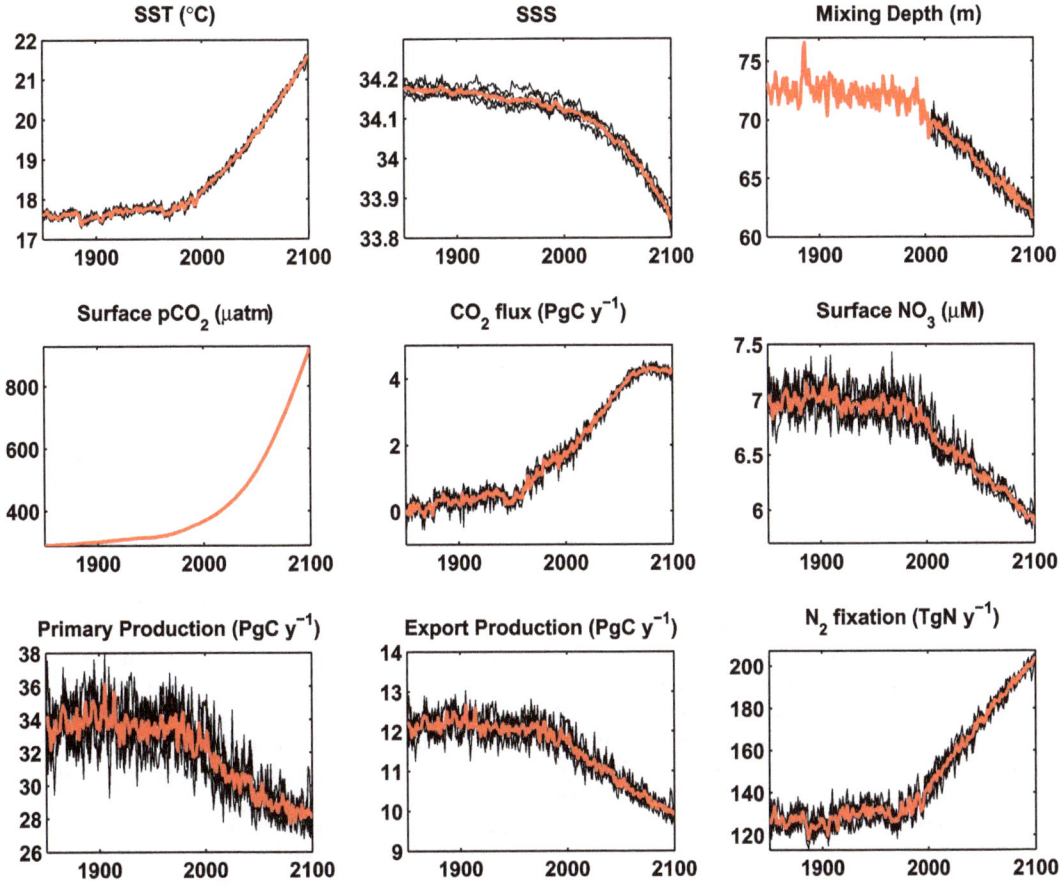

Figure 1. Global means or integrals of ocean surface fields from 1850–2100 under RCP8.5. Five ensemble members (thin black lines) are shown along with the ensemble mean (thick red line). Mixed layer depth has only a single realization for the historical run. Export is at 100 m. Primary production, export, CO_2 flux and N_2 fixation are global integrals; other fields are global mean surface values.

considered. The difference between the means for these regions for 2081–2100 of RCP8.5 and the means for 300 years of unforced control run was considered to be the 'fingerprint' of anthropogenic climate change. The projection of this fingerprint onto each year of the forced (historical + RCP) run, i.e.,

$$Y_i = P + aX \qquad (2)$$

where X is the anthropogenic fingerprint, Y_i is a vector of regional means for the current year of the forced run and P is a vector of regional means for the long-term mean of the unforced control run, was calculated by linear least squares.

$$a = \left(X^t C^{-1} X\right)^{-1} X^t C^{-1} (Y_i - P) \qquad (3)$$

where C is the covariance matrix for the control run. The same method was applied to (300) individual years (P_i) of the control run in order to generate a distribution of values (a_0) from which emergence can be determined. As for the global means, the year of emergence (YOE) was considered to be the year in which a exceeded three times the standard deviation of a_0 (σ_{a0}) and remained consistently in excess of $3\sigma_{a0}$ thereafter [31].

A 'combined' fingerprint experiment was conducted where the time series of regional means of all variables except pCO2, CO2 flux and mixing depth were superimposed, for a total of 92

variables/regions in the S02 case (sea surface nitrate concentration in the tropical Atlantic was excluded because its extremely low variance made the covariance matrix singular). pCO2 and CO2 flux were excluded because their much earlier emergence would dominate the result (pCO2 in particular has a much smaller preindustrial standard deviation than other fields and emerges very early); mixing depth was excluded because there was only a single realization of the historical run available. As noted above, high-latitude regions were excluded for DNF, because it only occurs where temperature exceeds 20°C.

Ocean observing station network

Model simulations were sampled at ten ocean observing stations, eight of the nine used by Moore et al. [28] and Station KNOT in the northeast Pacific (Table 1). The Ross Sea station used by Moore et al. was replaced, because of its shallow depth and proximity to the ice shelf, with a more oceanic location in the South Atlantic (SATL). These stations were chosen to represent most major regions of the world ocean, excluding the Arctic and the marginal seas, and to include actual sampling stations where observations have been made. Four of the stations are located in the tropics and subtropics and three each in the Southern Ocean and the northern midlatitudes (Table 1). Northern midlatitude stations range in latitude from 44–50°N, and Southern Ocean stations from 51–62°S.

Table 1. Locations of stations at which the model simulations were sampled for local emergence.

Station	Latitude	Longitude
BATS	32°N	64°W
HOT	23°N	158°W
PAPA	50°N	145°W
KNOT	44°N	155°E
ARAB	16°N	62°E
EQPAC	0°	140°W
SATL	51°S	19°W
NABE	47°N	19°W
KERFIX	51°S	68°E
PLRFR	62°S	170°W

At each station, the YOE was determined in the same manner as for the global means, except that the criterion for emergence was set to 2σ instead of 3σ. In addition, 20 and 30 year linear trends were calculated for individual ensemble members to identify the point at which these become consistently positive or negative, termed the "Last Zero Crossing" (LZC). In this case individual ensemble members are used because the influence of natural variability must be preserved. For a LZC to be recorded for N year trends, the last sign reversal must occur at least $N/2+10$ years prior to the end of the run for at least 4 of 5 ensemble members; LZC is then averaged over the realizations in which a LZC was recorded.

Results

Global mean trends

Global mean trends in ocean physical and biogeochemical fields show substantial alteration in the 21st century under the no-mitigation scenario (Figure 1), and in many cases these trends are well underway by the end of the 20th (which is scenario-independent). MLD declines by ~3 m by 2000 and 10 m by 2100 (Figure 1). Export production and dinitrogen fixation show more or less monotonic trends that are well underway by 2000 (Figure 1). Ocean CO_2 uptake continues to increase, but the rate of growth declines rapidly near the end of the 20th century (Figure 1). Some biogeochemical fields, such as primary production, have weak trends due to competing influences of e.g., temperature and stratification, as well as offsetting trends in different regions, so that the trend is relatively small compared to natural variability, at least initially (Figure 1).

The emergence of the global mean or integral values of the selected fields from the envelope of natural variability was tested by comparing ensemble means for historical + RCP2.6/4.5/8.5 simulations with the unforced control simulations, for each year of the simulation (1850–2100). Year of emergence was recorded as the first year that the ensemble mean differed from the preindustrial mean by at least 3 standard deviations of the control run and remained in excess of this threshold continuously thereafter. Surface ocean pCO_2 emerges in the 1870s, but air-sea CO_2 flux does not emerge until about a century later (Figure 2). Sea surface temperature (SST) emerges before sea surface salinity (SSS), as is expected for the global mean. SSS has offsetting trends in different regions, because the global surface net freshwater flux is close to zero, but anthropogenic warming increases the local flux in both net evaporative and net

precipitation regions [15,23]. Emergence for SST closely approximates the date at which detection of an anthropogenic contribution to ocean heat content change is deemed to have become statistically significant [9]. YOE for MLD was not determined but would be around 2005 (not shown).

Primary production does not emerge in RCPs 2.6 or 4.5, and even in RCP8.5 does not emerge until the 2040s. This is partly due to offsetting increasing and declining trends in different regions (see below sections 3.2 and 3.3). In RCP4.5 global mean primary production exceeds the 3σ threshold for about 10 years in the 2040s, but does not remain there because the overall trend is not monotonic under the mitigation scenarios (not shown). In RCP2.6, air-sea CO_2 flux, surface nitrate concentration, and export production all fall back within the 3σ window after initially emerging from it the early 21st century (not shown); in these cases the fields are recorded as emergent (Figure 2), because they remained outside the 3σ window continuously for>50 years. Note that emergence can occur slightly earlier in a lower emission scenario due to natural variability, which is reduced but not eliminated by using ensemble means (Figure 2). In fact, it is possible for emergence to occur after 2005 in one scenario while occurring before 2005 in the others, even though atmospheric CO_2 is identical up to 2005.

Detecting the spatial 'fingerprint' of anthropogenic change

Because some fields have offsetting increases and decreases in different regions [23], global means are not necessarily a good metric of whether the forced simulation has emerged from the envelope of variability of the unforced control. I have calculated a spatial 'fingerprint' of anthropogenic change by taking the difference between the mean for 2081–2100 of RCP8.5 and the mean of the preindustrial control run (equation 2). The projection of the anthropogenic fingerprint onto individual years of the transient run (a in equations 2 and 3) for individual data fields is shown in Figure 3 for two different choices of the averaging regions. The difference between the two is generally small and emergence from the 3σ window occurs by 2005 in all cases. Historical volcanic eruptions such as Krakatoa (1883), Agung (1963) and Pinatubo (1991) are visible in the trends for some fields (Figure 3). The eruption of Pinatubo occurs near the point of emergence for primary production, export production and dinitrogen fixation and delays emergence by 5–10 years (Figure 3).

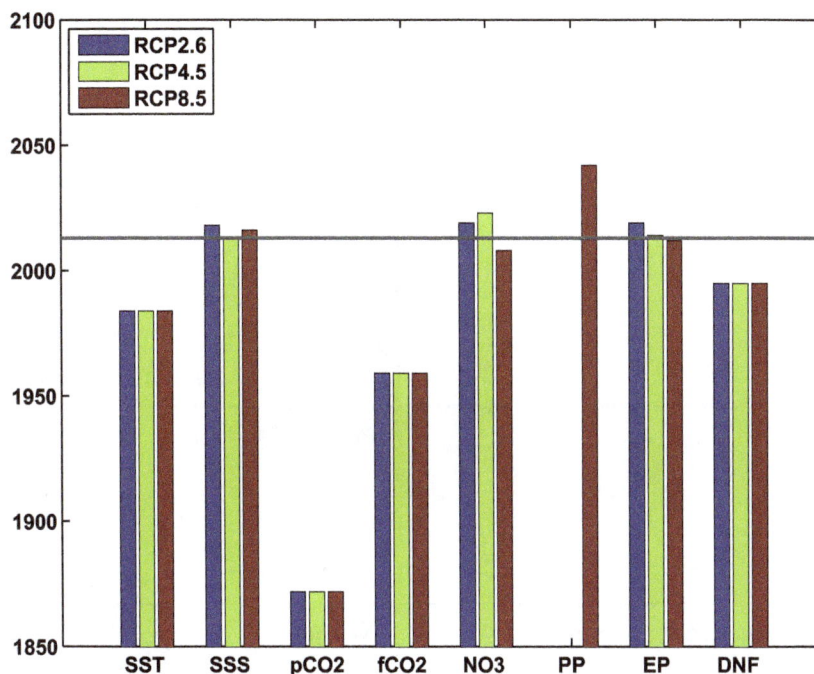

Figure 2. Year of emergence of global mean ocean surface fields from the range of natural variability for RCPs 2.6, 4.5, and 8.5.
Emergence is defined as the year that the ensemble mean of 5 forced (historical + RCP) runs exceeds three times the standard deviation (3σ) of 300 years of unforced control run and remains in excess of 3σ thereafter. Horizontal grey line indicates beginning of 2013. SST = sea surface temperature; SSS = sea surface salinity; pCO2 = surface ocean pCO_2; fCO2 = ocean-atmosphere CO_2 flux; NO3 = surface nitrate concentration; PP = primary production; EP = export production at 100 m; DNF = dinitrogen fixation.

Trends in the presence of the anthropogenic fingerprint do not in most cases differ much between the 14 regions of S02 or global zonal means for 10° latitude bands as in [31] (Figure 3). The variables that show the largest difference, i.e. the most sensitivity to basin-specific information, are sea-surface salinity and air-sea CO_2 flux (Figure 3). There are large interbasin differences in evaporation and precipitation [11], so it is not surprising that the spatial 'fingerprint' of anthropogenic warming for SSS has a larger component that is basin-specific than for most other fields. For CO_2 flux the anthropogenic fingerprint shows enhanced net uptake (which may be reduced outgassing in outgassing regions) in the equatorial Pacific upwelling zone, the Gulf Stream and Kuroshio termination regions, the high-latitude North Atlantic and much of the Southern Ocean (not shown). However, there is a fair amount of variation among regions of the Southern Ocean, with the strongest enhancement in the Atlantic sector and a mosaic of positive and negative trends in the Pacific and Indian sectors. By recalculating the fingerprint for a reduced set of regions, with different basins being combined for specific latitude ranges, the latitudes in which basin-specific information is important can be identified (not shown). For CO_2 flux about half of the total effect is in the Southern Ocean, with the balance in the tropics. For SSS the only region where there is sensitivity to basin-specific averaging is the subtropics (i.e., regions of net evaporation).

The combined fingerprint of all non-carbon fields (except MLD) shows early emergence, and is also insensitive to the choice of averaging regions (Figure 4, Table 2). As with some of individual fields, volcanic eruptions can delay emergence. In this case the eruption of Agung (1963) delays emergence by nearly half a century in the zonal means case, whereas the S02 case is unaffected because it remains slightly outside the window following the eruption (Figure 4, Table 2). This illustrates how

sensitive the exact date of emergence can be to the somewhat arbitrary criteria employed, but further serves to illustrate that an anthropogenic signal was present even in the first half of the twentieth century. The biological fields (surface nitrate concentration, primary production, export production and dinitrogen fixation) do not affect emergence time much compared to SST and SSS alone (Table 2), but this results in large part from the particular timing of the eruption of Agung (Figure 4). Inclusion of both sets of fields narrows the window of preindustrial variability substantially relative to either alone (Figure 4). Had the eruption not occurred when it did, the differences in emergence times among these three cases could be much larger.

Local emergence and ocean observing networks

An increasing number of ocean time series data span several decades [7,14]. But since these are largely localized observations, and interannual to interdecadal variability is also present in the data records, how can such data be used to detect a longer-term trend? And how clearly can such trends be associated with anthropogenic forcing? The following analysis of a network of 10 ocean observing stations explores these questions.

Year of emergence at 2σ for the ten station means is shown in Figure 5. YOE's are generally later for individual stations than for the global means, and are quite variable among stations (Figure 5). However, the ranges for different variables are generally consistent with YOE's estimated for global means or spatial fingerprints except for CO_2 flux (Figure 5). Surface ocean pCO_2 emerges much earlier than other fields (prior to ~1960), but the range among stations is almost 100 years. For all fields except pCO_2 and SST, there are some stations at which emergence does not occur by 2100, and a few (MLD, surface nitrate, and export production) emerge at less than half of the stations (Figure 5).

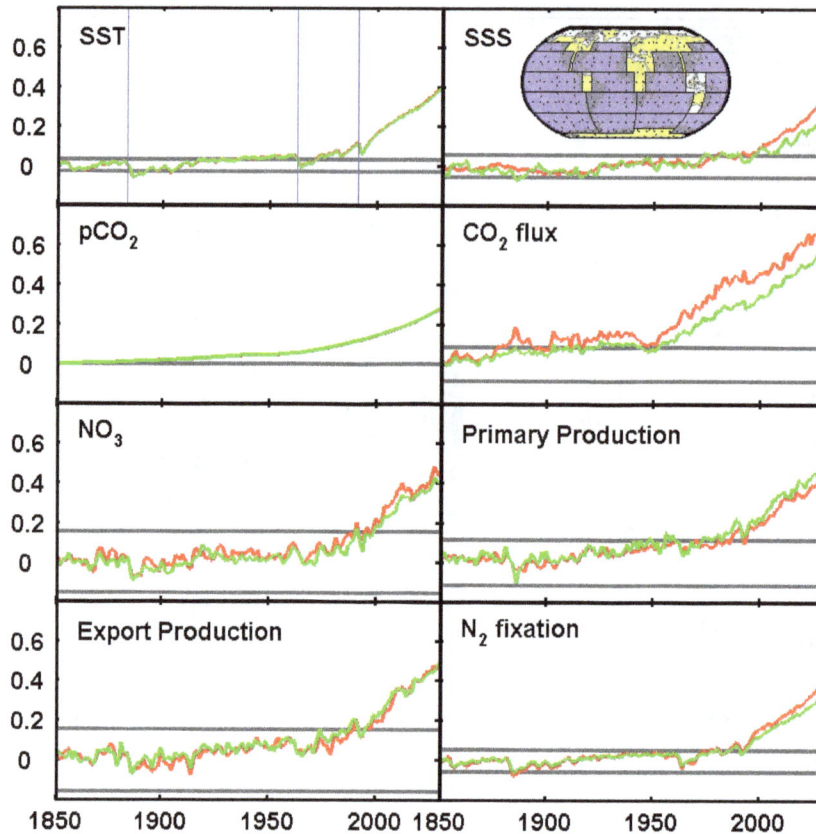

Figure 3. Contribution of 2081–2100 (RCP8.5) anthropogenic fingerprint to difference of current year from preindustrial, for individual fields. Fingerprint is based on areal means for the 14 ocean regions of Sarmiento et al. (2002) (S02, red) or global zonal means of 10° latitude bands (green). Vertical axes are normalized as shown in equation (1). Horizontal black lines are plus or minus three standard deviations of preindustrial values (3σ range is shown for S02 only; values are almost identical for the two methods). Vertical blue lines in first panel indicate eruptions of Krakatoa (1883), Agung (1963), and Pinatubo (1991). Inset map in second panel indicates S02 regions (blue rectangles); Pacific regions are single boxes spanning the basin.

Quasi-linear trends have been shown to occur over extended periods at ocean observing stations, particularly for carbon-related fields, for which the anthropogenic trend is large relative to the natural variability [7]. However, for other fields the time series are as yet too short for trends to be unambiguously associated with anthropogenic forcing. I have calculated 20 and 30 year linear trends from the model solution at the 10 observing stations in Table 1, and compared the time of emergence as estimated above with the time of the last sign reversal, or LZC.

The LZC for various fields and stations is shown in Table 3. The table shows the final year where the sign of the regression coefficient was opposite to the mean for the last 10 intervals. The LZC is not a statistically rigorous estimate of the time at which the anthropogenic trend becomes significant relative to natural variability, but it does give a rough indication (i.e., in the control climate the coefficients are equally often positive and negative), and it is shown below that its relation to the YOE is quite consistent. In quite a few cases sign reversals occur throughout the 20th century (blank entries in Table 3). In other cases the LZC occurs quite early, suggesting that the anthropogenic impact is present in the observations taken at these locations, which in most cases date from the 1980s or 1990s.

Because in this experiment the YOE (based on the ensemble mean and preindustrial standard deviation) is known (Figure 5), one can determine whether the LZC is correlated with the YOE

across different fields and locations and whether it tends to over- or underestimate the YOE. The LZC as an estimate of emergence time is strongly correlated with YOE for both 20 and 30 years trends, with r = 0.91 and 0.92 respectively (Figure 6). It provides consistently conservative estimates (exceeds the YOE) for 20 year trends, but not for 30 year trends, which tend to have an "early emergence" bias (Figure 6). This suggests that the proper trend length for which the LZC will approximate the YOE is ~25 years, but the exact value is sensitive to the characteristics of the specific data sets employed (see Discussion).

Discussion

A central result of this analysis is that if the model is a reasonable representation of the real world, most of the fields have already emerged from the preindustrial range (Figures 2 and 3, Table 2). As a result, the emission scenario does not matter much for the global YOE because it precedes the point where they start to diverge. However, for individual locations YOEs are generally later, sometimes much later (Figure 5) and therefore emergence may not occur, depending on the future trajectory of emissions. Global YOE is affected by mitigation only for primary production (Figure 2).

YOE is strongly affected by historical volcanic eruptions, and in some cases delayed by decades (Figures 3 and 4). The eruption of Mt Pinatubo (1991) delays emergence by about 10 years for

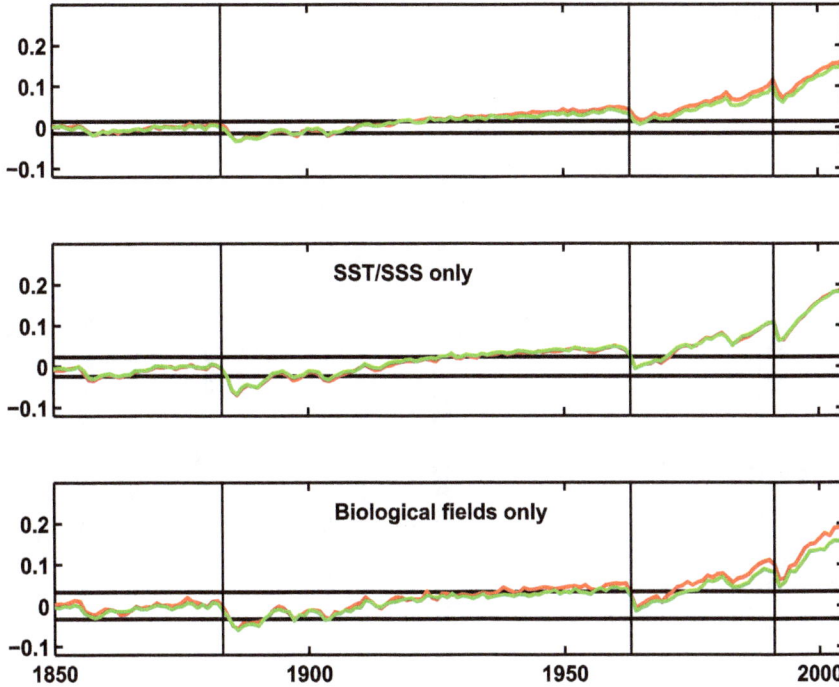

Figure 4. As Figure 3 but for combined fingerprint of all fields except pCO$_2$, CO$_2$ flux and mixing depth, for SST + SSS only, and for biological fields (surface nitrate concentration, primary production, export production and dinitrogen fixation) only. Red line indicates fingerprint based on the 14 ocean regions of S02; green line is for global zonal means of 10° latitude bands. Vertical black lines indicate eruptions of Krakatoa (1883), Agung (1963), and Pinatubo (1991).

several fields (Figure 3). Eruption of Agung in 1963 delayed emergence in the combined fingerprint experiment by more than 40 years in the zonal means case, while the S02 case is unaffected because it remains slightly outside the window following the eruption (Figure 4). Note that the downward trend at this time

appears to begin prior to the eruption: this may result in part from natural variability, in part from errors in the volcanic aerosol data set, and in part from a small increase in volcanism globally in the years 1960–1963; this downward trend is present in multiple models (see e.g., [22]).

Table 2. Year of emergence by estimation of single or multiple field anthropogenic fingerprint calculated for the 14 ocean regions of Sarmiento et al. (2002) or for global zonal means for 10° latitude bands, for two or three standard deviations of preindustrial values.

Regions	S02	S02	Zonal	Zonal
Emergence threshold	2σ	3σ	2σ	3σ
Individual				
SST	1972	1973	1972	1973
SSS	1976	1990	1997	1997
Surface pCO2	1859	1860	1859	1860
Air-sea CO2 flux	1877	1913	1918	1953
Surface [NO3]	1986	2000	1996	2000
Primary production	1969	1995	1973	1984
Export production	1988	1996	1981	1997
N2 fixation	1994	1995	1994	1995
Combined				
All, except pCO2 and CO2 flux	1919	1923	1967	1968
Biological fields	1971	1972	1971	1972
SST + SSS only	1971	1973	1973	1973

Combined fingerprint excludes pCO$_2$ and CO$_2$ flux; biological fields are surface nitrate concentration, primary production, export production and dinitrogen fixation.

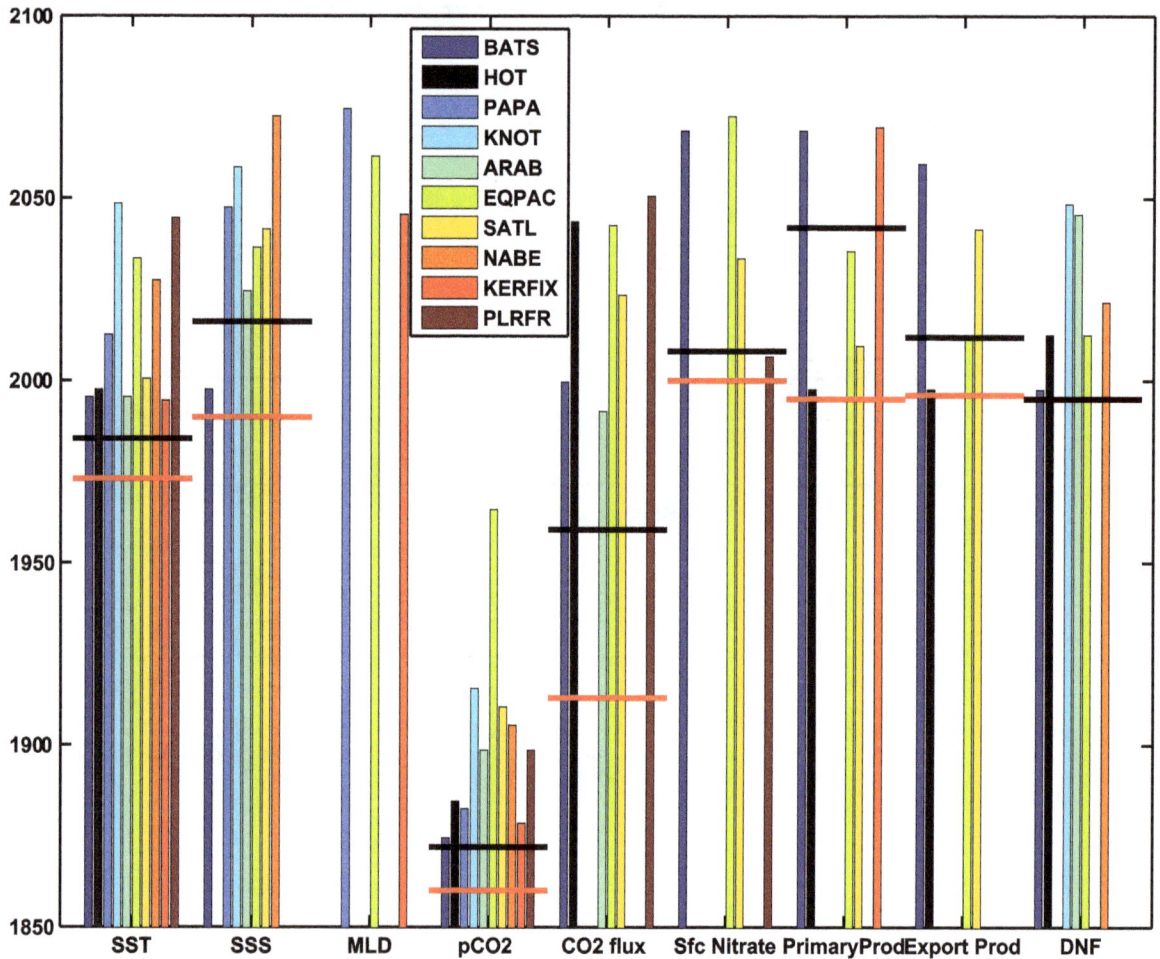

Figure 5. Year of emergence of mean ocean surface fields from the range of natural variability at 2σ for individual locations under RCP8.5 (see Table 1). Horizontal lines indicate YOE's for global means or spatial fingerprints from S02 (see Figures 2 and 3). Black indicates YOE for the global mean, red for the spatial fingerprint. For dinitrogen fixation red bars are not visible because YOE's are the same. No vertical bar indicates that the variable does not emerge by 2100 at that location.

The spatial fingerprint analysis provides a measure of the importance of diverse observations for detection of anthropogenic effects. The analysis presented here does not deal with detection directly, because only model simulations are used, but it is likely that earlier YOE implies earlier detection assuming sufficient observations are available. Spatial fingerprints generate earlier YOE than the global means for all fields except DNF (Figure 5). In many cases basin-specific information provides little additional information above what is provided by global zonal means (Figure 3, Table 2). A combined fingerprint of all fields results in very early emergence (~1920 were it not for the eruption of Agung in 1963). Inclusion of the biological fields has relatively little effect on time of emergence beyond what is available from SST and SSS alone, but does make the window smaller, i.e., the additional information increases confidence that the projection of the anthropogenic fingerprint on the preindustrial climate is close to zero (Figure 4). This is probably because the SST response to planetary heating is global and more or less instantaneous (disregarding modulation by natural variability), whereas impacts on ocean biology are more indirect, resulting from stratification that derives from surface heating. It is therefore unsurprising that fields like export production emerge later and have relatively little effect on the combined fingerprint YOE. DNF is (in the model) an

approximately linear function of SST, so it has strong trends but contains little information beyond what is present in SST alone. Nonetheless, emergence of the combined fingerprint for biological fields alone is quite similar to the total fingerprint, so there is likely to be an anthropogenic influence on most modern ocean biogeochemistry measurements.

Simplifying assumptions such as a single plankton species with fixed elemental ratios limit variability of the modelled ocean ecosystem. Plankton models with multiple species are better able to reproduce changes in ecosystem structure that occur under different physical forcing regimes [2]. There is some evidence that changes in plankton elemental stoichiometry can also 'amplify' the biological response to relatively small changes in physical forcing [26]. These biases are present in both the control and the forced runs, but they bias the model towards a 'damped' biological response to changing physical forcing and thus induce a bias towards later emergence. Excluding DNF at higher latitudes also potentially biases the method towards later emergence because a small amount of DNF occurs in regions where it is absent in the control run (as SST begins to exceed 20°C), but these rates are very low (not shown).

The use of ensembles to average out internal variability in the transient runs gives a statistically robust estimate of emergence

Table 3. Last year where 20 or 30 year trend had opposite sign to that recorded at the end of the 21st century.

	# years	BATS	HOT	PAPA	KNOT	ARAB	EQPAC	SATL	NABE	KERFIX	PLRFR
SST	20	1979	2027	2011	2041	1983	2058	2019	2049	1999	
	30	1968	1968	1974	1977	1967	1979	1968	2001	1957	2073
SSS	20	1980	2060	2074	2080	2039	2059	2050	2060	2067	
	30	1958		2038	2038	1997	2015	2021	2021	2064	
Mixing Depth	20						2076	2077	2077		2075
	30		2074	2070			2021	2056	2054		2070
Surface pCO_2	20	1869	1931	1883	1963	1947	1972	1943	1936	1909	1942
	30	*	1869	*	1940	1873	1941	1885	1867	*	1853
Air-sea CO_2 flux	20	2083		2062	2084	2044		2041	2078		2059
	30			2046		1993		1993	2057		2021
Surface $[NO_3]$	20		2081		2082			2053		2074	
	30	2029			2081		2069	2036	2046	2073	
Primary production	20				2082		2085	2051		2069	
	30	2030	2052		2072		2047	1976	2046	2047	2079
Export production	20	2019	2069		2082	2074	2085	2080		2075	2079
	30		1988		2081	2072	2059	2033	2044	2066	2075
N_2 fixation	20	2001	2055			2083	2078				
	30	1968	1968			2071	2008				

No data indicates sign reversals continue up until the end of the 21st century, except for N_2 fixation which only occurs at the stations where dates are listed.
* indicates that the trend was of a consistent sign from the outset. Mixing depth has only a single realization for the historical simulation but no values <2005 appear.

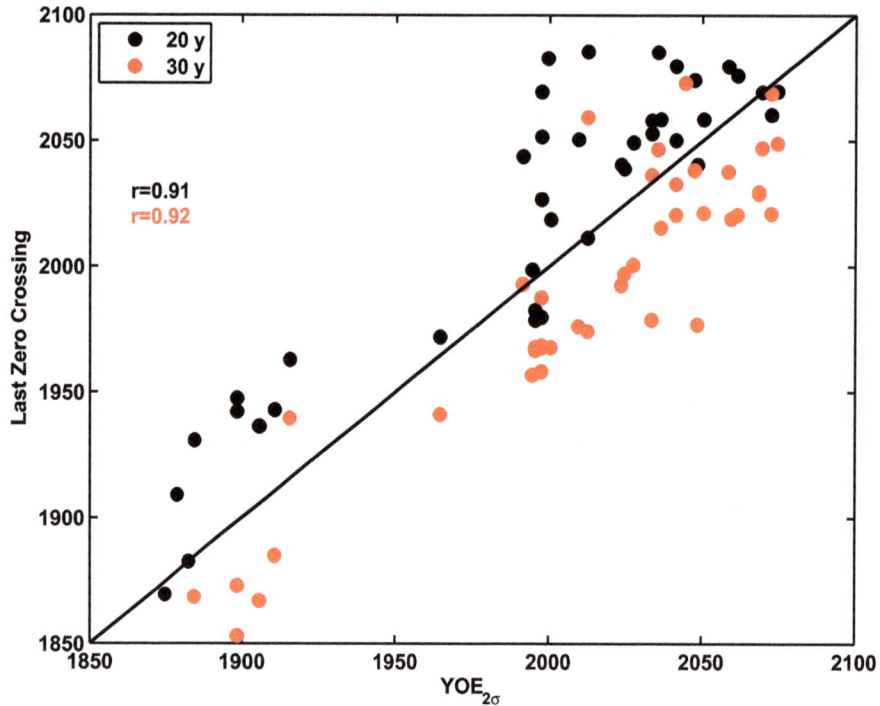

Figure 6. Last-zero-crossing for 20- and 30-year regression coefficients (Table 3) relative to YOE defined as in Figure 5, for all data fields at individual sampling stations where a last-zero-crossing was recorded.

relative to the variability of the control run. The time of emergence in an individual realization would depend strongly on internal variability, because natural variability superimposed on an overall trend will tend to produce periods where the record is flat or has a weak counter-trend (similar to the recent "warming hiatus", cf. [16]), followed by periods of rapid change. If there were no overall trend, the positive and negative trends would be symmetric (as in the unforced control run). But in the presence of a long term trend they are not, and there will be short periods with strong positive trends.

To fully understand the probability of medium-term trends occurring, one would ideally wish to know the frequency spectrum of natural variability. In the model, the trends are clearly forced by anthropogenic greenhouse gases; they do not occur in the unforced control run. But it is not known whether the model ocean underestimates (or overestimates, although that is less likely) natural variability [12,25]. In the forced model runs, the probability of a trend occurring is inversely proportional to the length, i.e., 20 year countertrends appear much more often than 30 year ones (see Table 3). But in a stationary climate it is not always true that shorter countertrends appear more frequently; it depends on the frequency spectrum. The more 'red' the spectrum (i.e., the stronger the autocorrelation), the more likely it is that a longer period with a trend will occur. If the spectrum is sufficiently 'red', longer periods with a consistent trend can occur more frequently than short ones [19]. In the model this does not occur, and there is a clear association of longer trends with external forcing [16], but it is not possible to know with certainty whether the model's internal variability has a less 'red' spectrum than the real ocean. At some locations, such as BATS in the subtropical Atlantic, the model's variance spectrum is red, and longer periods with consistent trends occur more frequently than shorter ones in the unforced control (not shown). In the 250 years of historical + RCP simulation, however, the anthropogenic trend is large

relative to the natural variability, so trends counter to the forced response are rare. This is at the heart of the detection problem: the instrumental record is gradually becoming long enough to resolve interdecadal variability, but there is already a large and growing anthropogenic component. As ocean time-series data accumulate, understanding of the spectrum of variability will increase, but separating out the anthropogenic component will remain extremely challenging.

Using the 2σ YOE estimated here as a benchmark (Figure 5), the last-zero-crossing for a 20-year trend (LZC_{20}) is a conservative estimate of emergence, while LZC_{30} has an early-emergence bias (Figure 6). Whether a particular length scale for this analysis is a conservative criterion depends on the specific data sets employed and can not readily be generalized across fields or emissions scenarios (although in these results there are consistent relationships among LZC_{20}, LZC_{30} and YOE for different data fields). In general, the faster the anthropogenic signal grows, the stronger the tendency for the LZC to give early emergence. This is a simple function of the signal-to-noise ratio with a nonlinearly growing signal ($d^2y/dt^2>0$) and a constant amount of noise. A noise-driven countertrend that exceeds the forced component is much more likely to occur early in the experiment, when dy/dt is small. Similar logic explains why the criterion becomes more conservative with redder noise. The greater the probability that a trend will occur over a period of e.g. 30 years, the more likely it is that a short-term trend counter to the anthropogenically forced trend will occur in the forced simulation after the 'true' emergence point has been passed. For the simulations considered here, the appropriate period appears to be about 25 years (Figure 6).

This analysis - using climate model simulations with future emissions scenarios - obviously benefits from hindsight that can never be available in ocean observations. If a 30 year secular trend is observed, and in the 31st year it does not reverse, nor in the 32nd, the observer is inclined to believe that a long-term trend is present. It is not the presence of a positive trend that is diagnostic, but the absence of counter-trends over shorter periods. Because in observations we lack the hindsight available in models, it is impossible to say with certainty how long a time since LZC is required to diagnose emergence, but model simulations can help us to estimate the probability of a counter-trend emerging in the future. In most locations it is unlikely that counter-trends will ever cease for periods less than about 15 years, even under very strong anthropogenic forcing, although the statistics are likely to be a latitude dependent [27]. For the stations and data fields examined here, LZC is a fairly reliable diagnostic of emergence for a period between 20 and 30 years, with the former being too conservative and the latter giving many false positives (Figure 6).

These results suggest that it will be difficult to detect an anthropogenic trend in localized ocean time-series data with confidence. The existence of a trend on timescales longer than interdecadal does not necessarily imply that it is anthropogenic, but as time series extend beyond the interdecadal range opportunities for detection and attribution will arise. Ultimately the case for anthropogenic forcing will have to rest on a

mechanistic understanding of the underlying processes [33]. It has only recently become possible to conduct an analysis of the kind presented here even with models, and the models will continue to improve. At the same time, ocean time series will be extended with modern methods that were novel when the time series began but are now mature and operationalized. Increased computational power also makes possible high resolution regional models that can aid in the interpretation of observations and elucidation of mechanisms.

Acknowledgments

Nathan Bindoff, Nathan Gillett, Debby Ianson, Slava Kharin and an anonymous reviewer made useful comments on an earlier draft of this manuscript.

Author Contributions

Conceived and designed the experiments: JC. Analyzed the data: JC. Wrote the paper: JC.

References

1. Andrews O, Bindoff N, Halloran P, Ilyina T, Le Quéré C (2013) Detecting an external influence on recent changes in oceanic oxygen using an optimal fingerprinting method. Biogeosciences 10: 1799–1813.
2. Armstrong R (1994) Grazing limitation and nutrient limitation in marine ecosystems - steady-state solutions of an ecosystem model with multiple food-chains. Limnol Oceanogr 39: 597–608.
3. Arora V, Boer G, Christian J, Curry C, Denman K, et al. (2009) The effect of terrestrial photosynthesis down regulation on the twentieth-century carbon budget simulated with the CCCma Earth System Model. J Clim 22: 6066–6088.
4. Arora V, Scinocca J, Boer G, Christian J, Denman K, et al. (2011) Carbon emission limits required to satisfy future representative concentration pathways of greenhouse gases. Geophys Res Lett 38 doi:10.1029/2010GL046270
5. Barnett T, Pierce D, Schnur R (2001) Detection of anthropogenic climate change in the world's oceans. Science 292: 270–274.
6. Barnett T, Zwiers F, Hegerl G, Allen M, Crowley T, et al. (2005) Detecting and attributing external influences on the climate system: a review of recent advances. J Clim 18: 1291–1314.
7. Bates N (2007) Interannual variability of the oceanic CO_2 sink in the subtropical gyre of the North Atlantic Ocean over the last 2 decades. J Geophys Res 112 doi:10.1029/2006JC003759
8. Beaulieu C, Henson S, Sarmiento J, Dunne J, Doney S, et al. (2013) Factors challenging our ability to detect long-term trends in ocean chlorophyll. Biogeosciences 10: 2711–2724.
9. Bindoff NL, Stott PA, AchutaRao KM, Allen MR, Gillett N, et al. (2013) Detection and attribution of climate change: from global to regional. In: Stocker, TF, Qin D, Plattner G-K, Tignor M, Allen SK, et el., editors. Climate Change 2013: The Physical Science Basis.Contribution of Working Group I to the Fifth Assessment Report of the Intergovernmental Panel on Climate Change. Cambridge: Cambridge University Press. pp. 867–952.
10. Chavez F, Ryan J, Lluch-Cota S, Niquen M (2003) From anchovies to sardines and back: Multidecadal change in the Pacific Ocean. Science 299: 217–221.
11. Chou S, Nelkin E, Ardizzone J, Atlas R, Shie C (2003) Surface turbulent heat and momentum fluxes over global oceans based on the Goddard satellite retrievals, version 2 (GSSTF2). J Clim 16: 3256–3273.
12. Christian J, Arora V, Boer G, Curry C, Zahariev K, et al. (2010) The global carbon cycle in the Canadian Earth System Model (CanESM1): Preindustrial control simulation. J Geophys Res 115 doi:10.1029/2008JG000920
13. Deutsch C, Brix H, Ito T, Frenzel H, Thompson L (2011) Climate-forced variability of ocean hypoxia. Science 333: 336–339.
14. Dore J, Lukas R, Sadler D, Church M, Karl D (2009) Physical and biogeochemical modulation of ocean acidification in the central North Pacific. Proc Natl Acad Sci USA 106: 12235–12240.
15. Durack P, Wijffels S, Matear R (2012) Ocean salinities reveal strong global water cycle intensification during 1950 to 2000. Science 336: 455–458.
16. Easterling D, Wehner M (2009) Is the climate warming or cooling? Geophys Res Lett 36 doi:10.1029/2009GL037810
17. Flato G, Marotzke J, Abiodun B, Braconnot P, Chou SC, et al. (2013) Evaluation of climate models. In: Stocker, TF, Qin D, Plattner G-K, Tignor M, Allen SK, et el., editors. Climate Change 2013: The Physical Science Basis. Contribution of Working Group I to the Fifth Assessment Report of the Intergovernmental Panel on Climate Change. Cambridge: Cambridge University Press. pp. 741–866.
18. Friedrich T, Timmermann A, Abe-Ouchi A, Bates N, Chikamoto M, et al. (2012) Detecting regional anthropogenic trends in ocean acidification against natural variability. Nat Clim Chang 2: 167–171.
19. Hasselmann K (1976) Stochastic climate models. 1. Theory. Tellus 28: 473–485.
20. Hasselmann K (1993) Optimal fingerprints for the detection of time-dependent climate-change. J Clim 6: 1957–1971.
21. Hegerl G, Hasselmann K, Cubasch U, Mitchell J, Roeckner E, et al. (1997) Multi-fingerprint detection and attribution analysis of greenhouse gas, greenhouse gas-plus-aerosol and solar forced climate change. Clim Dyn 13: 613–634.
22. Hegerl G, Zwiers F, Tebaldi C (2011) Patterns of change: whose fingerprint is seen in global warming? Environ Res Lett 6 doi:10.1088/1748-9326/6/4/044025
23. Helm K, Bindoff N, Church J (2010) Changes in the global hydrological-cycle inferred from ocean salinity. Geophys Res Lett 37 doi:10.1029/2010GL044222
24. Henson S, Sarmiento J, Dunne J, Bopp L, Lima I, et al. (2010) Detection of anthropogenic climate change in satellite records of ocean chlorophyll and productivity. Biogeosciences 7: 621–640.
25. Huybers P, Curry W (2006) Links between annual, Milankovitch and continuum temperature variability. Nature 441: 329–332.
26. Karl D, Letelier R, Hebel D, Tupas L, Dore J, et al. (1995) Ecosystem changes in the North Pacific subtropical gyre attributed to the 1991–92 El-Niño. Nature 373: 230–234.
27. Mahlstein I, Knutti R, Solomon S, Portmann R (2011) Early onset of significant local warming in low latitude countries. Environ Res Lett 6 doi:10.1088/1748-9326/6/3/034009
28. Moore J, Doney S, Kleypas J, Glover D, Fung I (2002) An intermediate complexity marine ecosystem model for the global domain. Deep Sea Res Part 2 Top Stud Oceanogr 49: 403–462.
29. Mora C, Frazier A, Longman R, Dacks R, Walton M, et al. (2013) The projected timing of climate departure from recent variability. Nature 502: 183–187.
30. Moss R, Edmonds J, Hibbard K, Manning M, Rose S, et al. (2010) The next generation of scenarios for climate change research and assessment. Nature 463: 747–756.
31. Muir L, Brown J, Risbey J, Wijffels S, sen Gupta A (2013) Determining the time of emergence of the climate change signal at regional scales. CAWCR Res Lett 10: 8–19.
32. Najjar R, Orr J (1998), Design of OCMIP-2 simulations of chlorofluorocarbons, the solubility pump and common biogeochemistry. http://ocmip5.ipsl.fr/documentation/OCMIP/phase2/simulations/Biotic/HOWTO-Biotic.html. Accessed 2014 Sep 19.
33. Rosenzweig C, Karoly D, Vicarelli M, Neofotis P, Wu Q, et al. (2008) Attributing physical and biological impacts to anthropogenic climate change. Nature 453: 353–357.
34. Santer BD, Wigley TML, Barnett TP, Anyamba E (1996) Detection of climate change and attribution of causes. In: Houghton JY, Meira Filho LG, Callander BA, Harris N, Kattenberg A, et al., editors. Climate Change 1995: The Science of Climate Change. Contribution of Working Group I to the Second Assessment Report of the Intergovernmental Panel on Climate Change. Cambridge: Cambridge University Press. pp. 406–443.
35. Santer B, Mikolajewicz U, Bruggemann W, Cubasch U, Hasselmann K, et al. (1995) Ocean variability and its influence on the detectability of greenhouse warming signals. J Geophys Res 100: 10693–10725.

36. Sarmiento J, Dunne J, Gnanadesikan A, Key R, Matsumoto K, et al. (2002) A new estimate of the CaCO₃ to organic carbon export ratio. Global Biogeochem Cycles 16 doi:10.1029/2002GB001919

37. Taucher J, Oschlies A (2011) Can we predict the direction of marine primary production change under global warming? Geophys Res Lett 38 doi:10.1029/2010GL045934

38. Taylor K (2001) Summarizing multiple aspects of model performance in a single diagram. J Geophys Res 106: 7183–7192.

39. Taylor K, Stouffer R, Meehl G (2012) An overview of CMIP5 and the experiment design. Bull Am Meteorol Soc 93: 485–498.

40. Wang M, Overland J, Bond N (2010) Climate projections for selected large marine ecosystems. J Mar Syst 79: 258–266.

41. Zahariev K, Christian J, Denman K (2008) Preindustrial, historical, and fertilization simulations using a global ocean carbon model with new parameterizations of iron limitation, calcification, and N₂ fixation. Prog Oceanogr 77: 56–82.

16

Defining Mediterranean and Black Sea Biogeochemical Subprovinces and Synthetic Ocean Indicators Using Mesoscale Oceanographic Features

Anne-Elise Nieblas[1]*, **Kyla Drushka**[2], **Gabriel Reygondeau**[3], **Vincent Rossi**[4], **Hervé Demarcq**[5], **Laurent Dubroca**[6], **Sylvain Bonhommeau**[1]

1 Unité Mixte Recherche Ecosystèmes Marins Exploités 212, Institut Français de Recherche pour l'Exploitation de la Mer (IFREMER), Sète, France, **2** Applied Physics Laboratory, University of Washington, Seattle, Washington, United States of America, **3** Center for Macroecology, Evolution and Climate, National Institute for Aquatic Resources, Technical University of Denmark (DTU Aqua), Charlottenlund, Copenhagen, Denmark, **4** Instituto de Física Interdisciplinary Sistemas Complejos, Institute for Cross-Disciplinary Physics and Complex Systems, (CSIC-UIB), Campus Universitat de les Illes Balears, Palma de Mallorca, Spain, **5** Unité Mixte de Recherche Ecosystèmes Marins Exploités 212, Institut de Recherche pour le Développement (IRD), Sète, France, **6** European Commission, Joint Research Center, Institute for Environment & Sustainability, Water Resources, Ispra, Italy

Abstract

The Mediterranean and Black Seas are semi-enclosed basins characterized by high environmental variability and growing anthropogenic pressure. This has led to an increasing need for a bioregionalization of the oceanic environment at local and regional scales that can be used for managerial applications as a geographical reference. We aim to identify biogeochemical subprovinces within this domain, and develop synthetic indices of the key oceanographic dynamics of each subprovince to quantify baselines from which to assess variability and change. To do this, we compile a data set of 101 months (2002–2010) of a variety of both "classical" (i.e., sea surface temperature, surface chlorophyll-a, and bathymetry) and "mesoscale" (i.e., eddy kinetic energy, finite-size Lyapunov exponents, and surface frontal gradients) ocean features that we use to characterize the surface ocean variability. We employ a k-means clustering algorithm to objectively define biogeochemical subprovinces based on classical features, and, for the first time, on mesoscale features, and on a combination of both classical and mesoscale features. Principal components analysis is then performed on the oceanographic variables to define integrative indices to monitor the environmental changes within each resultant subprovince at monthly resolutions. Using both the classical and mesoscale features, we find five biogeochemical subprovinces for the Mediterranean and Black Seas. Interestingly, the use of mesoscale variables contributes highly in the delineation of the open ocean. The first axis of the principal component analysis is explained primarily by classical ocean features and the second axis is explained by mesoscale features. Biogeochemical subprovinces identified by the present study can be useful within the European management framework as an objective geographical framework of the Mediterranean and Black Seas, and the synthetic ocean indicators developed here can be used to monitor variability and long-term change.

Editor: Silvia Mazzuca, Università della Calabria, Italy

Funding: A. -E. N. was supported by a joint grant from France Filière Pêche (http://www.francefilierepeche.fr/) and Institut Français de Recherche pour l'Exploitation de la Mer (http://wwz.ifremer.fr/). V. R. acknowledges support from Ministerio de Ciencia e Innovación (http://www.idi.mineco.gob.es/portal/site/MICINN/) and FEDER (http://www.europe-en-france.gouv.fr/Configuration-Generale-Pages-secondaires/FEDER) through the ESCOLA project (CTM2012-39025-C02-01). S. B. was supported by the Agence nationale de la recherche SEAS-ERA MERMAID project (ANR-12-SEAS-0003) and the French National Research project INSU-MERMEX. The funders had no role in study design, data collection and analysis, decision to publish, or preparation of the manuscript.

Competing Interests: The authors have declared that no competing interests exist.

* Email: anne.elise.nieblas@gmail.com

Introduction

Growing pressure on the European marine environment has led to an increasing demand for comprehensive evaluation and monitoring programs [1–4]. The Mediterranean and Black Seas are ecologically- and economically-important semi-enclosed seas characterized by highly specific biogeochemcial, oceanographic, and environmental conditions that have resulted in pronounced endemism of exploited marine species [5–7]. The Mediterranean Sea is commonly divided into two basins, east and west, which each have specific hydrological conditions and marked seasonal cycles [8]. Recently, the International Panel on Climate Change has designated the Mediterranean as one of the most perturbed marine ecosystems of the global ocean, as both deep and surface environments show significant change in the open seas, coastal, benthic and neritic areas [9–11]. In addition, it is undergoing increasing anthropogenic pressure, including pollution, overfishing, and habitat loss via coastal development [1,7,12].

In this context, the European Union has recently adopted the Integrated Maritime Policy framework for the protection of European Seas; the primary objective of which is to achieve environmentally healthy waters by 2020 [1–4]. The first step toward the goal of healthy waters and the aim of this study is to identify an objective spatial partitioning in the Mediterranean and Black Seas, where environmental conditions are homogeneous, to act as a framework for marine zoning [13,14], for ecological management [15,16], as well as to determine baseline conditions which can then be used to effectively monitor variability and change.

Marine bioregionalization aims to identify unique and homogeneous biogeochemical partitions delineated by observable frontiers, such as frontal structures. This discipline, recently redefined by [17], is based on objective statistical methodologies and has been applied in several regions of the global ocean at several different scales [18–21]. However, owing to the complex hydrodynamics [22,23] and the important influence of mesoscale activity on biogeochemcial processes [24–26], bioregionalization of the Mediterranean and Black Seas remains difficult.

Marine bioregionalizations are classically performed on oceanographic features that are thought to be representative of the oceanographic and biogeochemical structure of a region; for example, sea surface temperature (SST), bathymetry and surface chlorophyll-a (chl) [20,27,28]. However, mesoscale processes must also be important for defining biogeochemical partitions as they are known to impact ocean productivity, including the spatial distribution and stocks of chlorophyll-a [29,30]; and basin-scale circulation and its mesoscale variability have been shown to be crucial in delineating hydrodynamical regions [31].

Previous studies have used single or multivariate analyses to derive regions of similar features in the Mediterranean Sea, including classical oceanographic indicators (i.e., chl [8]; SST, chl, sea surface salinity, and bathymetry [32]), bio-physical indicators, such as Ekman pumping, nutrient concentration, euphotic depth, and stratification [32]; and exploited fish distributions and biodiversity [7]. Recently, the Mediterranean Sea was subdivided into several hydrodynamical provinces delineated by multi-scale oceanic frontal structures in order to assess the ecological connectivity of the whole basin [31].

In this study, we derive objective biogeochemical subprovinces (sensu [33]) of the Mediterranean and Black Seas based on multivariate analyses of classical oceanographic features (SST, chl, and bathymetry), mesoscale features (eddy kinetic energy (EKE), SST and chl surface fronts, finite-size Lyapunov exponents (FSLE), and the Okubo-Weiss (OW) parameter), and a combination of both classical and mesoscale features. We also quantify the stability of the boundaries between the biogeochemical subprovinces in time and space. Synthetic oceanographic indices for the subprovinces are then extracted to act as baseline indicators, using principal components analysis (PCA), similar to the multivariate ocean-climate indices recently developed by [34]. Finally, we examine the temporal variability of these indices and their relationships with large-scale climate indices. The biogeochemical subprovinces identified in this study and the time series of their synthetic indicators could become important tools within the European management framework for assessing the environmental variability and change within the Mediterranean Sea.

Materials and Methods

Data

We used daily 4-km, version 5 Advanced Very High Resolution Radiometer pathfinder SST (1982–2012) available at http://www.nodc.noaa.gov/sog/pathfinder4km/. Chl data were taken from the National Aeronautics and Space Administration's daily 4-km level-3 Moderate Resolution Imaging Spectroradiometer daily data set (2002–2010) available at http://oceancolor.gsfc.nasa.gov/. We extracted weekly 1/3° (i.e., about 33 km at these latitudes) Ssalto/Duacs sea level anomalies and geostrophic velocity anomalies (u,v) computed and distributed by Aviso (1992–2012), with support from the Centre National d'Études Spatiales (http://www.aviso.oceanobs.com/duacs/). The bathymetry of the Mediterranean basin was extracted from the ETOPO1 database hosted on the National Oceanic and Atmospheric Administration's website at ~4 km resolution using the *getNOAA.bathy* function (marmap package, http://cran.r-project.org/). For consistency between variables, analyses were performed for data between May 2002 and November 2010, totaling 101 months.

Oceanographic indices

Several features were derived by further processing the remotely-sensed data. SST and chl fronts were computed with the gradient method, using a common sobel operator (e.g., [35,36]). These continuous values indicate the frontal intensity between water masses.

Using geostrophic velocity anomalies, we calculated several indicators of mesoscale ocean features. These data were used to derive backward-calculated FSLEs, which measure the horizontal mixing and dispersion in the ocean [37] and help to detect mesoscale Lagrangian coherent structures of ecological significance (e.g., [38]). FSLEs are defined as $\lambda(\mathbf{x},t,\delta_0,\delta_f) = (1/\tau)\log(\delta_0/\delta_f)$ where $\lambda(\mathbf{x},t,\delta_0,\delta_f)$ is the FSLE at position $\mathbf{x}$ and time t with an initial separation distance from $\mathbf{x}$ of δ_0 and final separation distance from $\mathbf{x}$ of δ_f. Here, we assign δ_0 to be 0.04 degrees and δ_f to be 0.6 degrees, and a time interval, τ, of 200 days, following the FSLE parameterizations of the Center for Topographic studies of the Ocean and Hydrosphere (http://ctoh.legos.obs-mip.fr/products/submesoscale-filaments/fsle-description), allowing us to detect mesoscale structures of <100 km, an appropriate scale for these seas [37]. Geostrophic velocity anomalies were also used to compute EKE, $(u^2+v^2)/2$, which is an indicator of the intensity of the eddy activity. Finally, we use geostrophic velocity anomalies to compute the OW parameter [39,40], $W = s_n^2+s_s^2+\omega^2$, where s_n and s_s are the normal and shear components of strain, and ω is the relative vorticity. This parameter is used to identify regions of high vorticity (W<0), which are likely related to the cores of mesoscale ocean eddies [41].

Monthly mean time series (2002–2010) were derived for all variables (except bathymetry) for the region including the Mediterranean and the Black Seas (30°N to 47°N, −6°E to 42°E) and regridded at 4 km resolution. We natural log transformed chl, chl fronts and bathymetry in order to stabilize their variance as their values can span several orders of magnitude.

Spatio-temporal multivariate k-means cluster analysis

Multivariate arrays were created from time-averages of the monthly scaled oceanographic indices (Figure 1) and combined into "classical" (i.e., SST, chl, and bathymetry), "mesoscale" (i.e., FSLE, OW, EKE, and SST and chl surface fronts), and "full" arrays (i.e., all features). After initial tests, k-means (*kmeans*, stats package, http://cran.r-project.org/; [42]) was determined to be the most robust cluster analysis algorithm to objectively classify

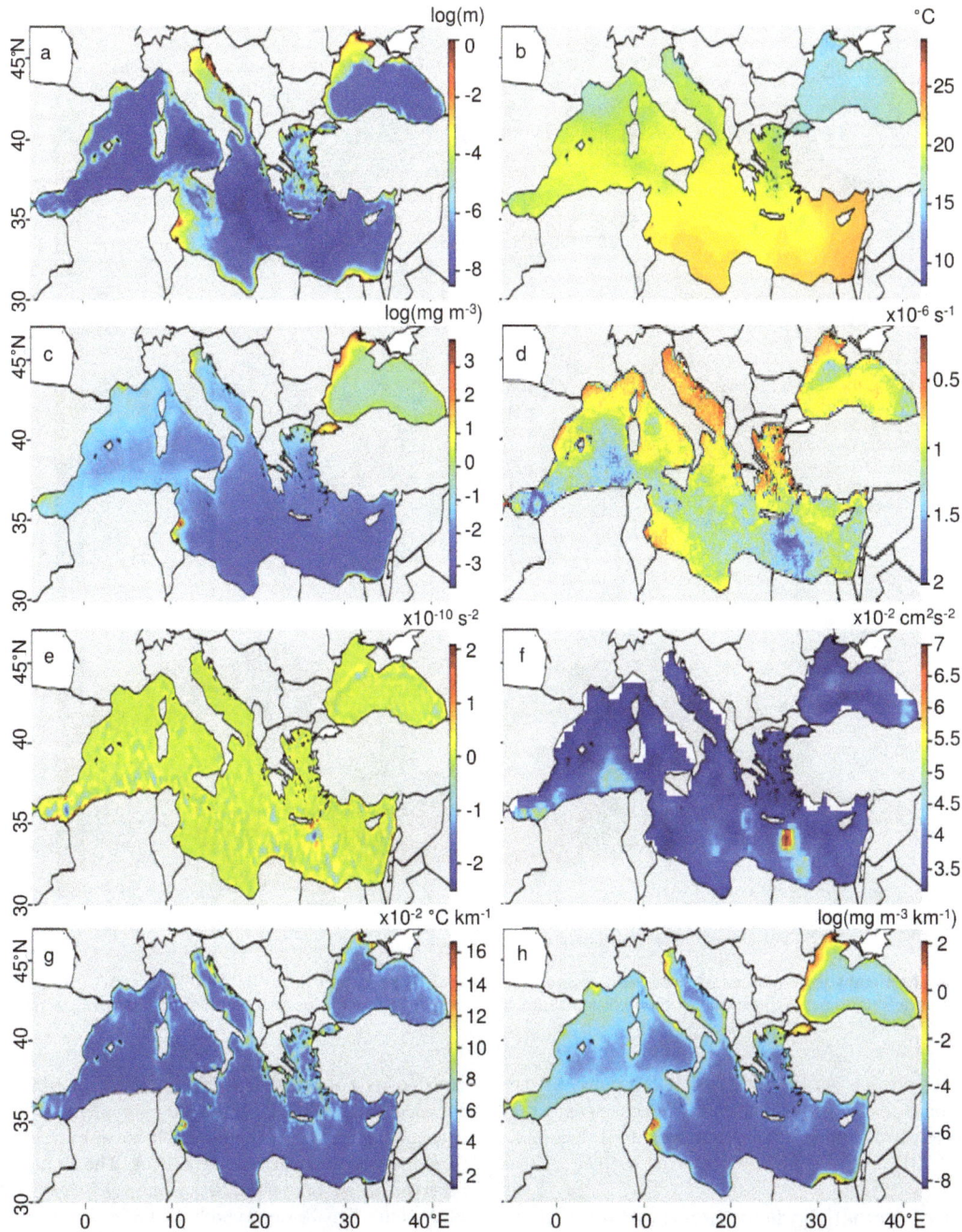

Figure 1. Time-averages of all oceanographic variables collected for the Mediterranean Sea. Variables include a) natural log-transformed bathymetry, b) sea surface temperature (SST), c) natural log-transformed chlorophyll-*a* (chl), d) finite-size Lyapunov exponents (FSLE), e) Okubo-Weiss parameter (OW), f) eddy kinetic energy (EKE), g) SST surface frontal gradients, and h) natural log-transformed chl surface frontal gradients.

biogeochemical subprovinces. This partitioning method, using Euclidean distances, assigns data points to k clusters and minimizes the sum of squares between the data points to cluster centre. With this algorithm, k must be defined *a priori*. In order to define k, we bootstrap (1000 times) k between 2 and 30. The between-clusters sum of squares is then divided by the total sum of squares to find the explained sum of squares. Arbitrary 1% and 5% thresholds are defined (Figure S1 in File S1), which we used to define the optimal k for the three multivariate arrays (Table 1), whereby the explained sum of squares for each additional k increases by less than 1% and 5%, respectively. K-means analyses were then

performed on each array using the optimal k for both threshold levels (1%; Figure S2 in File S1 and 5%; Figure 2). The resultant clusters were defined as the biogeochemical subprovinces of the Mediterranean and Black Seas as a subdivision of the Mediterranean provinces defined by [33].

To investigate the spatial stability of the subprovinces through time, we used the optimal k values found for each of the three multivariate arrays for both the 1% and 5% threshold levels, and performed a k-means analysis on each of the multivariate arrays for every month of the data set (n = 101 months). Then, based on an adaptation of the effectiveness test implemented by [43], the

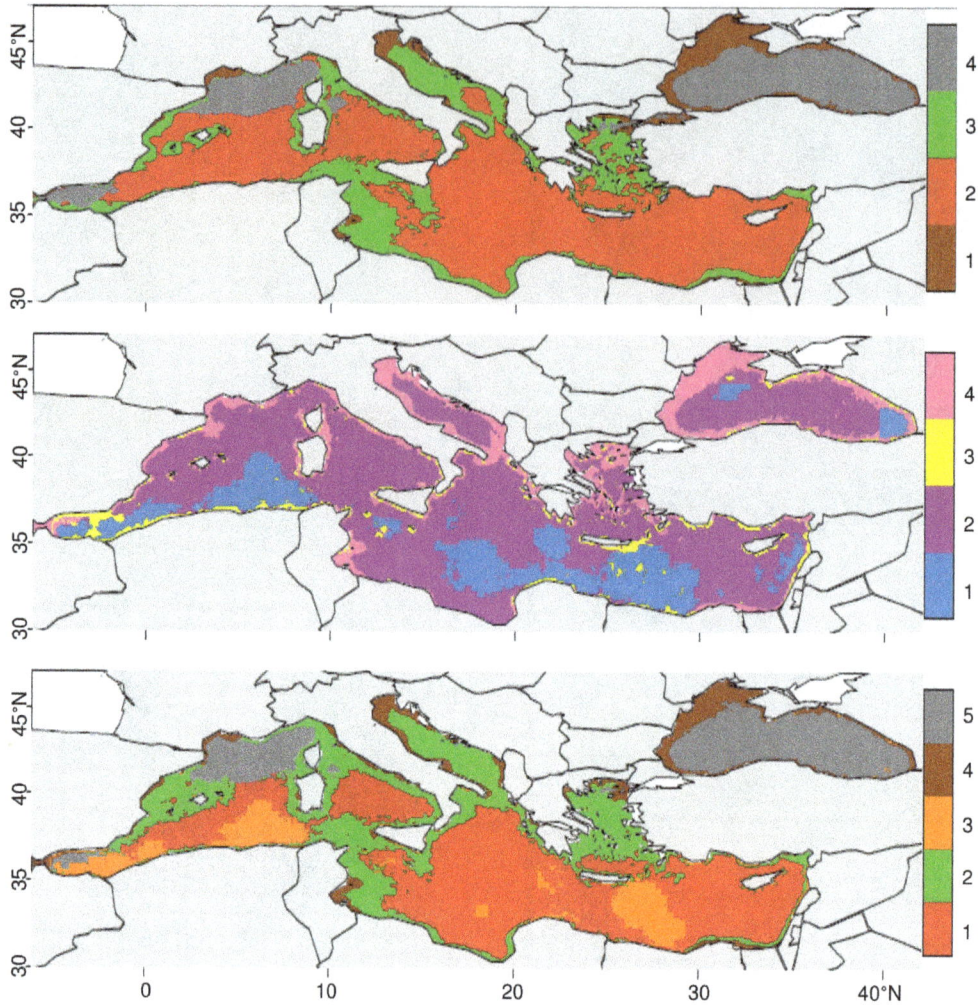

Figure 2. Biogeochemical subprovinces of the Mediterranean and Black Seas. Subprovinces for the (a) "classical", (b) "mesoscale", and (c) "full" multivariate arrays using a 5% threshold for the explained sum of squares to define the optimal number of subprovinces (see text).

temporal stability of each geographical cell is computed as the percentage of time that a geographical cell is considered as a boundary between two clusters at each temporal step (Figure 3, Figure S3 in File S1).

Development of synthetic indices through PCA

In order to develop synthetic indices of the oceanographic indicators for each biogeochemical subprovince, we extracted the scaled and centered monthly time series of each oceanographic variable (except bathymetry) for each pixel within each biogeochemical subprovince. Although bathymetry is important for determining the biogeochemical subprovinces, it does not vary in time and was not included in the PCA. The strong seasonal cycle observed in all time-series was removed before performing the PCA as this signal swamps both the lower- and higher-frequencies of the time series (e.g., [44]). We then performed a PCA for each biogeochemical subprovince with an individual being the monthly value of each oceanographic variable for each pixel. We used the

Table 1. Optimal number of clusters, k, for each multivariate array for the 1% and 5% threshold levels obtained by bootstrapping 1000 times the k-means analysis on k between 2 and 30.

Array	Threshold	
	1%	5%
Classical	9	4
Mesoscale	14	4
Full	13	5

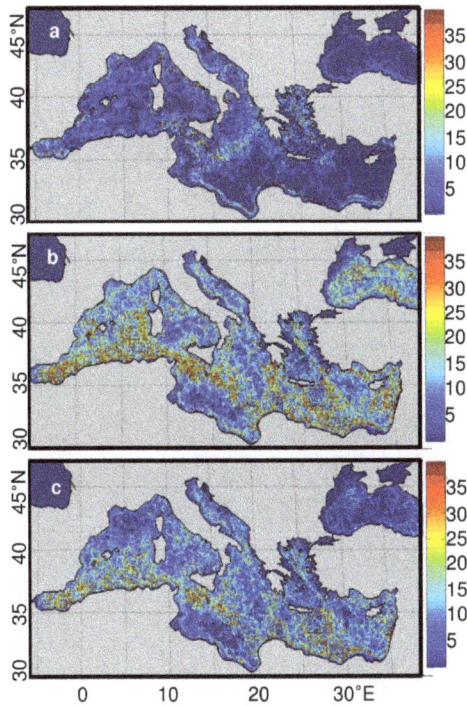

Figure 3. Spatial stability of the borders of biogeochemical subprovinces. Stability plots for the (a) classical, (b) mesoscale, and (c) full multivariate arrays. K-means analyses, using the k found in the time-averaged analyses (Table 1), are performed on the multivariate arrays at monthly time steps for the 101 months of the data set and using a 5% threshold of the explained sum of squares to define the optimal number of subprovinces (see text). Spatial stability is represented as the percentage of time that a boundary of the biogeochemical subprovinces is found at a particular pixel over the 101 months of the data set. Red colors indicate stable borders.

common cutoff of eigenvalues >1 to retain the unrotated principal components (PCs) (Table S1). We then took the monthly mean of the retained PCs over all the pixels, and used these as the synthetic indices of each biogeochemical subprovince.

Finally, we investigated the mode of temporal variability of these synthetic indices. Spectra were calculated to show the variability of each time-series. Lagged correlations were then investigated between time-series and monthly anomalies of four independent large-scale climate indices known to influence Mediterranean Sea dynamics [45,46]: North Atlantic Oscillation (NAO), the East Atlantic pattern (EA), the East Atlantic-West Russia pattern (EAWR), and the Scandinavian pattern (SCAND). These indices were computed by the National Oceanic and Atmospheric Administration/Climate Prediction Center.

Results

At the 5% threshold level, we find four subprovinces for the classical oceanographic features, four subprovinces for the mesoscale features and five subprovinces for the full combination of features (Table 1, Figure 2). Overall, the "classical" array (i.e., defined using the set of classical features) has the most stable boundaries in time and space (Figure 3a), while the boundaries for the "mesoscale" array (i.e., defined using the set of mesoscale features only) are highly variable (Figure 3b), as are the boundaries of the "full" array (i.e., defined using both classic and mesoscale features) (Figure 3c). This indicates that the apparent stability

found for the classical array is not representative of the "true" in-situ variability of the oceanic environment. The 5% threshold identifies fewer biogeochemical subprovinces than the 1% threshold (Table 1; Figure 2; Figure S2 in File S1), which, in addition to higher stability (Figure 3; Figure S3 in File S1), makes them easier to monitor in a management context. The full array at the 5% threshold was finally determined to be the most useful for management purposes, as the subprovinces are realistic and inclusive of both classical and mesoscale features (Figure 2, 3, 4). We perform a PCA for the subprovinces defined for the 5% threshold of the full array to derive synthetic time series (Table S1, Figure 5). The first two PCs are retained for full subprovinces 1–3 (explaining up to 39% and 21% of the variance for PC1 and PC2, respectively), and the first three PCs are retained for full subprovinces 4–5 (explaining up to 27%, 20%, and 18% of the variance for PC1, PC2 and PC3, respectively). We find strong low frequency (interannual) energy for the first two PCs in all full subprovinces (Figure 6), with the PCs of subprovince 5 being particularly different from the rest, especially PC2. We do not find any meaningful correlations to large-scale climate indices for any of the retained PCs (Table S2).

The stability of the boundaries for a given biogeochemical subprovince indicates whether the subprovince is representative of the local hydrological and biogeographical conditions, with higher stability giving greater credence. We found that the stability of the borders of the biogeochemical subprovinces defined by the 5% threshold (Figure 3) were more stable in terms of time and space than those of the 1% threshold (Figure S3 in File S1). The boundaries of the biogeochemical subprovinces of the classical array are very clearly defined and spatially stable through time (Figure 3a), with the greatest stability of boundaries found at the coastal isobaths (Figure 1a). Greater variability in the boundaries occurs through the Strait of Sicily and in the Aegean Seas, which correspond to high spatial bathymetric variability. The variability in biogeochemical subprovince boundaries observed in the Ionian Sea appears to correspond to a latitudinal transition of SST (Figure 1b). The boundaries for the mesoscale array are extremely variable (Figure 3b), with boundaries shifting significantly for each month of the data set. Especially high variability occurs in the southern parts of both the eastern and western basins, including the Alboran Sea, Algerian Basin, through the Strait of Sicily and south of Malta, where mesoscale activity is particularly high (Figure 1d,e,f).

Despite this spatial variability, we find that the subprovinces derived from the classical and mesoscale features are often grouped in the same manner. For example, the coastal Gulf of Lions, the northern Adriatic Sea, the Aegean Sea, the coastal Gulf of Gabes, the coast near the Suez Canal, and the western Black Sea are consistently grouped together in all three multivariate arrays (classical subprovince 1, mesoscale subprovince 4, full subprovince 4; Figure 2). These subprovinces are characterized by shallow bathymetry, generally low but variable SSTs, and high and variable chl (Figure 4a, c; Table 2). In terms of mesoscale features, these subprovinces have relatively consistently high FSLE, indicating low horizontal mixing, low EKE, and high SST and chl frontal gradients (Figure 4b, c; Table 2). Offshore of the Gulf of Lions and the offshore Black Sea are also commonly grouped by the three multivariate arrays (classical subprovince 4, mesoscale subprovince 2, full subprovince 5), and are characterized by deep bathymetry, low and relatively constant SSTs, and high chl (Figure 4a, c; Table 2). This subprovince has high FSLE, again indicating low horizontal mixing, low EKE (Figure 4b, c; Table 2), and high chl frontal gradients for the full subprovince 5 (Figure 4c; Table 2). The Strait of Gibraltar/Alboran Sea is

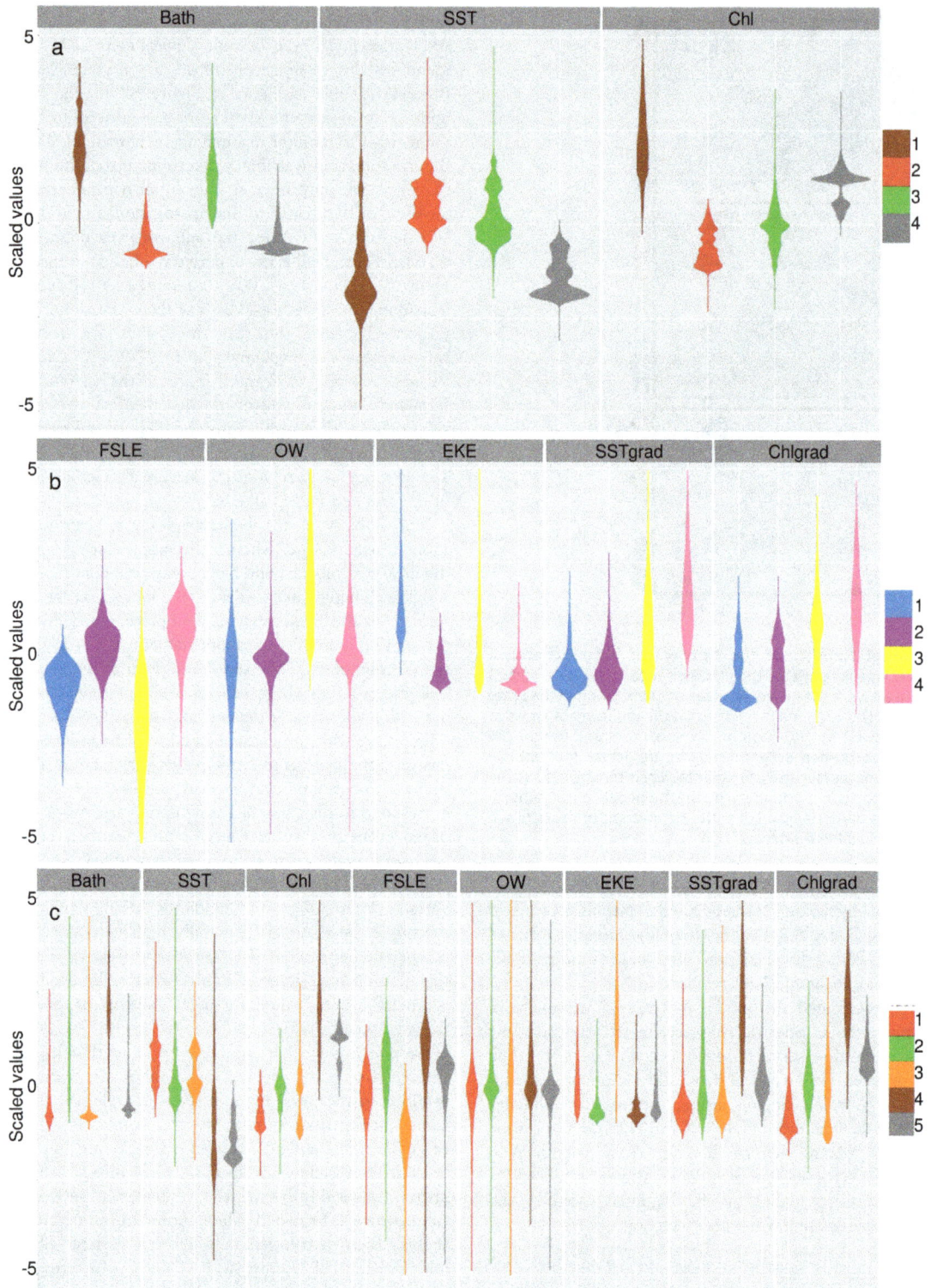

Figure 4. Violin plots of the scaled values for each oceanographic variable. Violin plots for the (a) classical, (b) mesoscale, and (c) full multivariate arrays. Colors indicate biogeochemical subprovinces. The mean of each variable is represented by the bulge in the violin and its variability is indicated by the tails. Here, positive bathymetry is shallow and negative bathymetry is deep. Variable abbreviations are as follows: sea surface temperature (SST), chlorophyll-*a* concentration (chl), finite-sized Lyapunov exponents (FSLE), the Okubo-Weiss parameter (OW), eddy kinetic energy (EKE), and SST and chl frontal gradients (SSTgrad, and chlgrad, respectively). Colorbars represent subprovinces.

grouped mainly by classical subprovinces 3 and 4 (Figure 2a, c), but is more variably characterized by mesoscale features (Figure 1, Figure 4). The majority of the southern parts of the eastern and

western basins are grouped together (classical subprovince 2, mesoscale subprovince 2, full subprovince 1), and are characterized by deeper bathymetry, higher SSTs, low chl, and low

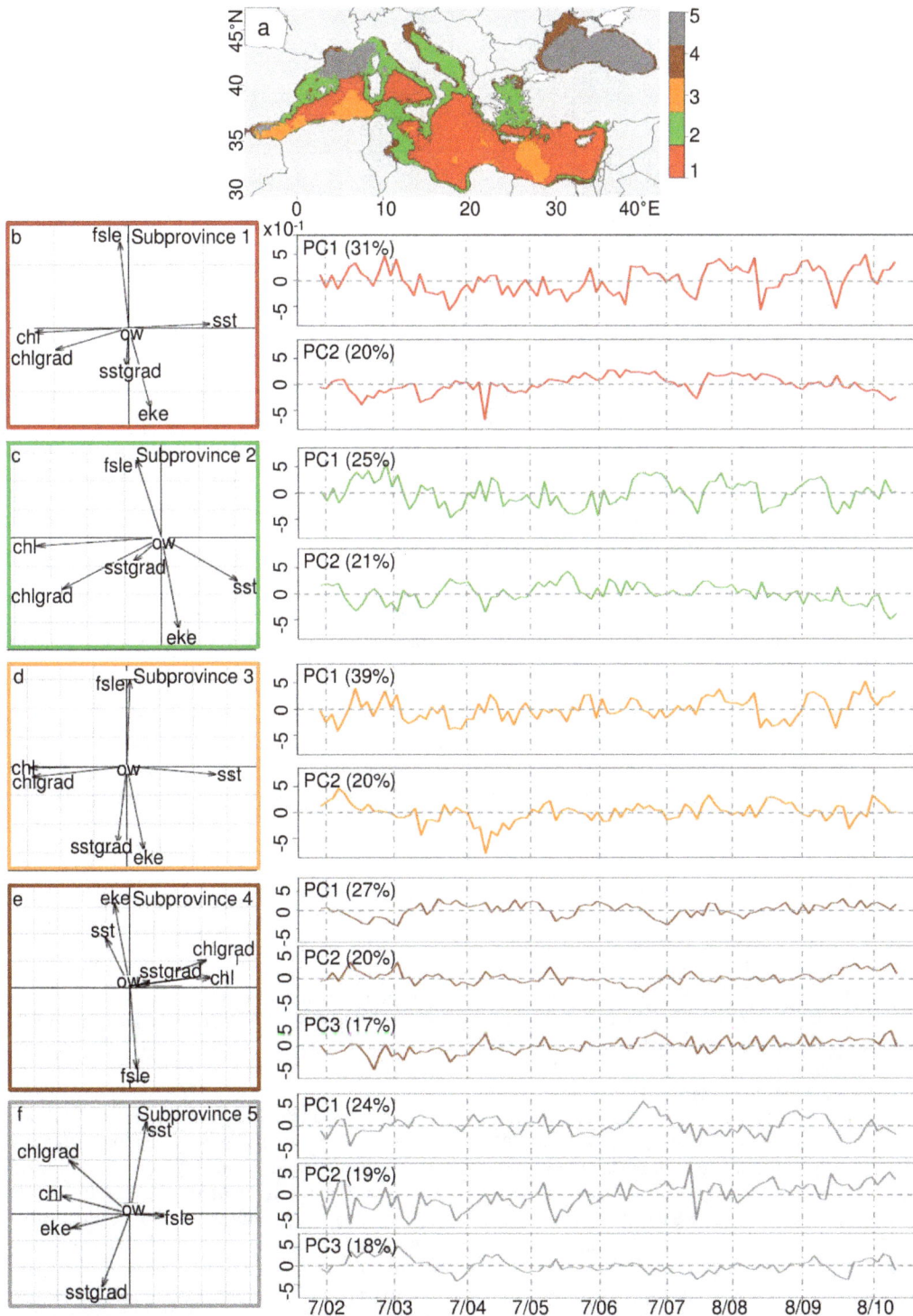

Figure 5. Principal component (PC) analysis for the biogeochemical subprovinces of the Mediterranean Sea. The (a) biogeochemical subprovinces of the Mediterranean Sea, as defined by the 5% threshold for the full multivariate array, and the (b–f) PC analysis arrow plots and time series of the retained PCs for each of the subprovinces, derived from the mean of all pixels in the subprovince for each month of the data set. Arrows that align well with an axis are well-explained by that axis. The longer the arrow, the more it contributes to explaining the variability of an axis. Variable abbreviations are as in Figure 4.

mesoscale activity (Figure 4; Table 2). The eastern and western basin are divided by the Strait of Sicily for the classical and full subprovinces (Figure 2a, c) due to relatively shallow bathymetry (Figure 4a,c; Table 2). Mesoscale subprovince 1 appears related to full subprovince 3, which both occupy the southern basin, and are characterized by low FSLE (strong mixing), low and variable OW (indicative of eddy cores), high EKE, and low SST and chl frontal gradients (Figure 4b; Table 2). This high EKE is especially

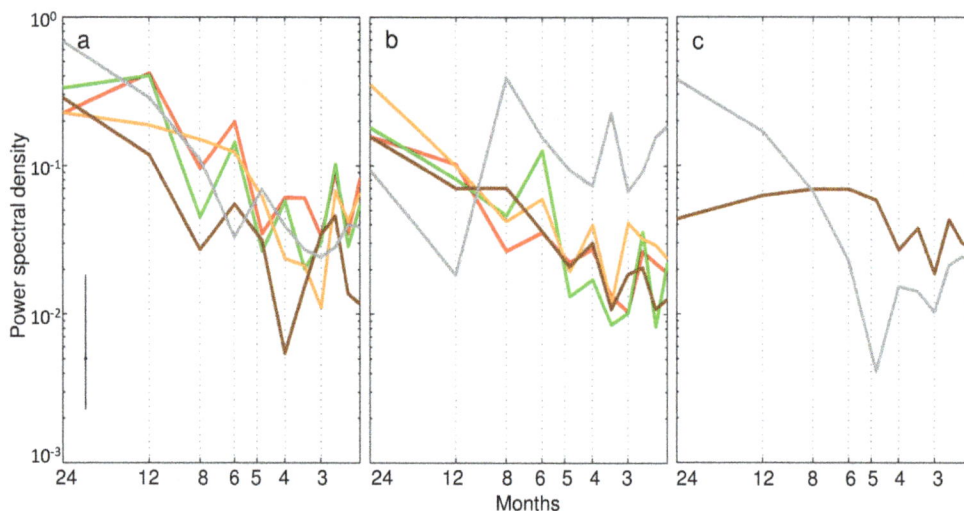

Figure 6. Spectral energy plots. Spectral energy plots for (a) principal component (PC) 1, (b) PC2, and (c) PC3 as retained for the different biogeochemical subprovinces computed for 101 months, having removed the seasonal signal. Peaks are indicated for all PCs at interannual frequencies. The error bar in the bottom-left indicates the 95% significance level.

apparent in full subprovince 3, south of Crete in the Levantine Sea (e.g., the wind-driven Ierapetra anticyclonic gyre), in the Algerian Basin, and in the Alboran Sea (Figure 1f, 2c).

In general, we find that classical variables for the full subprovinces primarily explain the first PC, and mesoscale variables for the full subprovinces generally explain the second PC (Figure 5b–f arrow plots, left panels; Table 3), with full subprovince 5 exhibiting much different patterns for all PCs than the other full subprovinces. For each PC, the variables having a correlation value >0.5 with a particular PC time series are deemed significant (a necessarily subjective cutoff value relevant only to this data set; Table 3). We find that the first two axes are typically divided between classical (chl and SST, as well as chl frontal gradients; PC1), and mesoscale features (typically FSLE and EKE, and SST frontal gradients for subprovince 3; PC2) (Figure 5b–d arrow plots), while OW does not explain any axis (Table 3). Chl and chl frontal gradients are highly correlated (r = 0.92), explaining the alignment of this mesoscale feature with classical features on the first axis. FSLE and EKE are also highly negatively correlated for each subprovince (Figure 5b–d arrow plots). PC1 for subprovinces 1, 2, and 3 are significantly correlated (p<0.001), with relatively high correlation coefficients (p<0.05, r = 0.61 to 0.8; Table S2), indicating that their classical features vary in the same manner. PC2 for subprovinces 1 and 2 are also significantly correlated (p<0.05, r = 0.58; Table S2).

Three axes are retained for subprovinces 4 and 5, as the eigenvalues for their PC3 were >1 (Figure 5e, f). The classical and mesoscale features are not as clearly divided by the retained axes for subprovinces 4 and 5 as they are for the first three subprovinces. The first axis for subprovince 4 is primarily explained by chl and chl frontal gradients and the second axis is explained by FSLE and EKE, similar to subprovinces 1–3, but the third axis is primarily explained by SST (Table 3). Subprovince 5 is even more different as the first and second axes are explained by a mix of classical and mesoscale features: PC1 is primarily explained by chl, chl frontal gradients and EKE, PC2 is explained by SST and SST frontal gradients, and PC3 is explained by FSLE.

The PC time series represent synthetic indices of oceanographic dynamics for each full subprovince (Figure 5b–f, right panels). After removing the dominant seasonal cycle, power spectra reveal

that low-frequency (interannual) variability dominates PC1 for all subprovinces (Figure 6a) as well as PC2 for all subprovinces except subprovince 5 (Figure 6b). Subprovince 5 has distinctive spectral characteristics, with no high-frequency peak for PC1 (Figure 6a) and no decrease in energy at high frequencies for PC2. Instead, PC2 has a strong peak at a period of 8 months (Figure 6b). Subprovince 5 also has a strong low-frequency signal for PC3, but no significant signals are found for PC3 of subprovince 4 (Figure 6c).

We find six out of 48 significant correlations between the retained PCs and large-scale climate indices (r = −0.2 to 0.27 with p-value <0.05). However, this could be due to multiple testing of time series. Since the correlations were low, we did not correct the p-values for multiple testing and we consider that there is no major links between the PCA axes and climatic indices. Lagged correlations between the PCs and the climate indices were also considered but did not render stronger relationships (Table S2).

Discussion

Our results synthetically characterize the hydrodynamics of the Mediterranean and Black Seas, complex and variable oceanic systems that function on multiple temporal and spatial scales [47]. Mesoscale features in the Mediterranean are an important source of variability [48], and we find that they are an important component to include in this bioregionalization. Indeed, the omission of mesoscale features and their temporal variability in such spatial analyses are misleading, as already suggested by [20], who promoted the use of dynamical biogeochemical subprovinces instead of their static equivalents. Classical features here are stable and are clearly representative of the biogeochemistry of the subprovince that they describe. Though highly variable, mesoscale features enable us to further discriminate additional regions. This is especially useful in the open ocean, which appears homogenous when considering only classical variables.

Overall, the biogeochemical subprovinces defined for the full array of variables (Figure 3c) can be organized into four broad categories that compare relatively well with previous studies. Subprovinces 1 and 2 represent the open ocean regions of the southern basin. These two subprovinces are highly correlated in

Table 2. Mean and standard deviations of the environmental variables for each biogeochemical subprovince defined by the 5% threshold of the classical, mesoscale, and full multivariate arrays over the 101 months of the data set, including sea surface temperature (SST, °C), chlorophyll-a concentration (chl; mg m^{-3}), finite-sized Lyapunov exponents (FSLE; s^{-1}), the Okubo-Weiss parameter (OW; s^{-2}), eddy kinetic energy (EKE; cm^2 s^{-2}), and SST and chl frontal gradients (SSTgrad; °C km^{-1}, and chlgrad; mg m^{-3} km^{-1}, respectively).

Subprovince	SST	Chl	Bath	EKE	FSLE	OW	SSTgrad	Chlgrad
Classical								
1	17.04±6.66	0.19±0.24	−56.75±73.12					
2	20.85±4.41	0.15±0.12	−2133.17±945.04					
3	20.19±4.55	0.24±0.26	−203.08±178.70					
4	16.98±6.01	0.65±0.59	−1786.03±658.50					
Mesoscale								
1				0.042±0.012	−1.32e-06±7.01e-07	−7.36e-12±8.69e-11	0.038±0.024	0.005±0.015
2				0.035±0.004	−1.00e-06±4.91e-07	1.44±3.59e-11	0.039±0.025	0.006±0.017
3				0.037±0.008	−1.84e-06±1.47e-06	4.99e-11±8.33e-11	0.052±0.035	0.026±0.094
4				0.034±0.003	−8.69e-07±5.67e-07	9.94e-12±3.34e-11	0.065±0.043	0.086±0.210
Full								
1	21.13±4.38	0.13±0.09	−2152.13±1021.32	0.036±0.005	−1.12e-06±5.5e-07	−7.62e-13±4.87e-11	0.03±0.02	0.003±0.004
2	19.95±4.57	0.25±0.005	−609.96±648.89	0.034±0.003	−9.39e-07±5.6e-07	8.09e-12±3.74 e-11	0.046±0.029	0.008±0.018
3	20.54±4.10	0.20±0.25	−2210.62±962.95	0.044±0.014	−1.57e-06±1.05e-06	4.00e-12±1.078e-10	0.040±0.026	0.008±0.026
4	17.80±6.48	1.91±2.84	−77.36±132.71	0.034±0.003	−8.95e-07±6.95e-07	1.18e-11±3.53e-11	0.072±0.047	0.155±0.279
5	16.87±6.09	0.65±0.49	−1774.19±722.23	0.035±0.005	−9.69e-07±4.66e-07	−3.76e-13±3.82e-11	0.045±0.030	0.017±0.045

Table 3. Correlation matrix of the first four principal components (PC) analyzed for the full multivariate array, as defined by the 5% threshold, including sea surface temperature (SST), chlorophyll-*a* concentration (chl), eddy kinetic energy (EKE), finite-sized Lyapunov exponents (FSLE), the Okubo-Weiss parameter (OW), and SST and chl frontal gradients (SSTgrad and chlgrad, respectively).

Subprovince	PC	Variance explained (%)	Eigenvalues of environmental variables						
			SST	Chl	EKE	FSLE	OW	SSTgrad	Chlgrad
1	1	31	**0.56**	**−0.64**	0.15	−0.06	0	−0.02	**−0.5**
	2	20	0.03	−0.04	**−0.64**	**0.68**	0	−0.31	−0.18
	3	16	−0.1	0.03	0.24	−0.2	0	−0.94	−0.02
2	1	25	0.42	**−0.69**	0.09	−0.14	0	−0.16	**−0.54**
	2	21	−0.31	−0.06	**−0.65**	**0.56**	0	−0.18	−0.37
	3	16	0.1	0.19	0.12	0.01	0	−0.96	0.14
3	1	39	**0.55**	**−0.6**	0.1	0.02	0	−0.06	**−0.58**
	2	20	−0.06	−0.01	**−0.58**	**0.6**	0	**−0.54**	−0.07
	3	15	0	0.04	0.12	−0.59	0	−0.79	0.04
4	1	27	−0.21	**0.69**	−0.13	0.06	0	0.19	**0.65**
	2	20	0.39	0.08	**0.64**	**−0.62**	0	0.05	0.21
	3	17	**0.8**	−0.15	−0.18	0.4	0	0.24	0.28
5	1	24	0.14	**−0.57**	**−0.5**	0.3	0	−0.24	**−0.51**
	2	19	**0.7**	0.14	−0.11	−0.02	0	**−0.56**	0.4
	3	18	0.23	−0.38	0.41	**−0.73**	0	−0.08	−0.3

Variables with correlation values >0.5 (bold) are deemed to significantly contribute to each PC.

terms of their PC1 and PC2 (Table S2), indicating that both their classical and mesoscale features vary similarly and appear to differ mostly in their bathymetry. Subprovince 3 is representative of regions of particularly high mesoscale activity. This is clearly the case for the Alboran Sea, the Algerian Basin, the Strait of Sicily and the Ierapetra, Rhodes and Mersa-Matrouh gyres, as confirmed in other studies (e.g., [8,22,37,47,49,50]). Mesoscale activity is also particularly high for the phenological regions in the Alboran Sea and the Strait of Sicily found by [8], who defined seven regions based on satellite ocean color. They found no apparent bloom pattern in these regions, which coincide primarily with full subprovince 3 (characterized by highly variable chl; Table 2). Subprovince 4 represents the coastal regions with both narrow and wide continental shelves. Finally, subprovince 5 represents oceanic gyres at high latitude, including the Lion Gyre, though the South Adriatic Gyre [50] is not included in this subprovince, as might be expected. The biodiversity hot spots identified by [7] using exploited fish distributions also show good spatial agreement with our subprovinces. They highlight the importance of the western Mediterranean shelves, the Alboran Sea, the Adriatic Sea, and the Levantine Basin, which coincide with our coastal and high mesoscale activity subprovinces. These subprovinces are characterized by high and highly variable chl, an indicator of primary productivity [51], which may be associated with the high ecological productivity and biodiversity here [52]. The coastal regions and the large oceanic gyres revealed in this study are consistent with the subdivisions found by [31] using only dynamical criteria, e.g., advection and dispersion schemes due to surface currents. This suggests that the horizontal circulation potentially explains a significant part of the basin-scale distribution of oceanic tracers such as SST and chl. The mesoscale variability of the oceanic circulation also controls their smaller-scale patterns (through the formation of SST and chl frontal gradients) and is thus responsible for the lower stability of the boundaries in regions where mesoscale features are particularly ubiquitous.

The synthetic indices developed in this study via PCA show temporal variability at seasonal and interannual time scales. The seasonal signal of oceanographic conditions in the Mediterranean that is the dominant signal in all PCs is generally related to changes of heat and momentum fluxes, which also vary at seasonal time scales driven by synoptic weather patterns [53]. However, interannual variability, as shown to be strong for PC1 and 2, is often more complex and puzzling. Our attempt to explain the interannual signal that we found in PCs 1 and 2 showed no clear relationship to large-scale climate indices, which supposedly represent external atmospheric forcing over the basin [45,46]. Though it is possible that our indices are not sufficiently long (101 months) to reveal any relationship to low-frequency signals, other potential drivers may help explain the interannual variability that we find. Among these are internal nonlinear ocean dynamics (e.g., unstable mesoscale eddy fields) that alter water properties and movements, the region and timing of deep-water formation, the extreme events and long-term variability of atmospheric forcing (e.g., winds, solar radiation, precipitation) that impact both the surface and sub-surface circulation, or interannual variations in the Gibraltar inflow [50].

The oceanographic variables defined here were included to capture as much of the ocean variability and dynamics as possible; however, our analysis indicated that some variables may be redundant. As noted, chl and chl frontal gradients are highly positively correlated (r = 0.92) and are almost always aligned with the same axis (Figure 5b–d arrow plots, Table S1, Table 3). For PC2, FSLE and EKE could also be redundant in that they are consistently aligned with the same axis for subprovinces 1–4 and

are significantly negatively correlated (r = −0.53, p<0.001). This is consistent with the compact relationship (power-law) found between FSLE and EKE in the global ocean [54], although it appears less robust in the Mediterranean [55]. OW does not appear to be a useful variable to include in this bioregionalization, as it does not significantly contribute to the explanation of any axis (Table 3). This may be because OW is related to EKE and FSLE [55] and may be redundant in this analysis. In addition, it may exist at a lower spatial resolution than that detected by the 5% threshold level. At the 1% threshold level, it appears that OW may play a more important role in the bioregionalization (Figure S2b, c in File S1). Finally, frontal structures themselves indicate the boundaries between two different water masses [56] and between macro- or meso-provinces [33], making these variables potentially redundant to a partitioning analysis. However, this is not obvious *a priori*, as there is clearly a complex relationship between chl and chl fronts as indicated by their high correlation. We find that the contribution of SST and chl fronts is not strong or consistent in the Mediterranean or Black Seas.

In this study, we characterize the biogeographical conditions of the Mediterranean Sea in a simple and synthetic approach. We develop objective biogeochemical subprovinces that agree with spatial characterizations made in previous studies and are specifically adapted to be useful as geographic reference points in a management context. Our results highlight the importance of mesoscale features to help delineate further regions in the seemingly homogeneous open ocean. We suggest that these subprovinces could be relevant for defining pelagic habitats for marine protected areas that are dynamic, yet predictable enough to be important for foraging and breeding aggregations [57]. In addition, the synthetic indices developed here represent a baseline from which variability and future changes of important classical and mesoscale features can be assessed. An understanding of the fundamental ocean processes of these heavily-impacted bodies of water has important implications for climatic studies, anthropogenic impact mitigation, and marine resources management.

Acknowledgments

We would like to thank NASA, NOAA, and AVISO for the freely available remotely-sensed data. We would also like to thank Francesco d'Ovidio for his expertise and FSLE code and Giorgio Caramanna for his help improving the manuscript.

Supporting Information

File S1   Supporting figures and tables. Figure S1 in File S1: Boxplots of the bootstrapped (1000 times) between-clusters sum of squares divided by the total sum of squares (i.e., y-axis represents the proportion of the explained sum of squares) for *k* between 2 and 30 for k-means analyses performed on the (a) "classical", (b) "mesoscale" and (c) "full" multivariate arrays. To identify the most appropriate *k* for each multivariate array, we define thresholds whereby the explained sum of squares for each additional *k* increases by less than 5% (red line) or less than 1% (blue line). Table S1 in File S1: The eigenvalues of each axis for the "full" multivariate array for biogeochemical subprovinces defined by the 5% threshold. To determine which principal components (PC) to retain, we used the common cutoff of eigenvalues ≥1. Figure S2 in File S1: Biogeochemical subprovinces of the Mediterranean Sea for the (a) "classical", (b) "mesoscale", and (c) "full" multivariate arrays using a 1% threshold on the explained sum of squares to define the optimal number of subprovinces (see text). Figure S3 in File S1: Spatial stability of the borders of biogeochemical subprovinces for the (a) classical, (b) mesoscale, and (c) full multivariate arrays. K-means analysis, using the k found in the time-averaged analyses (Table 1), are performed on the multivariate arrays at monthly time steps for the 101 months of the data set and using a 1% threshold on the explained sum of squares to define the optimal number of subprovinces (see text). Spatial stability is represented as the percentage of time that a boundary of the biogeochemical subprovinces is found at a particular pixel over the 101 months of the data set. Red colors indicate stable borders. Table S2 in File S1: Correlation coefficients between the retained principal components (PC) for each of the full biogeochemical subprovinces and the monthly anomalies of the large-scale climate indices: North Atlantic Oscillation (NAO), the East Atlantic pattern (EA), the East Atlantic-West Russia pattern (EAWR), and the

Scandinavian pattern (SCAND). Only correlations above the 95% significance level are included. Table S3 in File S1: Correlation coefficients between the retained principal components (PC) for each of the full biogeochemical subprovinces. Significance levels are represented as $p < 0.001$ '***', $p < 0.05$ '*', not significant ' '.

Author Contributions

Conceived and designed the experiments: AEN SB LD. Performed the experiments: AEN SB. Analyzed the data: AEN SB KD GR. Contributed reagents/materials/analysis tools: HD. Wrote the paper: AEN KD GR VR HD LD SB.

References

1. Durrieu de Madron X, Guieu C, Sempéré R, Conan P, Cossa D, et al. (2011) Marine ecosystems' responses to climatic and anthropogenic forcings in the Mediterranean. Prog Oceanogr 91: 97–166. doi:10.1016/j.pocean.2011.02.003.

2. Fraschetti S (2012) Mapping the marine environment: advances, relevance and limitations in ecological research and marine conservation and management. Biol Mar Mediterr 19: 79–83.

3. Micheli F, Levin N, Giakoumi S, Katsanevakis S, Abdulla A, et al. (2013) Setting Priorities for Regional Conservation Planning in the Mediterranean Sea. PLoS One 8: e59038. doi:10.1371/journal.pone.0059038.

4. Portman ME, Notarbartolo-di-Sciara G, Agardy T, Katsanevakis S, Possingham HP, et al. (2013) He who hesitates is lost: Why conservation in the Mediterranean Sea is necessary and possible now. Mar Policy 42: 270–279. doi:10.1016/j.marpol.2013.03.004.

5. Abdulla A, Gomei M, Hyrenbach D, Notarbartolo-di-Sciara G, Agardy T (2009) Challenges facing a network of representative marine protected areas in the Mediterranean: prioritizing the protection of underrepresented habitats. ICES J Mar Sci 66: 22–28.

6. Bianchi CN (2007) Biodiversity issues for the forthcoming tropical Mediterranean Sea. Hydrobiologia 580: 7–21.

7. Coll M, Piroddi C, Steenbeek J, Kaschner K, Lasram FBR, et al. (2010) The biodiversity of the Mediterranean Sea: estimates, patterns, and threats. PLoS One 5: e11842. doi:10.1371/journal.pone.0011842.

8. d'Ortenzio F, Ribera d'Alcalà M (2009) On the trophic regimes of the Mediterranean Sea: a satellite analysis. Biogeosciences 6: 139–148. doi:10.5194/bg-6-139-2009.

9. Parry ML, Canziani O, Palutikof J, van der Linden P, Hanson C (2007) Climate change 2007: impacts, adaptation and vulnerability: contribution of Working Group II to the fourth assessment report of the Intergovernmental Panel on Climate Change. Cambridge: Cambridge University Press. 982 p.

10. Pachauri RK, Reisenger A (2007) Climate change 2007 Synthesis report: Contribution of Working Groups I, II and III to the fourth assessment report of the Intergovernmental Panel on Climate Change. Geneva: Intergovernmental Panel on Climate Change.

11. Lejeusne C, Chevaldonne P, Pergent-Martini C, Boudouresque CF, Perez T (2012) Climate change effects on a miniature ocean: the highly diverse, highly impacted Mediterranean Sea, Trends Ecol Evol 25: 250–260. doi:10.1016/j.tree.2009.10.009.

12. Halpern BS, Walbridge S, Selkoe KA, Kappel CV, Micheli F, et al. (2008) A global map of human impact on marine ecosystems. Science 319: 948–952.

13. Spalding MD, Fox HE, Allen GR, Davidson N, Ferdaña ZA, et al. (2007) Marine Ecoregions of the World: A Bioregionalization of Coastal and Shelf Areas. BioScience 57: 573–583. doi:10.1641/B570707.

14. Rice J, Gjerde KM, Ardron J, Arico S, Cresswell I, et al. (2011) Policy relevance of biogeographic classification for conservation and management of marine biodiversity beyond national jurisdiction, and the GOODS biogeographic classification. Ocean Coast Manage 54: 110–122.

15. Gabrié C, Lagabrielle E, Bissery C, Crochelet E, Meola B, et al. (2012) The status of marine protected areas in the Mediterranean Sea. MedPAN and RAC/SPA. 256 p. Available: http://www.medpan.org/documents/10180/0/The+Status+of+the+Marine+Protected+Areas+in+the+Mediterranean+Sea+2012/069bb5c4-ce3f-4046-82cf-f72dbae29328. Accessed 2014 October 7.

16. Coll M, Cury P, Azzurro E, Bariche M, Bayadas G, et al. (2013) The scientific strategy needed to promote a regional ecosystem-based approach to fisheries in the Mediterranean and Black Seas. Rev Fish Biol Fish 23: 415–434.

17. Vierros M, Bianchi G, Skjoldal HR (2008) The Ecosystem approach of the convention on biological diversity. In: Bianchi G, Skjoldal, editors. The Ecosystem approach to fisheries. pp. 39–46.

18. Edgar GJ, Moverley J, Barrett NS, Peters D, Reed C (1997) The conservation-related benefits of a systematic marine biological sampling programme: the Tasmanian reef bioregionalisation as a case study. Biol Conserv 79: 227–240.

19. Grant S, Constable A, Raymond B, Doust S (2006) Bioregionalisation of the Southern Ocean: report of experts workshop (Hobart, September 2006). Sydney:

WWF-Australia and ACE CRC. 44 p. Available: http://awsassets.wwf.org.au/downloads/mo007_bioregionalisation_of_the_southern_ocean_8sep06.pdf. Accessed 2014 October 7.

20. Reygondeau G, Longhurst A, Martinez E, Beaugrand G, Antoine D, et al. (2013) Dynamic biogeochemical provinces in the global ocean. Global Biogeochem Cycles 27: 1046–1058.

21. Kavanaugh MT, Hales B, Saraceno M, Spitz YH, White AE, et al. (2014) Hierarchical and dynamic seascapes: A quantitative framework for scaling pelagic biogeochemistry and ecology. Prog Oceanogr 120: 291–304.

22. Millot C, Taupier-Letage I (2005) Circulation in the Mediterranean Sea. In: Saliot A, editor, The Mediterranean Sea, Handbook of Environmental Chemistry. Berlin Heidelberg: Springer, pp. 29–66.

23. Giakoumi S, Sini M, Gerovasileiou V, Mazor T, Beher J, et al. (2013) Ecoregion-based conservation planning in the Mediterranean: dealing with large-scale heterogeneity. PLoS One 8: e76449. doi:10.1371/journal.pone.0076449.

24. Millot C (1991) Mesoscale and seasonal variabilities of the circulation in the western Mediterranean. Dyn Atmospheres Oceans 15: 179–214. doi:10.1016/0377-0265(91)90020-G.

25. Millot C (1999) Circulation in the Western Mediterranean Sea. J Mar Syst 20: 423–442. doi:10.1016/S0924-7963(98)00078-5.

26. Robinson AR, Malanotte-Rizzoli P, Hecht A, Michelato A, Roether W, et al. (1992) General circulation of the Eastern Mediterranean. Earth-Science Reviews 32: 285–309. doi:10.1016/0012-8252(92)90002-B.

27. Sherman K, Alexander LM (1989) Biomass yields and geography of large marine ecosystems. Boulder: Westview Press. 493 p.

28. Sherman K, Sissenwine M, Christensen V, Duda A, Hempel G (2005) A global movement toward an ecosystem approach to management of marine resources: Politics and socio-economics of ecosystem-based management of marine resources. Mar Ecol Prog Ser 300: 275–279.

29. Lehahn Y, d'Ovidio F, Levy M, Heifetz E (2007) Stirring of the northeast Atlantic spring bloom: A Lagrangian analysis based on multisatellite data. J Geophys Res 112: C08005. doi:10.1029/2006JC003927.

30. Rossi V, Lopez C, Sudre J, Hernandez-Garcia E, Garcon V (2008) Comparative study of mixing and biological activity of the Benguela and Canary upwelling systems. Geophys Res Lett 35: L11602. doi:10.1029/2008GL033610.

31. Rossi V, Ser-Giacomi E, López C, Hernández-García E (2014) Hydrodynamic regions and oceanic connectivity from a transport network help designing marine reserves. Geophys Res Lett 41: 2883–2891. doi:10.1002/2014GL059540.

32. Reygondeau G, Olivier Irisson J, Guieu C, Gasparini S, Ayata S, et al. (April 2013) Toward a dynamic biogeochemical division of the Mediterranean Sea in a context of global climate change. EGU General Assembly Conference Abstracts 15: 10011. Available: http://scholar.google.fr/scholar?q=Toward+a+dynamic+biogeochemical+division+of+the+Mediterranean+Sea+in+a+context+of+global+climate+change&btnG=&hl=en&as_sdt=0%2C5. Accessed 2014 October 7.

33. Longhurst AR (2010) Ecological geography of the sea. San Diego: Academic Press. 560 p.

34. Sydeman WJ, Thompson SA, Garcia-Reyes M, Kahru M, Peterson WT, et al. (2014) Multivariate ocean-climate indicators (MOCI) for the central California Current: Environment change, 1990-2010. Prog Oceanogr 120: 352–369.

35. Canny J (1986). A computational approach to edge detection. IEEE Trans Pattern Anal Mach Intell 6: 67–698.

36. Nieto K, Demarcq H, McClatchie S (2012) Mesoscale frontal structures in the Canary Upwelling System: new front and filament detections algorithms applied to spatial and temporal patterns. Remote Sens Environ 123: 339–346.

37. d'Ovidio F, Fernández V, Hernández-García E, López C (2004) Mixing structures in the Mediterranean Sea from finite-size Lyapunov exponents. Geophys Res Lett 31: L17203. doi:10.1029/2004GL020328.

38. Tew-Kai E, Rossi V, Sudre J, Weimerskirch H, Lopez C, et al. (2010) Top marine predators track Lagrangian coherent structures. Proc Natl Acad Sci U S A 106: 8245–8250. doi:10.1073/pnas.0811034106.

39. Okubo A (1970) Horizontal dispersion of floatable particles in the vicinity of velocity singularities such as convergences. Deep Sea Res Part 1 Oceanogr Res Pap 17: 445–454. doi:10.1016/0011-7471(70)90059-8.

40. Weiss J (1991) The dynamics of enstrophy transfer in two-dimensional hydrodynamics. Physica D 48: 273–294. doi:10.1016/0167-2789(91)90088-Q.

41. Henson SA, Thomas AC (2008). A census of oceanic anticyclonic eddies in the Gulf of Alaska. Deep Sea Res Part 1 Oceanogr Res Pap 55: 163–176. doi:10.1016/j.dsr.2007.11.005.

42. Hartigan JA, Wong MA (1979) Algorithm AS 136: A k-means clustering algorithm. J R Stat Soc Ser C Appl Stat 28: 100–108.

43. Oliver MJ, Glenn S, Kohut JT, Irwin AJ, Schofield OM, et al. (2004) Bioinformatic approaches for objective detection of water masses on continental shelves. J Geophys Res 109: C07S04. doi:10.1029/2003JC002072.

44. Beaugrand G, Ibanez F (2004) Monitoring marine plankton ecosystems. II: Long-term changes in North Sea calanoid copepods in relation to hydro-climatic variability. Mar Ecol Prog Ser 284: 35–47.

45. Josey SA, Somot S, Tsimplis M (2011) Impacts of atmospheric modes of variability on Mediterranean Sea surface heat exchange. J Geophys Res 116: C02032. doi:10.1029/2010JC006685.

46. Papadopoulos VP, Josey SA, Bartzokas A, Somot S, Ruiz S, et al. (2012) Large-scale atmospheric circulation favoring deep- and intermediate-water formation in the Mediterranean Sea. J Clim 25: 6079–6091. doi:10.1175/JCLI-D-11-00657.1.

47. Fernández V, Dietrich DE, Haney RL, and Tintoré J (2005) Mesoscale, seasonal and interannual variability in the Mediterranean Sea using a numerical ocean model. Prog Oceanogr 66: 321–340.

48. Larnicol G, Ayoub N, Le Traon PY (2002) Major changes in Mediterranean Sea level variability from 7 years of TOPEX/Poseidon and ERS-1/2 data. J Mar Syst 33: 63–89.

49. Malanotte-Rizzoli P, Manca BB, d'Alcalà MR, Theocharis A, Bergamasco A, et al. (1997) A synthesis of the Ionian Sea hydrography, circulation and water mass pathways during POEM-Phase I. Prog Oceanogr 39: 153–204.

50. Pinardi N, Masetti E (2000) Variability of the large scale general circulation of the Mediterranean Sea from observations and modelling: a review. Palaeogeogr Palaeoclimatol Palaeoecol 158: 153–173.

51. Longhurst AR, Sathyendranath S, Platt T, Caverhill C (1995) An estimate of global primary production in the ocean from satellite radiometer data. J Plankton Res 17: 1245–1271.

52. Chase JM, Leibold MA (2002) Spatial scale dictates the productivity-biodiversity relationship. Nature 416: 427–430.

53. Zavatarelli M, Mellor GL (1995) A numerical study of the Mediterranean Sea circulation. J Phys Oceanogr 25: 1384–1414.

54. Hernández-Carrasco I, López C, Hernández-García E, Turiel A (2012) Seasonal and regional characterization of horizontal mixing in the global ocean. J Geophys Res 117: C10007. doi:10.1029/2012JC008222.

55. d'Ovidio F, Isern-Fontanet J, López C, Hernández-García E, García-Ladona E (2009) Comparison between Eulerian diagnostics and finite-size Lyapunov exponents computed from altimetry in the Algerian basin. Deep Sea Res Part 1 Oceanogr Res Pap 56: 15–31.

56. Cayula JF, Cornillon P (1992) Edge detection algorithm for SST images. J Atmos Oceanic Tech 9: 67–80.

57. Hyrenbach KD, Forney KA, Dayton PK (2000) Marine protected areas and ocean basin management. Aquatic Conserv 10: 437–458.

17

Spatial Scale, Means and Gradients of Hydrographic Variables Define Pelagic Seascapes of Bluefin and Bullet Tuna Spawning Distribution

Diego Alvarez-Berastegui[1]*, Lorenzo Ciannelli[2], Alberto Aparicio-Gonzalez[3], Patricia Reglero[3], Manuel Hidalgo[3], Jose Luis López-Jurado[3], Joaquín Tintoré[1,4], Francisco Alemany[3]

1 Balearic Islands Coastal Observing and Forecasting System (SOCIB), Palma de Mallorca, Balearic Islands, Spain, 2 College of Earth, Ocean, and Atmospheric Sciences (CEOAS), Oregon State University, Corvallis, Oregon, United States of America, 3 Instituto Español de Oceanografía - Centre Oceanogràfic de les Balears (COB-IEO), Palma de Mallorca, Balearic Islands, Spain, 4 Institute for Advanced Studies (IMEDEA), Consejo Superior de Investigaciones Científicas (CSIC) and the University of the Balearic Islands (UIB). Esporles, Spain

Abstract

Seascape ecology is an emerging discipline focused on understanding how features of the marine habitat influence the spatial distribution of marine species. However, there is still a gap in the development of concepts and techniques for its application in the marine pelagic realm, where there are no clear boundaries delimiting habitats. Here we demonstrate that pelagic seascape metrics defined as a combination of hydrographic variables and their spatial gradients calculated at an appropriate spatial scale, improve our ability to model pelagic fish distribution. We apply the analysis to study the spawning locations of two tuna species: Atlantic bluefin and bullet tuna. These two species represent a gradient in life history strategies. Bluefin tuna has a large body size and is a long-distant migrant, while bullet tuna has a small body size and lives year-round in coastal waters within the Mediterranean Sea. The results show that the models performance incorporating the proposed seascape metrics increases significantly when compared with models that do not consider these metrics. This improvement is more important for Atlantic bluefin, whose spawning ecology is dependent on the local oceanographic scenario, than it is for bullet tuna, which is less influenced by the hydrographic conditions. Our study advances our understanding of how species perceive their habitat and confirms that the spatial scale at which the seascape metrics provide information is related to the spawning ecology and life history strategy of each species.

Editor: Athanassios C. Tsikliras, Aristotle University of Thessaloniki, Greece

Funding: The development of this study was supported by the Project "PERSEUS" financed by the Seventh Framework Program for Research (FP7) under theme "Oceans Tomorrow", grant Agreement No. 287600 (http://www.perseus-fp7.eu/), the Spanish Foundation for Science and Technology, under the project "BALEARES" Grant number: CTM2009-07944/MAR (http://www.fecyt.es/fecyt/home.do) and the Science, Engineering and Education for Sustainability NSF-Wide Investment (SEES) Research Coordination Network, Grant 1140207 "Sustainability of Marine Renewable Resources in Subarctic Systems Under Incumbent Environmental Variability and Human Exploitation" (http://grantome.com/grant/NSF/OCE-1140207). Hydrographic data used in the study were provided by IBAMAR (http://www.ba.ieo.es/ibamar) a public data base founded by the Council for Economy and Competitiveness of the Balearic Islands Government (http://www.caib.es/govern/organigrama/area.do?coduo=6&lang=es), grant number AAEE0022/08. The research was developed under the framework of the BLUEFIN TUNA Project driven by the Balearic Island Observing and Forecasting System (www.socib.es) and the Spanish Institute of Oceanography (www.ieo.es). The funders had no role in study design, data collection and analysis, decision to publish, or preparation of the manuscript.

Competing Interests: The authors have declared that no competing interests exist.

* Email: dalvarez@socib.es

Introduction

Seascape ecology represents an emerging field in the study of how the habitat structure shapes the spatial distribution of marine species and influences key ecological processes [1],[2]. This discipline initiated applying techniques and metrics from the traditional landscape ecology to characterize and quantify spatial structure of benthic habitats, observed as a mosaic of patches of different habitat classes [1],[3],[4], [5],[6]. Nevertheless, there is still a gap in the development of concepts and techniques providing metrics to characterize the spatial structure of the seascape in the pelagic environments, where there are no clear boundaries delimiting different habitats [1],[2],[6]. In the framework of landscape ecology spatial gradients have been recently proposed as more appropriated metric than traditional categorical patch mosaic based metrics to characterize continuous habitats [7]. Accordingly, a location in a pelagic seascape would be better characterized by the combination of the value of a particular hydrographic variable and its spatial gradient.

Several studies have applied gradients of hydrographic parameters to characterize the spatial distribution of marine species during various life history stages, as nursery, foraging or spawning [8],[9],[10],[11],[12]. It is likely that the scale at which an individual perceive a change in the environment (i.e., a gradient) varies according to life history, ontogeny and to the hydrographic variable in exam. For instance, while large-scale gradients associated with the North Pacific transition zone drive the location of many pelagic predators including albacore tuna (*Thunnus alalunga*) during their feeding migratory stages [13], once off the west coast of the US, albacore tuna distribution is associated with smaller scale features linked to upwelling fronts [14]. In the Mediterranean Sea during spawning, bluefin tuna distribution is regulated by oceanographic variables that can change at relatively small scales [15],[16]. In spite of the expected importance of gradient scales, to our knowledge there are no studies that have evaluated the effect of changing the spatial scales at which environmental gradients are calculated to model the spatial distribution of fish. The goal of our study is to examine the distribution of large pelagic predators during spawning by explicitly considering mean, gradients and scale of gradients of hydrographic variables.

Atlantic bluefin tuna (*Thunnus thynnus*) and bullet tuna (*Auxis rochei rochei*) are two species of pelagic predators showing different spawning strategies. We target these species in the Balearic Sea (Figure 1), known as a recurrent spawning area for large pelagic species in the Western Mediterranean [17]. Bluefin tuna is a highly migratory species; the Eastern population enters in the Mediterranean Sea from the North Atlantic at the end of spring and early summer [18],[19]. Their spawning activity at the Balearic Sea is linked to the regional oceanography with spawning grounds located in the vicinity of frontal structures formed when the recent Atlantic water mass encounters the more saline resident surface Atlantic waters [15]. The area is characterized by highly dynamic processes that promote a seascape shaped by filaments, fronts and eddies whose location varies between years [20],[21]. Bullet tuna, by contrast, is smaller and more frequent in near coastal areas [22]. The spawning of bullet tuna is associated with the geography. Young larvae are found mainly in coastal areas and with little influence of the local oceanography in comparison with bluefin [16].

We expect that, when selecting spawning locations, a large-bodied and long-distance migratory pelagic fish, such as the bluefin tuna explores its environment at larger spatial scales than bullet tuna, a small-bodied and non-migratory pelagic fish. Therefore, we expect that pelagic seascape metrics based on the combination of hydrographic parameter values and their gradients calculated at appropriate spatial scales provide relevant information for bluefin tuna but not for bullet tuna, where we expect a greater reliance on geographic and hydrographic parameters, calculated at comparatively small-scale.

In this work we analyze the influence of the pelagic environment by depicting the spatial scales at which gradients of hydrographic variables are linked to the spawning ecology of these tuna species. We investigate the two most relevant hydrographic variables describing their spawning spatial distribution: salinity and geostrophic currents velocity [16], already determined in previous studies. Our analytical approach has two steps. Firstly, we identify the scale at which each hydrographic variable (Figure 2) maximizes the performance of a model fitted to larval distribution. Second, we evaluate whether components of the seascape (i.e., mean and gradients of oceanographic variables) are interactively affecting the spatial distribution of tuna larvae. By performing this analysis on two species we evaluate how fish with contrasting life history strategies perceive their environments when deciding for spawning locations.

Materials and Methods

Data acquisition

Bluefin and bullet tuna larvae were collected during ichthyoplankton surveys using Bongo nets from 2001 to 2005. The surveys were conducted by the Instituto Español de Oceanografía (www.ieo.es), a Spanish Government marine research institution. The sampling scheme was communicated and approved by the Spanish Directorate of Fisheries before the sampling was conducted. No specific ethical approval was required and the survey of biological data was conducted using Bongo nets, which are accepted standard

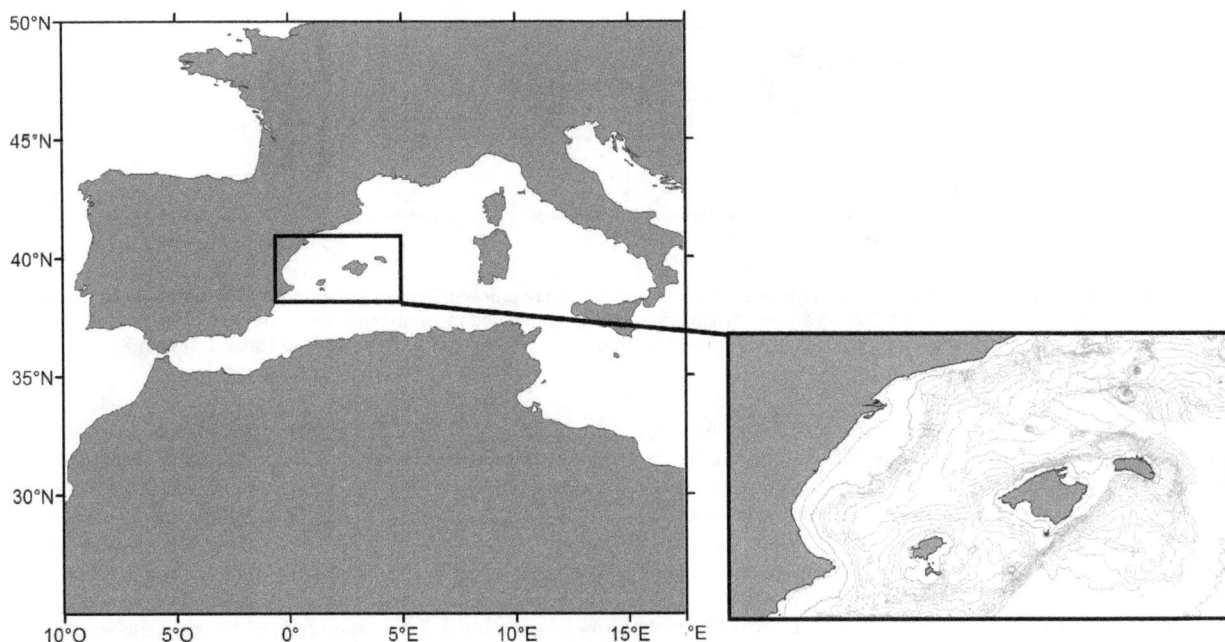

Figure 1. Location of the Balearic Islands, Western Mediterranean.

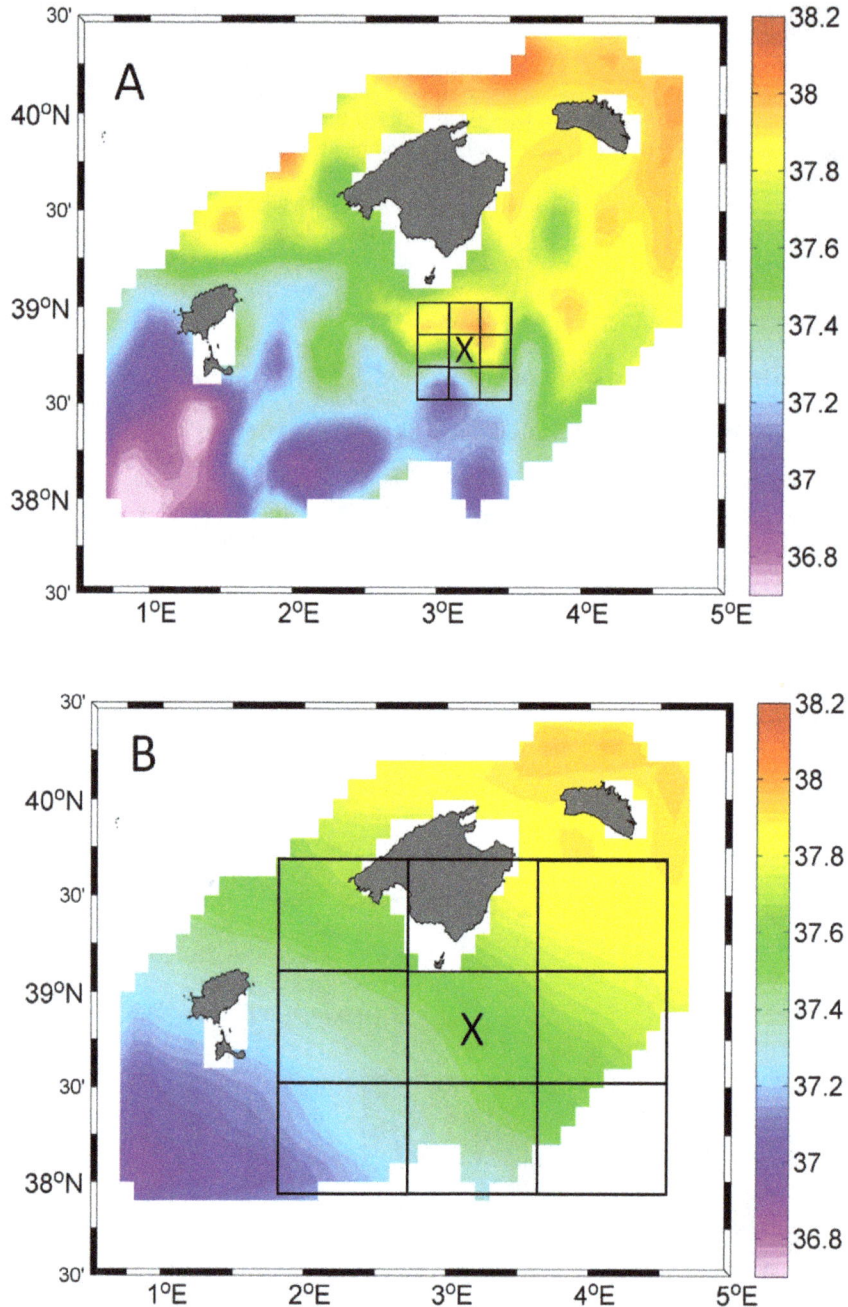

Figure 2. Sea surface salinity field in 2003. Spatial means of sea surface salinity processed at two different scales A) 0.15 degrees and B) 0.75 degrees. Spatial means were interpolated following an objective analysis onto a regular grid by using minimum error variance methods (Bretherton, 1976). The squares in each figure show the polygons used for the calculation of the spatial gradients at the two scales at station X.

techniques for this type of surveys, used worldwide for the collection of plankton samples, including billfish and tuna larvae [15],[17], [23], [24]. The nets were towed at low speeds, around 2 knots, during 8–10 minutes, and plankton samples were immediately preserved with 4% formalin buffered with borax onboard.

Around 200 stations were sampled every year, in a regular sampling grid of 10×10 nm located between $37.85°$–$40.35°$ N and $0.77°$–$4.91°$E, covering a total area of 101360 km2 ($= 280 \times 362$ km) around the Balearic archipelago. The sampling was conducted during June–July coinciding with the spawning period of bullet and bluefin tuna (see [15] for details of the

sampling procedures). Tuna larvae were identified to the species level and measured in standard length. All larvae identified as yolk sac and preflexion stages (<4.5 mm) were classified as "stage 1".

Mean and maximum age of larvae under this size are 6 days and 11 days old respectively, accounting for growth [25] and hatching time [26]. Displacement from spawning areas during these intervals in the study area are below 25 kilometers for the mean ages, which are in the range of 1.4 sampling stations, and 46 kilometers for the maximum ages, what is in the range of 2.6 sampling stations. These values have been calculated following methods in [27]. Considering these values, abundances of stage 1 larvae have been defined to get a proxy of spawning locations as in previous research in the study area [28].

Vertical profiles of conductivity, temperature and pressure data were recorded at all stations, by means of Sbe911 CTD. Sea surface salinity at each station (SAL) was calculated as the mean salinity over the mixed layer depth. Geostrophic velocities (GVEL) were calculated at sea surface from the first-derivative of the sea surface height between adjacent points, which was obtained by vertical integration of the specific volume, using 600 m as the level of no motion [21].

These two variables (SAL and GVEL) were selected, since they have been demonstrated to be the two most relevant environmental variables describing the spawning spatial distribution of tuna [16]. Sea surface temperature was also included in the models since in this area it is a secondary but relevant variable mainly related to the phenology of the spawning process [29]. However, the spatial gradient was not explored because sea surface temperature during the summer changes relatively fast due to solar irradiance [21].

Processing of spatial gradients along continuous spatial scales

Spatial gradients from the sea surface salinity field (gSAL) and geostrophic velocity field (gGVEL) within the sampled region were calculated at six spatial scales, from 0.15° to 0.90° with a spatial increment of 0.15°. The minimum (0.15°) and the maximum scale (0.90°) were in the range of the smallest (from 0.13° to 0.27°) and largest (up to 0.92°) mesoscale oceanographic structures in the area [21]. For the computation of the gradients, nine square polygons at every scale were delimited around each sampling position (see examples for scale 0.15° and 0.75° in figure 2A, 2B respectively). The gradient was then computed as the maximum absolute difference between the mean hydrographic variable at the center polygon and each of the eight surrounding polygons standardized to distance [9]. The software for the spatial processing was developed in R language [30].

Identification of spatial scales

Comparison of how models perform along scales allowed identifying the spatial scales at which information provided by gradients is maximized. The effect of gGVEL and gSAL at each scale on the abundance of bullet and bluefin tuna larvae was assessed using nonparametric regression statistical models (generalized additive models, GAMs, [31]). A base model was formulated to describe inter-annual variability (variable YEAR), sampling location (latitude and longitude variables), and the hour of the day on the catch of tuna larvae. Over-dispersed Poisson distribution family and a natural-log link were selected to model larval data. The volume of water filtered was included as an offset (after natural log transformation), to account for the effort expanded in catching the sample (Equation 1). The effects of these variables on the base model have been already analyzed in previous studies [16]. Here, the base model represented the null hypothesis of no gradient effect on tuna larvae distribution, against which all other more complex formulations will be compared.

Equation 1: Base model

$$Larvae\ abundance = offset(log(m3)) + factor(year)$$
$$+ sm_1(long, lat) + sm_2(hour)$$

m3 = volume filtered by the bongo nets (m3); long = longitude; lat = latitude; hour = hour of the day expressed from 0 to 1, sm_1 and sm_2 the smoothing functions.

At each spatial scale a GAM model was processed including the gradient of one hydrographic variable (gSAL, gGVEL) as a new additive term (s3) in the standardization model. The number of knots for the new smoother was always set to a maximum of three (i.e. two degrees of freedom) in order to avoid over fitting in the responses.

The identification of characteristic spatial scales (cgSAL, cgGVEL) was assessed with scalogram where the scale of the covariate is plotted against a measure of the model goodness of fit, which in our case were represented by the adjusted R-squared (Rsq, the higher the better), and the Generalized Cross Validation (GCV, the lower the better) [31]. We selected the scale that maximize Rsq and minimize the GCV. Results of the base model (when a seascape covariate was not included) were presented in the same graphics. Note that due to the greater complexity of the gradient model higher Rsq values do not necessarily imply an improvement in relation to the base model, while they do represent a better performance when compared to other gradient models.

Significant differences of Rsq values between models, or GCVs, were obtained from t-test of these parameters obtained from 500 iterations where 10% of the data was excluded. For all cases, alternative hypothesis (difference in means is not equal to 0) was accepted only if the P value was lower than 0.001, with a confidence level of 0.99. When one variable presented similar Rsq and GCV values at various scales, selection was assessed by inspection of the plot showing the response of the abundance in relation to the gradient processed at those scales.

Once the characteristic scale of the gradients was identified, we tackled the questions of whether the information provided by the gradients is different and complementary to the information provided by the hydrographic variables from which they were calculated, and in that case, how the information from these two variables (spatial mean and gradient) should be combined to maximize the goodness of fit of the models and the ecological information they provide. To assess these questions we analyzed the performance of models with different complexity:

i) The base model from equation 1.

ii) Hydrographic models combining the sea surface salinity, geostrophic velocity and sea surface temperature at the sampling station (stSAL, stGVEL, stSST).

iii) Seascape models combining the gradients at characteristic scales (cgSAL, cgGVEL) and the hydrographic variable at the sampling location (stSAL, stGVEL, stSST). Different seascape models were constructed including the two components of the seascape (values at stations and gradients) as additive and interactive terms. An interaction may be ecologically meaningful when a species is selecting its spawning habitat on a specific side of a frontal region, for example. In such case, it is the combination of both the gradient and the mean that provide the suitable conditions for spawning. The performance of different model configurations for each species was assessed by the delta AIC (ΔAIC), calculated as the difference between model AICs and the base model AIC. The AIC in this case is best suited for model comparisons because each model had different number of variables [32]. Rsq, GCV and explained deviances were used to compare how models perform between the two species, as AIC values among models with different dependent variables are not comparable.

Results

Identification of characteristic spatial scales

In all years considered, the recent Atlantic water masses encountered the more saline resident water masses forming an oceanic frontal zone inside the study area. The size of such frontal zone was bigger than other oceanographic phenomena as small eddies and meanders derived from the instabilities along the haline front and the effect of strong bathymetric changes (Figures S3, S4).

The scalogram of gGVEL for bluefin showed that Rsq values gradually improve as the spatial scales increased to a maximum at 0.6° (Rsq = 0.44, Figure 3A), which was chosen as the geostrophic velocity gradient characteristic scale for bluefin tuna. Values of GCV showed a similar pattern of model improvement, being significantly better than the base model at 0.6 degrees (Figure 3B). At this characteristic scale the response of the larvae abundance is positively related to the gradient of geostrophic velocity (Figure 3C).

The scalograms of gSAL for bluefin tuna showed also an increment of Rsq with higher values at 0.6° and 0.75° that also coincide with lower values of GCV (Figure 3D, 3E). Differences of R-sq between these two scales (0.6° and 0.75°) were not significant. The characteristic scale for gSAL was chosen at 0.6° as the model response at this scale presented a less ambiguous effect on larval abundance (Figure 3F). The gSAL at 0.75° spatial gradients displays a dome-shaped response with a less clear ecological

interpretation (Figure S1). At this scale GCV was lower than the base model.

In contrast to the results obtained for salinity, the gradients of geostrophic velocities (gGVEL) did not show any single scale that maximizes R-square and minimizes GCV for bullet tuna (Figure 4A, 4B). The Rsq scalogram showed a flat trend with the highest value at 0.15 degrees. The Rsq value at this scale (= 0.18) showed similar values than other scales (values between 0.170 and 0.173) or when compared to the base model (= 0.166). On the contrary the GCV scalogram showed significantly lower values than the base model at 0.45 and 0.6 degrees, scales at which Rsq were not even significantly higher than the base model. Therefore, the contradictory response of the model performance indicators, the flat trend of Rsq scalogram and their very low values (despite the higher complexity of the gradient models in relation to the base model), may indicate that the spatial gradient of geostrophic velocity is not a valid seascape metric for the spawning locations of bullet tuna. Consequently, gGVEL was excluded from further analysis in relation to this species.

The gSAL scalogram of bullet tuna showed a moderate effect of the scale at which gradients were calculated. The shape of the scalogram did not show a peak scale at which model performs better (Figure 4C). Scales from 0.45°, 0.6° and 0.75° for gSAL seemed to maximize Rsq, but not being different from each other. In this case GCV-scalograms showed the lowest at 0.75° (Figure 4D), being significantly lower than the base model, which

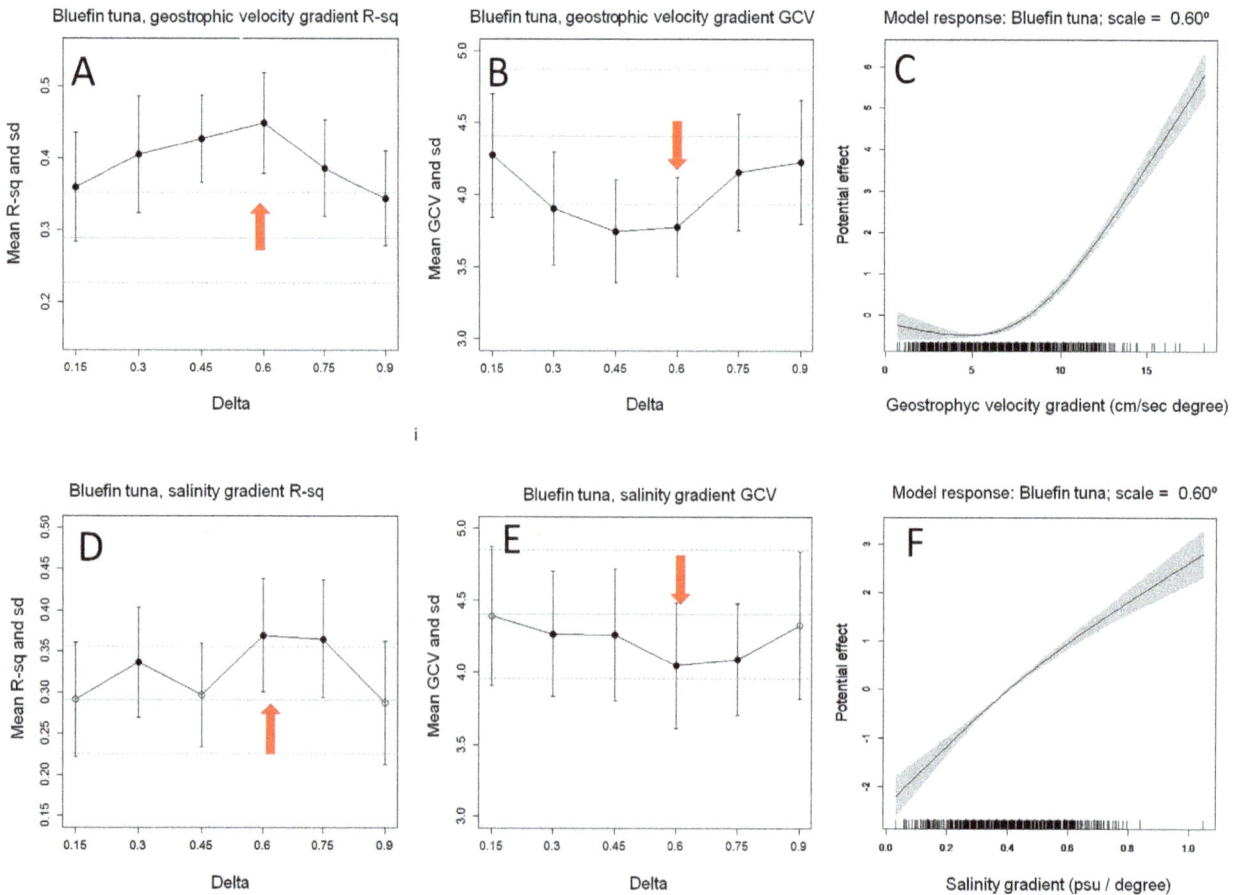

Figure 3. Rsq and GCV Scalograms of bluefin tuna larva abundance models along spatial scales, standard deviations. Horizontal grey lines indicate statistics from the base model (Straight line = mean, dashed line = Sd). Black dots show scales at which values are significantly different from the standardization model. Red arrows indicate the selected characteristic scale.

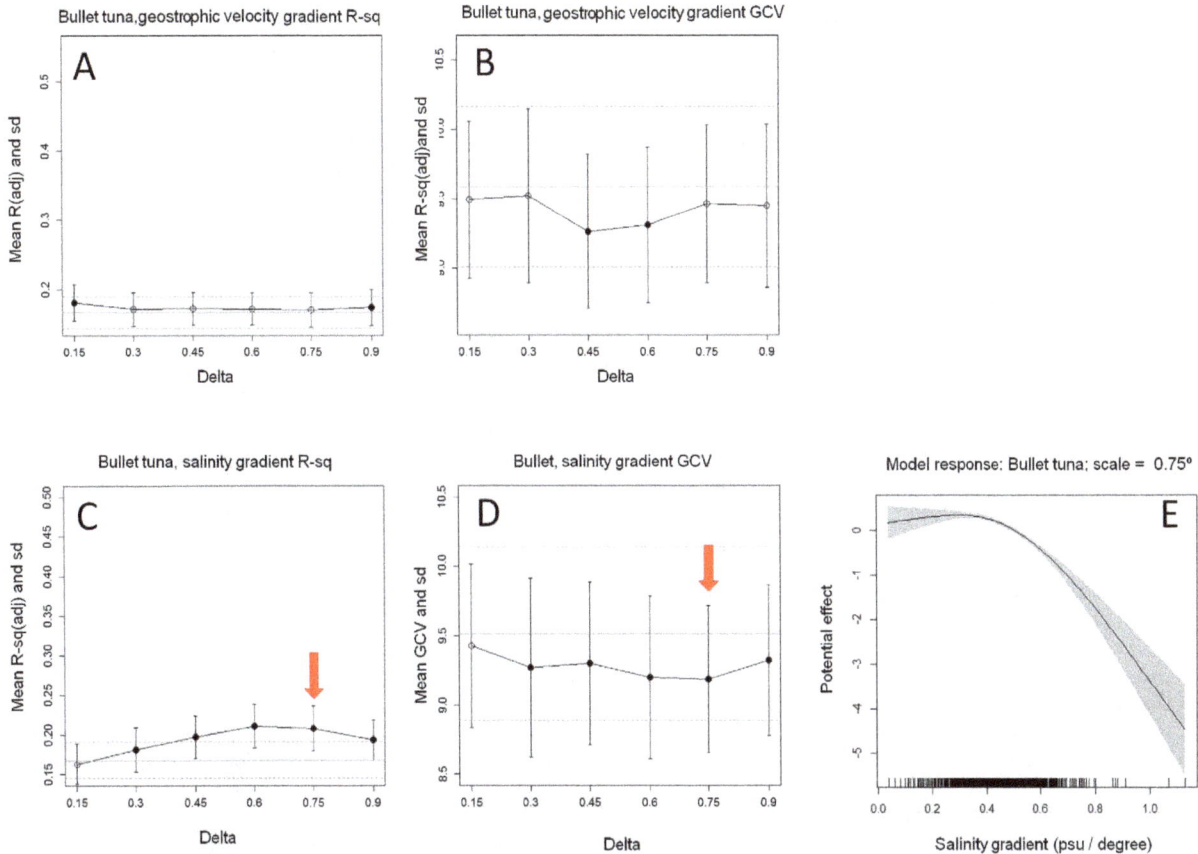

Figure 4. Rsq and GCV scalograms of bullet tuna larva abundance models along spatial scales, standard deviations. Horizontal grey lines indicate statistics from the standardization model (Strait line = mean, dashed line = Sd). Black dots show scales at which values are significantly different from the standardization model. Red arrows indicate selected characteristic scale.

was selected as characteristic scale. The Rsq at this scale was higher (= 0.21) than that of the base model (Rsq = 0.16). At this scale, gradients displayed a negative effect on the bullet tuna larvae abundance (Figure 4E) showing that bullet tuna spawning locations are found with more probability in areas where salinity is spatially homogeneous– a result that contrasted to that obtained for bluefin tuna.

Species-specific seascape characterization

The best model for each species included a gradient and a mean term (Tables 1 and 2). Note however that the hydrographic model already represents an improvement respect to the standardization model (Tables 1 and 2). For bluefin tuna the best seascape model had an improvement of 186% of the Rsq when compared to the base model (Rsq base mode = 0.23; Rsq best seascape model = 0.66, Table 1). The improvement for bullet was 61%, considerably lower compared to bluefin (Rsq base mode = 0.16; Rsq best seascape model = 0.25, Table 2).

Pearson correlation coefficients between hydrographical variables and their gradients at characteristic scales were in all cases below 0.50 and pair plots showed no clear tendencies on the correlations (Figure S2), indicating that selected gradients provided complementary information to that of the hydrographical variable. Models showed a better performance (i.e. lower GCV and higher ΔAIC; Tables 1 and 2) in all the cases when the gradient and the hydrographic value were considered as an interaction term, suggesting dependence in their effect on larvae abundance rather

than an additive response. However, larvae abundance of each species responded differently to the interaction of seascape components (Figures 5A, 5B, 5C).

For Bluefin tuna, higher probability of spawning was associated to higher values of geostrophic velocity gradients, but where velocities at station may present either high or low values (Figure 5A), (Figures S7, S8). Considering that a gradient is characterized by an area with high current speed near an area of low current speed, this result indicates that spawning locations were not associated to a particular side of the gradient, but in an area around the location where maximum velocities occurs. The extension of this area would be around a circle of 0.6 degrees of radius (aprox. 65 km in the study area), the characteristic scale at which the gradients were more relevant. In contrast, the interaction of the salinity seascape variables showed high larvae abundances in areas with high salinity gradients and intermediate-high salinity levels (Figure 5B), indicating an effect of the location of the main haline front and a preference for spawning at the high salinity values of that front (Figures S5, S6). The characteristic spatial correlation scale of local oceanographic structures in the area is around 18 nmi (aprox. 0.15 meridian degrees) [21] and therefore surface oceanographic structures at smaller spatial scales are ephemeral. The oceanographic structures at larger spatial scales, relevant for bluefin tuna, are linked to the Med-Atlantic salinity front and last longer.

The functional form of the interaction terms was different for bullet tuna. The relation between the bullet tuna larvae

Table 1. Summary of GAM models of larvae abundance for Atlantic bluefin tuna (*Thunnus thynnus*).

Model group	Model variables	R2	Dev(%)	GCV	AIC	delta AIC
Base model	(latitude, longitude) + filtered volume + hour	0.232	40.8	4,596	3985,54	0
One additive variable models	base model + stGVEL0.15	0,271	43,4	4,412	3827,79	157,75
	base model + gGVEL0.6	0,39	49,4	3,947	3465,03	520,50
	base model + stSAL0.15	0,222	41,8	4,534	3924,70	60,84
	base model + gSAL0.6	0,301	45,1	4,275	3723,53	262,01
Hydrographic model	base model + stGVEL + stSAL + stTEMP	0,472	51,6	3,814	3338,62	646,92
GVEL seascape models	Hydrographic model + stGVEL + gGVEL0.6	0,53	55,8	3,500	3087,14	898,40
	Hydrographic model + (stGVEL,gGVEL0.6)	0,666	59,3	3,251	2881,29	1104,25
SAL seascape models	Hydrographic model + stSAL + gSAL0.6	0,506	54,8	3,571	3145,80	839,74
	Hydrographic model + (stSAL,gSAL0.6)	0,533	57,1	3,417	3009,58	975,96

Interaction terms included in parenthesis.

abundance and the interaction of spatial distribution of sea surface salinity and its gradients is presented in Figure 5C. Spawning locations were associated to areas where the salinity at the station were lower and gradients presented intermediate values, but the interaction plot revealed that spawning also appears in areas of higher salinities associated to very low gradients. Areas defined by this twofold combination were located at both sides of the front avoiding more mixed waters. This spatial distribution was more evident in years 2001, 2003 and 2005 (Figures S9, S10)

Discussion

We have found that the combination of sea surface current velocities, salinities and their gradients calculated at characteristic spatial scales are relevant for the parameterization of the pelagic seascape affecting a key ecological process of bluefin tuna. For bullet tuna only salinity and their gradient provided a valid seascape metric not being relevant the gradients of sea surface current velocities. In agreement with our expectations, the importance of these metrics was much higher for bluefin, a large-bodied, long distance migratory and more dependent on local oceanography than were for bullet tuna a smaller coastal species with shorter migration distance.

Previous studies have documented the links between the frontal activity and the spawning of bluefin tuna [15],[16],[33]. In this study we add to these results by examining the effect of gradients

and their interactions with hydrographic mean. These metrics improved our understanding of the conditions for bluefin and bullet tuna spawning when compared to models using just the hydrographical values but not the gradients. Furthermore the identification of characteristic scales of gradients provided a new source of information for the interpretation of how local oceanography determines the selection of the site to spawn.

For bluefin tuna larvae, spatial salinity and geostrophic velocity gradients maximize spatial model performance when calculated at 0.6 degrees. The higher abundance of bluefin tuna larvae in areas with intermediate to high salinities and with high gradients of velocity is consistent among years. Higher abundance occurs around the location of the main frontal area, at the side of higher salinity of the front and where current speed presents high values. Higher salinity water likely has higher resident time near the islands than the less saline water, which may run along the front towards east getting farther from the archipelago. Spawning at the higher salinity side may favor spatial overlap with other larval species that are also located in this water mass [17],[34].

Results for bullet tuna showed that pelagic seascape metrics are not as relevant to explain the spawning distribution as they are for bluefin. In the western Mediterranean, bullet tuna spawning has been associated to near coastal areas [22], being less influenced by the local oceanography than bluefin tuna [16], which is consistent with our results.

Table 2. Summary of GAM models of larvae abundance for bullet tuna (*Auxis rochei rochei*). Interaction terms included in parenthesis.

Model group	Model variables	R2	Dev(%)	GCV	AIC	delta AIC
Base model	(latitude, longitude) + filtered volume + hour	0,158	32,8	9,615	8857,17	0
One additive variable models	base model + GVEL st0.15	0,177	35,3	9,281	8572,91	284,26
	base model + stSAL0.15	0,16	36,5	9,130	8440,06	417,11
	base model + gSAL0.75	0,207	39,6	8,741	8098,28	758,89
Hydrographic model	base model + stGVEL + stSAL + stTEMP	0,192	38,8	8,847	8183,39	673,78
SAL seascape models	Hydrographic model + stSAL + gSAL0.75	0,215	40,6	8,617	7984,18	872,99
	Hydrographic model + (stSAL,gSAL0.75)	0,255	43,1	8,327	7708,82	1148,35

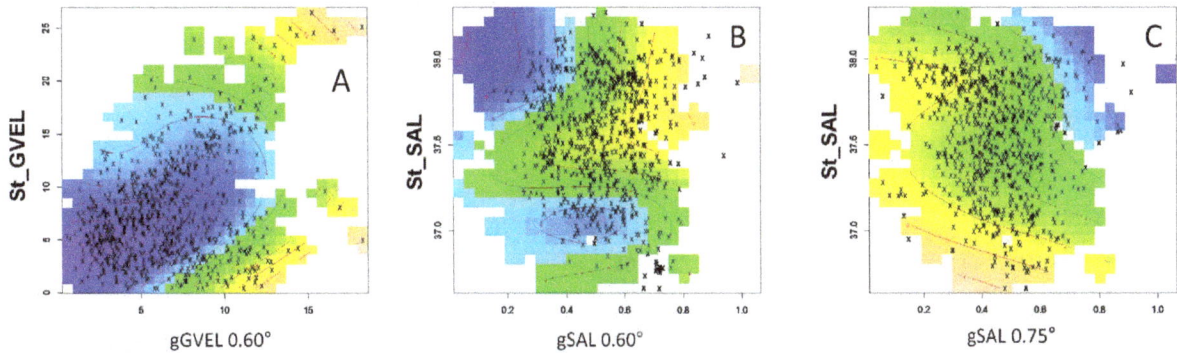

Figure 5. The effect of the interactions of the seascape components on the larval abundance as estimated from the seascape generalized additive model. The effects are shown for bluefin tuna (A–B) and bullet tuna (C). For bluefin tuna: A) the effect of the gGVEL and st_GVEL interaction. B) the effect f the gSAL and st_SAL interaction. For bullet tuna C) the effect of the gSAL and st_SAL interaction. Isolines indicate larval abundances predicted by the model. Peak of abundances are indicated in pink-yellow. Low and very low abundances are indicated in green and blue, respectively.

Despite the lower importance seascape metrics in bullet tuna, the inclusion of salinity gradients provided additional information for the identification of spawning sites. The analysis indicated that bullet tuna spawning areas are mostly found in areas where salinity gradients are low. Bullet tuna was found at both sides of the front but avoiding more mixed waters, located closer to the front. This was verified when observing the spatial distribution of larvae in relation to the salinity seascapes among the different years. For instance, in 2001, 2003 and 2005 high larvae abundances were observed North of the archipelago (high salinity waters with very low gradients), but intermediate abundances, indicating spawning, also occurs in Southern areas (low salinities and intermediate gradients). In 2002 and 2004 higher abundances were linked to low salinity and intermediate gradients shown in the south of the archipelago (Fig. S9 and S10). These results reinforce the theory of bullet tuna spawning occurs in widespread geographic areas, and not only close to the coast and suggest that the location of the main haline front negatively affects the spawning of this species.

Overall results related to bullet tuna point to the fact that, besides the avoidance of areas near strong surface haline gradients, other factors not considered in this study may also be relevant for spawning site selection in this species. It is also relevant that the spatial pattern in relation to the salinity is the opposite to that shown by bluefin tuna, located in areas near the front, suggesting possible avoidance of predators by bullet spawners [35].

The application of seascape metrics derived from salinity and geostrophic currents to characterize the spawning habitat provides new descriptors for environmental variables that improve model quality and predictions. This improvement allows a more precise identification of the relationships between the spatial location of the spawning grounds and the local oceanographic processes. Moreover, our study demonstrates that seascapes must be characterized at specific spatial scales to provide useful information as proposed in previous studies [36] and supporting results on terrestrial landscapes [37],[38] and bottom seascapes [39]. Therefore, the relations between the location of spawning sites and the mesoscale oceanographic processes may prove to be non significant if seascape metrics are not processed at the right spatial scales.

Seascape ecology is an emerging field generally being applied for the analysis of how benthic habitats pattern in coastal areas drives different aspects of marine species ecology [39]. Techniques are applied following categorical approach where the seascape is composed by a number of patches of different type of habitats

[40],[7]. However, very little attention has been paid to the techniques and concepts to investigate pelagic seascape ecology due to the complex spatiotemporal dynamics of this system [2]. Thus, the work presented here sheds new light to modeling spatial distribution and investigating key ecological processes of species highly dependent on the variability of the pelagic environment, like spawning ecology of many of the big tuna species are [41]. In areas as the Balearic Sea, for which new operational oceanography platforms provide near real time data of hydrography [42] and also in combination with remote sensing data (e.g. altimetry [43]) and modeling [44],[45] these metrics will improve the species spatial distribution forecast that has proved effective for management [46].

In contrast to seascapes, landscape metrics have a long history in terrestrial ecology, and have improved over time. For instance, the effect caused on the habitat analysis derived from the spatial definition of the input habitat maps or the extent of the study area are common studied topics, [47],[48]. Likewise, calculation of seascape metrics and the final results from their application in ecological studies may be affected by different issues, like the different ways of computing the hydrographic variables and their gradients, or the origin of the input data source like from in situ measurements, remote sensing or hydrodynamic models, each with different sources of uncertainty. A relevant question is how seascapes can provide information for other type of species and ecological processes. Addressing all these challenges and developing comparative studies between different data sources, processing methods, species and ecological processes will allow advancing towards the understanding of how seascape metrics can provide information about how ecological processes and oceanography are linked together.

In summary, pelagic seascapes based on gradients and characteristic scales allow improving spatial distribution models and the identification of essential fish habitat of pelagic species. They also provide a tool for analyzing the links between particular ecological processes and local oceanography going far beyond than stochastic models based on just hydrographic parameters as salinity, temperature or geostrophic velocities. As a consequence these metrics will provide an improvement in all the management approaches and tools pending on the capability of models to identify essential habitats as near real-time spatial management based on habitat predictions [46],[10], pelagic species distribution from deterministic models [49] or the standardization of larvae indices to assess adult stock, [50],[51],[52].

Supporting Information

Figure S1   Model response of bluefin tuna in relation to salinity gradient processed at 0.75 degrees. Fitted line (solid line) and 95% confidence intervals (grey shaded areas) are shown. Whiskers on the x-axis show the locations of measurements.

Figure S2   Correlation between the gradients at the characteristic scales and the hydrographical variables at the sampled station. A) Current velocity and B) salinity for Atlantic Bluefin tuna. C) Salinity for Bullet tuna.

Figure S3   Sea surface salinities in the area during the five years analyzed (2001 to 2005).

Figure S4   Sea surface geostrophic currents in the area during the five years analyzed (2001–2005).

Figure S5   Spatial distribution of bluefin tuna (*Thunnus thynnus*) larvae in relation to the salinity mean calculated at its characteristic scale (0.6 degrees). Relative stage-1 larval abundances are shown in the maps such as dots.

Figure S6   Spatial distribution of bluefin tuna (*Thunnus thynnus*) larvae in relation to the salinity gradient calculated at 0.6 degrees. Relative stage-1 larval abundances are shown in the maps such as dots.

Figure S7   Spatial distribution of bluefin tuna (*Thunnus thynnus*) larvae in relation to the geostrophic velocity mean calculated at its characteristic scale (0.6 degrees). Relative stage-1 larval abundances are shown in the maps such as dots.

Figure S8   Spatial distribution of bluefin tuna (*Thunnus thynnus*) larvae in relation to the geostrophic velocity gradient calculated at the characteristic scale (0.6 degrees). Relative stage-1 larval abundances are shown in the maps such as dots.

Figure S9   Spatial distribution of bullet tuna (*Auxis rochei rochei*) in relation to the salinity mean calculated at 0.75 degrees. Relative stage-1 larval abundances are shown in the maps such as dots.

Figure S10   Spatial distribution of bullet tuna (*Auxis rochei rochei*) in relation to the salinity gradient calculated at 0.75 degrees. Relative stage-1 larval abundances are shown in the maps such as dots.

Acknowledgments

We thank the scientists and crew participating in the TUNIBAL cruises. We also thank P. Tugores and L. Rueda for their help in software development, and R. Balbín for their comments about methodological approaches.

Author Contributions

Conceived and designed the experiments: DAB LC AAG PR FA MH JLLJ JT. Performed the experiments: DAB LC FA. Analyzed the data: DAB LC AAG PR FA MH. Contributed reagents/materials/analysis tools: DAB LC AAG PR FA MH JLLJ JT. Wrote the paper: DAB LC AAG PR FA MH JLLJ JT. Sample collection: FA JLLJ PR. Laboratory sample processing: FA JLLJ AAG. Software development: DAB LC AAG. Numerical modelling design: DAB LC PR. Environmental variables calculation: DAB LC AAG PR FA MH JLLJ JT. Relevant literature review in different fields: DAB LC AAG PR FA MH JLLJ JT. Final manuscript edition: DAB LC AAG PR FA MH JLLJ JT.

References

1. Hinchey EK, Nicholson MC, Zajac RN, Irlandi EA (2008) Preface: marine and coastal applications in landscape ecology. Landscape Ecology 23: 1–5.
2. Pittman SJ, Kneib RT, Simenstad CA (2011) Practicing coastal seascape ecology. Marine Ecology-Progress Series 427: 187–190.
3. Turner MG (1989) Landscape Ecology - the Effect of Pattern on Process. Annual Review of Ecology and Systematics 20: 171–197.
4. Pittman SJ, McAlpine CA, Pittman KM (2004) Linking fish and prawns to their environment: a hierarchical landscape approach. Marine Ecology-Progress Series 283: 233–254.
5. Boström C, Pittman SJ, Simenstad C, Kneib RT (2011) Seascape ecology of coastal biogenic habitats: advances, gaps, and challenges. Marine Ecology Progress Series 427: 191–217.
6. Wedding LM, Lepczyk CA, Pittman SJ, Friedlander AM, Jorgensen S (2011) Quantifying seascape structure: extending terrestrial spatial pattern metrics to the marine realm. Marine Ecology-Progress Series 427: 219–232.
7. Cushman SA, Gutzweiler K, Evans JS, McGarigal K (2010) The gradient paradigm: a conceptual and analytical framework for landscape ecology. In: Huettmann F, Cushman S, editors. Spatial complexity, informatics and wildlife conservation. Tokyo: Springer. pp. 83–110.
8. Mannocci L, Laran S, Monestiez P, Dorémus G, Van Canneyt O, et al. (2013) Predicting top predator habitats in the Southwest Indian Ocean. Ecography 37: 261–278.
9. Worm B, Sandow M, Oschlies A, Lotze HK, Myers RA (2005) Global patterns of predator diversity in the open oceans. Science 309: 1365–1369.
10. Druon JN, Fromentin JM, Aulanier F, Heikkonen J (2011) Potential feeding and spawning habitats of Atlantic bluefin tuna in the Mediterranean Sea. Marine Ecology-Progress Series 439: 223–240.
11. Louzao M, Pinaud D, Peron C, Delord K, Wiegand T, et al. (2011) Conserving pelagic habitats: seascape modelling of an oceanic top predator. Journal of Applied Ecology 48: 121–132.
12. Hidalgo M, Gusdal Y, Dingsor GE, Hjermann D, Ottersen G, Stige LC, et al. (2012) A combination of hydrodynamical and statistical modelling reveals non-stationary climate effects on fish larvae distributions. Proceedings of the Royal Society B: Biological Sciences 279: 275–283.
13. Block BA, Jonsen ID, Jorgensen SJ, Winship AJ, Shaffer SA, et al. (2011) Tracking apex marine predator movements in a dynamic ocean. Nature 475: 86–90.
14. Phillips AJ, Ciannelli L, Brodeur RD, Pearcy WG, Childers J (2014) Spatio-temporal associations of albacore CPUEs in the Northeastern Pacific with regional SST and climate environmental variables. ICES, Journal of Marine Science. In press.
15. Alemany F, Quintanilla L, Velez-Belchi P, Garcia A, Cortés D, et al. (2010) Characterization of the spawning habitat of Atlantic bluefin tuna and related species in the Balearic Sea (western Mediterranean). Progress in Oceanography 86: 21–38.
16. Reglero P, Ciannelli L, Álvarez-Berastegui D, Balbín R, López-Jurado JL, et al. (2012) Geographically and environmentally driven spawning distributions of tuna species in the western Mediterranean Sea. Marine Ecology Progress Series 463: 273–284.
17. Torres AP, Reglero P, Balbin R, Urtizberea A, Alemany F (2011) Coexistence of larvae of tuna species and other fish in the surface mixed layer in the NW Mediterranean. Journal of plankton research 33: 1793–1812.
18. Block BA, Teo SL, Walli A, Boustany A, Stokesbury MJ, et al. (2005) Electronic tagging and population structure of Atlantic bluefin tuna. Nature 434: 1121–1127.
19. Rooker JR, Secor DH, De Metrio G, Schloesser R, Block BA, et al. (2008) Natal homing and connectivity in Atlantic bluefin tuna populations. Science 322: 742–744.
20. La Violette PE, Tintoré J, Font J (1990) The surface circulation of the Balearic Sea. Journal of Geophysical Research 95: 1559–1568.
21. Balbín R, López-Jurado JL, Flexas MM, Reglero P, Vélez-Velchí P, et al. (2013) Interannual variability of the early summer circulation around the Balearic Islands: driving factors and potential effects on the marine ecosystem. Journal of Marine Systems In Press, DOI: 10.1016/j.jmarsys.2013.12.007.

22. Sabatés A, Recasens L (2001) Seasonal distribution and spawning of small tunas (*Auxis rochei* and *Sarda sarda*) in the northwestern Mediterranean. Scientia Marina 65: 95–100.

23. Muhling BA, Lamkin JT, Roffer MA (2010) Predicting the occurrence of Atlantic bluefin tuna (*Thunnus thynnus*) larvae in the northern Gulf of Mexico: building a classification model from archival data. Fisheries Oceanography 19: 526–539.

24. Rooker JR, Simms JR, Wells RD, Holt SA, Holt GJ, et al. (2012) Distribution and habitat associations of billfish and swordfish larvae across mesoscale features in the Gulf of Mexico. PloS one 7: e34180.

25. de la Gándara F, Ortega A, Blanco E, Viguri FJ, Reglero P (2013) La flexión de la notocorda en larvas de atún rojo,*Thunnus thynnus* (L, 1758) cultivadas a diferentes temperaturas. In: Sociedad Española de Acuicultura, editors. XIV Congreso Nacional de Acuicultura. Gijón: Universidad Laboral de Gijón. pp. 181–182. http://www.repositorio.ieo.es/e-ieo/bitstream/handle/10508/1559/5905A.pdf?sequence=2&isAllowed=y. Accessed 23 July 2014.

26. Gordoa A, Carreras G (2014) Determination of Temporal Spawning Patterns and Hatching Time in Response to Temperature of Atlantic Bluefin Tuna (*Thunnus thynnus*) in the Western Mediterranean. PloS one 9: e90691.

27. Reglero P, Balbín R, Ortega A, Alvarez-Berastegui D, Gordoa A, et al. (2013) First attempt to assess the viability of bluefin tuna spawning events in offshore cages located in an a priori favourable larval habitat. Scientia Marina 77: 585–594.

28. Reglero P, Urtizberea A, Torres AP, Alemany F, Fiksen Ö (2011) Cannibalism among size classes of larvae may be a substantial mortality component in tuna. Marine Ecology Progress Series 433: 205–219.

29. Blank JM, Morrissette JM, Landeira-Fernandez AM, Blackwell SB, Williams TD, et al. (2004) In situ cardiac performance of Pacific bluefin tuna hearts in response to acute temperature change. Journal of Experimental Biology 207: 881–890.

30. RDevelopment CORE (2011) TEAM. 2008. R: A language and environment for statistical computing. R Foundation for Statistical Computing, Vienna, Austria. ISBN 3-900051.

31. Wood, Simon N. (2006) Generalized additive models: an introduction with R. CRC, Boca Raton, Forida: Chapman & Hall.

32. Burnham, Kenneth P. and Anderson, David R. (2002) Model selection and multimodel inference: a practical information-theoretic approach. New York: Springer-Verlag. 488 p.

33. Muhling BA, Reglero P, Ciannelli L, Alvarez-Berastegui D, Alemany F, et al. (2013) Comparison between environmental characteristics of larval bluefin tuna *Thunnus thynnus* habitat in the Gulf of Mexico and western Mediterranean Sea. Marine Ecology Progress Series 486: 257–276.

34. Rodriguez JM, Alvarez I, Lopez-Jurado JL, Garcia A, Balbín R, et al. (2013) Environmental forcing and the larval fish community associated to the Atlantic bluefin tuna spawning habitat of the Balearic region (Western Mediterranean), in early summer 2005. Deep Sea Research Part I: Oceanographic Research Papers 77: 11–22.

35. Bakun A (2013) Ocean eddies, predator pits and bluefin tuna: implications of an inferred 'low risk-limited payoff' reproductive scheme of a (former) archetypical top predator test. Fish and Fisheries 14: 424–438.

36. Steele JH (1989) The ocean landscape. Landscape Ecology 3: 185–192.

37. Wiens JA (1989) Spatial Scaling in Ecology. Functional ecology 3: 385–397.

38. Wu JG, Li HB (2006) Perspectives and methods of scaling. In: Jianguo Wu, K.. Bruce ones, Harbin Li, Orie L.. Loucks, editors. Scaling and Uncertainty Analysis in Ecology. Netherlands: Springer. pp. 17–44.

39. Bostrom C, Pittman SJ, Simenstad C, Kneib RT (2011) Seascape ecology of coastal biogenic habitats: advances, gaps, and challenges. Marine Ecology-Progress Series 427: 191–217.

40. Forman, R. T. T. (1995) Land mosaics: the ecology of landscapes and regions. Cambridge, UK: Cambridge University Press.

41. Reglero P., Tittensor D.P., Álvarez-Berastegui D., Aparicio-González A., Worm B. (2014) Worldwide distributions of tuna larvae: revisiting hypotheses on environmental requirements for spawning habitats. Marine Ecology Progress Series 501: 207–224.

42. Tintore J, Vizoso G, Casas B, Heslop E, Pascual A, et al. (2013) SOCIB: The Balearic Islands Coastal Ocean Observing and Forecasting System, Responding to Science, Technology and Society Needs. Marine Technology Society Journal 47: 101–117.

43. Pascual A, Bouffard J, Ruiz S, Buongiorno Nardelli B, et al. (2013) Recent improvements in mesoscale characterization of the western Mediterranean Sea: synergy between satellite altimetry and other observational approaches. Scientia Marina 77: 19–36.

44. Juzza M, Mourre B, Renault L, Tintoré J (2014) Assessment and intercomparison of numerical simulations in the Western Mediterranean Sea. Geophysical Research Abstracts, Vol 16, EGU2014-6893, 2014, 27 April, 2014 Vienna, Austria. http://adsabs.harvard.edu/abs/2014EGUGA..16.6893J. Accessed 23 July 2014

45. Juza M, Renault L, Ruiz S, Tintoré J (2013) Origin and pathways of Winter Intermediate Water in the Northwestern Mediterranean Sea using observations and numerical simulation. Journal of Geophysical Research: Oceans 118: 6621–6633.

46. Hobday AJ, Hartmann K (2006) Near real-time spatial management based on habitat predictions for a longline bycatch species. Fisheries Management and Ecology 13: 365–380.

47. Cushman SA, McGarigal K, Neel MC (2008) Parsimony in landscape metrics: strength, universality, and consistency. Ecological Indicators 8: 691–703.

48. Wu JJ (2013) Landscape ecology. In: Rik Leemans, editors. Ecological Systems. New York: Springer. pp. 179–200.

49. Lehodey P, Senina I, Murtugudde R (2008) A spatial ecosystem and populations dynamics model (SEAPODYM)-Modeling of tuna and tuna-like populations. Progress in Oceanography 78: 304–318.

50. Ingram Jr GW, Richards WJ, Porch CE, Restrepo V, Lamkin JT, et al. (2008) Annual indices of bluefin tuna (*Thunnus thynnus*) spawning biomass in the Gulf of Mexico developed using delta-lognormal and multivariate models. ICCAT Working Document SCRS/2008/086.

51. Ingram Jr GW, Alemany F, Alvarez-Berastegui D, García A (2013) Development of indices of larval bluefin tuna (*Thunnus thynnus*) in the Western Mediterranean Sea. ICCAT,Collect Vol Sci Pap 69: 1057–1076.

52. Muhling BA, Lee SK, Lamkin JT, Liu Y (2011) Predicting the effects of climate change on bluefin tuna (*Thunnus thynnus*) spawning habitat in the Gulf of Mexico. ICES Journal of Marine Science: Journal du Conseil 68: 1051–1062.

18

The Relationship between the Distribution of Common Carp and Their Environmental DNA in a Small Lake

Jessica J. Eichmiller*, Przemyslaw G. Bajer, Peter W. Sorensen

Department of Fisheries, Wildlife, and Conservation Biology, Minnesota Aquatic Invasive Species Research Center, University of Minnesota, Twin Cities, St. Paul, Minnesota, United States of America

Abstract

Although environmental DNA (eDNA) has been used to infer the presence of rare aquatic species, many facets of this technique remain unresolved. In particular, the relationship between eDNA and fish distribution is not known. We examined the relationship between the distribution of fish and their eDNA (detection rate and concentration) in a lake. A quantitative PCR (qPCR) assay for a region within the cytochrome *b* gene of the common carp (*Cyprinus carpio* or 'carp'), an ubiquitous invasive fish, was developed and used to measure eDNA in Lake Staring (MN, USA), in which both the density of carp and their distribution have been closely monitored for several years. Surface water, sub-surface water, and sediment were sampled from 22 locations in the lake, including areas frequently used by carp. In water, areas of high carp use had a higher rate of detection and concentration of eDNA, but there was no effect of fish use on sediment eDNA. The detection rate and concentration of eDNA in surface and sub-surface water were not significantly different ($p \geq 0.5$), indicating that eDNA did not accumulate in surface water. The detection rate followed the trend: high-use water > low-use water > sediment. The concentration of eDNA in sediment samples that were above the limit of detection were several orders of magnitude greater than water on a per mass basis, but a poor limit of detection led to low detection rates. The patchy distribution of eDNA in the water of our study lake suggests that the mechanisms that remove eDNA from the water column, such as decay and sedimentation, are rapid. Taken together, these results indicate that effective eDNA sampling methods should be informed by fish distribution, as eDNA concentration was shown to vary dramatically between samples taken less than 100 m apart.

Editor: Arga Chandrashekar Anil, CSIR- National institute of oceanography, India

Funding: Funding for this project was provided by the Minnesota Environment and Natural Resources Trust Fund as recommended by the Legislative-Citizen Commission on Minnesota Resources (LCCMR). The funders had no role in study design, data collection and analysis, decision to publish, or preparation of the manuscript.

Competing Interests: The authors have declared that no competing interests exist.

* Email: eich0146@umn.edu

Introduction

Methods to quantify the abundance of fish populations, such as mark-recapture and electrofishing, are costly and time-consuming. In addition, fish are often difficult to capture and detect at low densities, and capture methods themselves can lead to behavioral changes of the target species [1–3]. Molecular methods to detect the DNA released by aquatic organisms into their environment are non-invasive, rapid, and potentially more sensitive than traditional census techniques [4–6]. This environmental DNA (eDNA) is released through processes such as cell sloughage, mucus excretions, and defecation [7]. Notably, eDNA is currently used to monitor the presence of invasive Bigheaded carps (often called 'Asian carps') (*Hypophthalmichthys* spp.) in the Chicago Area Waterway System and the Mississippi River [8]. Although initially developed as a detection tool, molecular techniques that utilize eDNA are evolving to answer more complex questions. For example, several studies have established relationships between eDNA concentration and biomass in aquatic habitats [9–11]. Next-generation sequencing approaches have successfully identified multiple species simultaneously [11,12].

Despite the immense potential for eDNA technology to revolutionize monitoring programs for fish and other aquatic species, little is known about the production, fate, and distribution of eDNA in the natural environment. The distribution of eDNA is of particular importance for development of effective monitoring methods [6]. Surprisingly, Pilliod et al. [9] found that time of day, sampling location, and distance from the target organism (salamanders) had no apparent effect on eDNA concentration in small streams. In contrast, eDNA from snails was more abundant in the middle of a river channel relative to the channel margins [13]. Surface water samples are widely used for eDNA studies [8,9,14]. The rationale for this approach has only been confirmed in one study done in experimental ponds [15]. The possibility that eDNA concentration within a water body may be influenced by fish distribution was initially posed by Takahara et al. [10]. In a lagoon in winter, the concentration of eDNA from common carp (*Cyprinus carpio*, hereafter 'carp') was positively correlated with water temperature and was spatially heterogeneous. The cause of this pattern, and in particular whether it was due to the distribution of carp or higher metabolic activity of fish in warmer waters, was not examined as the distribution of carp was not measured. From the few studies that address the question of eDNA distribution in water, it is clear that it varies among habitat types, and more conclusive explanations of eDNA distribution patterns are needed. Also of interest is the distribution of eDNA in sediments, as sediments likely retain eDNA for long periods of time [16].

To determine whether fish distribution affects eDNA concentration and detection rate in lake water and sediment, we examined the distribution of carp eDNA in a small, shallow lake and compared it to known patterns of carp distribution, which had been monitored for several years. We were interested both in detection rate (percentage of samples in which eDNA levels were present above detection threshold) as well as concentration, because the former is commonly used to assess the likely distribution of invasive Bigheaded carps while the latter measure, if understood, might add more resolution and value to the technique. First, a qPCR assay specific for *C. carpio* eDNA was developed and validated in the lab. Next, since eDNA is often assumed to accumulate in surface water and sediment, surface, sub-surface, and sediment samples were taken throughout the lake. Finally, the concentration and detection rate of eDNA was compared between areas of low- and high-fish use identified from radiotelemetry data. Results of this study provide insights into optimal eDNA sampling methods for small lakes as well as information on how eDNA is distributed in aquatic systems in relation to the distribution of target organisms.

Materials and Methods

Quantitative PCR marker development and validation

Although two *C. carpio* qPCR assays had been developed prior to this study [10,17], a screen against the NCBI database indicated potential non-specific amplification of non-target fish species (Table S1). Therefore, a qPCR assay was developed for the current study. Four genes were considered in the development of a novel qPCR marker specific to the common carp: (1) mitochondrial gene cytochrome *b*, (2) mitochondrial gene cytochrome *c* oxidase subunit 1, (3) mitochondrial gene control (D-loop) region, and (4) the nuclear gene recombination activating gene 1 (RAG1). Candidate primer sets were identified by NCBI Primer-BLAST using sequences under GenBank accession number X61010.1 [18] for mtDNA and EF458304.1 [19] for the RAG1 gene. Specificity was initially screened against the BLASTn database sequences for 15 fish species (Table S1). Minor groove binder (MGB) probes were manually designed using the Primer Probe Test Tool in Primer Express Software v3.0.1 (Life Technologies, Grand Island, NY). Assays with amplification efficiency outside the range of acceptable values of 90–110% or a limit of detection above 300 copies per reaction were not considered. We defined the limit of detection (LOD) as the lowest value at which three replicate reactions would successfully amplify with a quantification cycle (Cq) value of less than 40 cycles within the linear range of the standard curve.

Candidate markers were screened for specificity for carp by testing for amplification of 15 ng of fin clip DNA from carp and 34 native and non-native fish species (Table S1). Fin-clip samples for genetic marker specificity testing were extracted using the DNeasy Blood and Tissue Kit (Qiagen, Hilden, Germany) and assayed as described below.

Next, we tested markers using aqueous samples. Three 340 L flow-through tanks were set up to confirm the ability of the marker to detect carp eDNA. Prior to this experiment, all tanks were treated with 10% bleach for 30 minutes to remove all traces of DNA. The flow through rate was set at 600 mL/min, and

temperature was maintained at 18°C. The first tank was stocked with 10 carp (35 g), and the second tank was stocked with 10 goldfish (*Carassius auratus*) (50 g), while the third tank was stocked with five fish of both species. These stocking levels corresponded to a biomass of 0, 438, and 875 mg/L of carp. Fish were fed once daily *ad libitum* a combination of flake feed (Color Tropical Marine Flake, Pentair Aquatic Eco-systems, Inc., Apopka, FL) and 2.5 mm pellet feed (Oncor Fry, Skretting USA, Tooele, UT) that did not contain target genetic markers. After 6 days, 4 1 L water samples were collected from each tank, immediately stored at 4°C, and filtered within 4 h. Molecular analyses followed protocols described below. This study was carried out in strict accordance with the recommendations in the Guide for the Care and Use of Laboratory Animals of the National Institutes of Health. The protocol for care and holding of laboratory fish was approved by the University of Minnesota's Institutional Animal Care and Use Committee (IACUC) (Protocol: 1407-31659A). No anesthesia or euthanasia was required as part of this study.

Study site

The study site was Lake Staring, a small freshwater lake located in the Upper Mississippi River Basin (44°50'14" N, −93°27'18" W). Lake Staring is a small, shallow lake that experiences frequent mixing due to wind and is typical of high carp density lakes in this region [20]. The surface area of the lake is 65.7 ha, consisting mostly of littoral zone with a depth of less than 2 m. The maximum depth is 4.8 m, and the lake bed is composed of fine sediment. Due to high carp density, the lake lacks aquatic vegetation except for white water lily (*Nymphaea odorata*), which covers less than 10% of the lake area.

Carp population abundance in Lake Staring was estimated in 2011 using a mark-recapture analysis [20]. This analysis showed that the lake was inhabited by approximately 26,000 carp, 95% CI [21,000, 31,000], or approximately 400 carp/ha. The mean body length of carp was 444 mm, indicating that the population was primarily composed of adults [20]. Approximately 14,000 fish were removed from the lake in the winters of 2012 and 2013, and a new population estimate was generated in the fall of 2013 by conducting a mark-recapture analysis. For the mark-recapture analysis, 46 carp were marked with individually numbered tags in November 2013, of which 22 were recaptured during the following four months among 5,457 carp that were captured and examined for marks. Using these data we estimated that Lake Staring was inhabited by 11,153 carp, 95% CI [7,972, 14,334] in the fall of 2013. The biomass decreased only slightly, from 490 kg/ha in 2011 to 397 kg/ha in 2013 because the mean body length of carp increased to 559 mm over this time frame. The biomass at the time of this study was approximately 20 mg/L, assuming an average lake depth of 2 m.

Since 2011, the distribution of carp has been regularly assessed by locating 10–20 carp equipped with internal radio tags (F 1850, Advanced Telemetry Systems, Isanti, MN). Carp location was determined by identification of signal directionality (bearing) with a hand-held antenna and a compass while positioned within 200 m of the radiotagged fish. Two bearings were measured for each fish, each from a different location, and their intersection calculated (LOAS, Ecological Software Solutions, CA) to estimate fish location. Mean measurement error (30 m) was estimated using dummy tags.

Using previously determined carp locations, the pattern of carp habitat usage was examined for the warm season (June–October) when carp maintain stable summertime distributions [21]. A total of 12 radiotelemetry surveys within the 2011–2013 time frame were conducted. Individual locations (N = 135) were pulled across fish and years. Areas of high carp use were estimated by calculating kernel density (search radius = 35 m, approximately one SE of carp location estimate; output cell size = 4.2 m) using spatial analyst in ArcMap (10.0, Esri, Redlands, CA).

Carp showed well-defined areas of habitat usage in Lake Staring (Figure 1A). A density of 800 radiotagged carp/km^2 was used as the cut-off between high- and low-use areas. As such, areas with lower values were considered to be low-use areas, whereas areas with higher densities were classified as high-use areas. The value of 800 radiotagged carp/km^2 corresponded to approximately 1,248 carp/ha. This cut-off value was chosen because lower densities were associated with relatively isolated radiotagged carp observations. The high-use areas were located within or near the patches of lilies, which the carp most likely use for cover as the lake lacks other physical structure. In addition, all high-use areas were within 200 m of the shoreline and less than 2 m deep.

Field sampling

Field sampling took place on 8 October 2013. Average wind speed was 20 km/h from the S [22], while air and water temperatures were 17.8°C and 19.8°C, respectively. Water and sediment samples were collected from 24 locations within the lake. Samples were taken at 18 points at 4 to 5 locations along three N to S transects of the lake and at one location at the E and W ends. Six additional sampling points were added within three patches (two samples per patch) of N. odorata where we knew carp were generally found. At each sampling site, surface water, sub-surface (0.5 m depth) water, and sediment were sampled. A surface water sample was also taken in both the inflow and outflow of the lake.

Water samples were collected in 1-L HDPE bottles (Nalgene, Rochester, NY) that had been previously soaked in 10% bleach for at least 30 min to remove all traces of DNA. Bottles were subsequently rinsed with distilled water to remove residual bleach. Surface water samples were taken by partially submerging a

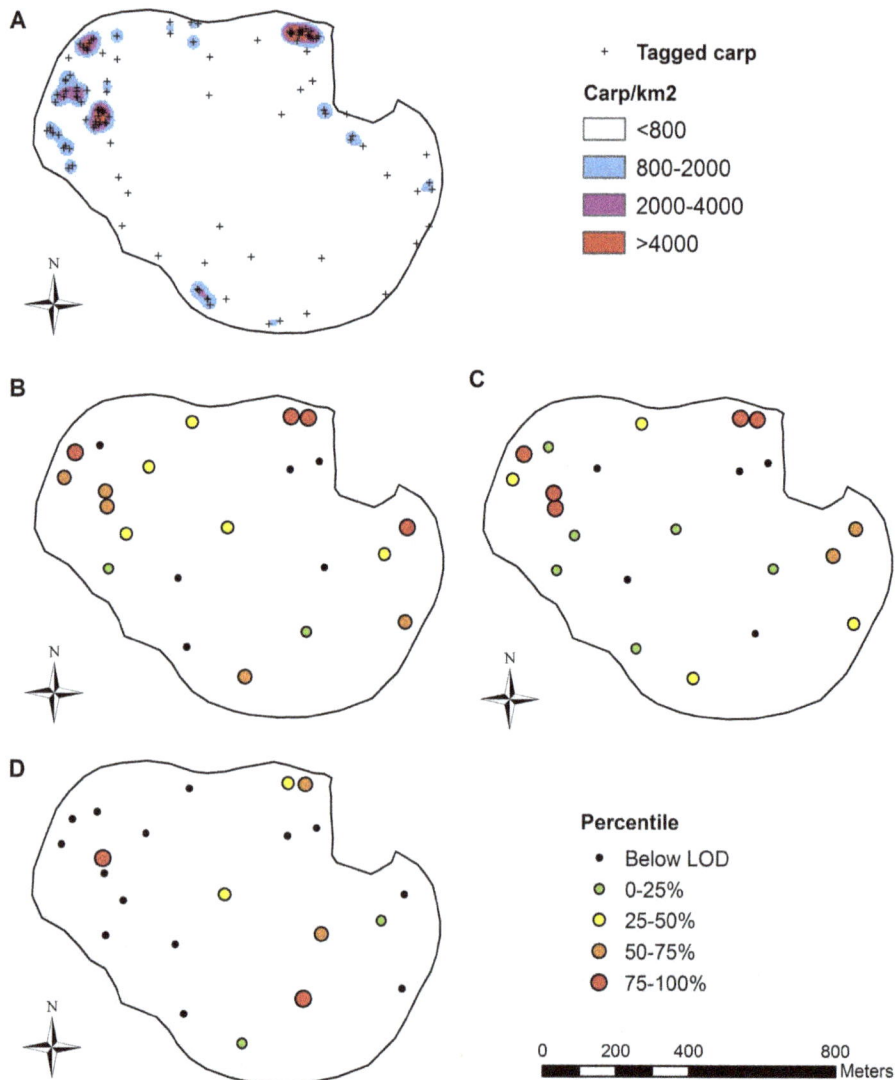

Figure 1. Carp use and distribution of eDNA in Lake Staring. Panel A shows locations of radiotagged carp and high- and low-use areas. Density categories represent the average number of locations of radiotagged carp/km^2. The high- and low-use area cut-off value of 800 radiotagged carp/km^2 corresponded to approximately 1,248 carp/ha. Panels B–D show the pattern of eDNA detection and concentration in surface water (B), sub-surface water (C), and sediment (D). All figures have the same scale. The symbol legend in the upper right refers to panel A, whereas lower right refers to panels B–D.

sample bottle to collect water from the top few cm. Sub-surface samples were taken using a stainless steel Van Dorn sampler (Wildlife Supply Company, Yulee, FL). Sediment samples were collected using a stainless steel Petite Ponar grab sampler (Wildlife Supply Company, Yulee, FL). Sediment was transferred to a sterile Whirl-Pak bag (Nasco, Fort Atkinson, WI) using a sterile polystyrene spatula (Bel-Art, Wayne, NJ). Once collected, samples were immediately placed on ice. Water and sediment samples were stored at 4°C and were filtered within 24 h. No specific permissions were required for access to the study site or collection of samples as part of this study. No animals were collected as part of this study.

Molecular Analyses

Water samples were filtered through Whatman 934-AH 1.5 μm glass microfiber filters (GE Whatman, Fairfield, CT) using a polyphenylsulfone filter funnel (Pall Corporation, Port Washington, NY). Filter funnels and forceps were soaked in 10% bleach and rinsed in distilled water prior to use and between samples. For tank samples, 1 L of water was filtered per sample. For field samples, only 200 mL could be filtered per sample due to clogging from the high amount of suspended solids. Filters were stored at − 80°C until DNA extraction.

Sediment samples were homogenized, and a 0.1 g subsample was stored at −80°C for DNA extraction. Preliminary experiments showed that extraction of greater than 0.1 g of sediment lead to reduction in eDNA yield and inhibition of qPCR, regardless of post-extraction inhibitor removal protocols or inclusion of PCR adjuvants. For moisture content analysis, a 10 g subsample of sediment was weighed and then dried at 100°C for 24 h.

DNA was extracted using the QIAamp DNA Stool Mini Kit (Qiagen, Hilden, Germany) using the human DNA analysis protocol. Frozen filters were sliced into 1 mm×5 mm fragments with a sterile razor blade and then transferred to extraction tubes. For sediment samples, extraction buffer was directly added to frozen sediment. Before extraction, 50 ng of UltraPure Salmon Sperm Solution (Life Technologies, Grand Island, NY) was added to adjust for extraction efficiency of DNA as previously described [23]. DNA was eluted in a final volume of 50 μL. To further remove potential inhibitors, all DNA extracts were processed with the Wizard Genomic DNA Purification Kit (Promega, Madison, WI).

A multiplex qPCR assay was designed to amplify both the CarpCyt*b* and the extraction control targets, and the oligonucleotide concentrations were optimized for this study (Table 1). CarpCyt*b* standard was created by cloning PCR product amplified from carp fin clip DNA for the CarpCyt*b* genetic marker (Table 1) using the StrataClone PCR kit (Stratagene, Santa Clara, CA). Purified plasmid DNA was quantified by using a QuantiFluor-ST Fluorometer (Promega, Madison, WI). For the extraction control, standards were created by diluting UltraPure Salmon Sperm Solution (Life Technologies, Grand Island, NY). CarpCyt*b* and extraction control standards were combined prior to preparation of five qPCR standards, ranging from 50 to 300,000 CarpCyt*b* copies and 1.6 to 10,000 pg control DNA per 5 μL.

The assay used iTaq Universal Probes Supermix (Bio-Rad, Hercules, CA). Reactions contained 12.5 μL mastermix, 10 μg bovine serum albumin (New England Biolabs Inc., Ipswich, MA) primers and probe, and water or sediment DNA in a final reaction volume of 25 μL. The volume of water and sediment DNA added to the qPCR reaction was adjusted by testing a dilution series of a subset of 5 samples from water and sediment to confirm that inhibition was not present. For sediment samples, 2.5 μL of DNA

extract was added to the reaction, and for water samples 5 μL of DNA extract was added. Reaction conditions consisted of an initial denaturation at 95°C for 3 min, followed by 40 cycles of denaturation at 95°C for 15 s and an annealing and extension step at 60°C for 1 min. Each qPCR run contained triplicate reactions of standards, non-transcript controls, and samples. Amplifications were performed using the StepOnePlus Real-Time PCR System (Life Technologies, Grand Island, NY), and C*q* values were automatically determined using the system software. Sample marker concentrations were calculated on a per-run basis. All sediment values are reported per dry g.

Statistical analyses

Detection rate of eDNA was defined as the proportion of samples that were above the qPCR assay LOD. To analyze the effects of water sample depth (surface, sub-surface), carp usage (low-use, high-use), and matrix type (water, sediment) on eDNA detection rate, the number of detections and non-detections were statistically compared using Fisher's exact test. Fisher's exact test was used due to low expected values (<5) in some cells.

Concentration of eDNA in water was analyzed using a three-way Analysis of Variance (ANOVA). Main effects and 2-way interactions of carp usage (low-use, high-use), water sample depth (surface, sub-surface), and lake depth (m) were examined. Lake depth was included to determine the potential for suspended sediment to affect water column eDNA concentration, with shallower depths more likely to be affected by sediment mixing into the water column. The 3-way ANOVA was restricted to sites greater than 0.5 m and less than 2 m depth (i.e. omitting 2 shallower and 7 deeper sampling sites) because of a partial confound (high-use areas were not found at depths greater than 2 m, and no low-use areas were sampled at depths less than 0.5 m). Finally, student's t-test was used to determine whether there was a significant difference between eDNA concentration of sites included in the ANOVA and those excluded for the low-use areas. Since only 2 sites within the high-use area were excluded, no statistical comparison was done.

For all parametric descriptive analyses and statistical tests, eDNA concentrations were log$_{10}$ transformed to achieve normal data distribution. Values below the LOD were given a value of half the LOD prior to analysis in order to reduce skewing of data. For graphical representation of eDNA concentrations across sampling points, data above the LOD were divided into four equal percentile categories. Percentiles were determined independently for the two sample types: (1) surface and sub-surface water and (2) sediment samples. All statistical tests were conducted in JMP, Version 10 (SAS Institute Inc., Cary, NC).

Results

Carp genetic marker development and laboratory validation

A carp-specific genetic marker (CarpCyt*b*) was developed for a 149 bp region in the cytochrome *b* gene (Table 1). This assay had an R^2 of over 0.99 and an average PCR efficiency of 92%. The lowest copy number that all three replicate reactions reliably and successfully amplified was determined to be the assay LOD. The assay LOD was 50 copies per reaction, which corresponded to 2.0×10^4 copies/L for water samples and 1.0×10^5 copies/g for sediment samples (calculated from the amount of DNA extract analyzed and the volume filtered or weight extracted). The sample LOD varied slightly for individual samples depending on extraction efficiency. Further details regarding qPCR calibration curves can be found in Table S2. The average extraction efficiency

Table 1. Primers and probes used for multiplex quantitative PCR.

Assay	Target	Locus	Primer/Probe	Sequence (5' to 3')[a]	Conc. (nM)	Ref.
CarpCytb	*Cyprinus carpio*	Cytochrome *b*	CCcytbF	CTAGCACTATTCTCCCCTAACTTAC	200	This study
			CCcytbR	ACACCTCCGAGTTTGTTTGGA	200	
			CCcytbP	(6FAM) CCCTCTAGTTACACCACC (MGBNFQ)	200	
Extraction control	*Oncorhynchus keta*	ITS[b] region 2	SketaF2	GGTTTCCGCAGCTGGG	200	[36]
			SketaR3	CCGAGCCGTCCTGGTCTA	200	
			SketaP2	(JOE) AGTCGCAGGCGGCCACCGT (BHQ-1)	100	

[a]BHQ1, black hole quencher-1; 6FAM, 6-carboxyfluorescein; JOE, 6-carboxy-4',5'-dichloro-2',7'-dimethoxyfluorescein.
[b]ITS, internal transcribed spacer.

was 11% for water samples and 6% for sediment samples. The assay reliably quantified up to 3.0×10^5 copies per reaction, the highest standard tested. No amplification of non-transcript controls was observed. The assay did not amplify DNA from a selection of 34 native and non-native fish species, including Bigheaded carps and other related Cyprinids (Table S1).

In validation tests using laboratory tanks which were stocked with combinations of carp and goldfish, no CarpCytb markers were detected in the tank that contained only goldfish. In the tank with 10 carp, the concentration of markers averaged 1.3×10^7 copies/L 95% CI [1.0×10^7, 1.6×10^7]. In the tank with 5 each of carp and goldfish, the concentration of markers was 4.5×10^6 copies/L, 95% CI [3.1×10^6, 6.7×10^6] which was significantly lower ($p = 0.007$, Student's t-test) than that of carp alone, indicating that the presence of more individuals yielded more eDNA. On average, individual carp contributed 3.1×10^{11} copies of CarpCytb in the mixed species tank and 4.4×10^{11} copies per individual in the carp only tank.

Detection rates of eDNA in Lake Staring

The overall detection rate of CarpCytb in water samples was 75% (Table 2). The distribution of eDNA was patchy, with carp eDNA detected tens of meters from sites where eDNA was not detected (Fig. 1B, C). The detection rate was not statistically different between surface and sub-surface samples ($p = 1.00$, Fisher's exact test). However, detection rate was significantly higher in water samples collected in high-use areas ($p = 0.009$, Fisher's exact test), a difference of nearly 40% (Table 2).

The overall detection rate of CarpCytb in sediment was only 36% (Table 3). Similar to the water samples, detection pattern was patchy (Fig. 1D). The detection rate of CarpCytb in sediment was slightly higher in high-use areas relative to low use areas, however, the difference between these use areas was not significant ($p = 1.00$, Fisher's exact test). The detection rate of CarpCytb in low-use areas in the sediment was approximately 30% less than for the average value for water samples, however, there was no difference between the detection rate ($p = 0.11$, Fisher's exact test). The detection rate of eDNA in sediment within high-use areas was nearly 60% less than water within high-use areas, and the difference was statistically significant ($p = 0.005$, Fisher's exact test).

Concentration of eDNA in Lake Staring

CarpCytb concentration in water ranged from below the LOD (2.0×10^4 copies/L) to 1.7×10^6 copies/L (Table S3). The mean CarpCytb concentration across all water samples was 5.7×10^4 copies/L, 95% CI [3.9×10^4, 8.3×10^4] (Table 2). Most samples

(84%) had less than 3.0×10^5 copies/L. Only three samples had a marker concentration above 5.0×10^5 copies/L, Surface water samples from the inflow and outflow streams had CarpCytb concentrations of 6.0×10^4 and 3.4×10^4 copies/L, respectively.

Three-way ANOVA showed that carp use pattern had a significant (main) effect on eDNA concentration (Table 4). There was no significant effect of lake depth or sample depth on eDNA, and no 2-way interactions were significant (Table 4). For the subset of sampling locations considered in the ANOVA (0.5 to 2 m lake depth), in low-use areas the average CarpCytb concentration was 3.3×10^4 copies/L, 95% CI [1.8×10^4, 5.9×10^4], and in high use areas CarpCytb concentration averaged 1.6×10^5 copies/L, 95% CI [1.1×10^5, 2.4×10^5]. There was no significant difference in eDNA concentration of the shallow water sites included in the ANOVA and those excluded for the low carp use areas ($p = 0.75$).

For sediment samples, the concentration of eDNA ranged from below the LOD (1.0×10^5 copies/g) to 5.4×10^5 copies/g (Table S3). For sediment samples above the LOD, the concentration of CarpCytb was slightly higher in high-use areas (Table 3), but the difference was not significant ($p = 0.3$, Student's t-test). On a per mass basis, the lowest measureable sediment concentration of CarpCytb was nearly two orders of magnitude greater than the water sample with the highest concentration of CarpCytb.

Discussion

This study found that both the detection rate and concentration of carp eDNA strongly correlated with the distribution of carp in lake water. In water, the concentration of the carp genetic marker CarpCytb was over 7 times greater in high-use areas as opposed to low-use areas, and detection rate rose from 63% to 100%. The detection rate and concentration of eDNA did not differ between surface and sub-surface water samples. Detection rate was comparably low in sediment, at 36%. The distribution of eDNA is fundamentally important in the design of eDNA sampling schemes and accurate interpretation of eDNA data [6]. Thus, we have shown that the distribution of a target organism must be carefully considered in the design of eDNA sampling schemes and accurate interpretation of eDNA data. Specifically, eDNA is patchily distributed in the environment, and the probability of detecting a target organism may drastically decline tens of meters from areas that are frequently inhabited. As part of this study, a highly-specific qPCR assay for common carp, an invasive and broadly distributed fish, was developed and validated.

The patchiness of eDNA distribution within water and sediment samples taken from Lake Staring was unexpected, given that the lake is small, shallow, and has a high biomass of carp. However,

Table 2. Concentration and detection rates of CarpCytb in water.

Carp usage	Surface			Sub-surface			Total		
	Mean (copies/L) [95% CI]	Detection rate (%)	N	Mean (copies/L) [95% CI]	Detection rate (%)	N	Mean (copies/L) [95% CI]	Detection rate (%)	N
Low-use	3.8×10^4 [2.2×10^4, 6.8×10^4]	60	15	2.7×10^4 [1.8×10^4, 4.2×10^4]	67	15	3.1×10^4 [2.1×10^4, 4.5×10^4]	63	30
High-use	2.1×10^5 [9.8×10^4, 3.7×10^5]	100	7	2.6×10^5 [1.3×10^5, 3.7×10^5]	100	7	2.4×10^5 [1.4×10^5, 3.8×10^5]	100	14
Total	6.6×10^4 [3.7×10^4, 1.2×10^5]	73	22	5.5×10^4 [3.1×10^4, 9.9×10^4]	77	22	5.7×10^4 [3.9×10^4, 8.3×10^4]	75	44

Pilliod et al. [9] also observed high variation in amphibian eDNA concentration of replicate water samples in small freshwater streams. The authors hypothesized that variation was due to downstream pulses of eDNA due to activity of the target organism or variation in the cell type or form (free or cellular) of eDNA. An alternative, and perhaps complimentary explanation, is that mechanisms of eDNA removal from the water column, such as sedimentation and decay, are rapid. Lake Staring is a eutrophic lake, with high productivity and turbid, nutrient rich water. Carp eDNA is primarily contained in the particle size fraction ranging from 1.0–10 μm [24]. Particulate eDNA is continuously settling into lake sediments, but suspended particles are also hot spots of microbial degradation in aquatic systems [25]. Although decay of eDNA was not measured in this study, microcosm studies suggest that eDNA decays rapidly in the environment, and eDNA degrades nearly 90% within several days in water [12,26,27,28]. Taken together, it is likely that rapid removal of eDNA from the water column through processes of decay and sedimentation prevented its accumulation and diffusion from release points in Lake Staring, leading to significantly higher concentrations of eDNA in areas where carp were present.

Regardless of the cause of eDNA's patchy distribution, it has implications for the optimal sampling of eDNA. In low-use areas, the detection rate of eDNA was 40% lower in low-use areas as opposed to high-use areas. Distances between high-use and low-use sampling sites were tens or hundreds of meters apart, and that small distance, in some instances, affected whether eDNA was detected or not. Differences on a fine spatial scale have been observed in streams [9,29], experimental ponds [15], and lakes [10]. Therefore, we conclude that eDNA sampling should be conducted on small spatial scales. Only after extensive testing and consideration should sampling intervals of greater distances be used.

To the authors' knowledge, only one other study has attempted to correlate fish eDNA distribution within a lentic system. A positive relationship between carp eDNA concentration and temperature was noted by Takahara et al. [10] within a Japanese lagoon using a different marker on a slightly coarser scale without explicit information of carp distribution. The authors posited that the distribution corresponded to temperature because carp prefer warmer water; however, the data were not compared to the actual distribution of carp, as in the present study. Therefore other factors, such as higher metabolic activity of fish within warmer waters, could not be ruled out. Nevertheless, there are concordances with patterns observed between these studies. For example, excluding the lagoon channel, areas with higher eDNA appeared to have been located near the shoreline [10]. Similarly, hot spots of eDNA near shore was observed within Lake Staring, due to carp aggregation.

The overall detection rate of carp eDNA in Lake Staring water was 77%, similar to other studies conducted in areas of high fish abundance. For example, the eDNA of Bigheaded carps in the surface waters of a reach of the Mississippi River with a high target population was detected in 64% of samples [30]. Similarly, a 90% of samples were positive for common carp eDNA in a lagoon used for breeding purposes by Takahara et al. [10].

Despite the assumption of eDNA accumulation in surface waters, water depth did not affect concentration or detection rate of carp eDNA in the study lake. Therefore, both surface and sub-surface samples were equally effective for eDNA sampling in the present study. Although most eDNA samples are taken from the water's surface [11,14,26,31], only one study, done in experimental ponds, has confirmed that eDNA is most frequently detected in surface waters [15]. The convention of surface water sampling

Table 3. Concentration and detection rates of CarpCytb in sediment.

Carp usage	Conc. of samples above LOD		Detection rate	
	Mean (copies/g)	N	(%)	N
	[95% CI]			
Low-use	1.2×10^5	5	33	15
	$[7.5 \times 10^4, 1.8 \times 10^5]$			
High-use	2.3×10^5	3	43	7
	$[1.1 \times 10^5, 4.8 \times 10^5]$			
Total	1.5×10^5	8	36	22
	$[1.1 \times 10^5, 2.1 \times 10^5]$			

may be a holdover from early methods, wherein eDNA was used to detect floating feces of large marine mammals [32]. Our results indicate that the level of eDNA accumulation in surface water may differ among species, and accumulation likely depends on the relative proportion of eDNA sources, buoyancy of fecal material, and site-specific factors.

Although eDNA is hypothesized to accumulate in sediment, the detection rate was unexpectedly low, at 36%. This is likely due to the high LOD of the CarpCytb marker in sediment. The LOD in sediment was 100,000 copies/g, nearly 4 orders of magnitude higher than for water samples on a per g basis. The high LOD is likely partly due to the limited amount of sediment that could be extracted. DNA extracts prepared with more than 0.1 g of sediment were observed to inhibit the qPCR reaction; therefore, greater amounts of sediment were not capable of being processed. Although the extraction efficiency of sediment was within the range previously observed for commercial DNA extraction kits [33], it was approximately half that of water samples. We do not know the cause of this discrepancy, but we hypothesize that sample chemistry can differentially affect extraction efficiency. Due to the difficulty of DNA recovery from sediment and the potential for qPCR inhibition, water sampling was more efficient for detection of carp in the study lake.

Regardless of the low detection rate of eDNA in sediment, its importance as a reservoir of carp eDNA cannot be disregarded. The concentration of CarpCytb was high in sediment locations where eDNA was detected, but as the majority of the samples (63%) were below the LOD, we were unable to reliably calculate a mean sediment eDNA concentration. The accumulation of eDNA

in sediment has been suspected based on the high concentration of microbial DNA in sediment [34,35], but measurements of fish eDNA in sediment have not been previously published. Hot spots of eDNA in sediment did not correlate with carp use. Therefore, there is a need for future studies to measure factors that may control eDNA distribution in sediment, such as deposition, resuspension, and degradation rates.

Conclusions

Sampling design has been previously identified as one of the four critical aspects that must be optimized in a DNA-based monitoring program [6]. The present study showed that common carp distribution led to spatial patterns in both eDNA concentration and detection rate in a small, shallow lake. Our results show that while eDNA is relatively evenly distributed in the water column, eDNA is patchily distributed horizontally. The large variation of eDNA on a small spatial scale, of tens to hundreds of meters, indicates that sampling for aquatic species using eDNA should use a similarly fine scale, at least for initial surveys. The results of this study also indirectly suggest that mechanisms of eDNA removal from the water column are rapid and may partially control eDNA distribution. Although the observations of the current study may not be universally applicable to all species and habitats, our results indicate that eDNA sampling schemes should be critically evaluated for the specific organism and the type of aquatic environment they inhabit. Future research is needed to examine the role of decay, sediment re-suspension, and eDNA release on eDNA distribution in aquatic habitats.

Supporting Information

Table S1 List of fish species tested for marker specificity.

Table S2 Quantitative PCR calibration data.

Table S3 Coordinates, lake depth, fish use, and eDNA concentration at sampling sites.

Acknowledgments

The authors thank Sendréa Best for field and laboratory assistance. Fish telemetry data was collected by Mary Headrick, Joseph Lechelt, Brett Miller, Robert Mollenheuer, and Tracy Szela. Erik Smith and Justine Koch provided assistance with ArcGIS. Mark Hove graciously provided fin

Table 4. Results of a 3-way ANOVA for CarpCytb marker in water samples.

Effects	df	F ratio	P value
Sample depth (surface, sub-surface)	1	0.07	0.79
Lake depth (m)	1	0.08	0.78
Carp use (low-use, high-use)	1	5.77	0.03*
Sample depth ×Lake depth	1	0.18	0.67
Lake depth ×Carp use	1	0.62	0.44
Sample depth ×Carp use	1	0.26	0.62
Error	19		

clips for use in specificity testing. Commercial fisher Tim Adams and crew also provided valuable assistance obtaining fin clips for specificity testing from the Mississippi River.

Author Contributions

Conceived and designed the experiments: JJE PGB PWS. Performed the experiments: JJE PGB. Analyzed the data: JJE PGB PWS. Contributed reagents/materials/analysis tools: JJE PGB PWS. Wrote the paper: JJE PGB PWS.

References

1. Mesa MG, Schreck CB (1989) Electrofishing mark-recapture and depletion methodologies evoke behavioral and physiological changes in cutthroat trout. Trans Amer Fish Soc 118: 644–658.
2. Cross DG, Stott B (1975) The effect of electric fishing on the subsequent capture of fish. J Fish Biol 7: 349–357.
3. Bayley PB, Austen DJ (2002) Capture efficiency of a boat electrofisher. Trans Amer Fish Soc 131: 435–451.
4. Lodge DM, Turner CR, Jerde CL, Barnes MA, Chadderton L, et al. (2012) Conservation in a cup of water: estimating biodiversity and population abundance from environmental DNA. Mol Ecol 21: 2555–2558.
5. Olson ZH, Briggler JT, Williams RN (2012) An eDNA approach to detect eastern hellbenders (Cryptobranchus a. alleganiensis) using samples of water. Wildl Res 39: 629–636.
6. Darling JA, Mahon AR (2011) From molecules to management: adopting DNA-based methods for monitoring biological invasions in aquatic environments. Env Res 111: 978–988.
7. Ficetola GF, Miaud C, Pompanon F, Taberlet P (2008) Species detection using environmental DNA from water samples. Biol Lett 4: 423–425.
8. Jerde CL, Mahon AR, Chadderton WL, Lodge DM (2011) "Sight-unseen" detection of rare aquatic species using environmental DNA. Conserv Lett 4: 150–157.
9. Pilliod DS, Goldberg CS, Arkle RS, Waits LP (2013) Estimating occupancy and abundance of stream amphibians using environmental DNA from filtered water samples. Can J Fish Aquat Sci 70: 1123–1130.
10. Takahara T, Minamoto T, Yamanaka H, Doi H, Kawabata Z (2012) Estimation of fish biomass using environmental DNA. PLoS One 7: e35868.
11. Thomsen PF, Kielgast J, Iversen LL, Wiuf C, Rasmussen M, et al. (2012) Monitoring endangered freshwater biodiversity using environmental DNA. Mol Ecol 21: 2555–2558.
12. Thomsen PF, Kielgast J, Iversen LL, Møller PR, Rasmussen M, et al. (2012) Detection of a diverse marine fish fauna using environmental DNA from seawater samples. PLoS One 7: e41732.
13. Goldberg CS, Sepulveda A, Ray A, Baumgardt J, Waits LP (2013) Environmental DNA as a new method for early detection of New Zealand mudsnails (Potamopyrgus antipodarum). Freshw Sci 32: 2555–2558.
14. U.S. Fish and Wildlife Service (2013) Quality assurance project plan (QAPP): eDNA monitoring of bighead and silver carps. 89 pp.
15. Moyer GR, Díaz-Ferguson E, Hill JE, Shea C (2014) Assessing environmental DNA detection in controlled lentic systems. PLoS One 9: e103767.
16. Bohmann K, Evans A, Gilbert TP, Carvalho GR, Creer S, et al. (2014) DNA for wildlife biology and biodiversity monitoring. Trends Ecol Evol 29: 358–357.
17. Mahon AR, Jerde CL, Galaska M, Bergner JL, Chadderton WL, et al. (2013) Validation of eDNA surveillance sensitivity for setection of Asian carps in controlled and field experiments. PLoS One 8: e58316.
18. Chang Y, Huang F, Lo UT (1994) The complete nucleotide sequence and gene organization of carp (Cyprinus carpio) mitochondrial genome. J Mol Evol 38: 138–155.
19. Mayden RL, Tang KL, Conway KW, Freyhof J, Sudkamp M, et al. (2007) Phylogenetic relationships of Danio within the order Cypriniformes: A framework for comparative and evolutionary studies of a model species. J Exp Zoo B Mol Dev Evol 308: 642–654.
20. Bajer PG, Sorensen PW (2012) Using boat electrofishing to estimate the abundance of invasive common carp in small midwestern lakes. N Amer J Fish Manag 32: 817–822.
21. Bajer PG, Chizinski CJ, Sorensen PW (2011) Using the Judas technique to locate and remove wintertime aggregations of invasive common carp. Fish Manag Ecol 18: 497–505.
22. (NOAA) National Oceanic and Atmospheric Administration Global Historical Climate Network. Available: http://www.ncdc.noaa.gov/cdo-web/datasets#GHCND. Accessed 11 February 2014.
23. Haugland RA, Siefring SC, Wymer LJ, Brenner KP, Dufour AP (2005) Comparison of Enterococcus measurements in freshwater at two recreational beaches by quantitative polymerase chain reaction and membrane filter culture analysis. Water Res 39: 559–568.
24. Turner CR, Barnes MA, Xu CCY, Jones SE, Jerde CL, et al. (2014) Particle size distribution and optimal capture of aqueous macrobial Edna. Method Ecol Evol doi: 10.1111/2041-210X.12206.
25. Simon M, Grossart H-P, Schweitzer B, Ploug H (2002) Microbial ecology of organic aggregates in aquatic ecosystems. Aquat Microb Ecol 28: 175–211.
26. Pilliod DS, Goldberg CS, Arkle RS, Waits LP (2014) Factors influencing detection of eDNA from a stream-dwelling amphibian. Mol Ecol Res 14: 109–116.
27. Dejean T, Valentini A, Duparc A, Pellier-Cuit S, Pompanon F, et al. (2011) Persistence of environmental DNA in freshwater ecosystems. PLoS One 6: e23398.
28. Barnes MA, Turner CR, Jerde CL, Renshaw MA, Chadderton WL, et al. (2014) Environmental conditions influence eDNA persistence in aquatic systems. Environ Sci Technol 48: 1819–1827.
29. Jane SF, Wilcox TM, McKelvey KS, Young MK, Schwartz MK, et al. (2014) Distance, flow, and PCR inhibition: eDNA dynamics in two headwater streams. Mol Ecol Res doi: 10.1111/1755-0998.12285.
30. Amberg JJ, McCalla SG, Miller L, Sorensen P, Gaikowski MP (2013) Detection of environmental DNA of Bigheaded carps in samples collected from selected locations in the St. Croix River and in the Mississippi River: U.S. Geological Survey Open-File Report 2013–1080, 44p.
31. Takahara T, Minamoto T, Doi H (2013) Using environmental DNA to estimate the distribution of an invasive fish species in ponds. PLoS One 8: e56584.
32. Tikel D, Blair D, Marsh HD (1996) Marine mammal faeces as a source of DNA. Mol Ecol 5: 456–457.
33. Mumy KL, Findlay RH (2004) Convenient determination of DNA extraction efficiency using an external DNA recovery standard and quantitative-competitive PCR. J Microbiol Meth 57: 259–268.
34. Anno AD, Corinaldesi C (2004) Degradation and turnover of extracellular DNA in marine sediments: ecological and methodological considerations. Appl Environ Microbiol 70: 4384–4386.
35. Pietramellara G, Ascher J, Borgogni F, Ceccherini MT, Guerri G, et al. (2008) Extracellular DNA in soil and sediment: fate and ecological relevance. Biol Fertil Soils 45: 219–235.
36. Domanico MJ, Phillips RB, Oakley TH (1997) Phylogenetic analysis of Pacific salmon (genus Oncorhynchus) using nuclear and mitochondrial DNA sequences. Can J Fish Aquat Sci 54: 1865–1872.

19

Carbon and Nitrogen Isotopes from Top Predator Amino Acids Reveal Rapidly Shifting Ocean Biochemistry in the Outer California Current

Rocio I. Ruiz-Cooley[1]*, Paul L. Koch[2], Paul C. Fiedler[3], Matthew D. McCarthy[1]

1 Ocean Sciences Department, University of California Santa Cruz, Santa Cruz, California, United States of America, 2 Earth and Planetary Sciences Department, University of California Santa Cruz, Santa Cruz, California, United States of America, 3 Southwest Fisheries Science Center, National Marine Fisheries Service, National Oceanic and Atmospheric Administration, La Jolla, California, United States of America

Abstract

Climatic variation alters biochemical and ecological processes, but it is difficult both to quantify the magnitude of such changes, and to differentiate long-term shifts from inter-annual variability. Here, we simultaneously quantify decade-scale isotopic variability at the lowest and highest trophic positions in the offshore California Current System (CCS) by measuring $\delta^{15}N$ and $\delta^{13}C$ values of amino acids in a top predator, the sperm whale (*Physeter macrocephalus*). Using a time series of skin tissue samples as a biological archive, isotopic records from individual amino acids (AAs) can reveal the proximate factors driving a temporal decline we observed in bulk isotope values (a decline of ≥ 1 ‰) by decoupling changes in primary producer isotope values from those linked to the trophic position of this toothed whale. A continuous decline in baseline (i.e., primary producer) $\delta^{15}N$ and $\delta^{13}C$ values was observed from 1993 to 2005 (a decrease of ~4‰ for $\delta^{15}N$ source-AAs and 3‰ for $\delta^{13}C$ essential-AAs), while the trophic position of whales was variable over time and it did not exhibit directional trends. The baseline $\delta^{15}N$ and $\delta^{13}C$ shifts suggest rapid ongoing changes in the carbon and nitrogen biogeochemical cycling in the offshore CCS, potentially occurring at faster rates than long-term shifts observed elsewhere in the Pacific. While the mechanisms forcing these biogeochemical shifts remain to be determined, our data suggest possible links to natural climate variability, and also corresponding shifts in surface nutrient availability. Our study demonstrates that isotopic analysis of individual amino acids from a top marine mammal predator can be a powerful new approach to reconstructing temporal variation in both biochemical cycling and trophic structure.

Editor: Wei-Chun Chin, University of California, Merced, United States of America

Funding: Funding was provided by Marine Mammal and Turtle Division, Southwest Fisheries Science Center-National Oceanographic Atmospheric Administration for data collection and isotope analysis and National Science Foundation(Division of Ocean Sciences(OCE)-1155728, and OCE-0623622) for analysis of amino acids and data. Funding for Open Access provided by the University of California, Santa Cruz, Open Access Fund. The funders had no role in study design, data collection and analysis, decision to publish, or preparation of the manuscript.

Competing Interests: The authors have declared that no competing interests exist.

* Email: rcooley@ucsc.edu

Introduction

The California Current System (CCS) contains one of the five major coastal upwelling zones in the world's oceans, and hosts a great diversity and abundance of marine life [1]. The oceanographic state of this large ecosystem is dynamic. Natural climate variation and anthropogenic stressors alter biochemical cycling, food web dynamics, and the fitness of species [1–3]. Known interannual and decadal changes are related both to the El Niño-Southern Oscillation (ENSO) and to basin-scale processes associated with the Pacific Decadal Oscillation (PDO) [1].The latter, is an index of interannual sea surface temperature (SST) variability in the North Pacific, that is related to physical and biochemical variations and influences community changes in plankton, fish and other taxa [4,5]. In addition to this natural variability, humans have perturbed climate by increasing atmospheric CO_2 concentrations, which have increased ocean temperatures, water column stratification, hypoxia, and water column

anoxia and have decreased surface ocean pH [6,7]. These environmental factors may negatively impact populations of species, increasing mortality and decreasing reproductive success due to habitat compression and metabolic constraints [8]. Other anthropogenic pressures, such as intensive fisheries and the past whaling industry (which principally targeted sperm whales, *Physeter macrocephalus*) might have triggered top-down effects. Given the lack of detailed proxy records to trace simultaneously biochemical baselines and length of food webs, assessing the extent to which biogeochemical cycling and community structure in pelagic ecosystems have changed over the past century is difficult, as is attributing change to natural cycles versus anthropogenic disturbances.

The isotopic values of marine primary producers are sensitive to environmental variation, such as change in temperature, and CO_2 or nitrate concentrations, as well as biological differences such as physiology and growth rate [9–11]. Hence, the carbon and nitrogen isotope values ($\delta^{13}C$ and $\delta^{15}N$ values, respectively) of primary producers, also known as "baseline isotope values", vary in space and time as a function of these fundamental ecosystem properties [12]. Baseline isotope values are then integrated into consumers' tissues through diet, typically with metabolic fractionation leading to enrichment in the heavier isotope (especially ^{15}N) in consumers [13,14]. Therefore, isotopic values of marine consumers could be used to reconstruct changes in diet and/or ecosystem biogeochemistry. The δ^{13}C and δ^{15}N values from a resident animal can potentially provide an integrated record of the biogeochemical characteristics of its habitat, as well as its trophic position [15]. However, because multiple factors influence the bulk δ^{13}C and δ^{15}N values ultimately recorded in consumer tissues, it is often difficult to disentangle the effects of changing trophic position from shifts in baseline values.

Studies in different ocean basins have shown that bulk tissue δ^{13}C or δ^{15}N values have declined over the last century, but interpretations of these trends have varied widely [16]. For example, declining bulk tissue δ^{15}N values are sometimes attributed to a drop in consumer trophic level [17,18] or to baseline shifts due to either changes in foraging zone or biogeochemical cycles [16]. In particular, two recent studies in the Pacific have revealed pervasive declines in δ^{15}N values in the offshore Central Pacific [18] and North Pacific Subtropical Gyre (NPSG) [19], but offered diametrically opposing interpretations as to underlying mechanism. In the highly productive CCS, despite accumulating evidence for oceanographic changes since the 1950s [2], isotopic data from plankton species have been contradictory. Bulk δ^{15}N values from three zooplakton species have exhibited no long-term trends, whereas data for a specialized zooplankton feeder decreased by approximately 3‰ [20,21]. Declines in δ^{13}C values over the 20th century are expected due to the combustion of fossil fuels (i.e., the Suess effect), and have been observed in many records and ecosystems [22]. However, variability in the magnitude and timing of δ^{13}C declines has suggested that other factors, such as declining primary productivity, could also contribute in some regions [16]. In the offshore CCS, there are currently no δ^{13}C time series for organic or inorganic material.

Isotopic analysis of individual amino acids (AAs) can effectively separate trophic effects from shifts in baseline isotope values [23,24]. Regardless of an animal's trophic position, the original δ^{15}N and δ^{13}C values from primary producers are relatively well preserved within the group of 'source-AAs' for nitrogen [25] and the 'essential-AAs' for carbon [26]. In contrast, isotopic values from the 'trophic-AAs' for nitrogen, and 'non-essential-AAs' for carbon, undergo significant metabolic fractionation, and vary in association with a consumer's diet [23,24], tissue turnover rates, and possibly metabolism [27]. Hence, isotopic analysis of amino acids from apex marine mammal predators offers a unique opportunity to simultaneously investigate temporal variation at the lowest and highest trophic levels of their food web. Sperm whales are top predators of the mesopelagic ocean. Mark-recapture studies, morphology, and acoustic analysis indicate that female sperm whales forage within the same oceanic region year round [28]. Consequently, they can function as natural biological samplers, broadly integrating biogeochemical information from their home ecosystem. In this study, we use sperm whale skin as a novel biological archive of time series data. Our data combine bulk tissue and AA isotope analysis to examine temporal variation in baseline values (reflecting ecosystem biogeochemistry) and whale trophic position (indicating trophic structure) from offshore waters of the California Current ecosystem.

Results and Discussion

Foraging zone of sperm whales sampled in CCS

In the CCS off the US west coast, sperm whales are found in oceanic waters from California to Washington [29]. Their habitat therefore excludes the coastal upwelling system that exhibits strong latitudinal isotopic gradients [30]. Mitochondrial and nuclear markers reveal that the CCS whales are an independent population and a single genetic stock [31]. Our isotopic data from skin biopsies (Figure 1) indicate that whales fed homogenously within the offshore northern and central CCS. First, the variation in bulk isotope values (n = 18; SD = 1.2‰ for δ^{13}C and 1.2‰ for δ^{15}N) is similar to the variation observed in other sperm whale populations that are considered to be resident (i.e., Gulf of Mexico and Gulf of California, SD≤0.8‰ for both δ^{15}N and δ^{13}C [15]; SE Pacific, SD = 3.5‰ for δ^{15}N and 0.7‰ for δ^{13}C [32]). In addition, the δ^{15}N values for phenylalanine (Phe; n = 12; mean (SD) = 10.9‰ (0.9)) are relatively consistent with expected nitrate and particulate organic matter δ^{15}N values from the oceanic northern CCS (~6 to 10 ‰) [12], and also with published Phe δ^{15}N values from muscle of the jumbo squid (*Dosidicus gigas*; potential prey of sperm whales) [33]. Phe δ^{15}N values are a proxy for primary producer values [25] as they exhibit only minor ^{15}N-enrichment with trophic transfer [23]. In top predators (such as sperm whales), this likely results in slightly higher Phe δ^{15}N values versus baseline inorganic N sources. Lastly, because latitudinal trends in the δ^{15}N values from predator source-AAs can indicate their geographic residency [24,33], the lack of any latitudinal variation in Phe δ^{15}N values (r^2 = 0; n = 12) strongly suggests that the individual sperm whales sampled here were not foraging in different localized regions, but rather foraged over a broad latitudinal range within the northern and central CCS. While the isotopic incorporation rate for extremely large animals like whales is not well known, the thick skin of sperm whales likely integrates information for at least three and possibly more than six months prior to sampling [34]. Our data set encompasses information mainly from the fall and winter, except for the samples collected in 2001 and 2003, which also integrate information from the summer.

Coupled decadal declines in δ^{15}N and δ^{13}C values

Bulk δ^{13}C and δ^{15}N values in whale skin decreased from 1993 to 2005 by 1.1‰ and 1.7‰, respectively. These decreases were statistically significant at the alpha = 0.05 level. Inclusion of a single sample available from 1972 further suggests possible longer-term temporal declines for both δ^{13}C and δ^{15}N values by ≥4‰ and >3‰, that are also statistically significant (Table 1). Together, these coupled time-series declines in bulk δ^{13}C and δ^{15}N values suggest coincident biogeochemical or trophic system perturbation (Table 1, Figure 2). In particular, the rate of decrease for bulk δ^{15}N values since the 1970's is at least five times greater than the rate for the long-term δ^{15}N decrease recently documented in the central Pacific from proteinaceous corals (2.3‰ in 150 years; annual decrease calculated at 0.015‰) [19], and it is more similar to the rate of change observed for a single zooplankton δ^{15}N record from southern California (~3‰ in 50 years) [21].

To disentangle the factors driving the declines in bulk isotope values, we analyzed individual AA isotope values, focusing on AAs that have been demonstrated to track baseline changes (as noted above, essential AA for δ^{13}C values, source AA for δ^{15}N values). Linear regression models for average δ^{13}C and δ^{15}N values from the most accurately measured essential- and source-AAs both exhibited strong negative temporal trends across all the data (i.e. for both 1972 and 1993 to 2005; Table 1, Figure 2), with drops of

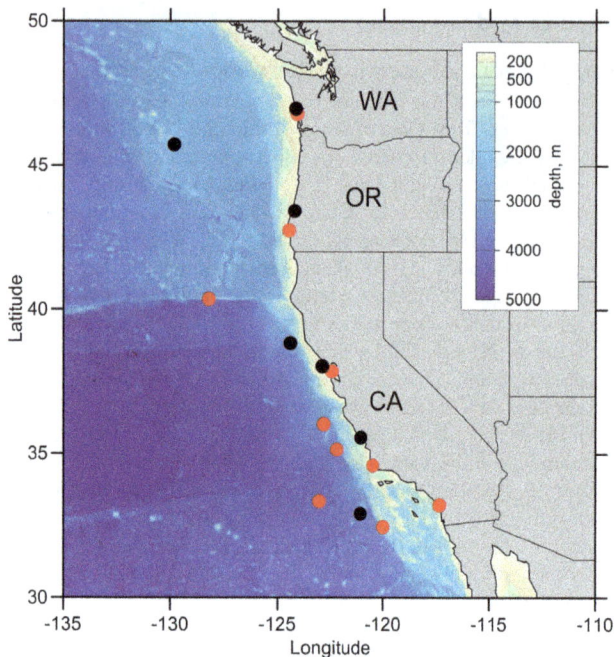

Figure 1. Sperm whales are distributed year-round in offshore deep waters (~>150 km off the US west coast [29]). Skin samples (○) from free-ranging sperm whales were collected together with skin from stranded individuals. Tissue samples were used for bulk (in black) and amino acid (in red) stable isotope analysis.

≥3‰ and >4‰ respectively indicated by compound-specific isotope data. Residuals for all regressions exhibited a random pattern. In contrast, average $\delta^{15}N$ values for the trophic-AAs were much variable, resulting in a lower r^2, but overall they paralleled the source-AA trend (Table 1, Figure 2A). These results are not consistent with any significant drop in sperm whale trophic level as the primary driver of decreases in bulk isotope ratios, and instead strongly implicate coupled changes in baseline $\delta^{15}N$ and $\delta^{13}C$ values.

These negative trends in baseline $\delta^{15}N$ and $\delta^{13}C$ values might relate to changes in biochemical cycling, rates of primary production, or primary producer species composition. In particular, the decline in average essential-AA $\delta^{13}C$ values (Figure 2B), which are a direct proxy for primary producers, is far too high to be explained solely by the Seuss effect (~0.2 ‰ per decade since 1960 [35]), and it also coincides with the decline in average source-AA $\delta^{15}N$ values. This suggests that the mechanism explaining a drop in primary producer $\delta^{15}N$ values should be consistent with a concurrent large decline in $\delta^{13}C$ values. One possiblity, which would represent a direct analogy to changes in other ocean regions, would be a shift towards more oligotrophic conditions for the outer CCS. This explanation would be consistent with coupled declines in both isotopes, linked to decreased primary production and a shift in species composition that is typically associated with warmer and more stratified ocean conditions [36]. Oligotrophy in the world ocean is increasing due to climate shifts [37] and is projected to continue increasing in the North Pacific [38]. Recent isotopic records from deep sea proteinacous corals, for example, provide strong support for such linked trends associated with warming of the NPSG [39]. The nitrogen isotope record from deep sea coral indicate that the long-term declines in baseline $\delta^{15}N$ values are likely linked to progressive increases in seasonal gyre

extent, leading to steady increases in N contribution from diazotrophy [19]. Therefore, an analogous explanation would imply that oceanographic conditions in the offshore CCS region (which have conditions more similar to the open ocean and represent the base of sperm whales' food web) might have shifted toward more "gyre-like" conditions, driving baseline isotope values toward those more typical of the oligotrophic open ocean.

However, to our knowledge, there is currently no evidence for substantially increasing SST and diazotrophy in the CCS itself. Instead, recent analyses suggest largely the opposite: overall, the thermocline weakened and shoaled in the offshore CCS between 1950 and 1993 [7], possibly increasing nutrient availability in the euphotic zone despite increased stratification [40]. Additionally, the offshore CCS has cooled (not heated) since the early 1990s (Figure 3), and this trend is also reflected in the present "cool" PDO regime. Furthermore, the generalization that global warming will universally increase stratification and thus decrease surface nutrient supply has been recently challenged for some regions including the CCS [41]. For example, one recent model projects increases in nitrate supply and productivity in the CCS during the 21st century despite increases in stratification and limited change in wind-driven upwelling [42]. In the southern CCS, coastal surface nutrients have increased possibly linked to a general shoaling of the nutricline [43]. In the Southern California Bight, the most intensively monitored region of the CCS, nutrients in source waters have also increased over the last three decades, but the N:P and Si:N ratios were greatly reduced, possibly shifting phytoplankton species composition and abundance [44]. Whether or not these trends in nutrient dynamics extend to other regions of the CCS is unclear, because the oceanographic state of this ecosystem varies regionally [1,45].

In particular, shifts in offshore and onshore oceanographic conditions appear to be decoupled. Coastal upwelling has recently increased, as expected for enhanced alongshore winds [46], but has decreased offshore where upwelling is driven by wind-stress curl [47]. Since 1997, trends in satellite chlorophyll estimates, an index of phytoplankton biomass, have been positive in coastal upwelling waters but tend to be zero or negative in offshore waters [48]. Together, this current evidence indicates cooling, but not increases in productivity, in the offshore CCS concurrent with the observed 1993–2006 trends in sperm whale $\delta^{15}N$ and $\delta^{13}C$ values. Lower temperatures increase the solubility of CO_2 and change the fractionation associated with carbon fixation, often resulting in lower phytoplankton $\delta^{13}C$ values [49]; lower temperatures might have also changed phytoplankton growth and species composition. If surface nitrate also increased along the outer CCS region sampled by these whales, then the degree of nitrate utilization by primary producers (and so their $\delta^{15}N$ values) could have also changed, since phytoplankton preferentially assimilate $^{14}NO_3^-$ [50]. In general, proportional nitrate utilization is lower where surface NO_3^- concentrations are higher [50]. Therefore, lower NO_3^- utilization during seasonal upwelling might also be expected to depress the $\delta^{15}N$ values of primary producers, propagating the ^{15}N-depleted signal into food webs during their most productive periods. At present, there simply are not enough detailed data on nutrient concentrations and other oceanographic factors in the outer CCS to deduce a mechanism. However, the observed declining baseline values revealed by sperm whales do indicate a recently progressive shift in primary producer dynamics, likely associated with changes in SST, average state of surface nutrients and/or primary production.

Table 1. Temporal variation in $\delta^{15}N$ and $\delta^{13}C$ values from the offshore California Current System in sperm whale skin samples.

Time period	Tracer	Linear Regression	n	r^2	p-value	Isotopic shift (‰)	Annual decrease
1993–2005	$\delta^{15}N$						
	Bulk	y = 302−0.143 * year	17	0.25	<0.05	1.7	0.14
	Mean Source-AA	y = 717−0.354 * year	11	0.52	= 0.01	4.2	0.35
	Mean Trophic-AA	y = 311−0.143 * year	11	0.12	>0.05	1.7	
1972–2005	$\delta^{15}N$						
	Bulk	y = 218−0.101 * year	18	0.37	<0.05	3.3	0.10
	Mean Source-AA	y = 298−0.145 * year	12	0.39	<0.05	4.7	0.14
	Mean Trophic-AA	y = 88−0.031 * year	12	0.03	>0.05	1.0	
1993–2005	$\delta^{13}C$						
	Bulk	y = 174−0.095 * year	17	0.24	<0.05	1.1	0.09
	Mean Essential-AA	y = 474−0.250 * year	8	0.62	<0.05	3.0	0.25
1972–2005	$\delta^{13}C$						
	Bulk	y = 242−0.129 * year	18	0.67	<0.01	4.2	0.12
	Mean Essential-AA	y = 184−0.105 * year	9	0.58	<0.05	3.4	0.10

For mean calculations: Source-AAs are phenylalanine, glycine, lysine, tyrosine; Trophic-AA: glutamic acid, alanine, isoleucine, leucine, proline; Essential-AA: phenylalanine, valine, leucine. Isotopic shifts were calculated using the corresponding linear regression equations listed in this table. The annual decrease was calculated for shifts that exhibited a p-value ≤ 0.05.

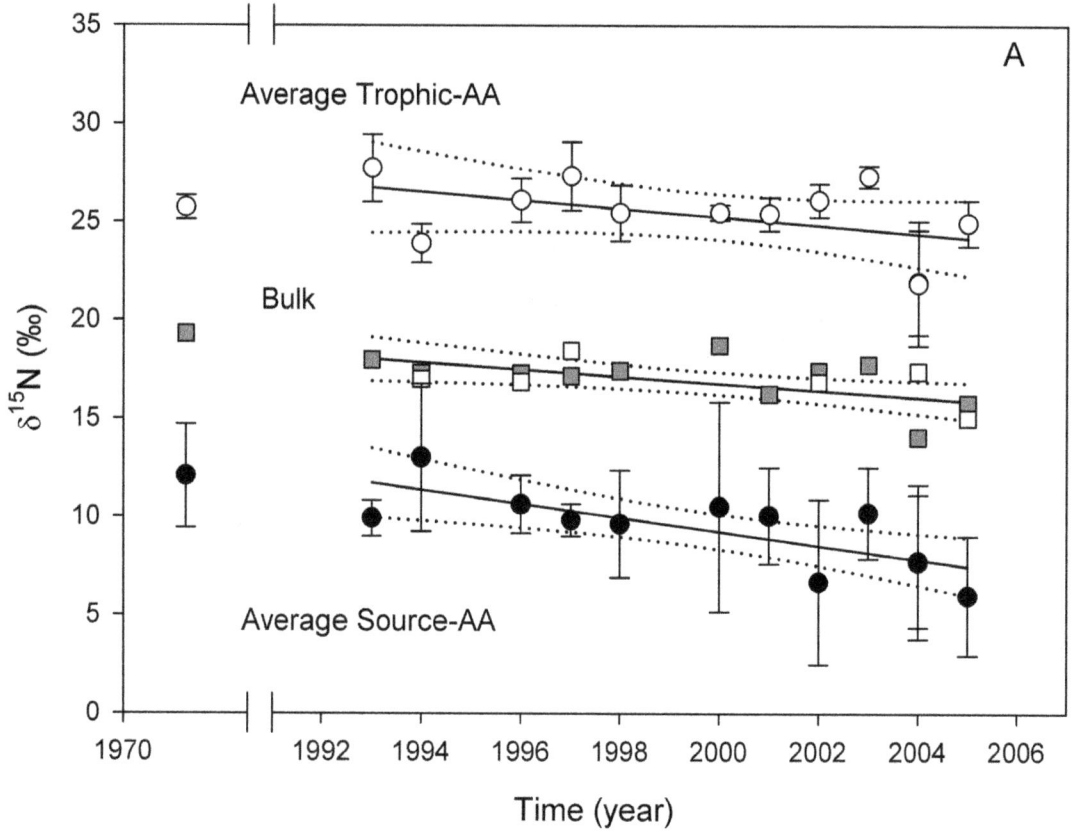

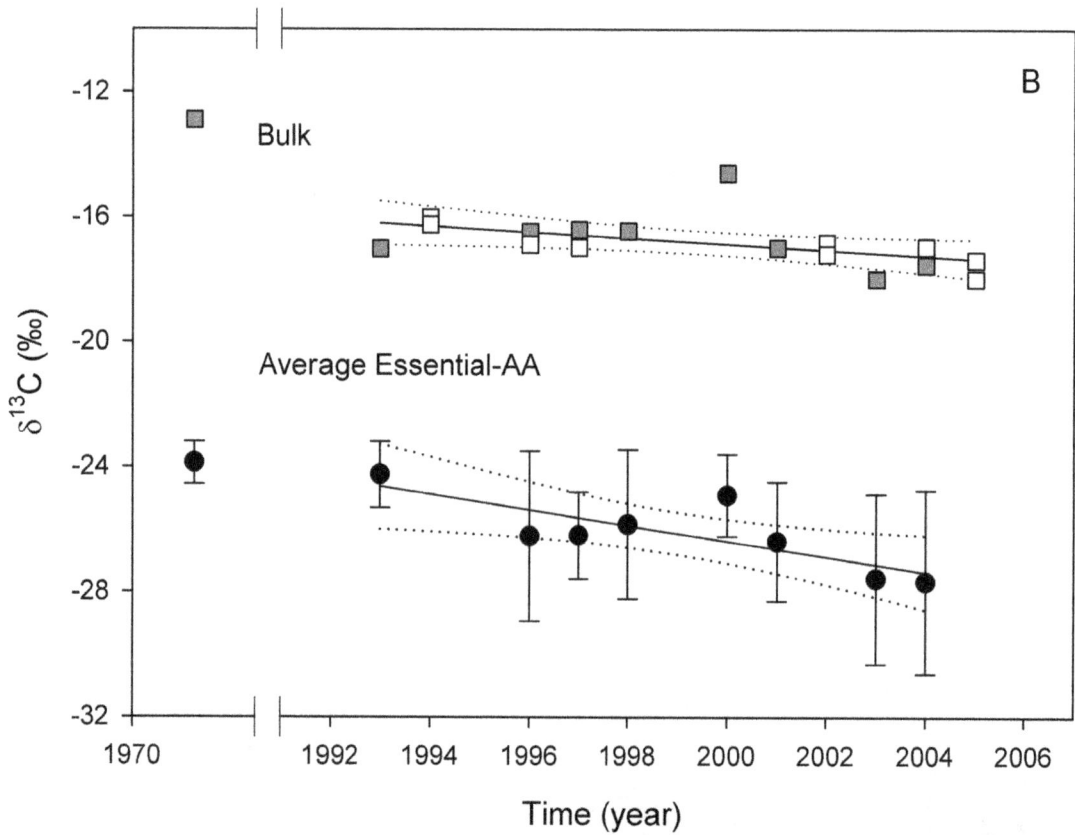

Figure 2. Time series of isotopic data from sperm whale skin. (A) $\delta^{15}N$ values from bulk skin, average source-AAs and average trophic- AAs ($\pm$ SD); and (B) $\delta^{13}C$ values from bulk skin and average essential-AAs ($\pm$SD). Bulk isotope data are plotted with a square symbol ($\square$), filled grey squares indicate the samples that were also analyzed for amino acid stable isotope analysis. The corresponding linear regression equations are provided in Table 1, as are the amino acids included within each AA-group.

Implications of Rapid Change for offshore CCS Biogeochemistry

Although our time series data are limited for both elements, the compound-specific AA data identify a parallel decline in both baseline $\delta^{15}N$ and $\delta^{13}C$ values in the outer CCS from 1992 to 2005, likely indicative of major recent shifts in biochemical cycling. At the same time, however, the overall similarity in whale trophic position signifies that the broad trophic structure is realtively unaffected. We note that in comparison with the recent deep sea coral data from the gyre offshore of this region [19], our data suggest that both the rate and scale of biochemical change on the CCS margin may be far greater than in the open Pacific Ocean. The coral record from the NPSG indicates a fairly steady $\delta^{15}N$ annual decrease of $\sim 0.015‰$ over the last 150 years with a total drop of 2.3 ‰ in exported primary production $\delta^{15}N$ values over that period. In contrast, our molecular-level proxies for $\delta^{15}N$ values at the base of the food chain (the source AAs) indicate more rapid annual declines of 0.35 ‰ since the 1990's. The independent molecular proxies for primary production $\delta^{13}C$ values (the essential AAs) indicate relatively similar declines.

Together with the CCS observations discussed above, the contrast with the NPSG coral data (while not directly comparable in terms of time scale), suggests that despite the fact that baseline $\delta^{15}N$ declines are observed in both data sets, different biogeochemical mechanisms may underlie the changes in these very different oceanographic regions. Climate variability likely affects the biochemistry of ecosystems differently depending on the oceanographic properties, microbial and phytoplankton communities, and species assemblages. In the eastern Pacific Ocean, the structure of the pycnocline varies strongly among the known biogeochemical provinces [51]. This likely influences geographic variation in surface nutrient availability, and therefore stable isotope ratios in POM, primary producers [12] and consumers [14]. Temporal trends in pycnocline depth, SST, stratification, and mixed layer depths also differ between these biogeochemical provinces [40]. For example, while SST decreased overall since 1958 in many parts of the California Current, SST increased in

the easternmost southern subtropical gyre and equatorial Pacific [40]. Ultimately, more detailed data that couple integrated measures of ecosystem baseline with oceanographic state will be required to understand the substantial biogeochemical changes our data indicate.

Our work highlights that detailed time-series of biochemical baseline and trophic structure records among different ecosystems will be crucial to identify rapid ecosystem shifts in response to climate change. In particular, in the face of uncertain coupling of natural and anthropogenic climate forcing, understanding the timing, extent and especially the mechanistic basis for baseline shifts now represents an urgent challenge. However, despite many efforts to unravel the linkage and feedback controls between the carbon and nitrogen cycles, and the effect of their variability on primary production and food-web dynamics, they are still not well understood. This study has demonstrated the great potential in coupling molecular isotopic tools with the unique bioarchive of sperm whales (or other top predators), as sentinels of offshore ecosystems. This may allow, for the first time, decoding of the factors that underlie temporal trends in bulk isotopic records, while simultaneosly monitoring changes at both the highest and lowest trophic levels. We suggest that integrating this approach with detailed oceanographic data will be a major new tool to identify the effects of natural climate variability versus anthropogenic global warming on ecosystem biochemistry and primary production. Elucidating such patterns from this and other ocean margin regions, in particular their relationships with oceanographic and climatic variations and shifts in primary production, will be an essential part of the critical task of predicting future trends in both ecosystem biochemistry and trophic dynamics.

Material and Methods

A total of 18 skin samples (Figure 1) were analyzed for bulk stable isotope analysis. Skin tissue samples with enough material (3.5 mg) were selected for CSIA-AA. Data from 12 samples were obtained for individual AA $\delta^{15}N$ values, and 9 samples for $\delta^{13}C$ values. The Southwest Fisheries Science Center/Pacific Islands

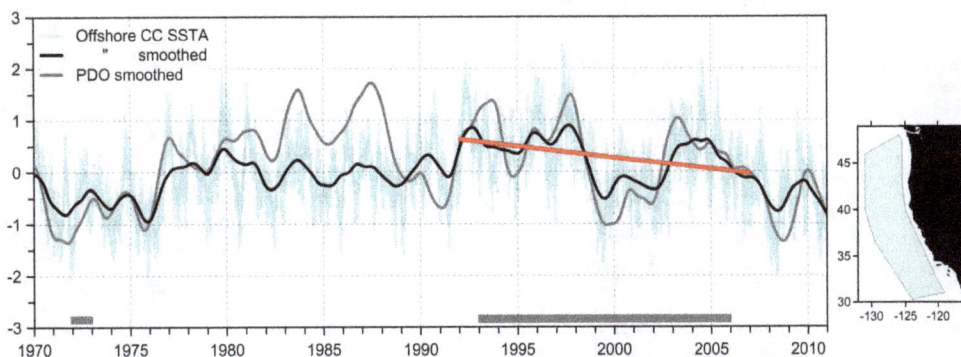

Figure 3. Time series data of sea surface temperature anomaly (SSTA) from the offshore California Current (inset map) and the Pacific Decadal Oscillation (PDO). Monthly SSTA was computed in 0.5-deg fields from the Simple Ocean Data Assimilation version 2.2.4 reanalysis (http://coastwatch.pfeg.noaa.gov/erddap/griddap/hawaii_d90f_20ee_c4cb.html), and then averaged in the offshore area (the plot shows ± 1sd). Monthly SSTA ($°C$) and PDO values were smoothed with a 25-month lowess smooth. The linear fit is for 1992–2006 (red line, slope $-0.044°C$ y-1). Sample periods are indicated along the time axis.

Fisheries Science Center Institutional Animal Care and Use Committee (IACUC) approved the original animal work that produced the samples. Sex was determined genetically using qPCR sexing assay by the PRD-Genetic Lab at NOAA [52]. These samples consisted of 5 females, 2 males and 2 unidentified individuals possibly corresponding to females or juvenile males. Large adult males were not included. Bulk isotope values were analyzed by continuous flow isotope ratio mass spectrometry (IRMS; Thermo Finnigan) and standardized relative to Vienna-Pee Belemnite (V-PDB) for carbon and atmospheric N_2 for nitrogen. Results are expressed in part per thousand (‰) and standard notation: $\delta^H X = [(R_{sample}/R_{standard})-1] \times 1000$, where H is the mass number of the heavy isotope, X is either C or N, and R_{sample} and $R_{standard}$ are the ratio of $^{13}C/^{12}C$ or $^{15}N/^{14}N$ in the sample and standard, respectively.

We hydrolyzed and prepared approximately 3.5 mg of skin as well as a control (Cyanno; bacteria tissue) [53] to quantify $\delta^{15}N$ values from source- and trophic-AAs and $\delta^{13}C$ values from essential- and non-essential-AAs. All derivatives were injected with an AA control, N-leucine, to verify accuracy during each run, and analyzed via gas chromatography-IRMS to obtain $\delta^{15}N$ and $\delta^{13}C$ values from individual AAs. Each sample was run 3–4 times to maximize accuracy among chromatograms. The associated analytical error among replicates was <1.0 ‰. For all samples, $\delta^{15}N$ values were obtained from a total of four source-AAs (phenylalanine, glycine, lysine, tyrosine), and five trophic-AAs (glutamic acid, alanine, isoleucine, leucine, proline) (Figure S1A). For $\delta^{13}C$ values, the essential-AAs that we consistently determined were phenylalanine, valine and leucine, and the non-essential-AA were alanine, proline, aspatic acid, glutamic acid and tyrosine (Figure S1B).

The relative pattern of AA $\delta^{15}N$ and $\delta^{13}C$ values was highly consistent with past work from other organisms and tissues [23,25,54]. We grouped data as source- or trophic-AAs for $\delta^{15}N$ values, and essential- or non-essential-AAs for $\delta^{13}C$ values to increase power in the analysis and evaluate temporal variation. We calculated average values for each AA group and they are reported in Table S1. Regression analyses were conducted to evaluate linear relationship between time and each isotopic tracer for both bulk and individual-AA $\delta^{15}N$ and $\delta^{13}C$ values (Table 1).

There was a weak correlation between average source-AA and trophic-AA ($r^2 = 0.13$; $p = 0.67$), indicating that trophic-AA $\delta^{15}N$ values could not be predicted by the variability in source-AAs, and vice versa. However, the correlation between average essential-AA and non-essential-AA $\delta^{13}C$ values was moderate ($r^2 = 0.63$, $p = 0.06$). Since the controls on isotopic patterns for non-essential-AA $\delta^{13}C$ values are complex and dependent on diet quality and quantity, including *de novo* synthesis and routing of AAs from diet-to-tissue, this group was not considered in the linear regression analysis.

Supporting Information

Figure S1 Stable isotope values of individual amino acids (AAs) in skin samples of sperm whales (*Physeter macrocephalus*). (A) Four $\delta^{15}N$ Source-AAs: phenylalanine (phe), glycine (gly), lysine (lys), tyrosine (tyr), and five Trophic-AAs: glutamic acid (glx), alanine (ala), isoleucine (ile), leucine (leu), proline (Pro); and (B) Three $\delta^{13}C$ essential-AAs: phe, leu, and valine (val).

Table S1 Average values and one standard deviations (SD) were calculated for Source-AAs (phenylalanine, glycine, lysine, tyrosine), Trophic-AAs (glutamic acid, alanine, isoleucine, leucine, proline) and Essential-AAs (phenylalanine, valine, leucine).

Acknowledgments

We thank L. T. Ballance, J. Barlow, K. Robertson (SWFSC/NMFS/NOAA) and J. Calambokidis (Cascadia Research) for facilitating the use of tissues samples, and the genetic SWFSC lab for molecular whale sex identification.

Author Contributions

Conceived and designed the experiments: RIRC MDM. Performed the experiments: RIRC MDM. Analyzed the data: RIRC PCF. Contributed reagents/materials/analysis tools: MDM. Contributed to the writing of the manuscript: RIRC PLK PCF MDM.

References

1. Checkley JDM, Barth JA (2009) Patterns and processes in the California Current System. Prog Oceanogr 83: 49–64.
2. Bograd SJ, William JS, Barlow J, Booth A, Brodeur RD, et al. (2010) Status and trends of the California Current region, 2003–2008. PICES Special Publication. 106–141 p.
3. McGowan JA, Bograd SJ, Lynn RJ, Miller AJ (2003) The biological response to the 1977 regime shift in the California Current. Deep Sea Res II 50: 2567–2582.
4. Brinton E, Townsend A (2003) Decadal variability in abundances of the dominant euphausiid species in southern sectors of the California Current. Deep Sea Res II 50: 2449–2472.
5. Chavez FP, Ryan J, Lluch-Cota SE, Ñiquen CM (2003) From anchovies to sardines and back: Multidecadal change in the Pacific Ocean. Science 299: 217–221.
6. Chan F, Barth JA, Lubchenco J, Kirincich A, Weeks H, et al. (2008) Emergence of anoxia in the California Current Large Marine Ecosystem. Science 319: 920.
7. Palacios DM, Bograd SJ, Mendelssohn R, Schwing FB (2004) Long-term and seasonal trends in stratification in the California Current, 1950–1993. J Geophy Res 109: C10016.
8. Bograd SJ, Castro CG, Di Lorenzo E, Palacios DM, Bailey H, et al. (2008) Oxygen declines and the shoaling of the hypoxic boundary in the California Current. Geophys Res Lett 35: L12607.
9. Farrell JW, Pedersen TF, Calvert SE, Nielsen B (1995) Glacial-interglacial changes in nutrient utilization in the equatorial Pacific Ocean. Nature 377: 514–517.
10. Rau GH, Sweeney RE, Kaplan IR (1982) Plankton $^{13}C:^{12}C$ ratio changes with latitude: differences between northern and southern oceans. Deep Sea Res 29: 1035–1039.

11. Goericke R, Fry B (1994) Variations of marine plankton $\delta^{13}C$ with latitude, temperature, and dissolved CO_2 in the World Ocean. Global Biogeochem Cy 8: 85–90.
12. Somes CJ, Schmittner A, Galbraith ED, Lehmann MF, Altabet MA, et al. (2010) Simulating the global distribution of nitrogen isotopes in the ocean. Global Biogeochem Cy 24.
13. Peterson BJ, Fry B (1987) Stable Isotopes in Ecosystem Studies. Annu Rev Ecol Evol Syst 18: 293–320.
14. Ruiz-Cooley RI, Gerrodette T (2012) Tracking large-scale latitudinal patterns of $\delta^{13}C$ and $\delta^{15}N$ along the eastern Pacific using epi-mesopelagic squid as indicators. Ecosphere 3: 63.
15. Ruiz-Cooley R, Engelhaupt D, Ortega-Ortiz J (2012) Contrasting C and N isotope ratios from sperm whale skin and squid between the Gulf of Mexico and Gulf of California: effect of habitat. Mar Biol: 1–14.
16. Schell DM (2001) Carbon isotope ratio variations in Bering Sea biota: The role of anthropogenic carbon dioxide. Limnol Oceanogr Methods 46: 999–1000.
17. Emslie SD, Patterson WP (2007) Abrupt recent shift in $\delta^{13}C$ and $\delta^{15}N$ values in Adélie penguin eggshell in Antarctica. Proc Natl Acad Sci USA 104: 11666–11669.
18. Wiley AE, Ostrom PH, Welch AJ, Fleischer RC, Gandhi H, et al. (2013) Millennial-scale isotope records from a wide-ranging predator show evidence of recent human impact to oceanic food webs. Proc Natl Acad Sci USA 110: 8972–8977.
19. Sherwood OA, Guilderson TP, Batista FC, Schiff JT, McCarthy MD (2013) Increasing subtropical North Pacific Ocean nitrogen fixation since the Little Ice Age. Nature 505: 78–81.

20. Rau GH, Ohman MD, Pierrot-Bults A (2003) Linking nitrogen dynamics to climate variability off central California: a 51 year record based on ^{15}N/^{14}N in CalCOFI zooplankton. Deep Sea Res Part II 50: 2431–2447.

21. Ohman MD, Rau GH, Hull PM (2012) Multi-decadal variations in stable N isotopes of California Current zooplankton. Deep Sea Res Part I 60: 46–55.

22. Sonnerup RE, Quay PD, McNichol AP, Bullister JL, Westby TA, et al. (1999) Reconstructing the oceanic [13]C Suess Effect. Global Biogeochem Cy 13: 857–872.

23. Chikaraishi Y, Ogawa NO, Kashiyama Y, Takano Y, Suga H, et al. (2009) Determination of aquatic food-web structure based on compound-specific nitrogen isotopic composition of amino acids. Limnol Oceanogr Methods 7 740–750.

24. Popp BN, Graham BS, Olson RJ, Hannides CCS, Lott MJ, et al. (2007) Insight into the trophic ecology of yellowfin tuna, Thunnus albacares, from compound-specific nitrogen isotope analysis of proteinaceous amino acids. In: Dawson TD, Siegwolf, R. T W., editor. Stable isotopes as indicators of ecological change. New York: Elsevier Academic Press. pp. 173–190.

25. McClelland JW, Montoya JP (2002) Trophic relationships and the nitrogen isotopic composition of amino acids in plankton. Ecology 83: 2173–2180.

26. O'Brien DM, Fogel ML, Boggs CL (2002) Renewable and nonrenewable resources: Amino acid turnover and allocation to reproduction in Lepidoptera. Proc Natl Acad Sci USA 99: 4413–4418.

27. Germain LR, Koch PL, Harvey JT, McCarthy MD (2013) Nitrogen isotopic fractionation of amino acids in harbor seals (Phoca vitulina): Differential trophic enrichment factors based on ammonia vs. urea excretion. Mar Ecol Prog Ser 482: 265–277.

28. Default S, Whitehead H, Dillon M (1999) An examination of the current knowledge on the stock structure of sperm whales (Physeter macrocephalus) worldwide. J Cetac Res Manage 1: 1–10.

29. Carretta JV, Forney KA, Lowry MS, Barlow J, Baker J, et al. (2010) U.S. Pacific marine mammal stock assessments: 2009. California, USA. 336 p.

30. Sigman DM, Casciotti KL (2001) Nitrogen Isotopes in the Ocean. In: Editor-in-Chief: John HS, editor. Encyclopedia of Ocean Sciences. Oxford: Academic Press. pp. 1884–1894.

31. Mesnick SL, Taylor BL, Archer FI, Martien KK, Treviño SE, et al. (2011) Sperm whale population structure in the eastern and central North Pacific inferred by the use of single-nucleotide polymorphisms, microsatellites and mitochondrial DNA. Mol Ecol Resour 11: 278–298.

32. Marcoux M, Whitehead H, Rendell L (2007) Sperm whale feeding variation by location, year, social group and clan: Evidence from stable isotopes. Mar Ecol Prog Ser 333: 309–314.

33. Ruiz-Cooley RI, Ballance LT, McCarthy MD (2013) Range expansion of the jumbo squid in the NE Pacific: δ^{15}N decrypts multiple origins, migration and habitat Use. PLoS ONE 8: e59651.

34. Ruiz-Cooley RI, Gendron D, Aguiniga S, Mesnick S, Carriquiry JD (2004) Trophic relationships between sperm whales and jumbo squid using stable isotopes of C and N. Mar Ecol Prog Ser 277: 275–283.

35. Francey RJ, Allison CE, Etheridge DM, Trudinger CM, Enting IG, et al. (1999) A 1000-year high precision record of δ^{13}C in atmospheric CO₂. Tellus B 51: 170–193.

36. Karl DM, Bidigare RR, Letelier RM (2001) Long-term changes in plankton community structure and productivity in the North Pacific Subtropical Gyre: The domain shift hypothesis. Deep Sea Res Part II 48: 1449–1470.

37. Polovina JJ, Howell EA, Abecassis M (2008) Ocean's least productive waters are expanding. Geophys Res Lett 35: L03618.

38. Polovina JJ, Dunne JP, Woodworth PA, Howell EA (2011) Projected expansion of the subtropical biome and contraction of the temperate and equatorial upwelling biomes in the North Pacific under global warming. ICES J Mar Sci 68: 986–995.

39. Guilderson TP, McCarthy MD, Dunbar RB, Englebrecht A, Roark EB (2013) Late Holocene variations in Pacific surface circulation and biogeochemistry inferred from proteinaceous deep-sea corals. Biogeosciences 10: 3925–3949.

40. Fiedler PC, Mendelssohn R, Palacios DM, Bograd SJ (2012) Pycnocline Variations in the Eastern Tropical and North Pacific, 1958–2008. J Climate 26: 583–599.

41. Dave AC, Lozier MS (2013) Examining the global record of interannual variability in stratification and marine productivity in the low-latitude and mid-latitude ocean. J Geophysi Res-Oceans 118: 3114–3127.

42. Rykaczewski RR, Dunne JP (2010) Enhanced nutrient supply to the California Current Ecosystem with global warming and increased stratification in an earth system model. Geophys Res Lett 37: L21606.

43. Aksnes DL, Ohman MD (2009) Multi-decadal shoaling of the euphotic zone in the southern sector of the California Current System. Limonol Oceanogr 54: 1272–1281.

44. Bograd SJ, Buil MP, Lorenzo ED, Castro CG, Schroeder ID, et al. (2014) Changes in source waters to the Southern California Bight. Deep Sea R Part II. Available: http://dx.doi.org/10.1016/j.dsr2.2014.04.009.

45. McClatchie S (2013) Regional fisheries oceanography of the California Current System: the CalCOFI Program. Dordrecht: Springer. 253 p.

46. García-Reyes M, Largier J (2010) Observations of increased wind-driven coastal upwelling off central California. J Geophy Res 115.

47. Jacox MG, Moore AM, Edwards CA, Fiechter J (2014) Spatially resolved upwelling in the California Current System and its connections to climate variability. Geophysl Res Lett 41: 3189–3196.

48. Kahru M, Kudela RM, Manzano-Sarabia M, Greg Mitchell B (2012) Trends in the surface chlorophyll of the California Current: Merging data from multiple ocean color satellites. Deep Sea Res Part II 77–80: 89–98.

49. Rau GH, Takahashi T, Marais DJD (1989) Latitudinal variations in plankton δ^{13}C: implications for CO₂ and productivity in past oceans. Nature 341: 516–518.

50. Wada E, Hattori A (1991) Nitrogen in the sea: forms, abundances, and rate processes. Boca Raton: CRC Press. 208 p.

51. Longhurst AR (2007) Ecological Geography of the Sea; Press. EA, editor. 542 p.

52. Morin PA, Nestler A, Rubio-Cisneros NT, Robertson KM, Mesnick S (2005) Interfamilial characterization of a region of the ZFX and ZFY genes facilitates sex determination in cetaceans and other mammals. Mol Ecol 14: 3275–3286.

53. McCarthy MD, Benner R, Lee C, Fogel M (2007) Amino acid nitrogen isotopic fractionation patterns as indicators of heterotrophy in plankton, particulate, and dissolved organic matter. Geochim Cosmochim Acta 71: 4727–4744.

54. Sherwood OA, Lehmann MF, Schubert CJ, Scott DB, McCarthy MD (2011) Nutrient regime shift in the western North Atlantic indicated by compound-specific δ^{15}N of deep-sea gorgonian corals. Proc Natl Acad Sci USA.

20

Estimating Trans-Seasonal Variability in Water Column Biomass for a Highly Migratory, Deep Diving Predator

Malcolm D. O'Toole[1]*, **Mary-Anne Lea**[1], **Christophe Guinet**[2], **Mark A. Hindell**[1]

1 Institute of Marine and Antarctic Studies, University of Tasmania, Hobart, Australia, **2** Marine Predator Department, Centre détudes biologiques de Chizé, Villiers-en-Bois, France

Abstract

The deployment of animal-borne electronic tags is revolutionizing our understanding of how pelagic species respond to their environment by providing *in situ* oceanographic information such as temperature, salinity, and light measurements. These tags, deployed on pelagic animals, provide data that can be used to study the ecological context of their foraging behaviour and surrounding environment. Satellite-derived measures of ocean colour reveal temporal and spatial variability of surface chlorophyll-a (a useful proxy for phytoplankton distribution). However, this information can be patchy in space and time resulting in poor correspondence with marine animal behaviour. Alternatively, light data collected by animal-borne tag sensors can be used to estimate chlorophyll-a distribution. Here, we use light level and depth data to generate a phytoplankton index that matches daily seal movements. Time-depth-light recorders (TDLRs) were deployed on 89 southern elephant seals (*Mirounga leonina*) over a period of 6 years (1999–2005). TDLR data were used to calculate integrated light attenuation of the top 250 m of the water column (LA_{250}), which provided an index of phytoplankton density at the daily scale that was concurrent with the movement and behaviour of seals throughout their entire foraging trip. These index values were consistent with typical seasonal *chl-a* patterns as measured from 8-daySea-viewing Wide Field-of-view Sensor (SeaWiFs) images. The availability of data recorded by the TDLRs was far greater than concurrent remotely sensed *chl-a* at higher latitudes and during winter months. Improving the spatial and temporal availability of phytoplankton information concurrent with animal behaviour has ecological implications for understanding the movement of deep diving predators in relation to lower trophic levels in the Southern Ocean. Light attenuation profiles recorded by animal-borne electronic tags can be used more broadly and routinely to estimate lower trophic distribution at sea in relation to deep diving predator foraging behaviour.

Editor: Z. Daniel Deng, Pacific Northwest National Laboratory, United States of America

Funding: This research was funded by the Australian Research Council, the Natural Sciences and Engineering Research Council of Canada, Sea World Research and Rescue Foundation Inc., and the Australian Antarctic Science Programme. MO was supported by an Australian Postgraduate Award. The funders had no role in study design, data collection and analysis, decision to publish, or preparation of the manuscript.

Competing Interests: The authors have declared that no competing interests exist.

* Email: otoolem@utas.edu.au

Introduction

Chlorophyll-a is an important biological parameter in the Southern Ocean and is considered a useful indicator of spatial and temporal variability of primary productivity [1–3]. To understand the foraging behaviour and habitat utilisation of higher trophic organisms requires knowledge of lower trophic dynamics, coupled with information on how organisms respond to these changes. Indeed, satellite measurements of ocean colour have revealed the complex temporal and spatial variability of weighted average near-surface chlorophyll-a concentration [4], but the quantity and quality of information obtained in this way is affected by cloud cover. Consequently, information from high latitudes and during the winter months is often sparse [5,6] and correspond poorly with marine animal behaviour. Moreover, to improve data availability, these patchy satellite data are often merged at spatio-temporal scales not necessarily relevant to marine animal behaviour. While fluorometers and water samples from ship-based surveys are the only in-vivo and in-vitro measurements to determine chlorophyll-a concentration, it is both costly and logistically difficult if collecting simultaneously with animal behaviour. In recent years, additional

ocean data recorded by animal-borne electronic tags have been used to supplement other data from buoys and satellites (e.g., [7,8]) and have improved our understanding of the relationship between marine predator distribution and environmental parameters, including chlorophyll-a [9,10]. Indeed, miniaturised fluorometers have now been deployed, in some instances simultaneously with light sensors, on elephant seals to estimate chlorophyll-a in the water column [11,12] but are costly and available data are scarce. Therefore, understanding lower trophic variability (i.e. phytoplankton) and its influence on marine predators in the Southern Ocean is still hampered by a lack of concurrent data.

Time-depth-light recorders (TDLRs) provide detailed information on dive behaviour of a wide range of animals over extensive areas [13,14], and are often coupled with sensors that record environmental data (*e.g.* temperature and salinity). Southern elephant seals (*Mirounga leonina*) are ideal platforms for these oceanographic sensors due to their circumpolar distribution extensive foraging across the Southern Ocean [9]. They are also a deep diving animals, diving up to 2000 m [15] while performing on average 60 dives per day (Hindell et al. 1991). Elephant seals can be used to measure *in situ* environmental conditions and provide important habitat information for the seals [9]. Seals equipped with sensors that collect information such as temperature, salinity can cover areas not sampled by conventional techniques (e.g. ship-based survey, satellite images), including within the sea-ice zone (*i.e.* south of 60°S) where it is particularly difficult to sample physical parameters of the ocean [7]. Furthermore, post-moult elephant seals are also at sea throughout winter when data collected by conventional techniques is scarce.

Light levels recorded by animal-borne sensors are commonly used to infer day length as a means of estimating geographical position [16,17], and can also be used as means of recording light levels at depth during animal diving [18–20]. Experiments have demonstrated the concept of estimating chlorophyll-a distribution from light-depth data compared to fluorescence (e.g., [10,12]). Fluorometers estimate chlorophyll-a by measuring its fluorescence intensity. Light sensors instead measures ambient light, which is attenuated throughout the water column for two reasons: (1) physical properties of the seawater and (2) quantity of inorganic and organic particles suspended (or dissolved) in the water column [21]. The Southern Ocean is typically characterised by Case I waters, whereby phytoplankton comprises the main source of particles suspended within the euphotic zone [21,22], and is consequently the main cause of light attenuation if we assume related coloured dissolved organic matter (CDOM) and detritus degradation products covary with phytoplankton [23] and physical properties are constant [24]. Indeed, it was Smith and Baker [1] that introduced the concept of measuring the bio-optical properties of the water column to estimate the concentration of chlorophyll-a in the ocean. A study by Teo et al. [10], one of the first to use light levels collected by Pacific blue fin tuna (*Thunnus orientalis*) to estimate chlorophyll-a distribution, found a positive relationship between light attenuation at depth and *in situ* chlorophyll-a collected by both water samples and fluorometers. Light levels collected by elephant seals equipped with light sensors were also strongly correlated with concurrent *in situ* fluorometer data [12]. Despite their findings, these studies were not performed over multiple seasons; instead tested over a much shorter time scale. Nor was light attenuation compared with satellite-derived chlorophyll-a estimates (hereafter *chl-a*). More recently, Guinet et al. [11] used a multi-seasonal dataset over several years and found *chl-a* to be related to surface chlorophyll-a estimates from seal-borne fluorometers. The bio-optical relationship between chlorophyll-a and phytoplankton and does vary according to phytoplankton taxonomic composition [25] but are still considered to correlate well with each other. To our knowledge, no study has used light level and depth data to generate a phytoplankton index that matches daily seal movements while at sea.

This study examined the feasibility of using light collected from TDLRs to calculate an index of phytoplankton distribution that is concurrent with marine animal behaviour in the Southern Ocean. We also highlight the advantages of a phytoplankton index recorded simultaneously with the foraging behaviour of a top marine predator, particularly at times of the year where *chl-a* data is lacking. Analyses were performed in Case 1 waters over multiple seasons between 1999 and 2005. Our primary objectives included:

(i)   providing an index of phytoplankton density at the daily scale that is concurrent with the movement and behaviour of seals throughout their entire foraging trip;

(ii) demonstrating that our index is consistent with typical seasonal chl-a patterns;

(iii) examining the efficacy of using a light-based index to estimate phytoplankton distribution.

Materials and Methods

Ethics statement

All necessary permits were obtained for the described field studies. Elephant seal research was sanctioned by the University of Tasmania Animal Ethics Committee (permit A6738) and the Australian Antarctic Science Advisory Council Ethics Committee (project 2794). Permits and permission to carry out research on Macquarie Island was obtained from Parks and Wildlife Service Tasmania.

TDLRs (Mk6, Mk7, Mk8 and Mk9; Wildlife Computers, Redmond, WA, USA) were attached to both post-breeding and post-moult adult female southern elephant seals (n = 89) at Macquarie Island (54°35'S, 158°58'E, Table 1) from 1999 to 2005. The seals were approached by foot and temporarily restrained with a head bag and anaesthetised intravenously with a 1:1 mixture of tiletamine and zolazepam (0.5 mg kg^{-1}) [26,27]. TDLRs were attached to the pelage above the shoulders using a two component industrial epoxy (Araldite AW 2101) [28]. Seals were observed during recovery from anaesthesia and allowed to enter the water when no longer sedated. TDLRs were retrieved at the end of the foraging trip once the seal had hauled out on land by repeating the above restraint procedures. These tracking devices or attachment method did not adversely affect individual performance and fitness over the short (seal growth) or long (seal survival) term [29].

Tag data

TDLRs measured time, depth and light at 30 s intervals for the duration of each foraging trip. Mk6–Mk8 tags used uncorrected watch crystals to measure time. They were offset to spread the time error (TE) over the likely range of seawater temperatures (T) (TE = $(1 \times 10^{-5} - 3.5 \times 10^{-8} \times (T-25)^2) \times 10^6$ µs). Mk9 tags used a temperature correction algorithm to keep the time error within 1 ppm. Depth measurements were made by a pressure transducer calibrated by the manufacturer (± 6 m). Light values are converted on-board the logger via a log treatment (see Figure S1) to compress the light measurements to a three digit value, thereby giving a linear relationship and increase the resolution at lower light levels. The light sensor data can be used to identify dawn/dusk events down to 300 m in clear waters and is temperature-compensated for the entire light level range (Wildlife Computers). The wavelength at the centre of the light sensor parabolic-shaped pass-band filter is ~430 nm and consequently the sensor only reads the violet/blue light band (370 nm–470 nm). All other bands of light are rejected and not measured. The light sensor measures on a scale of 20 readings per decade, so the light level error is considered to be 1/20th of a decade. Tags also recorded temperature (± 0.1°C). The lag in temperature measurement (inherent in the design of the TDLRs) was accounted for (see [30,31]).

Data extraction

Twice daily at-sea location estimates were derived from the recorded light levels for sunrise and sunset using the geo-location procedure outlined inThums et al. [32]. Geo-location by light enables animal movement estimation, based on measurements of light intensity over time recorded by the in-built light sensor of each TDLR [17]. However, an inherent problem with this approach is that an array of factors may change the natural light intensity pattern, thereby affecting the accuracy and precision of location estimates calculated from these light patterns [33–35]. With the incorporation of the 'tripEstimation' method (see [36]) geo-location mean longitudinal and latitudinal error is shown to be

Table 1. Summary of tag deployments for 89 female elephant seals: year, trip, tag type used, number of individuals tagged, and period of data records (start to finish dates).

Year	Trip	Tag type	No. individuals	Period of data records	
				Start	Finish
1999	PB	mk6	4	23-Oct-1999	12-Jan-2000
1999	PB	mk7	15	15-Oct-1999	17-Jun-2000
2000	PM	mk7	9	26-Jan-2000	15-Oct-2000
2000	PB	mk7	11	21-Oct-2000	14-Jun-2001
2001	PM	mk7	3	9-Feb-2001	3-Oct-2001
2001	PM	mk8	3	15-Jan-2001	13-Oct-2001
2002	PM	mk7	3	30-Jan-2002	22-Sep-2002
2002	PM	mk8	12	26-Jul-2001	7-Nov-2002
2004	PM	mk8	12	30-Jan-2004	19-Oct-2004
2004	PB	mk8	7	18-Oct-2004	31-Jan-2005
2004	PM	mk9	3	24-Jan-2004	15-Dec-2004
2004	PB	mk9	2	22-Oct-2004	5-May-2005
2005	PM	mk8	3	11-Jan-2005	13-Oct-2005
2005	PM	mk9	2	19-Jan-2005	15-Oct-2005

PB - post-breeding, PM - post-moult.

estimated at $\sim 57 \pm 9$ km (i.e. 0.83°) and $\sim 54 \pm 8$ km (i.e. 0.49°) respectively (Chew unpublished). All dive recorders were corrected for drift in the pressure sensor using a customised zero-offset correction routine. We then identified individual dive cycles, defined as the first sub-surface record until the last surface interval of the subsequent post-dive surface interval below 10 m. The surface interval encompassed depth values between 0 and 10 m. This tolerance accounted for subsurface movements of seals between dives.

Environmental data

Satellite-derived chlorophyll-a estimates. The *chl-a* data (mg m^{-3}) was estimated from Sea-viewing Wide Field-of-view Sensor (SeaWiFs) images [37]. Because of the patchy nature of these data at high latitudes, particularly during winter, we used 8-day *chl-a* composites at 0.1° resolution (http://oceancolor.gsfc.nasa.gov/).

Sea ice. We extracted sea ice data from daily satellite images (grid cell size of 25 km×25 km) [38]. Satellite *chl-a* data in regions with >20% sea ice coverage were excluded from analyses as reflective irradiance from the ice may affect the accuracy of satellite imagery (P. Strutton, Personal Communication). Sea ice data were also used to calculate the seasonal mean sea ice extent between 1999 and 2005. The sea ice extent was defined by the open ocean (i.e. ice-free pelagic region) – sea ice (>50% concentration) interface.

Bathymetry. We aggregated bathymetry data, derived from the ETOPO2 bathymetry data set at 2′ resolution (http://www.ngdc.noaa.gov/mgg/global/global-.html), to calculate mean bathymetric depth of each 1°×1° grid cell associated with each seal location.

Mixed layer. In order to assess changes to phytoplankton density using integrated light attenuation we consider total phytoplankton in the water column, the bulk of which is found within the mixed layer [39]. Temperature and profiles recorded by the TDLRs were used to identify the mixed layer depth (hereafter MLD) for each dive to establish the vertical extent of phytoplank-

ton in the water column. A custom broken stick method was used to find the greatest inflection point along each temperature-depth profile to a depth of 350 m (limit of light sensor sensitivity is ~ 300 m). The inflection point was considered the MLD if the difference between temperature at the surface (~ 10 m) and temperature inflection point was greater than 0.2°C [40].

The same procedure was applied to light profiles, also recorded by the TDLRs, to identify the depth of the most significant light inflection point for each dive. It is important to note consistent light-depth profile differences between the descent and ascent phase of each dive (Figure 1), owing to a time-response lag inherent in the light sensor that is greatest at low light levels (Wildlife Computers), as well as possible changes in water properties, surface irradiance and animal behaviour. We calculated the average depth of the most significant light inflection point for each dive to account for this bias. From the surface to the depth of the most significant light inflection point was considered the section of the water column that incorporated the bulk of phytoplankton.

These analyses showed the proportion of dives with a given temperature and light inflection depth closely corresponds with each other (Figure 2). According to these results we conclude that the mixed layer depth and bulk of phytoplankton were frequently above 250 m (82.6% and 74.3% of dives respectively), and fewer dives encountered mixed layer depths or the bulk of phytoplankton exceeding 300 m (17.4% and 25.7%) (Figure 2).

Frontal zones. Frontal structures in the Southern Ocean are sharp, horizontal gradients in water properties that mark the boundaries between different frontal zones (FZ) (Figure 3 – represented by historical mean front positions). The general position for each front can be marked using representative values of temperature and salinity at approximately 200 m depth. The FZ occupied by the seal was identified by water temperatures at 200 m depth (T_{200}), as indicated by Park et al. [41] and Orsi et al. [42]. The sub-Antarctic Front (SAF) limit was defined by sub-surface values of 7°C, the Polar Front (PF) was defined by the northern limit of 2.8°C, and the Southern Antarctic Circumpolar

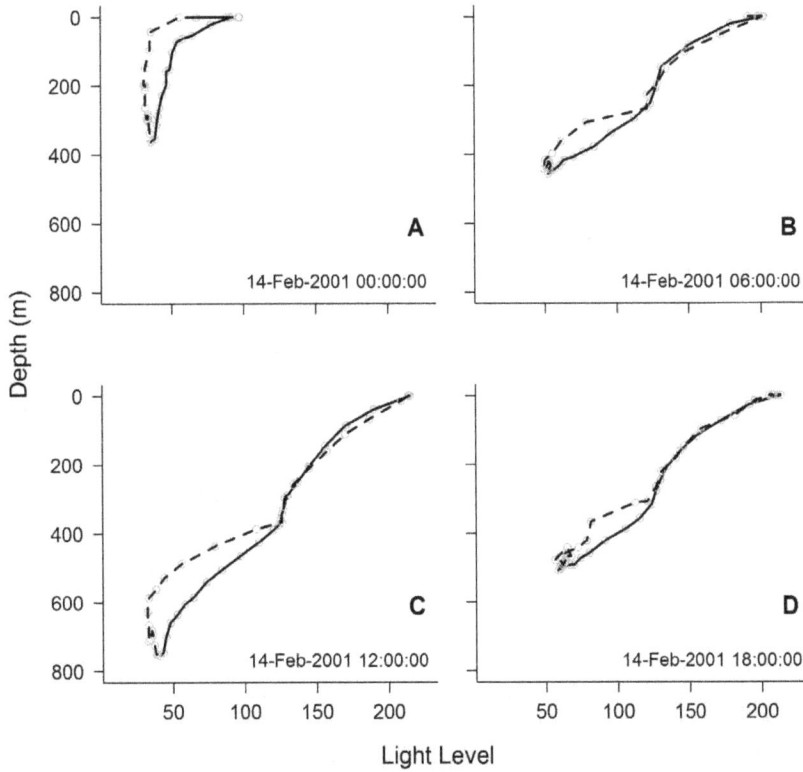

Figure 1. Examples of light-depth profiles collected from the descent (solid) and ascent (dashed) phases of dives. Profiles recorded at local (A) midnight, (B) 6am, (C) noon and (D) 6pm on 14 February 2001. Light level values are related to blue light intensity (W cm^{-2}). Calibrations are checked at levels 10^{-5}, 10^{-7} and 10^{-9} W cm^{-2}, which correlated to light level values around 150, 110 and 70 respectively (see Figure S1).

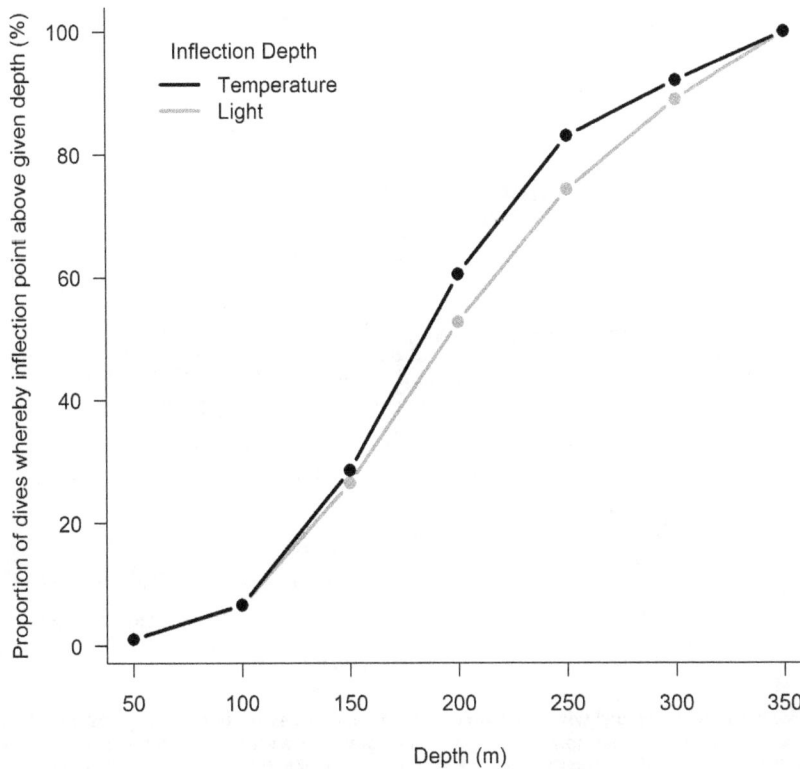

Figure 2. Proportion of dives whereby greatest temperature and light inflection points are above a given depth.

Current Front (SACCF) was defined by a temperature of 1.6°C [43]. In this study we use these subsurface boundaries to distinguish between three major FZ: the Polar Frontal Zone (PFZ) was where seals encountered temperatures greater than 2.8°C; north of the southern Antarctic Circumpolar Current Front (SACCF-N) was where seals encountered temperatures between 2.8°C and 1.6°C; and south of the SACCF (SACCF-S) was where seals encountered temperatures below 1.6°C.

Temperature recorded for both the descent and ascent phase of individual dives were used and a temperature value at 200 m (T_{200}) was derived for both the descent and ascent phase of each dive using a linear interpolation between the non-regular series of depths and temperature. Temperature values from the two phases were averaged. Because we only retained local noon light attenuation values for our analyses only local noon temperature estimates were used to calculate the mean daily noon T_{200}. Each daily noon light level profile was assigned to a given FZ based on these mean daily noon T_{200} values (Figure S2).

Light attenuation

By examining the mixed-layer depths and light-depth profiles encountered by the seals we determined that the bulk of phytoplankton was likely found in the top 250 m of the water column (see mixed layer section). Moreover, light levels recorded at depths of 300 m or more become unreliable as the light sensors reach their sensitivity limit. Consequently, integrated light attenuation between the surface and a depth of 250 m (LA_{250}) was used as an index of phytoplankton in the water column (based on the assumptions outlined in our introduction).

To calculate LA_{250}, light data were first interpolated linearly between the non-regular series of depths to estimate light levels at 250 m for each dive (LL_{250}). We used light levels recorded for both the descent and ascent phase due to sensor and measurement error (see mixed layer section). The surface light level for each dive was estimated from the mean sub-surface light levels in the top 10 m of the water column at the end of the ascent phase (LL_0) [12]. Light levels above the surface (indicated by the wet/dry sensor on the tag) were excluded from LL_0 estimates. For each

dive, LL_{250} was subtracted from LL_0 and divided by the depth (z) of LL_{250} (i.e. 250 m) to calculate the LA_{250} (m^{-1}):

$$LA_{250} = \frac{LL_0 - LL_{250}}{z}$$

Only LA_{250} values 1 h either side of local noon (1100–1300) were used in order to minimise variability in the ambient light field (see discussion in [10]). The interpolation of geo-locations was done to attempt best correspondence with noon dives and chl-a (see maps – Figure S2). This was based on the assumption that the seals' trajectory between consecutive locations was straight. Data recorded 1 h either side of local noon encompassed, on average, 4.3 ± 1.3 light profiles per seal per day. Since we assume Southern Ocean waters are Case-1, attenuation will be dominated by phytoplankton (see introduction for details).

Statistical analysis

As part of this study we aimed to demonstrate that LA_{250} values (our phytoplankton index) are consistent with typical seasonal chl-a patterns. However, spatial error associated with positions derived by geo-location [17] impart uncertainty in the true position of the recorded light attenuation. If then compared to chl-a values, for which the spatial errors were considerably less, any resulting correlation may be subsequently weakened. We attempt to account for spatial bias in geo-location position errors by spatial averaging of the data into 1°×1° grid cells. Grid cells with less than 3 dive profiles were excluded from the analysis as these were likely to give unreliable estimates of the resulting mean LA_{250} and mean chl-a per grid cell. Moreover, chl-a data were sparse at high latitudes (>64°S) and during winter months due to elevated cloud cover (Table 2). Conversely, the seal light data were sparse at low latitudes (<52°S) as few seals travelled north of this region (Table 2). Focal analysis was therefore based on data collected between 52°S and 64°S, and excluded winter months; for two reasons: (i) it is not possible to establish interaction effects when there are missing data in the dataset; and (ii) low data frequency may result in interaction effect bias.

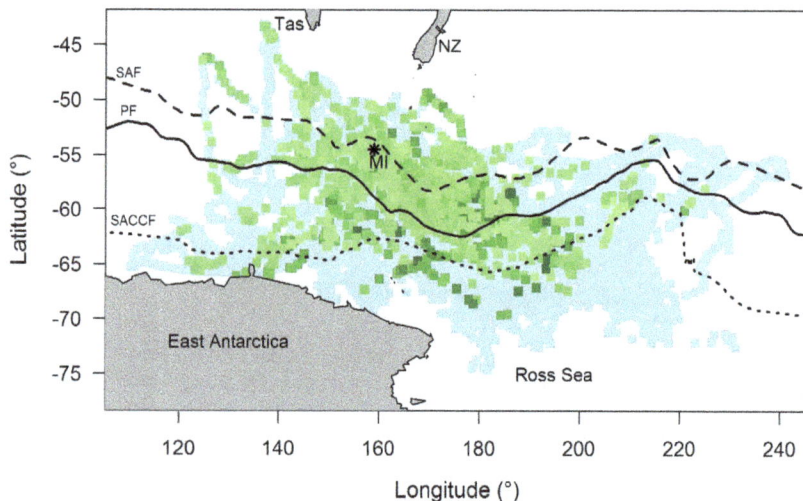

Figure 3. Noon locations for all seals with and without concurrent chl-a. Includes locations during post-moult and post-breeding foraging trips collectively (light blue), and of these, all that correspond with chl-a (green). Scale of chl-a values denoted by grading from light green (0.03 mg m^{-3}) to dark green (2.48 mg m^{-3}). Map shows the bottom of Tasmania (Tas) and New Zealand (NZ) and the coast of East Antarctica and Ross Sea (bottom). The black asterisk shows Macquarie Island (MI). Lines represent the historical mean positions of the Sub-Antarctic Front (SAF - dashed), Polar Front (PF - solid) and Southern Antarctic Circumpolar Current Front (SACCF - dotted).

Table 2. Frequency of daily locations for each season by 1° latitudinal bins.

Latitude (°S)	Season			
	Autumn	Spring	Summer	Winter
44	-	-	-	6
45	-	-	-	4
47	4	-	-	1
48	1	-	-	2
49	1	4	-	1
50	2	7	3	1
51	1	8	5	6
52	3	6	8	2
53	5	10	5	-
54	2	18	11	-
55	4	63	38	-
56	6	72	44	1
57	3	51	50	-
58	12	41	43	-
59	14	21	53	-
60	21	27	46	-
61	22	26	47	-
62	27	17	40	-
63	24	12	27	-
64	33	2	15	-
65	40	-	13	-
66	28	-	14	-
67	10	-	4	-
68	5	-	-	-
69	3	-	-	-
70	2	-	-	-

We investigated the relationship between *chl-a* and LA_{250} aggregated at 1° resolution. We used the mixed effect model (*nlme*) package in R [44] to assess this relationship with and without the random intercept term and slope effect to determine whether individual seals were contributing to the model fit. Season and latitude (and their interaction terms) were included in our analysis because of their likely effect on phytoplankton abundance in the water column (for details see discussion). Season was divided according to the austral seasonal cycle: summer (Dec–Feb); autumn (Mar–May); winter (Jun–Aug); and spring (Sep–Nov). Because FZ is largely influenced by latitude in the Southern Ocean (e.g., [45,46]) we expect the inclusion of FZ and latitude in our mixed model to have a confounding effect on *chl-a* distribution. For that reason we assessed the inclusion of each of these effects in our mixed model relative to each other and found that latitude was more useful for the purpose of this study (Table S1). We therefore tested the individual fixed effects (including LA_{250}, season, latitude and their interactions) by sequentially removing non-significant terms from the model according to Zuur et al. [47]. In all cases, models were ranked via Akaike Information Criterion [48], the most parsimonious model having the lowest AIC value. Model selection was carried out using Maximum Likelihood (ML) estimation. In addition, we used F and t statistics to examine the significance of individual fixed effects. The final model is presented using restricted maximum likelihood (REML) methods. Both *chl-a* and LA_{250} values were log-transformed to ensure a normal distribution.

Results

We used data from entire foraging trips for 67 (75%) of the 89 deployments (31 post-breeding/36 post-moult trips). Twenty one trips were excluded due to light sensor failure at some point during the time at sea. Data for one seal were also omitted due to unrealistic track estimates (*i.e.* the track passed over land). Data were obtained over 1561 days from 22 Oct 1999 through to 8 Oct 2005. A total of 31614 light profiles at 9552 noon locations were recorded during this period (Table 3)). There were 7212 noon locations available that included 3 or more light profiles (*i.e.* LA_{250} values) and did not coincide with heavy sea-ice, of which only 1461 noon locations coincided with *chl-a* values (20.3%) (Table 3, Figure 3). This showed approximately one-fifth of seal locations (with daily LA_{250} values) coincided with *chl-a* values. Filtered data (*i.e.* included 3 or more light profiles, did not coincide with heavy sea-ice) were then gridded into 3940 1°×1° cells for model analysis, of which only 1066 cells (25.1%) corresponded with gridded *chl-a* data (Table 3). Each cell incorporated 1.26±0.02 locations (4.77±0.09 light profiles). Seals travelled either to the sea ice zone in the north of the Ross Sea and off the coast of East

Table 3. Data summary for each deployment (*i.e.* trip) by year: number of seals (*n*); total light profiles and locations at noon and concurrent chl-a; filtered[†] light profiles and locations at noon and concurrent chl-a; number of 1° grid cell locations and concurrent chl-a.

Year	Trip	Seals (n)	Total			Filtered[†]			Grid Cells	
			Light profiles	Locations	Concurrent chl-a	Light profiles	Locations	Concurrent chl-a	Locations	Concurrent chl-a
1999	PB	14	4319	984	307	4096	939	306	514	207
2000	PB	10	2954	713	167	2805	681	165	365	115
2001	PB	-	-	-	-	-	-	-	-	-
2002	PB	-	-	-	-	-	-	-	-	-
2004	PB	7	1733	433	180	1644	415	177	255	129
2005	PB	-	-	-	-	-	-	-	-	-
		31	*9006*	*2130*	*654*	*8545*	*2035*	*648*	*1134*	*451*
1999	PM	-	-	-	-	-	-	-	-	-
2000	PM	6	4120	1328	113	3385	1051	111	544	89
2001	PM	5	3091	951	93	2659	808	85	433	69
2002	PM	10	6199	2001	269	4241	1262	254	689	188
2004	PM	11	6832	2292	297	5034	1632	283	892	205
2005	PM	4	2366	850	81	1312	424	80	248	64
		36	*22608*	*7422*	*853*	*16631*	*5177*	*813*	*2806*	*615*
Total		**67**	**31614**	**9552**	**1507**	**25176**	**7212**	**1461**	**3940**	**1066**
%		-	-	-	**15.8**	-	-	**20.3**	-	**27.1**

[†]locations with >3 light profiles that do not coincide with heavy sea-ice.

Table 4. Ranked mixed models.

Candidate models	df	AIC	ΔAIC	logLik
LA$_{250}$+lat+S+LA$_{250}$: lat+S: lat	**12**	**547.8**	**0**	**−261.9**
LA$_{250}$+lat+S+S: lat	11	550.7	2.9	−264.3
LA$_{250}$+lat+S+LA$_{250}$: S+LA$_{250}$: lat+S: lat	14	551.7	4	−261.9
LA$_{250}$+lat+S+LA$_{250}$: S+S: lat	13	553.9	6.2	−264
LA$_{250}$+lat+S+LA$_{250}$: S+LA$_{250}$: lat+S: lat+LA$_{250}$: S: lat	16	554.2	6.5	−261.1
LA$_{250}$+lat+S+LA$_{250}$: lat	10	554.5	6.7	−267.2
LA$_{250}$+lat+S	9	557.1	9.3	−269.5
LA$_{250}$+lat+S+LA$_{250}$: S+LA$_{250}$: lat	12	558.5	10.7	−267.2
LA$_{250}$+lat+S+LA$_{250}$: S	11	560.3	12.5	−269.1
LA$_{250}$+S	8	568.1	20.4	−276.1
LA$_{250}$	6	662	114.2	−325
LA$_{250}$+lat	7	662.7	115	−324.4
S+lat	6	664.8	117.1	−326.4
S	5	685.2	137.5	−337.6
lat	4	787.1	239.3	−389.5
~1	3	794.5	246.7	−394.2

The *chl-a* explained by light attenuation at 250 m (LA$_{250}$), season (S) and latitude (lat) ($n=67$ seals). Mixed models are ranked by decreasing Akaike's Information Criterion (AIC) and change in AIC (ΔAIC). The most parsimonious model is in bold.

Antarctica, or to the shelf break of East Antarctica (Figure 3). Most seals travelled to areas south of the SACCF.

Relationship between light attenuation and *chl-a*

The best model relating LA_{250} to *chl-a* included individual variability (random intercept term *seal*), the random slope term (LA_{250}); most parsimonious model included fixed effects LA_{250}, season, latitude, and the 2-way interaction terms LA_{250}: latitude and latitude: season (Table 4). The LA_{250} was positively related to *chl-a* (estimated coefficient = 0.76±0.13, p<0.0001, Table 5, Figure 4). Predicted *chl-a* values from our model show no obvious latitudinal or longitudinal error pattern over the study region (Figure 5A, B). However, results indicated that predicted *chl-a* values largely overestimated *chl-a* by 10–30%, particularly over water depths greater than 4000 m, those most frequented by seals (Figure 5C).

Distribution of light-based *chl-a* estimates

Fitted values from the mixed model results (*i.e.* phytoplankton index) were used to calculate the spatial distribution of light-based *chl-a* collected by TDLRs (hereafter TDLR$_{chl}$) encountered by the focal seals (Figure 6A), revealing different seasonal patterns in relation to latitude (Figure 6B). During summer months, seals encountered generally higher TDLR$_{chl}$ compared to other times of the year, particularly at latitudes between 60°S and 65°S, southeast of Macquarie Island (north of the Ross Sea). Conversely, seals encountered uniformly low TDLR$_{chl}$ across latitudes during autumn. In spring, TDLR$_{chl}$ encountered by seals were marginally greater (Figure 6B), and levels gradually elevated toward the mean spring-time sea ice extent.

These same fitted values were also used to calculate inter-annual TDLR$_{chl}$ variability and compared with *chl-a* within the 55–65°S latitudinal band (Figure 7). Mean monthly TDLR$_{chl}$ agreed well with *chl-a* inter-annual variability, despite large differences for January 2002 and 2004, and to a lesser extent, December 2005.

Table 5. Results from the most parsimonious mixed model: relating *chl-a* to integrated light attenuation in the top 250 m of the water column (LA_{250}), latitude and season and their significant interactions.

	Coefficient ± SE	Coefficient p
LA$_{250}$	**0.76±0.13**	**<0.0001**
Season (Summer)	**0.33±0.04**	**<0.0001**
Season (Spring)	**0.21±0.04**	**<0.0001**
Latitude	0.06±0.03	0.0654
LA$_{250}$: Latitude	**0.06±0.03**	**0.0264**
Season (Summer): Latitude	**0.03±0.01**	**0.0071**
Season (Spring): Latitude	**0.03±0.01**	**0.0040**

Term coefficients are presents ± SE and p-values for each coefficient are also shown. Significant terms (p<0.05) are denoted by bold characters. For the season variable that was a factor in the model, coefficients are given in reference to autumn.

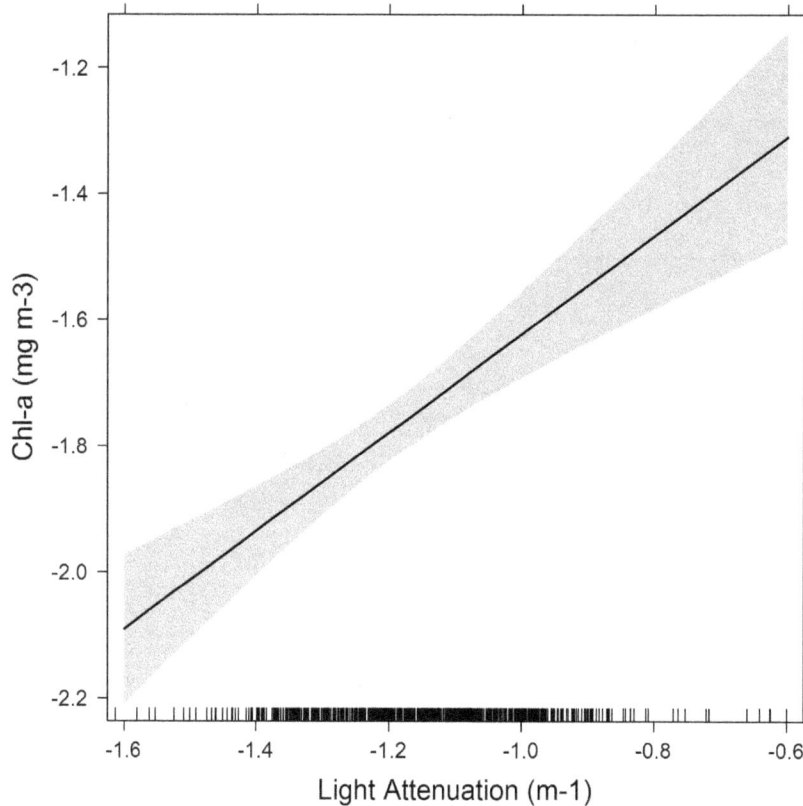

Figure 4. Relationship between *chl-a* **and light attenuation (LA₂₅₀) from our mixed model.** Shaded area indicates the confidence level. Both axes are log transformed.

These large differences correspond well with the few available data (Figure 7).

Light verses *chl-a* data coverage

The TDLR$_{chl}$ data sets provided more information than the *chl-a* data that corresponded with seal locations (hereafter corresponding *chl-a*), but both followed similar spatial and temporal trends (Figure 8A and Figure 8B respectively). However, coverage of TDLR$_{chl}$ and the overall *chl-a* available within the focal study region (hereafter overall *chl-a*) each followed different spatial and temporal trends (Figure 8C and Figure 8D respectively).

Spatial coverage of TDLR$_{chl}$ and corresponding *chl-a* peaked at latitudes between 55°S and 64°S; however peak corresponding *chl-a* coverage was considerably less than TDLR$_{chl}$ coverage (Figure 8A). In general, peak TDLR$_{chl}$ coverage increased with latitude up to 64°S only to drop with increasing proximity to the Antarctic Continent. However, the extent of TDLR$_{chl}$ coverage was still considerable at latitudes as high as 67°S. Conversely, peak corresponding *chl-a* data coverage steadily decreased from 56°S, becoming virtually negligible at 66°S. Overall, *chl-a* data coverage was greatest at 44°S, but was inversely related to latitude; virtually negligible at latitudes greater than ~67°S (Figure 8C).

Temporal coverage of TDLR$_{chl}$ and corresponding *chl-a* data peaked twice over a 12-month period; the largest peak during March, the other during spring with the exception of a sharp drop of coverage in October (Figure 8B). However, the two peaks in TDLR$_{chl}$ coverage were considerably greater than that of corresponding *chl-a* data (Figure 8B). Less coverage of TDLR$_{chl}$ and corresponding *chl-a* data was evident at the beginning of summer, during winter and in October. Specifically, minimal

coverage of TDLR$_{chl}$ occurred in October during the breeding season, although still maintained moderate-to-low coverage at this time compared to virtually nil coverage of corresponding *chl-a* data throughout winter. Overall *chl-a* data coverage was poor during the winter months, particularly in July when overall *chl-a* data coverage was completely unavailable (Figure 8D).

Discussion

This is the first multi-year dataset (67 elephant seals) used to provide a light-based index of phytoplankton density that is concurrent with a marine animal's entire foraging trip over multiple seasons. Of the 3940 LA_{250} gridded cells recorded over 5 years, only 25.1% cells coincided with *chl-a* measurements demonstrating the deficiency of remotely sensed data sources concurrent with animal behaviour. Model output also revealed that seasonal trends detected by our phytoplankton index were in agreement with data collected by remote sensing. It demonstrates how our phytoplankton index is consistent with near-surface *chl-a* values in the Southern Ocean, and that phytoplankton changes at depth generally reflect near-surface primary producer conditions. This opens the way for the use of simple light data as a bio-optical index for phytoplankton in the Southern Ocean that is concurrent with animal at-sea behaviour.

Relationship between water column light and *chl-a*

The Southern Ocean is characterised by Case I waters, where phytoplankton organisms are the most optically significant components of the water column [22]. It is therefore likely that bio-optical differences detected by TDLR light sensors at depth

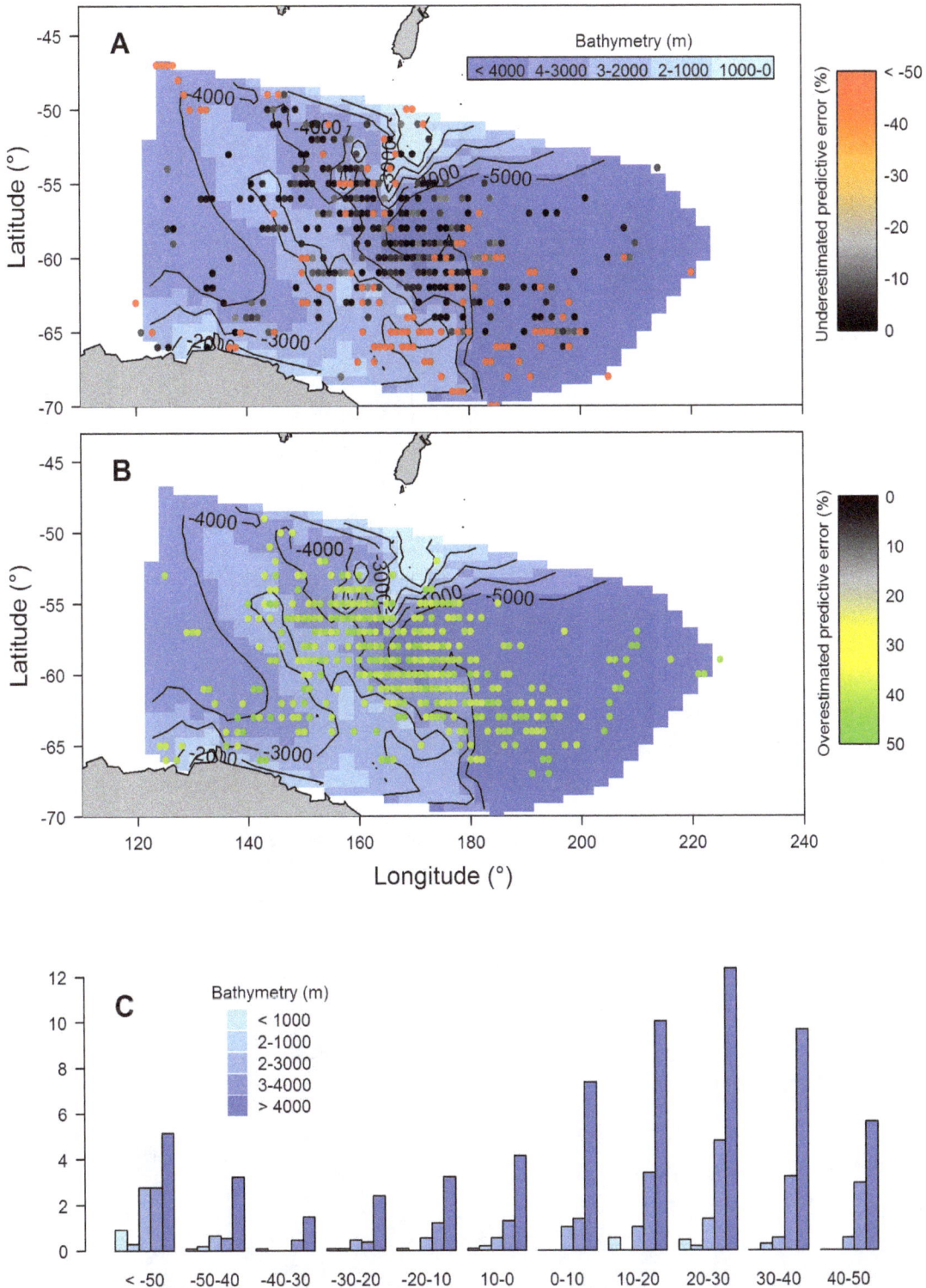

Figure 5. Spatial distribution of predictive chlorophyll-a error from our mixed model. Plots show locations associated with (A) underestimated predictive error (%), (B) overestimated predictive error (%), and (C) proportion (%) of locations with associated predictive error (%) in relation to bathymetric bands.

are representative of plankton densities. Our results show that *chl-a* (derived from satellite images) is significantly related to our phytoplankton index estimated from the integrated light attenuation recorded by TDLRs in the top 250 m of the water column.

Light at 250 m generally coincides with the limit of the euphotic zone, so all photosynthetic organisms in the water column influence light attenuation at this depth. In general, there is a good relationship between *chl-a* concentration within the top 30 m

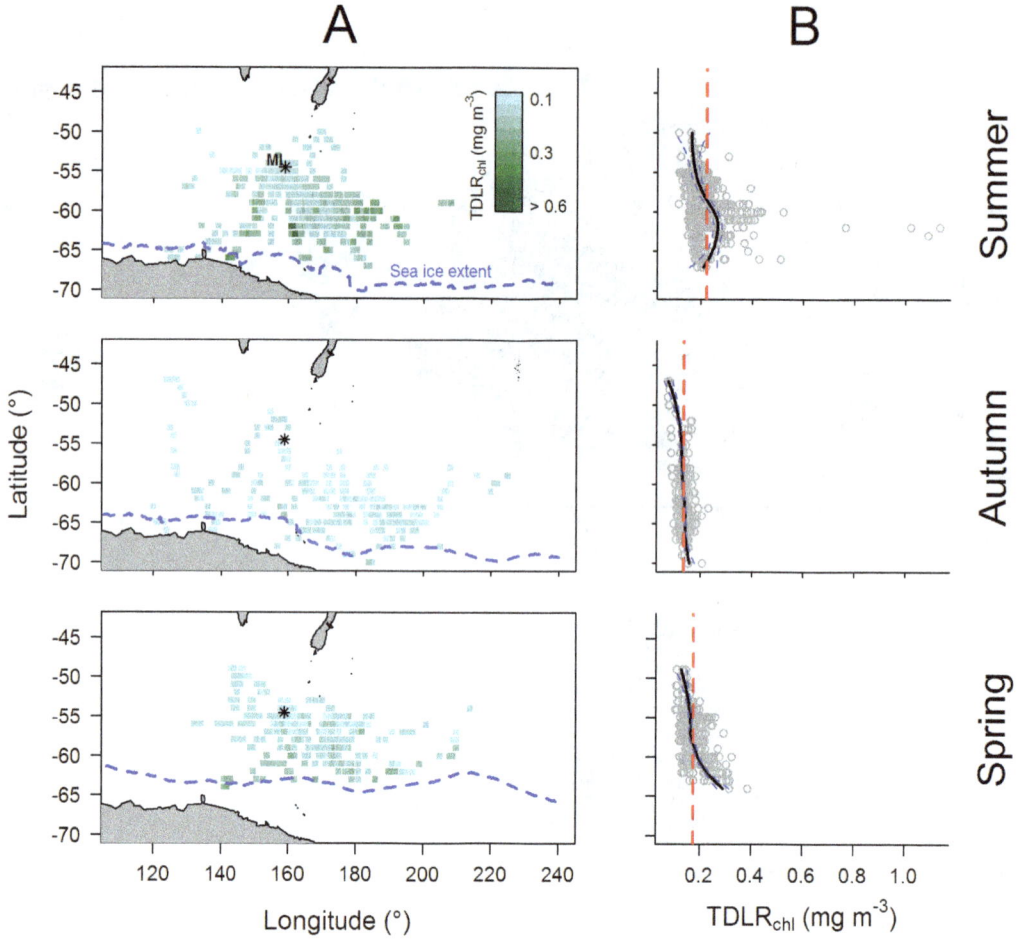

Figure 6. (A) Seasonal spatial distribution and (B) latitudinal patterns of TDLR$_{chl}$† from the final mixed model. Each map shows the bottom of Tasmania and New Zealand (top), the coast of East Antarctica and Ross Sea (bottom), and the sea ice extent (blue dashed line). The black asterisk shows Macquarie Island. For each corresponding plot (B) the black line represents a loess fit and blue dashed lines represent the 95% confidence level, and the vertical red dashed line represents the mean TDLR$_{chl}$. †Light-based *chl-a* estimates from our final mixed model collected by TDLRs.

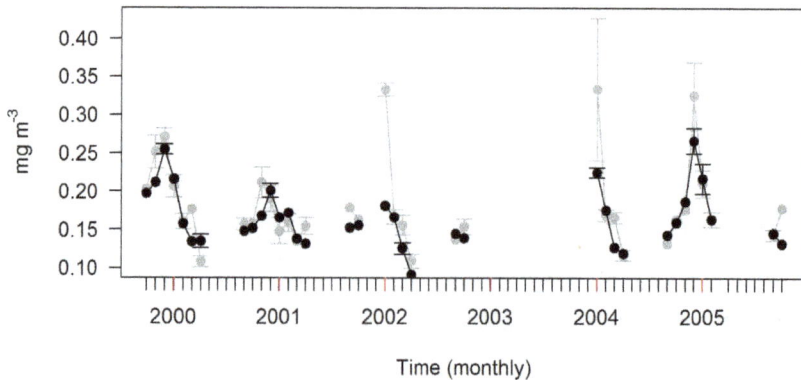

Figure 7. Mean inter-annual cycles. The *chl-a* (grey) and light-based *chl-a* estimates from our final mixed model collected by TDLRs (TDLR$_{chl}$ – black) mean inter-annual cycles within a 55°S to 60°S latitudinal band of the study site (*i.e.* where seal density is highest – see Table 3). Values include standard error bars. Red ticks on the x-axis represent January of each year. N.B. mean inter-annual trends are incomplete because study lacked PB deployments for 2001, 2002 and 2003, and PM deployments for 2003 (see Table 1). Furthermore, mixed model analysis excluded winter months (see Results).

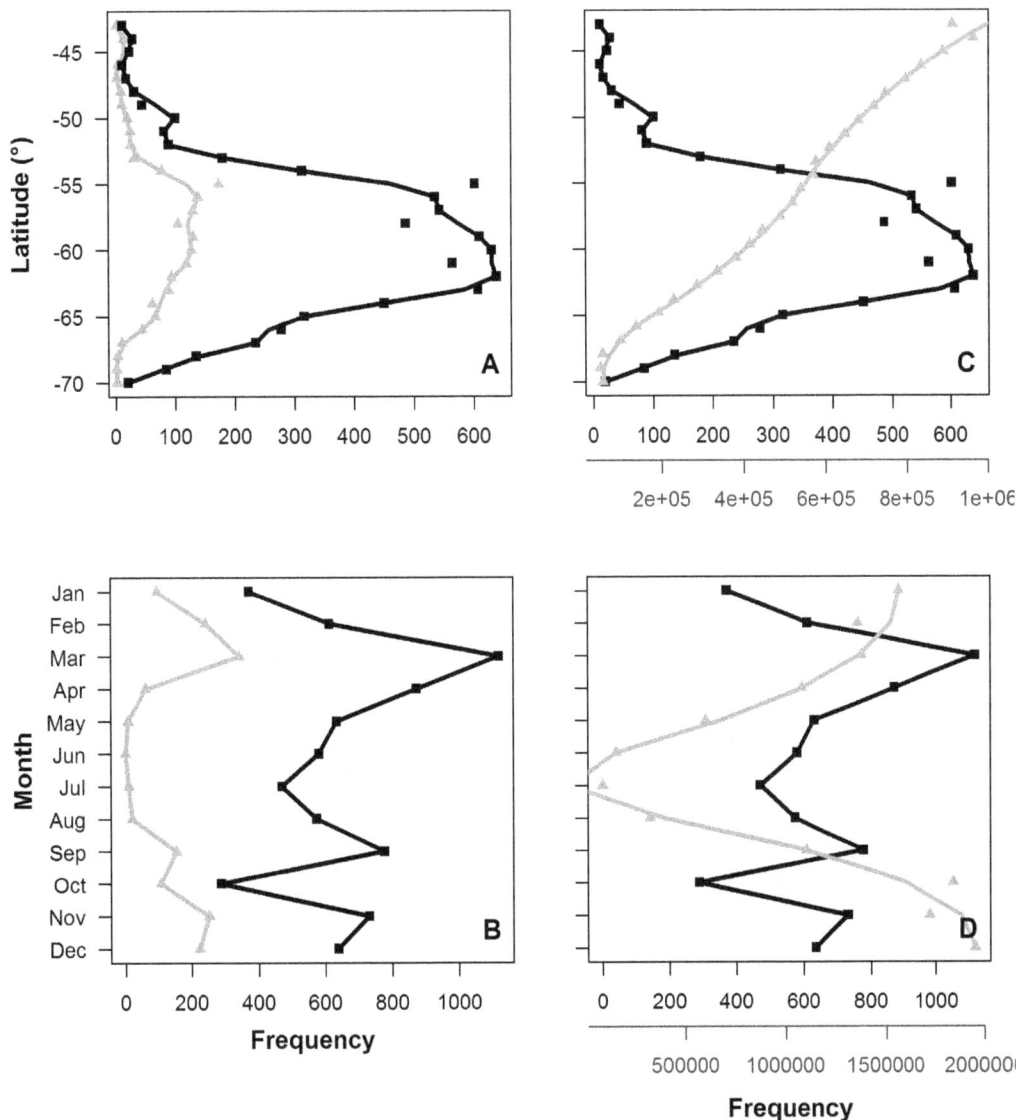

Figure 8. Data frequency coverage of *chl-a** and TDRL$_{chl}$†: data coverage of the study region is shown by (**A**) latitude (at 1° increments) and (**B**) months (between 1999 and 2005). Black represents TDRL$_{chl}$ coverage and grey represents *chl-a* coverage. Lines represent a loess fit. * *chl-a* data. †Light-based phytoplankton index calculated from TDLR data.

(as detected by satellite images) and chlorophyll-a integrated over the entire euphotic zone [11]. Nevertheless, because the density of phytoplankton particulates is particularly low in the Southern Ocean [1,49] perhaps the compounding effect of phytoplankton cells on LA_{250} enables TDLR sensors to detect differences that correlate well with *chl-a*. Preliminary analysis in this study suggests that most phytoplankton, which is retained within the mixed layer, is often found in the top 250 m of the water column. Furthermore, because phytoplankton is virtually negligible below the euphotic zone [50] no further information is likely to be gained by considering depths greater than 250 m.

We only used data from a 2 h period around the local noon to reduce the influence of ambient light field variability, thereby improving the accuracy and reducing the variability of light attenuation between dives (for details see [10]). However, solar elevation angle at local noon is affected by latitude and time of year (*i.e.* season), invariably altering light penetration at depth, and ultimately, light attenuation at 250 m. This would consistently

affect the relationship between *chl-a* and LA_{250} as seals travel extensively across the Southern Ocean. We suggest that the interactions between LA_{250} and latitude, as well as latitude and season, were retained in our final model to account for changes to the solar elevation angle at local noon. Variability of light attenuation in the water column is also due partly to differences in optical properties between phytoplankton species [25,51]. Different phytoplankton groups (based on their bio-optical characteristics) can be highly influenced by latitude (e.g., haptophytes and diatoms are found mostly in high latitudes [52]) and season (e.g., diatoms blooms dominate during spring and summer [52]). Moreover, different FZ can influence distribution of phytoplankton groups [53–55], and therefore, light attenuation variability, although we suggest that latitude can account for this effect in the Southern Ocean. It is possible, however, that distinct phytoplankton assemblages are not closely associated with our defined FZs. Perhaps a better understanding of FZ and their associated phytoplankton assemblages may show, in fact, that FZ is a useful contributing predictor to our light-based index of phytoplankton distribution. Finally, one of the reasons for using LA_{250} (*i.e.* relative decrease in irradiance) is that it normalises out small variations in sensor sensitivity or calibration. Nonetheless, seal was included as a random term in mixed model analyses despite Wildlife Computers checking light sensor calibrations (see Figure S1). We expect light values still vary between individual seals and potentially influence the relationship between light and *chl-a* if not accounted for.

These findings show the value of using existing datasets collected from animal-borne light sensors to calculate an index for phytoplankton density in the water column. Indeed, our results revealed that seasonal trends detected by our phytoplankton index were in agreement with data collected by remote sensing. This is despite temporal and spatial accuracy issues associated with both *chl-a* and LA_{250} data that were a potential source of persistent error in our analysis. First, typically dense cloud cover in the Southern Ocean (particularly during winter and at high latitudes) required use of 8-day composite SeaWiFS data (rather than 1-day) to improve data coverage, thereby compromising temporal resolution. Second, we expect spatial error inherent in our geo-location estimates (see methods) to result in spatial mismatch between LA_{250} and *chl-a*. Analyses were performed at 1° degree resolution to minimise location error bias. Third, it is possible that body position of the diving seals affects detection of irradiance by the light sensor. Indeed Sala et al. [56] have shown how body roll is incorporated into typical diving bouts throughout a seal's entire foraging trip, although we see little evidence of body position affecting light profiles (for examples refer to Figure 1) and expect error due to roll to be minimal. Although these issues may exist in our analysis we were still able to show how our phytoplankton index revealed seasonal trends consistent with data from *chl-a*.

It is likely that much of the discrepancy in our model between *chl-a* and LA_{250} largely originates from our data sources. Satellites do not provide a direct measure of *chl-a* and instead measure radiance and use empirically derived algorithms to estimate values. The SeaWiFS algorithm used for estimating *chl-a* tends to underestimate values in the Southern Ocean [57,58]. Surface prey aggregation may also contribute to the overestimation of *chl-a* detected during spring when zooplankton in particular become more abundant [59]. However, it is also possible that these prey aggregations could consistently coincide with elevated *chl-a* and therefore still correlate well with surface (or shallow subsurface) *chl-a* in any case. Nonetheless, we would expect that subsurface biology accounts for some of the discrepancy between our phytoplankton index and *chl-a*. Prey aggregations (*e.g.* zooplankton, fish), for instance, can affect light attenuation [10,12], which become increasingly likely with depth. Moreover, deep chlorophyll-a maxima (DCM) can be more than 30% that of surface values in some regions [11] and may cause further decoupling of *chl-a* and LA_{250}. Holm-Hansen et al. [60] showed that DCMs are located predominately over the deep ocean basins, regions regularly frequented by the focal elephant seals. Indeed, model predictions were more likely to overestimate *chl-a* by 10–30% when light was recorded over bathymetry greater than 4000 m (Figure 5). Our light-based phytoplankton index may therefore be useful for estimating total phytoplankton densities in the water column, rather than only providing near-surface *chl-a* information where seals dive.

Ecological significance

The light-based phytoplankton index from our model (hereafter phytoplankton index) produced seasonal patterns typical of *chl-a* distribution in the Southern Ocean south of Australia and New Zealand [61]. Summer values were consistent with Sokolov and Rintoul [61] that showed relatively high phytoplankton south of the Antarctic Circumpolar Current, and where the Polar Front interacted with the Mid-Ocean Ridge (*i.e.* regions between 60°S and 65°S). Particulate density estimates also show typical seasonal patterns that are in agreement with *chl-a* values for the same region (Sokolov and Rintoul [61]: low *chl-a* across the entire Southern Ocean leading into austral winter and a rise in early spring in the vicinity of the sea ice extent. It is important to consider, however, that locations visited by the seals may result in biased phytoplankton distributional trends. For example, in spring, seals may target slightly elevated phytoplankton patches at high latitudes and therefore not sample the relatively low phytoplankton densities of surrounding areas. We did, however, show that an inter-annual trend in mean monthly phytoplankton estimates corresponded well with *chl-a* within a latitudinal band most frequented by the focal seals; further validating that light-based estimates are detecting biological activity.

This study also shows that TDLRs record data in areas where satellite coverage is limited or completely absent. Specifically, satellite coverage is poor at high latitudes and during winter where cloud and ice coverage is more prevalent. This lack of data potentially limits our understanding of resource distribution in the Southern Ocean in relation to seal movement and their foraging behaviour [5]. Electronic tags deployed on animals have already been used to collect *in situ* temperature and salinity data along its track to improve our understanding of habitat utilisation [62]. Using light to estimate relative phytoplankton distribution may prove a useful covariate recorded simultaneously with elephant seal behaviour in future studies, particularly at the large scale and where *chl-a* data is sparse as described here. Indeed, our light-based phytoplankton index recorded at depth could be more relevant to a deep diving apex predator rather than *chl-a* data taken at the near-surface, although this is beyond the scope of this study. Regardless, these light data are already widely available, for a range of marine species, as light is traditionally recorded for estimating geo-location. This provides an opportunity to augment the application of light data in this study with data collected by multiple species.

Phytoplankton blooms typically support high zooplankton densities [59,63,64], and this in turn provides an important food resource for pelagic fish and higher predators. Traditionally, *chl-a* data has been the primary source of resource information in the marine environment, but are often limited by cloud cover at high latitudes and lack information at depth. Studies have often not found any significant relationship between *chl-a* and top predators foraging movements (e.g., [65]), unless at large scales, where general associations are apparent (e.g., [66]). In some instances foraging behaviour has even been shown to be inversely related to *chl-a* (e.g., [5]). However such studies cite either a lack of *chl-a* data [5,6], limited satellite resolution [6,67] or "downstream" effects decoupling phytoplankton from its physical conditions of origin [67] as possible explanations. It is therefore crucial that concurrent data is used where possible in efforts to model and understand trophic linkages in Southern Ocean ecosystems. Recording animal behaviour and light data simultaneously may enable researchers to help improve the predictive capacity of ecological models. Light data may provide important biological context in regions of the Southern Ocean and during specific months of the year that are historically poorly understood. Tag configuration that incorporates both fluorometer and light sensors could improve our ability to disseminate between phytoplankton and zooplankton distribution in the 3D marine environment. These data could give new insight into the biology of foraging habitat and/or oceanographic structures (e.g. upwelling eddies, ocean fronts) visited by tagged animals by providing information on resource distribution.

Supporting Information

Figure S1 The relationship between relative light level and blue light intensity (W cm^{-2}) for a typical tag. Calibrations are checked by Wildlife Computers at levels 10^{-5}, 10^{-7} and 10^{-9} W cm^{-2}, which correlates to light level values around 150, 110 and 70 respectively. Furthermore, these light level values roughly equate to specific daylight conditions ranging from full sunlight to overcast night. Source: Wildlife Computers, USA.

Figure S2 Locations of each annual deployment cohort. Each location is assigned to one of three frontal zones: Polar Frontal Zone (PFZ – orange); north of the southern Antarctic Circumpolar Current (SACCF-N - green); south of the Southern Antarctic Circumpolar Current Front (SACCF-S - blue). Maps show the bottom of Tasmania (Tas) and New Zealand (NZ, top)

and the coast of East Antarctica and Ross Sea (bottom). The black asterisks show Macquarie Island (MI).

Table S1 Ranked mixed models at 1° resolution. Satellite-derived chlorophyll (*chl-a*) explained by integrated light attenuation above 250 m (LA$_{250}$), season, latitude and frontal zone (FZ) ($n = 67$ seals)[†]. Mixed models are ranked by decreasing Akaike's Information Criterion (AIC) and change in AIC (ΔAIC) [41]; the most parsimonious model having the lowest AIC.

Acknowledgments

We thank Michele Thums, Steve Wall and Corey Bradshaw for field assistance. We thank the Australian Antarctic Division for providing logistical support. We wish to thank Pete Strutton for his advice.

Author Contributions

Conceived and designed the experiments: MO MAH ML CG. Analyzed the data: MO. Wrote the paper: MO.

References

1. Smith RC, Baker KS (1978) The bio-optical state of ocean waters and remote sensing. Limnology and Oceanography 23: 247–259.
2. Behrenfeld MJ, Falkowski PG (1997) Photosynthetic rates derived from satellite-based chlorophyll concentration. Limnology and Oceanography 42: 1–20.
3. Arrigo KR, van Dijken GL, Bushinsky S (2008) Primary production in the Southern Ocean, 1997–2006. Journal of Geophysical Research C: Oceans 113.
4. Moore JK, Abbott MR (2002) Surface chlorophyll concentrations in relation to the Antarctic Polar Front: Seasonal and spatial patterns from satellite observations. Journal of Marine Systems 37: 69–86.
5. Bradshaw CJA, Higgins J, Michael KJ, Wotherspoon SJ, Hindell MA (2004) At-sea distribution of female southern elephant seals relative to variation in ocean surface properties. Ices Journal of Marine Science 61: 1014–1027.
6. Sumner MD, Michael KJ, Bradshaw CJA, Hindell MA (2003) Remote sensing of Southern Ocean sea surface temperature: implications for marine biophysical models. Remote Sensing of Environment 84: 161–173.
7. Charrassin J-B, Hindell M, Rintoul SR, Roquet F, Sokolov S, et al. (2008) Southern Ocean frontal structure and sea-ice formation rates revealed by elephant seals. Proceedings of the National Academy of Sciences 105: 11634–11639.
8. Fedak M (2013) The impact of animal platforms on polar ocean observation. Deep Sea Research Part II: Topical Studies in Oceanography 88–89: 7–13.
9. Biuw M, Boehme L, Guinet C, Hindell M, Costa D, et al. (2007) Variations in behavior and condition of a Southern Ocean top predator in relation to in situ oceanographic conditions. Proceedings of the National Academy of Sciences 104: 13705–13710.
10. Teo SLH, Kudela RM, Rais A, Perle C, Costa DP, et al. (2009) Estimating chlorophyll profiles from electronic tags deployed on pelagic animals. Aquatic Biology 5: 195–207.
11. Guinet C, Xing X, Walker E, Monestiez P, Marchand S, et al. (2013) Calibration procedures and first dataset of Southern Ocean chlorophyll a profiles collected by elephant seals equipped with a newly developed CTD-fluorescence tags. Earth System Science Data 5: 15–29.
12. Jaud T, Dragon A-C, Garcia JV, Guinet C (2012) Relationship between Chlorophyll a Concentration, Light Attenuation and Diving Depth of the Southern Elephant Seal Mirounga leonina. PLoS One 7: e47444.
13. Boyd IL, Croxall JP (1996) Dive durations in pinnipeds and seabirds. Canadian Journal of Zoology-Revue Canadienne De Zoologie 74: 1696–1705.
14. Hindell M, Crocker D, Mori Y, Tyack P (2010) Foraging Behaviour. In: Boyd IL, Bowen WD, Iverson SJ, editors. Marine Mammal Ecology and Conservation: A Handbook of Techniques. Oxford: Oxford University Press. pp. 241–262.
15. McIntyre T, Bornemann H, Plötz J, Tosh CA, Bester MN (2012) Deep divers in even deeper seas: Habitat use of male southern elephant seals from Marion Island. Antarctic Science 24: 561–570.
16. Delong RL (1992) Documenting migrations of northern elephant seals using day length. Marine Mammal Science 8: 155–159.
17. Sumner MD, Wotherspoon SJ, Hindell MA (2009) Bayesian estimation of animal movement from archival and satellite tags. PLoS One 4: e7324.
18. McCafferty DJ, Walker TR, Boyd IL (2004) Using time-depth-light recorders to measure light levels experienced by a diving marine mammal. Marine Biology 146: 191–199.
19. Vacquié-Garcia J, Royer F, Dragon A-C, Viviant M, Bailleul F, et al. (2012) Foraging in the Darkness of the Southern Ocean: Influence of Bioluminescence on a Deep Diving Predator. PLoS One 7: e43565.
20. Campagna C (2001) Detecting bioluminescence with an irradiance time-depth recorder deployed on southern elephant seals. Marine Mammal Science 17: 402–414.
21. Morel A, Maritorena S (2001) Bio-optical properties of oceanic waters: A reappraisal. Journal of Geophysical Research 106: 7163–7180.
22. Morel A, Prieur L (1977) Analysis of variations in ocean color. Limnology and Oceanography 22: 709–722.
23. Bricaud A, Morel A, Prieur L (1981) Absorption by dissolved organic matter of the sea (yellow substance) in the UV and visible domains. Limnol Oceanogr 26: 43–53.
24. Bricaud A, Morel A, Babin M, Allali K, Claustre H (1998) Variations of light absorption by suspended particles with chlorophyll a concentration in oceanic (case 1) waters: Analysis and implications for bio-optical models. Journal of Geophysical Research: Oceans 103: 31033–31044.
25. Stramski D, Bricaud A, Morel A (2001) Modeling the inherent optical properties of the ocean based on the detailed composition of the planktonic community. Applied Optics 40: 2929–2945.
26. Field IC, McMahon CR, Burton HR, Bradshaw CJA, Harrington J (2002) Effects of age, size and condition of elephant seals (Mirounga leonina) on their intravenous anaesthesia with tiletamine and zolazepam. Veterinary Record 151: 235–240.
27. McMahon CR, Burton H, McLean S, Slip D, Bester M (2000) Field immobilisation of southern elephant seals with intravenous tiletamine and zolazepam. Veterinary Record 146: 251–254.
28. Hindell MA, Slip DJ (1997) The importance of being fat: maternal expenditure in the southern elephant seal Mirounga leonina. In: Hindell MA, Kemper CM, editors. Marine mammal research in the Southern Hemisphere: Status, ecology and medicine: Surrey Beatty & Sons. pp. 72–77.
29. McMahon CR, Field IC, Bradshaw CJA, White GC, Hindell MA (2008) Tracking and data-logging devices attached to elephant seals do not affect individual mass gain or survival. Journal of Experimental Marine Biology and Ecology 360: 71–77.
30. Bradshaw CJA, Hindell MA, Michael KJ, Sumner MD (2002) The optimal spatial scale for the analysis of elephant seal foraging as determined by geo-location in relation to sea surface temperatures. ICES Journal of Marine Science 59: 770–781.
31. Boyd IL, Walker TR, Taylor RI, McCafferty DJ (1999) Can marine mammals be used to monitor oceanographic conditions? Marine biology 134: 387–395.
32. Thums M, Bradshaw CJA, Hindell MA (2008) Tracking changes in relative body composition of southern elephant seals using swim speed data. Marine Ecology Progress Series 370: 249–261.
33. Ekstrom PA (2004) An advance in geolocation by light. Memoirs of the National Insitute of Polar Research, Special Issue 58: 210–226.
34. Hill RD (1994) Theory of geolocation by light levels. Elephant seals: population ecology, behavior, and physiology University of California Press, Berkeley: 227–236.
35. Hill RD, Braun MJ (2001) Geolocation by light level. Electronic tagging and tracking in marine fisheries: Springer. pp. 315–330.

36. Sumner M, Wotherspoon S (2010) Metropolis sampler and supporting functions for estimating animal movement from archival tags and satellite fixes. Available: http://CRAN.R-project.org/package=tripEstimation%3E.

37. McClain CR, Cleave ML, Feldman GC, Gregg WW, et al. (1998) Science quality sea WiFS data for global biosphere research. Sea Technology 39: 10.

38. Cavalieri DJ, Parkinson CL, Gloersen P, Zwally H (2012) Sea Ice Concentrations from Nimbus-7 SMMR and DMSP SSM/I-SSMIS Passive Microwave Data. Boulder, Colorado USA: NASA DAAC at the National Snow and Ice Data Center.

39. de Baar HJW, Boyd PW, Coale KH, Landry MR, Tsuda A, et al. (2005) Synthesis of iron fertilization experiments: From the iron age in the age of enlightenment. Journal of Geophysical Research C: Oceans 110: 1–24.

40. Thomalla SJ, Fauchereau N, Swart S, Monteiro PMS (2011) Regional scale characteristics of the seasonal cycle of chlorophyll in the Southern Ocean. Biogeosciences 8: 2849–2866.

41. Park YH, Charriaud E, Pino DR, Jeandel C (1998) Seasonal and interannual variability of the mixed layer properties and steric height at station KERFIX, southwest of Kerguelen. Journal of Marine Systems 17: 571–586.

42. Orsi AH, Whitworth Iii T, Nowlin WD Jr (1995) On the meridional extent and fronts of the Antarctic Circumpolar Current. Deep Sea Research Part I: Oceanographic Research Papers 42: 641–673.

43. Guinet C, Vacquié-Garcia J, Picard B, Bessigneul G, Lebras Y, et al. (2014) Southern elephant seal foraging success in relation to temperature and light conditions: insight into prey distribution. Marine Ecology Progress Series 499: 285–301.

44. Pinheiro J, Bates D, DebRoy S, Sarkar D, R Development Core Team (2012) nlme: Linear and Nonlinear Mixed Effects Models.

45. Field I, Hindell M, Slip D, Michael K (2001) Foraging strategies of southern elephant seals (*Mirounga leonina*) in relation to frontal zones and water masses. Antarctic Science 13: 371–379.

46. Bost CA, Cotte C, Bailleul F, Cherel Y, Charrassin JB, et al. (2009) The importance of oceanographic fronts to marine birds and mammals of the southern oceans. Journal of Marine Systems 78: 363–376.

47. Zuur AF, Leno EN, Walker NJ, Saveliev AA, Smith GM (2009) Mixed effects models and extensions in ecology with R: Springer.

48. Burnham KP, Anderson DR (2002) Model Selection and Multi-Model Inference: A Practical Information-Theoretic Approach: Springer.

49. Fenton N, Priddle J, Tett P (1994) Regional variations in bio-optical properties of the surface waters in the Southern Ocean. Antarctic Science 6: 443–448.

50. Kirk JTO (1994) Light and Photosynthesis in Aquatic Ecosystems: Cambridge University Press.

51. Loisel H, Nicolas JM, Deschamps PY, Frouin R (2002) Seasonal and inter-annual variability of particulate organic matter in the global ocean. Geophysical Research Letters 29.

52. Alvain S, Moulin C, Dandonneau Y, Breon FM (2005) Remote sensing of phytoplankton groups in case 1 waters from global SeaWiFS imagery. Deep-Sea Research Part I-Oceanographic Research Papers 52: 1989–2004.

53. Garcia-Munoz C, Lubian LM, Garcia CM, Marrero-Diaz A, Sangra P, et al. (2013) A mesoscale study of phytoplankton assemblages around the South Shetland Islands (Antarctica). Polar Biology 36: 1107–1123.

54. Zhao L, Zhao Y, Zhang WC, Zhou F, Zhang CX, et al. (2013) Picoplankton distribution in different water masses of the East China Sea in autumn and winter. Chinese Journal of Oceanology and Limnology 31: 247–266.

55. D'Ovidio F, De Monte S, Alvain S, Dandonneau Y, Lévy M (2010) Fluid dynamical niches of phytoplankton types. Proceedings of the National Academy of Sciences of the United States of America 107: 18366–18370.

56. Sala J, Quintana F, Wilson R, Dignani J, Lewis M, et al. (2011) Pitching a new angle on elephant seal dive patterns. Polar Biology 34: 1197–1209.

57. Hirawake T, Satoh H, Ishimaru T, Yamaguchi Y, Kishino M (2000) Bio-optical relationship of case I waters: the difference between the low- and mid-latitude waters and the Southern Ocean. Journal of Oceanography 56: 245–260.

58. Johnson R, Strutton PG, Wright SW, McMinn A, Meiners KM (2013) Three improved satellite chlorophyll algorithms for the Southern Ocean. Journal of Geophysical Research: Oceans 118: 3694–3703.

59. Robins DB, Harris RP, Bedo AW, Fernandez E, Fileman TW, et al. (1995) The relationship between suspended particulate material, phytoplankton and zooplankton during the retreat of the marginal ice zone in the Bellingshausen Sea. Deep Sea Research Part II: Topical Studies in Oceanography 42: 1137–1158.

60. Holm-Hansen O, Kahru M, Hewes CD (2005) Deep chlorophyll a maxima (DCMs) in pelagic Antarctic waters. II. Relation to bathymetric features and dissolved iron concentrations. Marine Ecology Progress Series 297: 71–81.

61. Sokolov S, Rintoul SR (2007) On the relationship between fronts of the Antarctic Circumpolar Current and surface chlorophyll concentrations in the Southern Ocean. Journal of Geophysical Research 112: C07030.

62. Costa DP, Huckstadt LA, Crocker DE, McDonald BI, Goebel ME, et al. (2010) Approaches to Studying Climatic Change and its Role on the Habitat Selection of Antarctic Pinnipeds. Integrative and Comparative Biology 50: 1018–1030.

63. Burghart SE, Hopkins TL, Vargo GA, Torres JJ (1999) Effects of a rapidly receding ice edge on the abundance, age structure and feeding of three dominant calanoid copepods in the Weddell Sea, Antarctica. Polar Biology 22: 279–288.

64. Lizorre MP (2001) The contributions of sea ice algae to antarctic marine primary production. American Zoologist 41: 57–73.

65. Bost C, Thiebot J, Pinaud D, Cherel Y, Trathan PN (2009) Where do penguins go during the inter-breeding period? Using geolocation to track the winter dispersion of the macaroni penguin. Biology Letters 5: 473–476.

66. Lea M-A, Guinet C, Cherel Y, Duhamel G, Dubroca L, et al. (2006) Impacts of climatic anomalies on provisioning strategies of a Southern Ocean predator. Marine Ecology Progress Series 310: 77–94.

67. Guinet C, Dubroca L, Lea MA, Goldsworthy S, Cherel Y, et al. (2001) Spatial distribution of foraging in female Antarctic fur seals Arctocephalus gazella in relation to oceanographic variables: a scale-dependent approach using geographic information systems. Marine Ecology Progress Series 219: 251–264.

21

Linking Environmental Forcing and Trophic Supply to Benthic Communities in the Vercelli Seamount Area (Tyrrhenian Sea)

Anabella Covazzi Harriague[1]*, **Giorgio Bavestrello**[1], **Marzia Bo**[1], **Mireno Borghini**[2], **Michela Castellano**[1], **Margherita Majorana**[1], **Francesco Massa**[1], **Alessandro Montella**[1], **Paolo Povero**[1], **Cristina Misic**[1]

1 Dipartimento di Scienze della Terra, dell'Ambiente e della Vita - DiSTAV – University of Genoa, Italy, 2 CNR-ISMAR, Institute of Marine Sciences, National Research Council, Section of La Spezia, Pozzuolo di Lerici, Italy

Abstract

Seamounts and their influence on the surrounding environment are currently being extensively debated but, surprisingly, scant information is available for the Mediterranean area. Furthermore, although the deep Tyrrhenian Sea is characterised by a complex bottom morphology and peculiar hydrodynamic features, which would suggest a variable influence on the benthic domain, few studies have been carried out there, especially for soft-bottom macrofaunal assemblages. In order to fill this gap, the structure of the meio-and macrofaunal assemblages of the Vercelli Seamount and the surrounding deep area (northern Tyrrhenian Sea – western Mediterranean) were studied in relation to environmental features. Sediment was collected with a box-corer from the seamount summit and flanks and at two far-field sites in spring 2009, in order to analyse the metazoan communities, the sediment texture and the sedimentary organic matter. At the summit station, the heterogeneity of the habitat, the shallowness of the site and the higher trophic supply (water column phytopigments and macroalgal detritus, for instance) supported a very rich macrofaunal community, with high abundance, biomass and diversity. In fact, its trophic features resembled those observed in coastal environments next to seagrass meadows. At the flank and far-field stations, sediment heterogeneity and depth especially influenced the meiofaunal distribution. From a trophic point of view, the low content of the valuable sedimentary proteins that was found confirmed the general oligotrophy of the Tyrrhenian Sea, and exerted a limiting influence on the abundance and biomass of the assemblages. In this scenario, the rather refractory sedimentary carbohydrates became a food source for metazoans, which increased their abundance and biomass at the stations where the hydrolytic-enzyme-mediated turnover of carbohydrates was faster, highlighting high lability.

Editor: Erik V. Thuesen, The Evergreen State College, United States of America

Funding: This work was undertaken within the PRIN (Progetti di Rilevante Interesse Nazionale) project "Tyrrhenian Seamount Ecosystems: An Integrated Study (TySEc)", financed by the Italian Ministry of Research and Instruction. The funders had no role in study design, data collection and analysis, decision to publish, or preparation of the manuscript.

Competing Interests: The authors have declared that no competing interests exist.

* Email: anabella7@hotmail.com

Introduction

The deep-sea communities of the Mediterranean Sea (namely those living below a 200 m depth) have been investigated rather intensively, but these studies have typically been characterised by a limited spatial or temporal scale of investigation [1,2,3,4,5,6]. Focusing on the deep Tyrrhenian Sea, very few papers have been published on microbial benthic communities [7] and meiofaunal communities [8] and studies on deep, soft-bottom macrofauna are lacking. This is surprising as the Tyrrhenian Sea hosts a number of morphological peculiarities (seep, vent, slope, and abyssal plain habitats, and seamounts etc.) that have led us to suppose that an interesting part of the Mediterranean's biodiversity could be hidden there.

The northwestern Tyrrhenian Sea is characterised by a complex hydrology [9,10], which responds to the bottom morphology. A submerged ridge, called the Vercelli Seamount, with its main axis SW-NE, reaches the photic layer from the bathyal plain [11,12].

Almost permanent frontal zones exist on the main Vercelli Seamount axis, modifying the pelagic-benthic coupling, while cyclonic and anticyclonic gyres move the water masses around it [13].

In this complex scenario, the forcings that usually shape benthic communities (depth, sediment texture, trophic supply) may change their roles suddenly. The presence of a Taylor column [14] and the impinging of water circulation on the seamount's flanks have been suggested as pivotal factors for the development of the benthic communities, for instance where the presence of a rather stable Taylor column isolates the summit and limits foraging of the down-current zones [15,16]. Higher and lower current speeds have been invoked to explain community trophic differences on different flanks of the seamount [11]. In such systems a constant pelagic-benthic coupling, providing the aphotic sediment with food, is hardly possible, strongly influencing the benthic assemblages [14,17,18,19]. All these variables and the variable slope of the flanks, that may exert a certain influence on the sediment texture, influence the characteristics of the benthic community, and have encouraged the hypothesis that seamounts and the surrounding area are peculiar hotspots of marine life [20,21,22].

This study aims to link the metazoan benthic community (meiofauna and macrofauna) to the environmental constraints, highlighting how the parameters considered (morphological as well as trophic) may have a role in community characterisation.

Material and Methods

Study area, sampling sites and sampling strategy

The sampling area lies in the NW Tyrrhenian Sea (NW Mediterranean), near Sardinia (Fig. 1), and it is located within the following coordinates: 40°46'N/10°39'E, 40°47'N/11°34'E, 41°24'N/11°34'E and 41°24'N/10°38'E. It is centred on the Vercelli Seamount, an elongated, chain seamount whose axis is oriented SW-NE, and whose summit (41°06'N/10°54'E) rises from the bathyal plain to a depth of 55 m. The seamount summit covers an area of about 0.36 km^2, characterised by alternating rocky and sandy surfaces of variable depths. About half of the area lies between 100 and 120 m, while only 15% is shallower than 80 m.

This area has been studied previously for its hydrodynamic features, within the framework of research on the Tyrrhenian Sea circulation and water mass fluxes. The previous studies showed the presence of a large cyclonic structure (the Bonifacio gyre) [23] that displayed permanent features, centred NW of the sampling area and crossing it in its W sector [9,13]. Krivosheya [24] noted the presence of an anticyclonic companion of the Bonifacio gyre to the southeast. The Vercelli Seamount is placed within the transition area between the two gyres [9], whose boundaries are frontal zones [10]. In addition, Vetrano et al. [13]noted the presence of another mesoscale structure in the NE section of the sampling area, although less stable than the others. All these structures extend from the surface to considerable depths [13].

Sediment and water samples were collected from the R/V *Urania* during May 2009. During our field studies endangered or protected species were not involved and the permit for sampling activity was issued by the Italian Ministry of Defence (Military Navy) and the Italian Ministry of Communications. Nine stations were visited (Fig. 1). Station 28 was only sampled for organic matter (OM), and enzymatic and granulometric parameters due to the steepness of the seafloor, which prevented proper closure of the sampling device, and station 0 only for macrofauna due to the risk of damage to the sampling device on the irregular sandy-rocky bottom.

The stations (Table 1) were located on the summit (station 0) and around the seamount at different distances from the summit. Stations 14 and 28 were located on the upper-flank area, NE and NW of the summit respectively, where the flanks were rather steep. The other stations were located at different depths on the seamount's flanks, stations 9 and 16 at medium depths (1100–1200 m) and stations 25 and 32 at lower depths (1700–1800 m), close to the bathyal plain. Station 41 was located in the extreme SE sector of the study area and up-current from the seamount. Station 53 was, instead, placed in the extreme NW sector, influenced by the permanent Bonifacio gyre. Both stations were considered control sites, being about 50 km from the seamount summit, although having very different depths and features.

Undisturbed sediment cores were collected with independent deployments using a box-corer with a 29 cm internal diameter. Immediately after the arrival of the box-corer on board, the overlaying bottom water was gently removed without notably

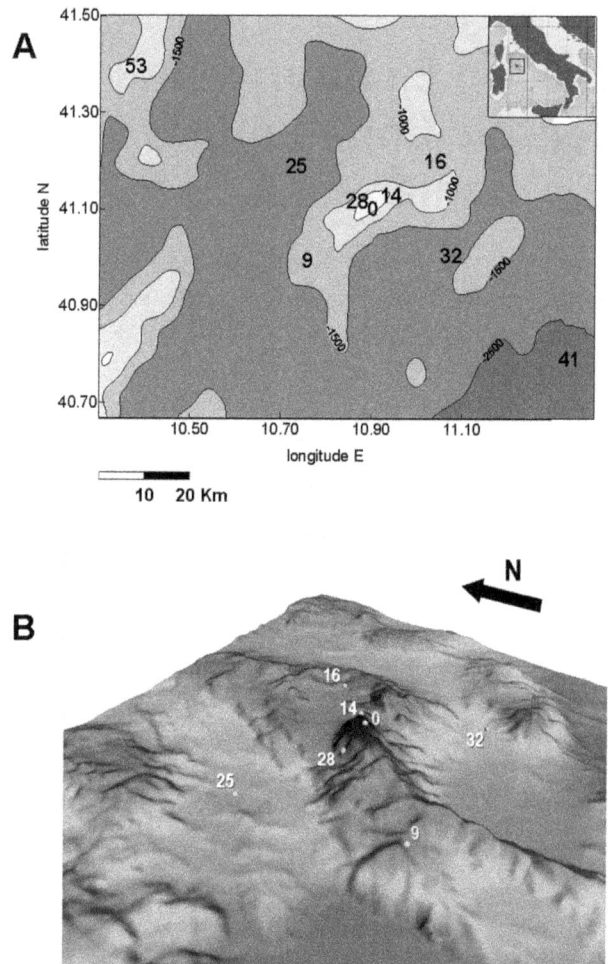

Figure 1. View of the Vercelli Seamount area (Tyrrhenian Sea). The bathymetric map (A) shows that the seamount has a SW-NE oriented axis and the summit (41°06'N/10°54'E) rises from the bathyal plain to a depth of 55 m. All the stations where sediment has been sampled are shown (see Table 1 for further details). The three-dimensional detail of the seamount (map B) is centred on the summit station 0 (at 115 m depth), the upper flank stations 14 and 28 (at 400 and 877 m depth), the medium-flank stations 9 and 16 (at 1232 and 1166 m depth), and the deeper-flank stations 25 and 32 (at 1833 and 1728 m depth).

disturbing the sediment surface. For the granulometric, OM, enzymatic activity and meiofaunal analyses, the sediment in the box-corer was subsampled by pushing PVC cores (5 cm internal diameter) into it, in duplicate for each analysis except meiofaunal, which was collected in triplicate. Each core was then aseptically sliced (at 0–2 and 2–10 cm depths). Each sediment slice was frozen until analysis, except for the enzymatic activity determinations, that were performed immediately. For the macrofaunal analyses, 3 deployments were performed for each station and entirely sorted with a 500 µm mesh net. The limited sampling for meiofauna could have reduced the representativeness of the data. In particular, the rare species could have been underestimated, thus leading only to preliminary considerations on meiofaunal diversity.

In addition to the sediment sampling, we also performed surface-water layer sampling at 25 stations distributed throughout the sampling area. Four depths were sampled with Niskin bottles: surface, oxygen minimum (between 30 and 50 m), fluorescence

Table 1. Location of sampling stations with respect to the seamount morphology and to the general current flow and evaluated variables.

station	latitude N	longitude E	depth (m)	location	bottom inclination (°)	exposition (°N)	current from (a)	sampling parameters
0	41.108	10.907	115	summit	1,70	77	SE	Gr, Ma
9 *	40.993	10.759	1232	medium flank	8,49	128	SW	Gr, OM, E, Me, Ma
14	41.128	10.940	400	upper flank	14,61	20	SE	Gr, OM, E, Me, Ma
16 *	41.200	11.038	1166	medium flank	4,91	261	N-NW	Gr, OM, E, Me, Ma
25	41.188	10.733	1833	deeper flank	0,36	352	SE	Gr, OM, E, Me, Ma
28	41.117	10.867	877	upper flank	17,46	332	SE	Gr, OM, E
32 *	41.003	11.074	1728	deeper flank	2,84	287	SE	Gr, OM, E, Me, Ma
41 *	40.790	11.339	2646	far-field	5,23	302	SE	Gr, OM, E, Me, Ma
53 *	41.394	10.378	887	far-field	5,84	238	W-SW	Gr, OM, E, Me, Ma

Asterisks denote those stations whose bottom inclination doesn't follow the seamount flank general pendency.
Gr: granulometry, OM: organic matter, E: enzymatic activity, Me: meiofauna, Ma: macrofauna.
(a): general current flows as reported in Vetrano et al. [13]

maximum (between 55 and 75 m), depth of extinction of the fluorescence signal (at ca. 120 m). Seawater was filtered through Whatman GF/F filters (in triplicate) for the chlorophyll-a concentration determination.

Analytical procedures

Benthic community. Meiofauna was extracted from the sediment by sieving through 500 μm and 45 μm mesh nets. The fraction retained was resuspended and processed according to the protocol reported by Danovaro [25]. All meiofaunal organisms were counted and classified to phylum or class level under a stereomicroscope. In order to obtain the functional parameter (biomass) of the meiofaunal component, the organisms were weighed after drying at 60°C for 24 h.

The macrofaunal specimens were recognised, when possible, down to species level. All the organisms, divided by species, were weighed to obtain the biomass value expressed as DW (drying at 60°C for 24 h). Molluscs and echinoderms were treated with 30% HCl prior to weighing.

Environmental variables. The grain size analysis was performed following Buchanan and Kain [26]; briefly, sediments were sieved (9 mesh sizes from 3.35 to 0.063 mm) after H_2O_2 treatment and drying (60°C, 48 h) and each fraction was weighed. Sediment particle-size diversity (Sed-H) was calculated from the percent dry weight of 5 size classes (<0.063 mm, 0.063–0.212 mm, 0.212–0.5 mm, 0.5–2 mm and >2 mm) using the Shannon–Wiener diversity index [27].

The chlorophyll-a concentrations in the water column were measured on board following the method of Holm-Hansen et al. [28], using a Perkin Elmer LS50B spectrofluorometer calibrated with chlorophyll-a from spinach (Sigma C5753). The specific standard deviation of the replicates was on average 4%.

The protein content of the sediment was determined following Hartree [29], the carbohydrate content was determined following Dubois et al. [30] and the lipid content was determined following Bligh and Dyer [31] and Marsh and Weinstein [32]. A Jasco V-500 spectrophotometer was calibrated with bovine serum albumin, glucose and tripalmitine solutions, respectively. Labile organic phosphorus was determined following the first step of the sequential extraction (SEDEX) proposed by Ruttenberg [33]. Briefly, sediment samples (3 to 5 g) were shaken for 2 h at 50°C in a 1 M $MgCl_2$ solution in order to detach the loosely absorbed P from the sediment. The supernatant was then treated with an oxidising solution ($K_2S_2O_8$) [34] in order to transform all the P into inorganic phosphates, which were then detected following Hansen and Grasshoff [35]with a SYSTEA Nutrient Probe Analyser. Organic carbon and total nitrogen were determined following Hedges and Stern [36] with a Carlo Erba Mod. 1110 CHN Elemental Analyser after acidification with hydrochloric acid to remove the inorganic carbonate fraction. Cyclohexanone-2,4-dinitrophenyl hydrazone was chosen as standard.

The hydrolytic enzymatic activities (β-glucosidase – BG, alkaline phosphatase – AP and leucine aminopeptidase – LA) were determined following Hoppe [37], using artificial substrates: 4-methylumbelliferyl β-D glucopyranoside and 4-methylumbelliferyl phosphate (excitation at 365 nm and emission at 460 nm) for BG and AP, respectively, and L-leucine 7-amido-4-methylcoumarin hydrochloride (excitation at 380 nm and emission at 440 nm) for LA. The samples and controls (sample sediment boiled as a blank for accidental contamination due to handling and for abiotic cleavage of the artificial substrates) were incubated in duplicate with 0.5 ml of substrate solution for 3 h. Incubations in the dark respected in-situ temperatures. Fluorescence was measured with a Perkin-Elmer 50 L spectrofluorometer previously calibrated with 4-methylumbelliferone and 7-amino-4-methylcoumarin solutions. The LA and BG activities were converted into equivalents of mobilised C assuming that 1 nmol of substrate hydrolysed enzymatically corresponded to 72 ng of mobilised C [38].

The two degradative enzymes were associated to their respective OM component: the LA with proteins, because the enzymatic hydrolysis of polypeptides is the preliminary step to amino-acid mineralisation [39], and the BG with carbohydrates, because the enzyme is involved in cellulose degradation [40]. AP activity has generally been associated with remineralisation of dissolved inorganic P, but it has bi-functional features [41]. We related it to labile P. The OM turnover times were calculated by converting the proteins and carbohydrates into C equivalents (factors of 0.49 and 0.40 for proteins and carbohydrates, respectively, according to the C content of the standard) and then dividing by the LA and BG activities transformed into their equivalents of mobilised C. The labile phosphorus turnover time was calculated following the same procedure, but assuming that 1 nmol of substrate hydrolysed enzymatically by AP corresponded to 31 ng of potentially released phosphate [38].

Statistics

We tested the differences in the same variable between different samplings with the one-way ANOVA test followed by the Newman-Kneuls post-hoc test (ANOVA+NK test) (Statistica software). To test the relationships between the various parameters, a Spearman-rank correlation analysis was performed. We used the PRIMER 6β programme package to perform SIMPER analysis on the metazoan abundance data, separately for meiofauna and macrofauna. The data matrices have been transformed using presence/absence. DISTML (distance-based linear model) routines were performed with the PERMANOVA+ programme package for PRIMER to analyse and model the relationship between the meiofaunal abundance and the environmental variables. Of the original set of environmental parameters, 9 were retained for further analysis. The variables with correlation R^2 values >0.9 (considered redundant)were omitted for the DISTLM procedures. The meiofaunal DISTML was constructed using the step-wise selection procedure and the adjusted R^2 as selection criterion to enable the fitting of the best explanatory environmental variables in the model. Euclidean distance was used as resemblance measure.

A PCA analysis, based on to the trophic quality of the OM, was performed of the 4th-root-transformed data (reported in Table 2) to reveal similarities between stations.

Results

Benthic communities

The meiofaunal communities had their highest densities at the stations situated on the upper and medium flanks of the seamount (Fig. 2A). The total abundance, although it has to be considered as preliminary data, decreased significantly with water depth (R = −0.82, n = 7, p<0.05), a trend confirmed also by the DISTLM analysis (Table 3). The first 2 cm depth was the layer mainly inhabited by the organisms (85–99% of the total abundance) at most of the stations, although in the northwestern area (stations 53 and 25) the vertical distribution was more homogeneous: only 62% and 58%, respectively, preferred the surface sediments (Fig. 2A). Overall 13 taxa were found in the study area, 11 at the seamount stations (Table 4), although the limited sampling procedure could have not highlighted all the rare species in the different sites. The number of taxa ranged between 5 (stations 9

Table 2. Quality indexes of the OM: C/N ratio, protein/carbohydrate ratio, turnover times (days) in the two sediment layers.

station	area	C/N ratio		protein/carbohydrate ratio		carbohydrate turnover		protein turnover		labile P turnover	
		0–2 cm	2–10 cm	0–2 cm	2–10 cm	0–2 cm	2–10 cm	0–2 cm	2–10 cm	0–2 cm	2–10 cm
53	far-field	14.8±0.8	13.5±1.6	0.23	0.15	379	823	2.6	1.6	0.09	0.19
25	deeper flank	20.7±2.8	10.6±0.1	0.34	0.24	218	840	1.9	3.2	0.04	0.25
9	medium flank	19.7±2.7	15.2±0.4	0.11	0.30	549	653	1.5	3.9	0.02	0.10
28	upper flank	12.4±1.1	15.0±2.9	0.16	0.32	382	768	0.8	2.1	nd	0.09
14	upper flank	11.7±0.9	10.5±1.8	0.11	0.10	179	997	3.7	3.0	0.01	0.06
16	medium flank	11.8±0.7	11.0±1.0	0.14	0.20	223	264	1.2	1.6	0.03	0.07
32	deeper flank	13.6±0.8	13.6±0.2	0.20	0.24	826	1332	1.8	2.4	0.08	0.18
41	far-field	12.0±0.8	12.6±0.3	0.43	0.27	427	854	7.6	8.2	0.09	0.11

and 25) and 8 (stations 14 and 16, Fig. 2A). The assemblage abundances were dominated by nematodes and copepods with nauplii (61–76% for nematodes and 5–11% for copepods). The community structure was similar in the different parts of the seamount and far field (SIMPER: dissimilarities from 33.6% to 23.8%). The biomass patterns were similar to the abundance ones (Fig. 2B), except at station 53, which had a higher biomass in the deeper sediments due to the high number of polychaetes found. The higher biomass contributions were given by nematodes (26–71%) and polychaetes (0–69%), while copepods represented only 4–9%.

Macrofauna was only found at four stations (53, 9, 16 and 0, Fig. 3A), characterised by depths shallower than 1500 m. It seems that the presence of macrofauna was related to depth, but actually station 14 (390 m depth) showed no macrofaunal organisms, indicating that features other than depth must be involved. Macrofaunal densities were lower than 50 ind m^{-2} at stations 53, 9 and 16, but at station 0 the value was greater than 3000 ind m^{-2}. The number of taxa was also low at stations 53, 9 and 16, with four species of polychaetes, one species of crustacean, and two species of sipunculans. Station 0, instead, showed a high diversification (Table 5) and was dominated by crustaceans. Molluscs, polychaetes and others showed similar contributions; only echinoderms were scarce (Fig. 3A). The SIMPER analysis showed higher dissimilarities between assemblages of the seamount flanks and of the far field area (85.7%), between the assemblages of the seamount flanks and of the summit (100%) and between the assemblages of the far field and of the summit

(94.7%). The macrofaunal biomass showed patterns similar to the density, but the taxa contribution was different: at the summit station crustaceans accounted for 94%, while at the other stations polychaetes made higher contributions (Fig. 3B).

Environmental variables

Seawater autotrophic biomass. The distribution of the autotrophic biomass in the water column (integrated values in the 0–120 m layer for the chlorophyll-a concentration) is shown in Fig. 4. The highest values in the area studied were found at station 0 (seamount summit, 0.43 µg l^{-1}) and in the SW and NW sectors (0.41 µg l^{-1} at station 53). The waters surrounding the summit had low (0.14 µg l^{-1} at station 14) and rather low (from 0.22 to 0.24 µg l^{-1} at stations 9, 16, 28 and 32) values. The SE sector also displayed rather low values (0.20 µg l^{-1} at station 41) while values higher than 0.24 µg l^{-1} were observed in the NE sector.

Sediment texture. The mean grain size of the sediment (Fig. 5) was in the range of silt & clay for half of the sampled stations, while only the shallowest station 0 had a medium-sand texture. The coarser grain sizes (gravel, >2 mm) were found only at the shallowest stations 0 and 14, while the silt & clay fraction was highly represented in the 0–2 cm layer of station 16 (40%). On average, the most represented fraction was that ranging from 0.064 to 0.212 mm (52%), followed by the fraction ranging from 0.212 to 0.5 mm (30%) and by the fraction lower than 0.064 mm (15%).

The Sed-H results (Fig. 5) highlighted the highest sediment diversity at stations 0 and 14, and the lowest at stations 25 and 32.

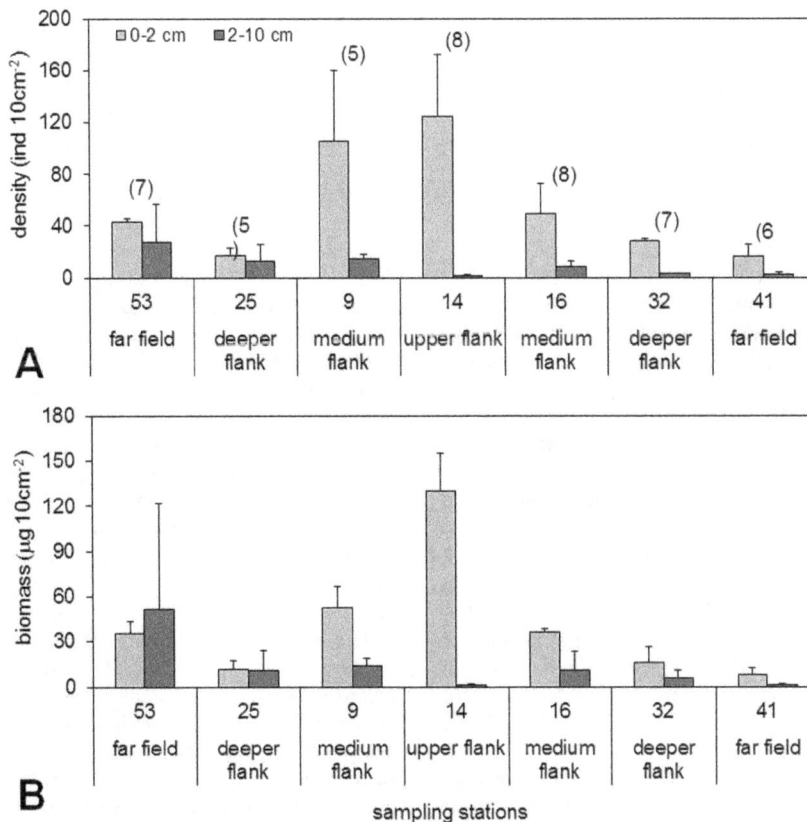

Figure 2. Preliminary meiofaunal abundance and biomass data for the two sediment layers of each station. Values averaged over replicates collected in the same deployment: bars denote standard error. A: abundance (individuals 10 cm^{-2}), in brackets the number of taxa, B: biomass (µg 10 cm^{-2}, the values have been calculated with the gravimetric method).

Table 3. Distance-based linear model (DISTLM) for meiofaunal abundance and selected environmental variables.

MARGINAL TESTS					
Variable	SS(trace)	Pseudo-F	P	Prop.	
H sed	2323.2	3.788	0.052	0.43	
water depth	3688.4	10.840	0.016	0.68	
protein turnover time	418.6	0.421	0.597	0.08	
carbohydrate turnover time	471.5	0.479	0.542	0.09	
labile P turnover time	2295.3	3.709	0.107	0.43	
proteins	1468.1	1.872	0.240	0.27	
carbohydrates	4276.6	19.212	0.002	0.79	
lipid	469.6	0.477	0.510	0.09	
labile P	1154.5	1.363	0.277	0.21	
SEQUENTIAL TESTS					
Variable	Adjust.R^2	SS(trace)	Pseudo-F	P	Prop.
+carbohydrates	0.7522	4276.6	19.212	0.003	0.79
+labile P	0.7561	236.6	1.080	0.364	0.04
+water depth	0.8380	439.8	3.022	0.140	0.08
+carbohydrate turnover time	0.9452	338.2	6.870	0.060	0.06
+protein turnover time	0.9857	85.6	6.660	0.172	0.02

Marginal tests: explanation of variation for each variable taken alone. Sequential tests: conditional tests of individual variables in constructing the model. Each test examines whether adding the variable contributes significantly to the explained variation. Selection procedure: step-wise, selection criterion: adjusted R^2. Prop.: % variation explained.

The sediment diversity was significantly correlated to the meiofaunal biomass (R = 0.77, n = 7, p<0.05), and also the meiofaunal density was directly proportional to the sediment diversity, although not significantly (Table 3).

Sedimentary organic matter (OM). The carbohydrate content (Fig. 6A) had the highest values of the OM components (on average 583.1±232.2 µg g^{-1} for the two layers). All the upper- and medium-flank stations of the seamount showed 0–2 cm layer values significantly higher than the 2–10 cm layer ones (one-way ANOVA and NK post-hoc test, p<0.05), with stations 9, 14 and 16 reaching the highest absolute values (from 776.9±125.6 to 1169.3±43.4 µg g^{-1} for stations 16 and 14, respectively). Considering the 0–2 cm layer, the correlation between meiofauna abundance and carbohydrates was highly significant (R = 0.95, n = 7, p<0.01). The DISTLM analysis confirmed that, within the environmental trophic features, the carbohydrate content was the variable mainly linked to the meiofaunal abundance.

Lower values were observed for the lipid content (on average 123.7±106.2 µg g^{-1} for the two layers), even lower if the upper- and medium-flank stations of the seamount were considered (on average 47.9±21.0 µg g^{-1}) (Fig. 6B). This difference was significant (one-way ANOVA and NK post-hoc test, p<0.01). On the other hand, no significant differences were observed between the two sediment layers of the upper- and medium-flank stations of the seamount, while the other stations showed a significant accumulation in the 2–10 cm layer (one-way ANOVA and NK post-hoc test, p<0.01).

The protein content values (Fig. 6C) were more homogeneous throughout area (on average 114.0±27.1 µg g^{-1}), without significant differences except for the two layers of station 41 (one-way ANOVA and NK post-hoc test, p<0.05). The labile P values (Fig. 6D) did not show significant differences between the stations. The OC and N contents (Figs. 6E and 6F) of the two layers of the upper- and medium-flank stations of the seamount,

considered together, showed significantly lower values than the other stations (one-way ANOVA and NK post-hoc test, p<0.01). Significant differences between the two layers were rare for protein, P labile, OC and N contents.

Sedimentary enzymatic activity and OM turnover. The enzymatic activities are shown in Fig. 7 and the related OM turnover times in Table 2.

The LA activity was on average 29.7±12.8 nmol g^{-1}h^{-1} for the whole area and the stations didn't show significant differences. The surface sediment layer always had higher values than the deep one, sometimes significant (one-way ANOVA and NK post-hoc test, p<0.05). The protein turnover times were, on average, 2.9±2.1 days for the whole area and the surface layer showed generally lower values than the deep one, except for stations 53 and 14. The deepest station, station 41, showed the highest value, followed by the shallowest, station 14.

The AP activity showed the highest values in the 0–2 cm layer of station 14, but station 9 also showed notable activity in its surface layer. The turnover times were significantly (one-way ANOVA and NK post-hoc test, p<0.05) lower for the upper- and medium-flank stations than for the other stations (on average 0.05±0.03 vs 0.13±0.07 days, respectively).

The BG activity was higher at stations 14 and 16, lower at the southernmost stations 32 and 41, and rather high variability was recorded for the upper- and medium-flank stations. The surface sediment layer showed significantly higher BG values than the deep layer (one-way ANOVA and NK post-hoc test, p<0.05). The carbohydrate turnover times were very high (on average 607±332 days) and the surface layer showed lower values than the deep one.

Keeping in mind that the meiofaunal data were preliminary due to the limited sampling procedure, the DISTML analysis (Table 3) indicated that the best model for the meiofaunal abundance and the environmental features included, together with the water depth and the carbohydrate sedimentary content, the turnover times for

Table 4. List of the meiofaunal taxa found in the sampling stations and position of each station with respect to the seamount morphology.

stations taxa	53 far-field	25 deeper flank	9 medium flank	14 upper flank	16 medium flank	32 deeper flank	41 far-field
Amphipoda						1.3	
Copepoda	6.1	10.0	8.2	9.2	4.2	11.3	7.4
Nauplii		2.5	0.4	1.7	1.2		
Gastrotricha					0.6		
Kinorhyncha	0.5						
Nematoda	79.7	81.2	82.6	78.1	85.5	77.5	83.3
Oligochaeta					1.2		
Ostracoda	1.0		1.7	0.6	0.6		1.8
Polychaeta	11.7	3.8	3.2	8.3	4.2	3.7	
Rotifera	0.5		3.9	1.2	1.9	2.5	1.9
Sipuncula		1.3		0.3			
Tanaidacea				0.3			
Thermosbaenacea	0.5			0.3		2.5	1.9
Turbellaria		1.3			0.6	1.2	3.7

Percentage contributions of the taxa to the meiofaunal total abundance for each station are presented.

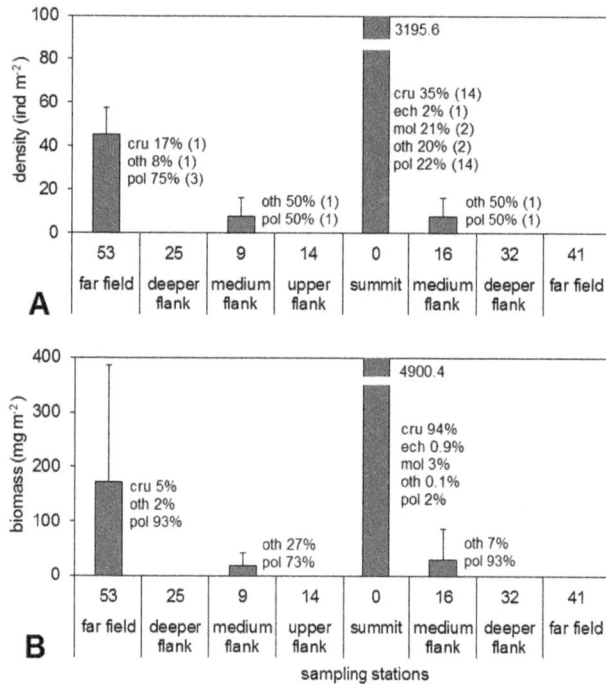

Figure 3. Average macrofaunal abundance and biomass and the contribution of each taxonomical group (cru: crustacean, pol: polychaetes, mol: molluscs, ech: echinoderms and oth: others) for each station. Values averaged over independent replicates: bars denote standard error. A: abundance (individuals m^{-2}), in brackets the number of taxa, B: biomass(μg m^{-2}, the values have been calculated with the gravimetric method).

proteins and carbohydrates. The PCA analysis (Fig. 8), focused on the indexes of OM lability, showed thatPCA1 and PCA2 explained 78.5% of the variation. The superimposed bubbles are proportional to the meiofaunal abundances and the asterisks indicate where macrofauna were found.

The statistical analyses indicated that the C/N and the protein/carbohydrate ratios failed to give a reasonable picture of the OM lability, at least for the sediments. For instance, the highest protein turnover times of station 41, indicating the poor trophic quality of the protein content, disagrees with the high protein/carbohydrate ratio values, thus indicating refractory protein accumulation rather than a large pool of trophic resources. This may be related to an analytical bias, because the protein method overestimates the concentration of proteins because of the presence of humic substances [42].

Discussion

The Tyrrhenian basin has been considered as one of the more oligotrophic of the western Mediterranean in terms of water column features [43]. In agreement with this scenario, previous studies on meiofauna [8] highlighted rather low abundances compared to extra-Mediterranean ones (in the Atlantic Ocean, for instance) [16], while benthic prokaryote abundance and biomass fell within the range of values reported in the literature for deep-sea sediments worldwide [7]. However, no exhaustive information has been provided for the deep Tyrrhenian Sea macrofaunal organisms, nor for the seamount-summit macrofauna of this area.

The summit station, 0, is a world apart for macrofauna, due firstly to the shallow depth (within the photic zone and enriched by

good trophic-quality phytopigments) and to a coarse granulometry that allows for deep oxygenation of the sediment. The heterogeneity of the substrata at the summit (rocky substrate alternating with coarse, biogenic debris) may provide different habitats for a high number of species [15]. In addition, the presence of a *Laminaria rodriguezii* meadow at the seamount summit [11]would provide the macrofauna with a surplus of food resources through the detrital food chain. In fact, as reported for the coastal *Posidonia oceanica* meadows of the Ligurian Sea [44], the macrobenthic community at the summit is dominated by carnivores (94% of the total abundance) such as Aciculata families (89% of polychaetes), Decapoda, Turbellaria, Lyssianassidae and the amphipod *Leucothoe occulta* (Krapp & Schickel, 1975) (Table 5). The good quality of the organic particulate matter [45] favours suspension feeders and mixed suspension-deposit feeders (ca. 6%) such as the echinoderm *Acrocnida brachiata* (Montagu, 1804), the bivalve *Lucinella divaricata* (Linnaeus, 1758) and cypridinid ostracods (Table 5). On the contrary, only deposit and mixed suspended-deposit feeders, such as Scolecida and Sipuncula, were found at the medium-flank stations (Table 5).

In our study we observed that, with the exception of the summit station, the benthic communities of the Vercelli Seamount area were poor in macrofauna (for instance, lower than at the Atlantic Condor Seamount [15]) and moderately rich in meiofauna (abundance and number of taxa higher than those reported for the sediments surrounding the Tyrrhenian Palinuro and Marsili seamounts[8] but lower than the abundances found at the Atlantic Condor Seamount [16]). Benthic communities are generally influenced by water depth [16,46]. In accordance with the above, a significant correlation between depth and our preliminary data of total meiofaunal abundance was found, and the macrofaunal organisms have only been found at some shallower stations.

Leduc et al. [47] showed that the sediment texture, transformed into a sediment-diversity index (Sed-H) by means of the Shannon-Wiener diversity index, deeply influences soft-bottom communities, especially those groups, such as meiofauna, that have a limited expansion rate due to their small size. In our study the preliminary data on meiofaunal abundance agreed with these observations, showing the highest abundances and biomasses where the Sed-H was the highest (station 14) and the lowest where the Sed-H was the lowest (stations 25 and 32).

Macrofauna did not directly respond to the Sed-H as the meiofauna did. Considering other environmental features that may be involved in the macrofauna distribution, we observed that three stations (9, 16 and 28, although the latter was not sampled for benthic communities) had a particular exposition degree, opposite that of the regular seamount flank slope. This implies the presence of irregularities along the flank that interfere with the incoming current, modifying the local hydrodynamic processes, allowing slowing of the current speed and increasing deposition. This would explain the fine granulometry of these stations, finer than the other flank-stations. Higher sedimentation rates on the northern flank of the seamount were also revealed by the composition and trophic strategies of the megafauna assemblages found there by Bo et al. [11]. The peculiar environment generated by these morphological features seems to favour macrofaunal development. Except for the summit station, only stations 9 and 16 within the very poor area of the Vercelli Seamount showed the presence of macrofaunal organisms. The lack of macrofauna at shallow station 14 may, then, be related to the high bottom slope and to an excessive hydrodynamic forcing, as this station lacks appropriate aprons. In addition, the high meiofaunal density and biomass at this site may have exerted more efficient competition

Table 5. List of the macrofaunal taxa found on the Vercelli Seamount summit (station 0), medium flank (stations 9 and 16) and in the far-field area (station 53).

Phylum	Class	Family	Species	0	9–16	53
					stations	
Annelida	**Polychaeta**					
	Scolecida	Capitellidae sp1			X	
		Capitellidae sp2		X		
		Opheliidae		X		
		Paraonidae sp1				X
		Paraonidae sp2		X		
	Aciculata	Chrysopetalidae sp1		X		
		Chrysopetalidae sp2		X		
		Exogoninaejuv.		X		
		Goniadidae		X		
		Hesionidae		X		
		Nereididae		X		
		Phyllodocidae sp1				X
		Phyllodocidae sp2		X		
		Sigalionidae		X		
		Syllidae sp1		X		X
		Syllidae sp2		X		
Artropoda	**Malacostraca**					
	Decapoda	Leucosiidae	*Ebalia* sp	X		
		Parthenopidae	*Parthenopoides massena*	X		
	Amphipoda	Ampithoidae	*Ampithoe helleri*	X		
		Leucothoidae	*Leucothoe occulta*	X		
		Lyssianassidae		X		
	Tanaidacea	Apseudidae	*Apseudopsis elisae*	X		
			Apseudes sp			X
	Isopoda sp1			X		
	Isopoda sp2			X		
	Isopoda sp3			X		
	Isopoda sp4			X		
	Maxillopoda					
	Copepoda			X		
	Ostracoda	Cypridinidae		X		
		Bythocytheridae		X		
Mollusca	**Bivalvia**	Limidae	*Limaria hians*	X		
		Lucinidae	*Lucinella divaricata*	X		
Echinodermata	**Ophiuroidea**	Amphiuridae	*Acrocnida brachiata*	X		
Sipuncula	**Sipunculidea**	Phascolionidae	*Phascolion (Phascolion) strombus*		X	
		Phascolosomatidae	*Phascolosoma (Phascolosoma) granulatum*		X	X
Platyhelminthes	**Turbellaria**			X		
Nematoda				X		

with macrofauna and/or a top-down control of macrofaunal juveniles [48].

However, water depth, sediment texture and sediment aspect are not enough to explain benthic community distribution [17,19], especially in a variable environment such as the Vercelli Seamount area. The quantitative features of the trophic supply in deep sediments have been indicated as a limiting factor for the growth

and activity of microbes [39] and small metazoan [49], and their qualitative features (bioavailability) have been invoked to explain community distribution as well [1,50].

Surprisingly, but in agreement with previous literature data [8], our preliminary data on meiofaunal abundance were related to a rather refractory fraction of the organic matter (OM), namely the carbohydrates, indicating that in this peculiar system the

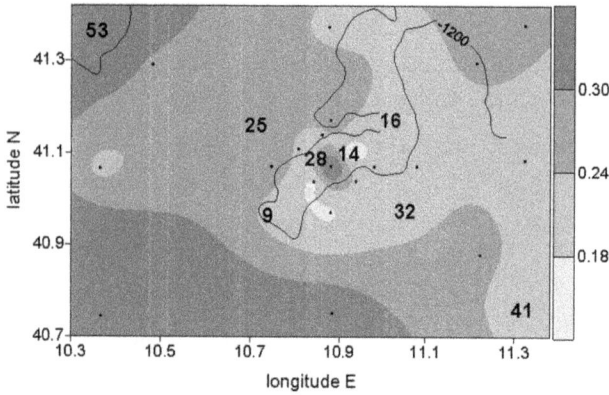

Figure 4. Seawater autotrophic pigment distribution. Chlorophyll-a concentrations (µg l^{-1}) in the water column have been averaged for the 0–120 m layer. The numbers indicate the stations also sampled for the sedimentary parameters, the dots indicate all the stations sampled for the autotrophic biomass determination in the water column. The line reports the 1200 m depth isobath.

carbohydrate trophic supply would play a role in the distribution of the meiofaunal organisms. The absence of significant relationships for the 2–10 cm layer suggested that these carbohydrates may derive from deposition via pelagic-benthic coupling.

In fact, the shallower seamount stations (9, 28, 14 and 16) showed a significant enrichment of the upper sediment layer, indicating that this carbohydrate distribution was related to the flux of OM from the surface layer. Frontal zones, as observed in the Vercelli Seamount area, are known to induce the sinking of particulate material towards the deeper layers [39]. Therefore, most of the organic bulk in these benthic areas is composed of high-molecular-weight compounds, so that the benthic zone is enriched with surface OM, which usually does not reach the seafloor. The phytoplanktonic debris, the main OM source in pelagic environments, was probably highly involved in these processes, as suggested by Speicher et al. [51]. While the more labile components of the OM flux (proteins and lipids, for instance) could be consumed and re-cycled during their sinking to the

sediment; structural carbohydrates (cellulose, for instance) were preserved and reached the sediments at a higher rate.

The complex hydrodynamic assessment of the studied area led to a carbohydrate accumulation that did not exactly match the chlorophyll-a distribution in the water column recorded during the same sampling cruise. The pelagic-benthic coupling was based on oblique fluxes rather than on vertical ones; the fluxes (and the accumulation) being higher in the shallower stations where the dilution is proportionally lower. In the southernmost section of the sampling area, the Bonifacio gyre flows from the SW [13], reaching the seamount area at station 9. The favourable conditions to phytoplanktonic biomass accumulation in the surface layer of the SW sector, that led to the very high chlorophyll-a signal recorded, could have generated a high OM flux, carrying high carbohydrate amounts to the sediments of station 9. During our cruise high chlorophyll-a values were also measured in the water column of the NE section. Sinking phytoplanktonic biomass, pushed by the gyre previously reported by Vetrano et al. [13], could have reached our station 16, increasing the carbohydrate content of the surface sediments. Station 14 was probably the only one directly subjected to the "seamount effect", indicated by the increase of phytoplanktonic biomass centred on the seamount summit that we recorded. Lateral transport of this OM could have reached shallow station 14, generating the significantly higher carbohydrate contents of the surface sediment.

Although the carbohydrate contents of the entire Vercelli Seamount area showed rather high values, the other biochemical fractions were lower than those reported for other Mediterranean and Atlantic areas [8,52], studied using the same analytical methods. The C-OM (namely the sum of the C contribution of proteins, carbohydrates and lipids) of the Portuguese, Adriatic and Cretan margins was three times the average value we found in the Vercelli Seamount area. Our results confirmed the general oligotrophy of the Tyrrhenian Sea and the tendency to ultraoligotrophy of the Vercelli area, highlighted by a very low protein content [53].

The depositional features of the studied area may have a role in these low values. Speicher et al. [51] highlighted a rather low particulate organic carbon (POC) export flux in the northern Tyrrhenian Sea (where the Vercelli Seamount is placed) during late May-early June compared to the middle and southern

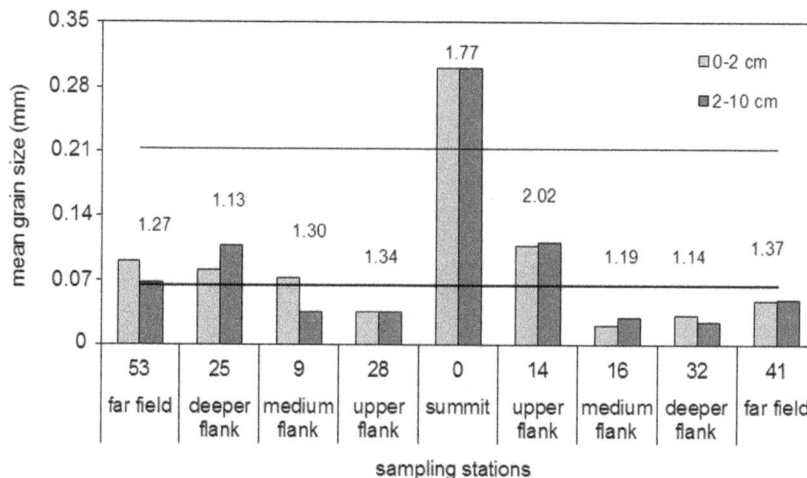

Figure 5. Sediment texture. Mean grain size (mm) for the sediment 0–2 cm layer and 2–10 cm layer. The black horizontal line highlights the upper limit of the silt & clay fraction (0.064 mm), the grey line the division between fine and medium sand (0.212 mm). The numbers above the bars report the sedimentary diversity index (Sed-H) for each station.

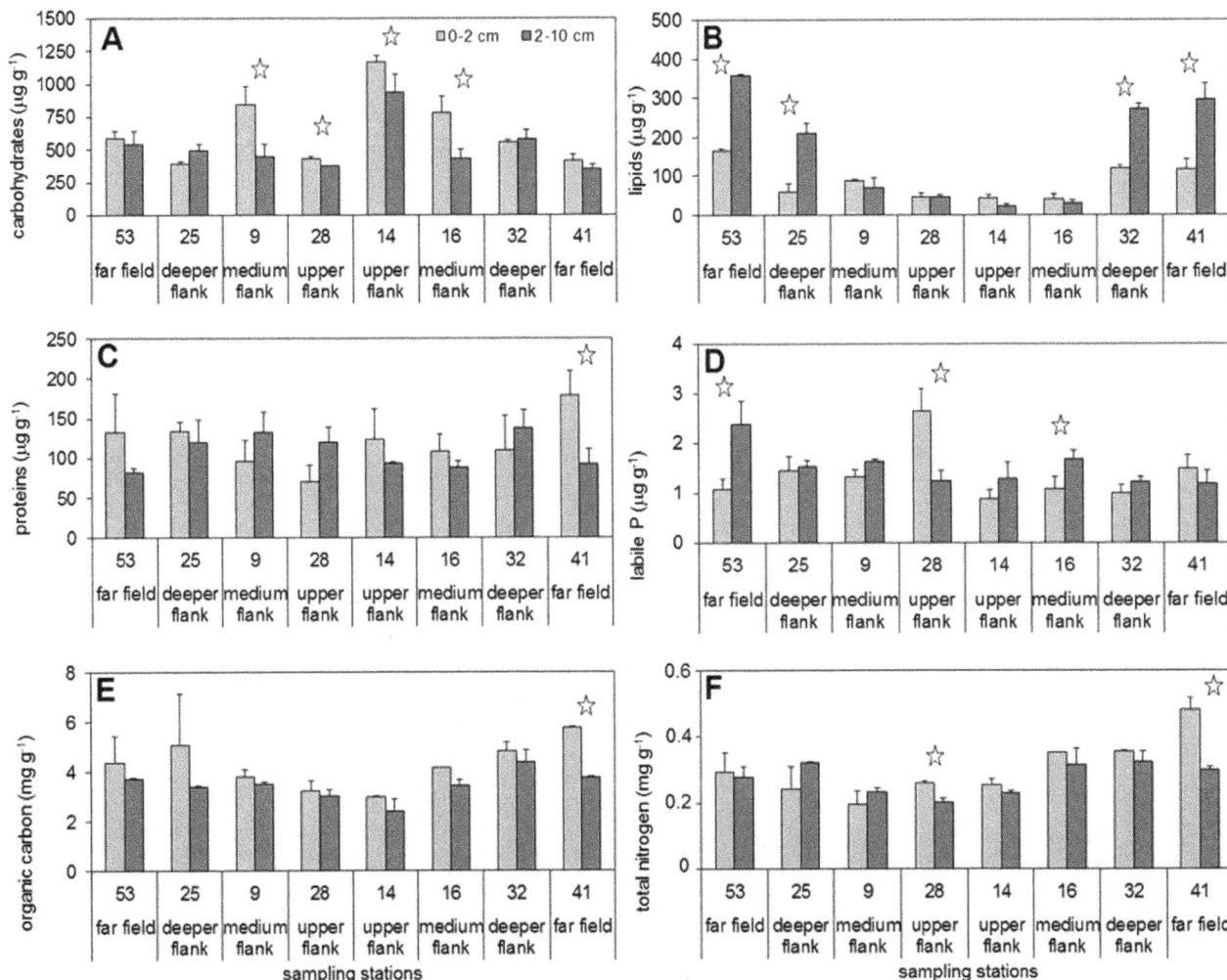

Figure 6. Organic matter contents for the two sediment layers. Bars denote standard deviations. A: carbohydrates ($\mu g\ g^{-1}$), B: lipids ($\mu g\ g^{-1}$), C: proteins ($\mu g\ g^{-1}$), D: labile P ($\mu g\ g^{-1}$), E: organic carbon (OC, $mg\ g^{-1}$), F: total nitrogen (N, $mg\ g^{-1}$). Stars denote significantly different values for the 0–2 cm and the 2–10 cm sediment layers at each station (one-way ANOVA, NK post-hoc test, $p < 0.05$).

Tyrrhenian. In addition, it is well known that the sediment texture regulates the OM accumulation, the coarser sediments providing a lower surface to be coated by OM. The generally low contribution of the silt & clay fraction (grain size $<63\ \mu m$) in the Vercelli area (on average 15%) compared to the 22–98% of the slope areas reported by Pusceddu et al. [8] agrees with this statement.

Together with the common quality indexes (C/N ratio and protein/carbohydrate ratio), the hydrolytic enzymatic activity may give clues on the potential lability of the OM [41], following the optimal resource allocation model of Sinsabaugh et al. [54], who stated that osmotrophic assemblages will optimize their energy expenditures by expressing high levels of particular hydrolases only if polymeric substrates are abundant and if the monomeric hydrolysates required for growth are scarce. The low values of the substrate quantity: enzyme activity ratio, namely low turnover times, highlight high trophic value OM. A major role of the OM turnover times was observed, and the highest meiofaunal abundances were found in stations where the OM was more labile (surface layer of stations 9, 14, 16) as indicated by multivariate analysis.

The BG activity and the carbohydrate turnover time may, therefore, be useful to explain meiofaunal as well as macrofaunal distribution. BG activity was always higher in the surface sediment layer, probably becoming too energy expensive for the trophically-limited deeper sediment layer. This matches the vertical meiofaunal distribution, with more than 80% of the total abundance found in the surface layer. Considering, instead, the macrofauna, its biomass value at station 16, higher than that of station 9, was in agreement with the higher trophic quality of the sediment, characterised by rather low turnover times for carbohydrates and proteins. In fact station 9, although showing a notable carbohydrate content, showed higher turnover times, indicating a lower lability for the OM reaching the bottom, due probably to the longer time the OM spent in the water column.

The deepest station, station 41, showed the highest protein turnover time values in both layers, indicating that the accumulated proteins did not induce LA activity as they did at the other stations of our study area. This may be because peptidase activity has been described as being negatively correlated to depth [55] but, from a trophic point of view, it indicates the lower lability of station 41 protein content. At the seamount-influenced sites, although the fluxes and/or in situ production were low, the proteins were generally more labile than those of the non-seamount station 41.

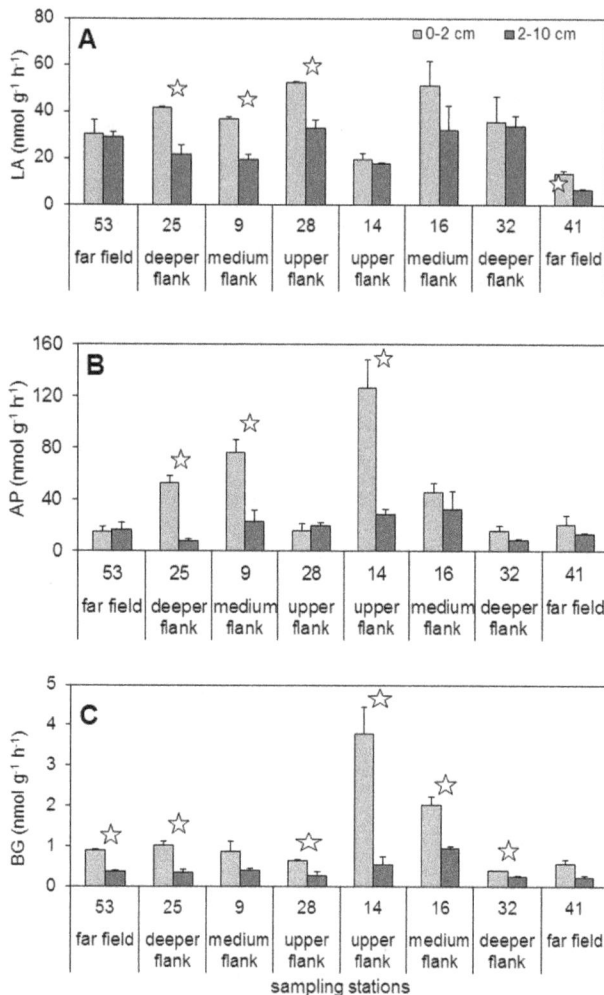

Figure 7. Enzymatic activities (nmol g⁻¹ h⁻¹) for the two sediment layers. Bars denote standard deviations. A: leucine aminopeptidase (LA), B: alkaline phosphatase (AP), C: β-glucosidase (BG). Stars denote significantly different values for the 0–2 cm and the 2–10 cm sediment layers at each station (one-way ANOVA, NK post-hoc test, p<0.05).

The relationship between benthic organisms and OM lability, expressed as a short turnover time, is complex. In fact, macrofaunal presence and also the preliminary information on meiofaunal abundance may explain the anomalous turnover times for proteins we detected for stations 53 and 14, which showed higher values in the surface sediment than in the deeper layer. If the coarser grain size of these stations allowed for higher mixing of the OM of the surface and deep sediment layers, station 53 showed high meiofaunal and rather high macrofaunal abundances, and the diffusion of OM is strongly stimulated by macrofaunal galleries [56]. At station 14, instead, the bioturbation was limited to the action of meiofauna, which showed the highest densities, due to the lack of macrofaunal organisms.

References

1. Danovaro R, Company JB, Corinaldesi C, D'Onghia G, Galil B, et al. (2010) Deep-Sea Biodiversity in the Mediterranean Sea: The Known, the Unknown,

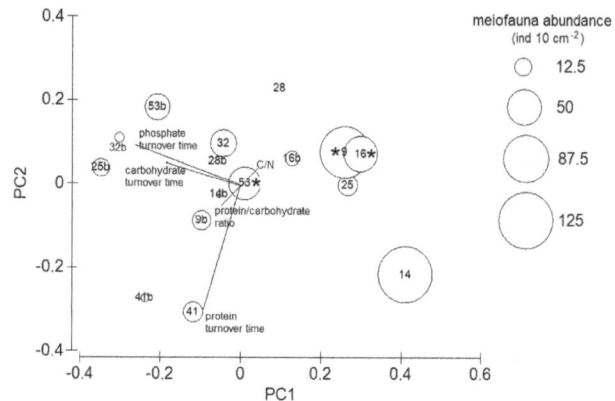

Figure 8. PCA analysis of the OM trophic-quality features (see text and Table 2 for details) for the upper flank stations 14 and 28 (at 400 and 877 m depth), the medium-flank stations 9 and 16 (at 1232 and 1166 m depth), the deeper-flank stations 25 and 32 (at 1833 and 1728 m depth), the far-field stations 53 and 41 (at 887 and 2646 m depth). C/N: (carbon/nitrogen ratio) the lower the ratio the higher the trophic quality, protein/carbohydrate: the higher the ratio the higher the trophic quality, protein-carbohydrate-labile P turnover times: the lower the time the higher the trophic quality. Meiofaunal total biomass is superimposed as bubbles and asterisks denote where macrofauna was found. "b" after the station number denotes the 2–10 cm layer of the station.

Conclusions

The Mediterranean Sea has been studied intensively, but its deep areas need more research to increase our understanding of its ecosystem functions. In particular, the studies on the deep areas of the Tyrrhenian Sea are scarce, although its morphological features (seamounts, for instance) suggest the presence of diversified habitats and ecosystems. In the present paper we observed that in the Vercelli Seamount area the peculiar environmental features (physical, morphological and trophic) differently shaped the benthic assemblages. Macrofauna showed different community composition comparing the seamount flanks and the far-field stations, and the summit station was a world apart in terms of density, diversity and biomass. Bottom inclination and aspect allowed the presence of aprons against the hydrodynamic forcing, favourable to macrofauna development. Our preliminary data on meiofaunal abundance showed, instead, a link to water depth and to trophic supply, especially to the lability of the OM. A variable pelagic-benthic coupling links the seawater with the bottom indicating that, irrespectively of being far from the coast and/or placed in the deep sea, these areas may be sensitive to global processes.

Acknowledgments

We would like to thank the captain and crew of the R/V *Urania* for their unstinting assistance during the cruise.

Author Contributions

Conceived and designed the experiments: ACH CM. Performed the experiments: ACH CM M. Bo M. Borghini MM FM AM. Analyzed the data: ACH CM FM. Contributed reagents/materials/analysis tools: ACH CM MC PP. Wrote the paper: ACH CM. Project planning: GB PP.

and the Unknowable. PLoS ONE 5(8): e11832. doi: 10.1371/journal.pone. 0011832.

2. Danovaro R, Dinet A, Duineveld G, Tselepides A (1999) Benthic response to particulate fluxes in different trophic environments: A comparison between the Gulf of Lions-Catalan Sea (Western Mediterranean) and the Cretan Sea (Eastern Mediterranean). Progr Oceanogr 44(1–3): 287–312.

3. Psarra S, Tselepides A, Ignatiades L (2000) Primary productivity in the oligotrophic Cretan Sea (NE Mediterranean): Seasonal and interannual variability. Progr Oceanogr 46: 187–204.

4. Tselepides A, Papadopoulou N, Podaras D, Plaiti W, Koutsoubas D (2000) Macrobenthic community structure over the continental margin of Crete (South Aegean Sea, NE Mediterranean). Progr Oceanogr 46(2–4): 401–428.

5. Galil BS, Goren M (1994) The deep sea Levantine fauna, new records and rare occurrences. Senckenb Marit 25 (1–3): 41–52.

6. Galil BS, Zibrowius H (1998) First benthos samples from Eratosthenes Seamount, Eastern Mediterranean. Senckenb Marit 28 (4–6): 111–121.

7. Danovaro R, Corinaldesi C, Luna GM, Magagnini M, Manini E, et al. (2009) Prokaryote diversity and viral production in deep-sea sediments and seamounts. Deep-Sea Res II 56: 738–747.

8. Pusceddu A, Gambi C, Zeppilli D, Bianchelli S, Danovaro R (2009) Organic matter composition, metazoan meiofauna and nematode biodiversity in Mediterranean deep-sea sediments. Deep-Sea Res II 56: 755–762.

9. Artale V, Astraldi M, Buffoni G, Gasparini GP (1994) Seasonal variability of gyre-scale circulation in the northern Tyrrhenian Sea. J Geophys Res 99 (C7): 14,127–14,137.

10. Nair R, Cattini E, Gasparini GP, Rossi G (1994) Circolazione ciclonica e distribuzione dei nutrienti nel Tirreno settentrionale. Proceedings of the X Symposium of the Italian Association of Limnology and Oceanology (AIOL). pp. 65–76.

11. Bo M, Bertolino M, Borghini M, Castellano M, Covazzi Harriague A, et al. (2011) Characteristics of the mesophotic megabenthic assemblages of the Vercelli Seamount (North Tyrrhenian Sea). PLoS ONE 6(2): e16357. doi: 10.1371/journal.pone.0016357.

12. Wezel FC (1985) Structural features and basin tectonics of the Tyrrhenian Sea. In: Stanley DJ, Wezel FC, editors.Geological evolution of the Mediterranean Basin.Springer-Verlag, New York. pp.153–194.

13. Vetrano A, Napolitano E, Iacono R, Schroeder K, Gasparini GP (2010) Tyrrhenian Sea circulation and water mass fluxes in spring 2004: Observations and model results. J Geophys Res. 115 (C06023), 18 pp.

14. Dower J, Freeland H, Juniper K (1992) A strong biological response to oceanic flow past Cobb Seamount. Deep-Sea Res A 39: 1139–1145.

15. Bongiorni L, Ravara A, Parretti P, Santos RS, Rodrigues CF, et al. (2013) Organic matter composition and macrofaunal diversity in sediments of the Condor Seamount (Azores, NE Atlantic). Deep-Sea Res II 98: 75–86.

16. Zeppilli D, Bongiorni L, Cattaneo A, Danovaro R, Serrão Santos R (2013) Meiofauna assemblages of the Condor Seamount (North-East Atlantic Ocean) and adjacent deep-sea sediments. Deep-Sea Res 98: 87–100.

17. Clark MR, Rowden A, Schlacher T, Williams A, Consalvey M, Stocks KI, Rogers AD, O'Hara TD, White M, Shank TM, Hall-Spencer JM (2010) The ecology of seamounts: structure, function and human impacts. Ann Rev Mar Sci 2: 253–278.

18. Gooday AJ (2002) Biological responses to seasonally varying fluxes of organic matter to the ocean floor: a review. J Oceanogr 58(2): 305–332.

19. Piepenburg D, Müller B (2004) Distribution of epibenthic communities on the Great Meteor Seamount (North-east Atlantic) mirrors pelagic processes. Arch Fish Mar Res 51: 55–70.

20. Gad G, Schminke HK (2004) How important are seamounts for the dispersal of interstitial meiofauna. Arch Fish Mar Res51: 43–54.

21. George KH (2004) Description of two new species of Bodinia, a new genus incertae sedis in Argestidae Por, 1986 (Copepoda, Harpacticoida), with reflections on argestid colonization of the Great Meteor Seamount plateau. Org Divers Evol 4 (4): 241–254.

22. Christiansen B, Wolff G (2009) The oceanography, biogeochemistry and ecology of two NE Atlantic seamounts: The OASIS project. Deep-Sea Res II 56 (25): 2579–2581.

23. Moen J (1984) Variability and mixing of the surface layer in the Tyrrhenian Sea: MILEX-80, Final Report. Saclancen Report SR-75, 128 pp.

24. Krivosheya VG (1983) Water circulation and structure in the Tyrrhenian Sea. Oceanology 23: 166–171.

25. Danovaro R (2010) Methods for the Study of Deep-sea Sediments their Functioning and Biodiversity. CRC Press, Boca Raton, FL. pp. 420.

26. Buchanan J B, Kain JM (1971) Measurement of the physical and chemical environment. In Holme NA, Mc Intyre AD, editors. Methods for the Study of Marine Benthos. Blackwell Scientific Publications, Oxford, Edinburgh. pp. 30–52.

27. Etter RJ, Grassle JF (1992) Patterns of species diversity in the deep sea as a function of sediment particle size diversity. Nature 369: 576–578.

28. Holm-Hansen O, Lorenzen CJ, Holmes RW, Strickland JDH (1965) Fluorometric determination of chlorophyll. J Cons – Cons Perm Int Explor Mer 30: 3–15.

29. Hartree EF (1972) Determination of proteins: a modification of the Lowry method that give a linear photometric response. Anal Biochem 48: 422–427.

30. Dubois M, Gilles K, Hamilton JK, Rebers PA, Smith F(1956) Colorimetric method for determination of sugars and related substances. Anal Chem 28: 350–356.

31. Bligh EG, Dyer WJ (1959) A rapid method for total lipid extraction and purification. Canadian J Biochem Physiol 37: 911–917.

32. Marsh BJ, Weinstein DB (1966) Simple charring method for determination of lipids. J Lipid Res 7: 574–576.

33. Ruttenberg KC (1992) Development of a sequential extraction method for different forms of phosphorous in marine sediments. Limnol Oceanogr 37: 1460–1482.

34. Koroleff F (1983) Determination of phosphorus. In: Grassoff, K, Ehrhardt, M, Kremling, K, editors.Methods of Seawater Analysis. Verlag Chemie, Weinheim, Haugland. pp.125–139.

35. Hansen HP, Grasshoff K (1983) Automated chemical analysis. In: Grassoff K, Ehrhardt M, Kremling K, editors.Methods of Seawater Analysis. Verlag Chemie, Weinheim, Haugland. pp. 347–379.

36. Hedges JI, Stern JH (1984) Carbon and nitrogen determination of carbonate-containing solids. Limnol Oceanogr 29: 657–663.

37. Hoppe HG (1983) Significance of exoenzymatic activities in the ecology of brackish water: measurements by means of methylumbelliferyl substrates. Mar Ecol-Prog Ser 11: 299–308.

38. Caruso G, Monticelli L, Azzaro F, Azzaro M, Decembrini F, et al. (2005) Dynamics of extracellular enzymatic activities in a shallow Mediterranean ecosystem (Tindari Ponds, Sicily). Mar Freshw Res 56: 173–188.

39. Bianchi A, Calafat A, De Wit R, Garcin J, Tholosan O, et al. (2003) Microbial activity at the deep water sediment boundary layer in two highly productive systems in the Western Mediterranean: the Almeria-Oran front and the Malaga upwelling. Oceanol Acta 25: 315–324.

40. Bhaskar PV, Bhosle NB(2008) Bacterial production, glucosidase activity and particle-associated carbohydrates in Dona Paula bay, west coast of India. Estuar Coast Shelf Sci 3: 413–424.

41. Mudryk ZJ, Skórczewski P (2004) Extracellular enzyme activity at the air-water interface of an estuarine lake. Estuar Coast Shelf Sci 59: 59–67.

42. Vakondios N, Koukouraki EE, Diamadopoulos E (2014) Effluent organic matter (EfOM) characterization by simultaneous measurement of proteins and humic matter. Water Res 63: 62–70.

43. Bosc E, Bricaud A, Antoine D (2004) Seasonal and interannual variability in algal biomass and primary production in the Mediterranean Sea, as derived from 4 years of SeaWiFS observations. Global Biogeochemical Cycles 18 (1).

44. Covazzi Harriague A, Bianchi CN, Albertelli G (2006) Soft-bottom macro-benthic community composition and biomass in a Posidonia oceanica meadow in the Ligurian Sea (NW Mediterranean). Estuar Coast Shelf Sci 70: 251–258.

45. Misic C., Bavestrello G, Bo M, Borghini M, Castellano M, et al. (2012) The "seamount effect" as revealed by organic matter dynamics around a shallow seamount in the Tyrrhenian Sea (Vercelli Seamount, western Mediterranean). Deep-Sea Res I 67: 1–11.

46. Rex MA, Etter RJ, Morris JS, Crouse J, McClain CR, et al. (2006) Global bathymetric patterns of standing stock and body size in the deep-sea benthos. Mar Ecol-Prog Ser 317: 1–8.

47. Leduc D, Rowden AA, Probert PK, Pilditch CA, Nodder SD, et al. (2012) Further evidence for the effect of particle-size diversity on deep-sea benthic biodiversity. Deep-Sea Res I 63: 164–169.

48. Zobrist EC, Coull BC (1992) Meiobenthic interactions with macrobenthic larvae and juveniles: an experimental assessment of the meiofaunal bottleneck. Mar Ecol-Prog Ser 88: 1–8.

49. Fonseca G, Soltwedel T (2009) Regional patterns of nematode assemblages in the Arctic deep seas. Polar Biol 32: 1345–1357.

50. Heinz P, Rueff D, Hemleben C (2004) Benthic foraminifera assemblages at Great Meteor Seamount. Mar Biol 144(5): 985–998.

51. Speicher EA, Moran SB, Burd AB, Delfanti R, Kaberi H, et al. (2006) Particulate organic carbon export fluxes and size-fractionated POC/^{234}Th ratios in the Ligurian, Tyrrhenian and Aegean Seas. Deep-Sea Res I 53: 1810–1830.

52. Pusceddu A, Bianchelli S, Canals M, Sanchez-Vidal A, Durrieu De Madron X, et al. (2010) Organic matter in sediments of canyons and open slopes of the Portuguese, Catalan, Southern Adriatic and Cretan Sea margins. Deep-Sea Res I 57: 441–457.

53. Rossi S, Grémare A, Gili J-M, Amouroux J-M, Jordana E, et al. (2003) Biochemical characteristics of settling particulate organic matter at two north-western Mediterranean sites: a seasonal comparison. Estuar Coast Shelf Sci 58: 423–434.

54. Sinsabaugh RL, Findlay S, Franchini P, Fischer D (1997) Enzymatic analysis of riverine bacterioplankton production. Limnol Oceanogr 42: 29–38.

55. Tamburini C, Garcin J, Ragot M, Bianchi A (2002) Biopolymer hydrolysis and bacterial production under ambient hydrostatic pressure through a 2000 m water column in the NW Mediterranean. Deep-Sea Res II 49: 2109–2123.

56. De Wit R, Bouloubassi I (1998) Oxygen penetration depth and aerobic microbial respiration in sediments of the Western Mediterranean. Third MTP-II Workshop on the Variability of the Mediterranean Sea. Rhodos (Greece). pp. 207.

PERMISSIONS

All chapters in this book were first published in PLOS ONE, by The Public Library of Science; hereby published with permission under the Creative Commons Attribution License or equivalent. Every chapter published in this book has been scrutinized by our experts. Their significance has been extensively debated. The topics covered herein carry significant findings which will fuel the growth of the discipline. They may even be implemented as practical applications or may be referred to as a beginning point for another development.

The contributors of this book come from diverse backgrounds, making this book a truly international effort. This book will bring forth new frontiers with its revolutionizing research information and detailed analysis of the nascent developments around the world.

We would like to thank all the contributing authors for lending their expertise to make the book truly unique. They have played a crucial role in the development of this book. Without their invaluable contributions this book wouldn't have been possible. They have made vital efforts to compile up to date information on the varied aspects of this subject to make this book a valuable addition to the collection of many professionals and students.

This book was conceptualized with the vision of imparting up-to-date information and advanced data in this field. To ensure the same, a matchless editorial board was set up. Every individual on the board went through rigorous rounds of assessment to prove their worth. After which they invested a large part of their time researching and compiling the most relevant data for our readers.

The editorial board has been involved in producing this book since its inception. They have spent rigorous hours researching and exploring the diverse topics which have resulted in the successful publishing of this book. They have passed on their knowledge of decades through this book. To expedite this challenging task, the publisher supported the team at every step. A small team of assistant editors was also appointed to further simplify the editing procedure and attain best results for the readers.

Apart from the editorial board, the designing team has also invested a significant amount of their time in understanding the subject and creating the most relevant covers. They scrutinized every image to scout for the most suitable representation of the subject and create an appropriate cover for the book.

The publishing team has been an ardent support to the editorial, designing and production team. Their endless efforts to recruit the best for this project, has resulted in the accomplishment of this book. They are a veteran in the field of academics and their pool of knowledge is as vast as their experience in printing. Their expertise and guidance has proved useful at every step. Their uncompromising quality standards have made this book an exceptional effort. Their encouragement from time to time has been an inspiration for everyone.

The publisher and the editorial board hope that this book will prove to be a valuable piece of knowledge for researchers, students, practitioners and scholars across the globe.

LIST OF CONTRIBUTORS

Benjamin A. Cash
Center for Ocean-Land-Atmosphere Studies, Fairfax, Virginia, United States of America

Xavier Rodó
Institut Catalá de Ciències del Clima (IC3), Barcelona, Catalunya, Spain
Institució Catalana de Recerca i Estudis Avanc̦ats, Barcelona, Catalunya, Spain

Michael Emch
University of North Carolina Chapel Hill, Chapel Hill, North Carolina, United States of America

Md. Yunus and Abu S. G. Faruque
International Centre for Diarrheal Disease Research, Dhaka, Bangladesh

Mercedes Pascual
Department of Ecology and Evolutionary Biology University of Michigan, Ann Arbor, Michigan, United States of America
Howard Hughes Medical Institute, Chevy Chase, Maryland, United States of America

Camerron M. Crowder and Virginia M. Weis
Department of Integrative Biology, Oregon State University, Corvallis, Oregon, United States of America

Wei-Lo Liang
Institute of Marine Biology, National Dong Hwa University, Pingtung, Taiwan, R.O.C.

Tung-Yung Fan
Institute of Marine Biology, National Dong Hwa University, Pingtung, Taiwan, R.O.C.
National Museum of Marine Biology and Aquarium, Pingtung, Taiwan, R.O.C.

John D. Hedley
Environmental Computer Science Ltd., Tiverton, Devon, United Kingdom

Kathryn McMahon
School of Natural Sciences and Centre for Marine Ecosystems Research, Edith Cowan University, Joondalup, Western Australia

Peter Fearns
Department of Imaging and Applied Physics, Curtin University of Technology, Perth, Western Australia

Paul G. Matson and Gretchen E. Hofmann
Department of Ecology, Evolution, and Marine Biology, University of California Santa Barbara, Santa Barbara, California, United States of America

Libe Washburn
Department of Geography, University of California Santa Barbara, Santa Barbara, California, United States of America

Todd R. Martz
Geosciences Research Division, Scripps Institution of Oceanography, University of California San Diego, La Jolla, California, United States of America

Sergei Katsev
Large Lakes Observatory, University of Minnesota Duluth, Duluth, Minnesota, United States of America
Department of Physics, University of Minnesota Duluth, Duluth, Minnesota, United States of America

Arthur A. Aaberg
Department of Physics, University of Minnesota Duluth, Duluth, Minnesota, United States of America

Sean A. Crowe
Department of Earth, Ocean, and Atmospheric Sciences, University of British Columbia, Vancouver, British Columbia, Canada

Robert E. Hecky
Large Lakes Observatory, University of Minnesota Duluth, Duluth, Minnesota, United States of America
Department of Biology, University of Minnesota Duluth, Duluth, Minnesota, United States of America

Teba Gil-Díaz, Ricardo Haroun, Fernando Tuya, Séfora Betancor and María A. Viera-Rodríguez
Centro de Biodiversidad y Gestión Ambiental, Universidad de Las Palmas de Gran Canaria, Las Palmas de Gran Canaria, Spain

Anne Molcard
Université du Sud Toulon-Var, Aix-Marseille Université CNRS/INSU/IRD, Mediterranean Institute of Oceanography (MIO), La Garde, France

Anna-Maria Rammou, Andrea Doglioli and Anne Petrenko
Aix-Marseille Université , CNRS/INSU/IRD, Mediterranean Institute of Oceanography (MIO), Marseille, France

Léo Berline
Université du Sud Toulon-Var, Aix-Marseille Université ,CNRS/INSU/IRD, Mediterranean Institute of Oceanography (MIO), La Garde, France
CNRS, Laboratoire d'Océanographie de Villefranche, Villefranche-sur-Mer, France
Université Pierre et Marie Curie, Paris 6, Laboratoire d'Océanographie de Villefranche, Villefranche-sur-Mer, France

Ulrike Braeckman, Carl Van Colen, Katja Guilini, Dirk Van Gansbeke, Magda Vincx and Jan Vanaverbeke
Ghent University, Department of Biology, Marine Biology Research Group, Ghent, Belgium

Karline Soetaert
Netherlands Institute for Sea Research, Department of Ecosystem Studies, Yerseke, The Netherlands

Chenyi Zhao and Fengqing Jiang
State Key Laboratory of Desert and Oasis Ecology, Xinjiang Institute of Ecology and Geography, Chinese Academy of Sciences, Urumqi, Xinjiang, China

Dongmei Peng
State Key Laboratory of Desert and Oasis Ecology, Xinjiang Institute of Ecology and Geography, Chinese Academy of Sciences, Urumqi, Xinjiang, China
Graduate University of Chinese Academy of Sciences, Beijing, China
Information Centre of Xinjiang Xingnong-Net, Urumqi, China

Xiujun Wang
State Key Laboratory of Desert and Oasis Ecology, Xinjiang Institute of Ecology and Geography, Chinese Academy of Sciences, Urumqi, Xinjiang, China
Earth System Science Interdisciplinary Center, University of Maryland, College Park, Maryland, United States of America

Xingren Wu
IM System Group and Environmental Modeling Center, National Centers for Environmental Prediction, National Oceanic and Atmospheric Administration, College Park, Maryland, United States of America

Pengxiang Chen
Xinjiang Meteorological Observatory, Urumqi, Xinjiang, China

Gilberto F. Barroso and Fábio da C. Garcia
Department of Oceanography and Ecology, Federal University of Espírito Santo, Vitória, Espírito Santo, Brazil

Monica A. Gonçalves
Espírito Santo State Water Resources Agency, Vitória, Espírito Santo, Brazil

Malcolm T. McCulloch
School of Earth and Environment, The University of Western Australia and ARC Centre of Excellence for Coral Reef Studies, Crawley, WA, Australia

Verena Schoepf
School of Earth and Environment, The University of Western Australia and ARC Centre of Excellence for Coral Reef Studies, Crawley, WA, Australia
School of Earth Sciences, The Ohio State University, Columbus, Ohio, United States of America

Stephen J. Levas, Yohei Matsui and Andréa G. Grottoli
School of Earth Sciences, The Ohio State University, Columbus, Ohio, United States of America

Mark E. Warner and Matthew D. Aschaffenburg
School of Marine Science and Policy, University of Delaware, Lewes, Delaware, Unite States of America

Tracy L. Perkins and James E. McDonald
School of Biological Sciences, Bangor University, Bangor, United Kingdom

Katie Clements, Jaco H. Baas and Colin F. Jago
School of Ocean Sciences, Bangor University, Bangor, United Kingdom

Shelagh K. Malham
School of Ocean Sciences, Bangor University, Bangor, United Kingdom
Centre for Applied Marine Sciences, Bangor University, Bangor, United Kingdom

Davey L. Jones
School of Environment, Natural Resources and Geography, Bangor University, Bangor, United Kingdom

Shalin Seebah and Matthias S. Ullrich
Molecular Life Science Research Center, Jacobs University Bremen, Bremen, Germany,

Caitlin Fairfield and Uta Passow
Marine Science Institute, University of California Santa Barbara, Santa Barbara, California, United States of America

Richard A. Feely
NOAA Pacific Marine Environmental Laboratory, Seattle, Washington, United States of America

Geraint A. Tarling and Sophie Fielding
British Antarctic Survey, High Cross, Cambridge, United Kingdom

Nina Bednaršek
NOAA Pacific Marine Environmental Laboratory, Seattle, Washington, United States of America
British Antarctic Survey, High Cross, Cambridge, United Kingdom
Centre for Ocean and Atmospheric Sciences, School of Environmental Sciences, University of East Anglia, Norwich Research Park, Norwich, United Kingdom

Dorothee C. E. Bakker
Centre for Ocean and Atmospheric Sciences, School of Environmental Sciences, University of East Anglia, Norwich Research Park, Norwich, United Kingdom

James R. Christian
Canadian Centre for Climate Modelling and Analysis, Victoria, B.C., Canada
Fisheries and Oceans Canada, Institute of Ocean Sciences, Sidney, BC, Canada

Anne-Elise Nieblas and Sylvain Bonhommeau
Unité Mixte Recherche Ecosystémes Marins Exploités 212, Institut Français de Recherche pour l'Exploitation de la Mer (IFREMER), Sète, France

Kyla Drushka
Applied Physics Laboratory, University of Washington, Seattle, Washington, United States of America

Gabriel Reygondeau
Center for Macroecology, Evolution and Climate, National Institute for Aquatic Resources, Technical University of Denmark (DTU Aqua), Charlottenlund, Copenhagen, Denmark

Vincent Rossi
Instituto de Física Interdisciplinary Sistemas Complejos, Institute for Cross-Disciplinary Physics and Complex Systems, (CSIC-UIB), Campus Universitat de les Illes Balears, Palma de Mallorca, Spain

Hervé Demarcq
Unité Mixte de Recherche Ecosystémes Marins Exploités 212, Institut de Recherche pour le Développement (IRD), Séte, France

Laurent Dubroca
European Commission, Joint Research Center, Institute for Environment & Sustainability, Water Resources, Ispra, Italy

Diego Alvarez-Berastegui
Balearic Islands Coastal Observing and Forecasting System (SOCIB), Palma de Mallorca, Balearic Islands, Spain

Lorenzo Ciannelli
College of Earth, Ocean, and Atmospheric Sciences (CEOAS), Oregon State University, Corvallis, Oregon, United States of America

Alberto Aparicio-Gonzalez, Patricia Reglero, Manuel Hidalgo, Jose Luis López-Jurado and Francisco Alemany
Instituto Español de Oceanografía - Centre Oceanogràfic de les Balears (COB-IEO), Palma de Mallorca, Balearic Islands, Spain

Joaquín Tintoré
Balearic Islands Coastal Observing and Forecasting System (SOCIB), Palma de Mallorca, Balearic Islands, Spain

Institute for Advanced Studies (IMEDEA), Consejo Superior de Investigaciones Científicas (CSIC) and the University of the Balearic Islands (UIB). Esporles, Spain

Jessica J. Eichmiller, Przemyslaw G. Bajer and Peter W. Sorensen
Department of Fisheries, Wildlife, and Conservation Biology, Minnesota Aquatic Invasive Species Research Center, University of Minnesota, Twin Cities, St. Paul, Minnesota,
United States of America

Rocio I. Ruiz-Cooley and Matthew D. McCarthy
Ocean Sciences Department, University of California Santa Cruz, Santa Cruz, California, United States of America

Paul L. Koch
Earth and Planetary Sciences Department, University of California Santa Cruz, Santa Cruz, California, United States of America

Paul C. Fiedler
Southwest Fisheries Science Center, National Marine Fisheries Service, National Oceanic and Atmospheric Administration, La Jolla, California, United States of America

Malcolm D. O'Toole, Mary-Anne Lea and Mark A. Hindell
Institute of Marine and Antarctic Studies, University of Tasmania, Hobart, Australia

Christophe Guinet
Marine Predator Department, Centre détudes biologiques de Chizé, Villiers-en-Bois,
France

Anabella Covazzi Harriague, Giorgio Bavestrello, Marzia Bo, Michela Castellano,
Margherita Majorana, Francesco Massa, Alessandro Montella, Paolo Povero and Cristina Misic
Dipartimento di Scienze della Terra, dell'Ambiente e della Vita - DiSTAV – University of Genoa, Italy

Mireno Borghini
CNR-ISMAR, Institute of Marine Sciences, National Research Council, Section of La Spezia, Pozzuolo di Lerici, Italy

Index

www.ingramcontent.com/pod-product-compliance
Lightning Source LLC
Chambersburg PA
CBHW061243190326
41458CB00011B/3566